Drawing Theories Apart

Drawing Theories Apart

The Dispersion of Feynman Diagrams in Postwar Physics

DAVID KAISER

The University of Chicago Press CHICAGO AND LONDON

David Kaiser is associate professor in the Program in Science, Technology, and Society and lecturer in the Department of Physics at the Massachusetts Institute of Technology.

The University of Chicago Press, Chicago 60637
The University of Chicago Press, Ltd., London

Printed in the United States of America

14 13 12 11 10 09 08 07 06   2 3 4 5 6 7 8 9 10

ISBN: 0-226-42266-6 (cloth)
ISBN: 0-226-42267-4 (paper)

Library of Congress Cataloging-in-Publication Data

Kaiser, David.
Drawing theories apart : the dispersion of Feynman diagrams in postwar physics / David Kaiser.
p. cm.
Includes bibliographical references and index.
ISBN 0-226-42266-6 (alk. paper)—ISBN 0-226-42267-4 (pbk. : alk. paper)
1. Feynman diagrams. 2. Physics—United States—History—20th century. 3. Physics—History—20th century. I. Title.
QC794.6.F4K35 2005
530′.0973′0904—dc22
2004023335

⊗ The paper used in this publication meets the minimum requirements of the American National Standard for Information Sciences—Permanence of Paper for Printed Library Materials, ANSI Z39.48-1992.

He teaches his theory, not as a body of fact, but as a set of tools,
to be used, and which he has actually used in his work.

John Clarke Slater describing Percy Bridgman after
Bridgman won the Nobel Prize in Physics, 11 January 1947

Contents

Preface and Acknowledgments

This book is about what theoretical physicists do and how their daily routines shifted in the wake of the Second World War. I focus on the crafting of theoretical tools, techniques, and practices—the everyday labor of theory—to complement the usual treatments in intellectual and conceptual histories. At stake are our assumptions about how theorists have spent their time and structured their activities. Have their primary concerns centered narrowly on the construction and selection of theories, or have they had other goals in mind when taking pencil and paper in hand? How should historians account for their work?

If we assume (as usual) that the quest for distinct scientific theories—delimited by the philosophers' anemic labels T_1, T_2, T_3—and the selection of one best theory have dominated theorists' work, then we might be inclined to organize our historical analyses around the birth and death of particular theories. Indeed, today we have huge literatures charting the development of Maxwell's theory of electrodynamics, Einstein's theories of special and general relativity, Heisenberg's matrix mechanics, Schrödinger's wave mechanics, their union in a theory of quantum mechanics, and so on. Yet as the best of these historical studies have shown, the objects of study in such cases—"special relativity," say, or "quantum mechanics"—are rarely single or unitary objects at all. Scientific theories are always open to divergent interpretations and uses, even decades after their construction and selection have supposedly been settled. Moreover, theoretical physicists have had much more on their plates than just theories. Theirs has been a world littered with *calculations*—calculations performed with an ever-changing toolbox of techniques.

These tools and techniques never apply themselves. Apprentice physicists must rehearse using theoretical tools until such calculational skills become second nature. These skills are rarely transmitted by formal, written instructions alone. Therefore, we must interrogate the wide range of pedagogical means by which students have become working theorists, eventually wielding the tools of theory with ease.

Few tools have meant more to theoretical physicists during the past half century than Feynman diagrams, named for the American theorist Richard Feynman. Feynman diagrams provide a rich map for charting larger transformations in the training of theoretical physicists during the decades after World War II. Even when drawing their diagrams in a similar fashion, members of different research groups gave them meaning and put them to work in distinct ways. Physicists carved out a wide variety of uses for the diagrams between the late 1940s and late 1960s. Historians may add one more: the diagrams demarcate the boundaries between research groups and chart their pedagogical reproduction over time.

Much like the diagrams it describes, this book is meant to contribute to constructive discussions among heterogeneous communities—historians, sociologists, philosophers, and physicists. With these diverse groups in mind, my goal has been to supply specialists with sufficient technical detail to demonstrate my arguments without overwhelming readers from broader backgrounds. Toward this end, the first and last sections of the book are meant to be broadly accessible, while most of the technical details appear in chapters 5, 6, and 8. Even in these chapters, I have attempted to provide guideposts for readers who last saw an integral sign in high school (and didn't like it then).

* * *

I have incurred a great many debts while working on this study. This book grew out of my doctoral dissertation in the Department of the History of Science at Harvard University, and it is a pleasure to thank my advisers, Peter Galison, Silvan S. Schweber, and Mario Biagioli, for untiring advice, encouragement, and friendship. Kathryn Olesko and Andrew Warwick functioned as de facto advisers every step of the way, and the book is better because of their engagement. While at Harvard I also completed doctoral work in theoretical physics; my thanks to Alan Guth, Robert Brandenberger, and Sidney Coleman for all their aid on my physics dissertation.

Many physicists who were active in the period I describe have graciously shared their memories, correspondence, and lecture notes with me, and I am deeply grateful to them: Stephen Adler, the late Louis Balàzs, Michel Baranger, E. E. Bergmann, James Bjorken, John Bronzan, Laurie Brown, Kenneth Case, Denyse Chew, Geoffrey Chew, Roy Chisholm, Carleton DeTar, Freeman Dyson, Gordon Feldman, Jerry Finkelstein, William Frazer, Marvin Goldberger, the late A. Carl Helmholtz, Macalaster Hull, C. Angas Hurst, the late Morton Kaplon, Tom Kinoshita, Abraham Klein, Norman Kroll, Andrew Lenard, Joseph Levinger, Peter Lindenfeld, Francis Low, Stanley Mandelstam, Gordon Moorhouse, Yoichiro Nambu, Roger Phillips, Silvan S. Schweber, Andrew Sessler, Howard Shugart, Henry Stapp, M. K. Sundaresan, Chung-I Tan, Kip Thorne, and Eyvind Wichmann.

I have also benefited from the comments of many people who read part or all of this book in manuscript: Ron Anderson, James Bjorken, Alexander Brown, Laurie Brown, Cathryn Carson, Patrick Catt, Deborah Coen, Denyse Chew, Geoffrey Chew, Roy Chisholm, the late James Cushing, Carleton DeTar, Joe Dumit, Freeman Dyson, James Elkins, Guy Emery, Peter Galison, Jean-François Gauvin, Tracy Gleason, Michael Gordin, Stephen Gordon, Loren Graham, Karl Hall, Kristen Haring, Kenji Ito, Robert Jaffe, Matthew Jones, Tom Kinoshita, Robert Kohler, Alexei Kojevnikov, Michael Lynch, Shawn Mullet, Yoichiro Nambu, Mary Jo Nye, Kathryn Olesko, Naomi Oreskes, Elizabeth Paris, Silvan S. Schweber, Sameer Shah, Peter Shulman, Robert Silbey, Susan Silbey, Rebecca Slayton, Chung-I Tan, Will Thomas, Jessica Wang, Andrew Warwick, Lambert Williams, Rosalind Williams, and Chen-Pang Yeang. Though we might not agree on all points, the book is incomparably better because of their generous input.

This study became feasible only with the aid of several research assistants, who tirelessly tracked down published articles, biographical information, secondary literature, and sundry other details. I am grateful to the following for all their help: Alexander Brown, Shane Hamilton, Rachel Prentice, Sameer Shah, and Peter Shulman. Sameer Shah in particular rose above the call of duty time and time again over the course of three years; Sameer, my thanks. My friends and colleagues Kenji Ito and Karl Hall also proved invaluable, especially for their help with sources in Japanese and Russian, respectively. Without working closely with Kenji and Karl, I could never have extended my original study (which had focused only upon the United States) to include the crucial international comparisons.

Archivists at the following institutions proved tremendously helpful: the Niels Bohr Library of the American Institute of Physics, in College Park, Maryland; the Bancroft Library of the University of California at Berkeley (especially David Farrell); the Rare and Manuscript Collections in Kroch Library at Cornell University; the Harvard University Archives in Pusey Library at Harvard University; the Institute for Advanced Study Archives in Princeton, New Jersey (especially Lisa Coats); the Seeley G. Mudd Manuscript Library at Princeton University; the Physics-Optics-Astronomy Library at the University of Rochester (especially Patricia Souloff); and the Syracuse University Archives in Bird Library at Syracuse University. Professor Freeman Dyson graciously shared with me his voluminous private correspondence, which is housed in his office at the Institute for Advanced Study.

The Program in Science, Technology, and Society at MIT has been a wonderful setting in which to expand my dissertation into this book. My heartfelt thanks to my colleagues and students. The many months I spent as a visitor in the Office for the History of Science and Technology at Berkeley during 1998 were made both productive and happy by the warm welcome extended by Cathryn Carson, Diana Wear, Roger Hahn, Sonja Amadae, Christopher Ritter, Paula DeVos, and Peter Westwick. More recently, this project has benefited from the kind hospitality of MIT's Center for Theoretical Physics; I would like to thank Robert Jaffe for making me an affiliate of the CTP, and Scott Morley and Joyce Berggren for their assistance. My physics collaborators over the years have helped me keep critical skills sharp; I have learned many things from working with Bruce Bassett, Joanne Cohn, Alan Guth, Roy Maartens, Philip Mannheim, and Ali Nayeri.

Research grants from the National Science Foundation (SES-0135615) and the Spencer Foundation (grant 200200081) are gratefully acknowledged, as are the funds provided by the Levitan Prize in the Humanities at MIT and the Leo Marx Career Development Assistant Professorship at MIT. I would also like to thank Christie Henry, Jennifer Howard, and Erik Carlson at the University of Chicago Press for all their help along the way.

Portions of this book have appeared previously in various articles, and I thank the editors of each journal for permission to use the material here. Sections of chapter 4 originally appeared in David Kaiser, Kenji Ito, and Karl Hall, "Spreading the tools of theory: Feynman diagrams in the United States, Japan, and the Soviet Union," *Social Studies of Science* 34 (December 2004): 879–922; most of chapter 9 originally appeared in David Kaiser, "Nuclear democracy: Political engagement, pedagogical reform, and particle physics in postwar America," *Isis* 93 (June 2002): 229–68; and portions of chapter 10 originally appeared in David Kaiser, "Stick-figure realism: Conventions, reification, and the persistence of Feynman diagrams, 1948–1964," *Representations* 70 (Spring 2000): 49–86.

Permission to reproduce figures is gratefully acknowledged as follows: figure 1.1 (Apple Computer, Inc.); figures 1.3–4, 2.5–6, 4.2, 5.1–3, 5.5, 5.7, 5.10, 6.2, 6.6, 6.9–11, 7.6, 8.3, 8.6–7, 8.11, 10.2, 10.8, and 10.9 (*left*) (American Physical Society); figures 2.4, 4.3, 6.4, 6.8, and 9.1 (Emilio Segrè Visual Archives, American Institute of Physics); figure 4.1 (Professor J. S. R. Chisholm); figures 4.4–9 (Japanese Physical Society); figure 4.12 (Sovfoto/Eastfoto); figures 5.4, 7.2, and 10.10 (McGraw-Hill); figures 5.8–9 (Springer-Verlag); figure 6.1 (Professor Freeman Dyson); figure 6.3 (*right*) (Professor Joseph Weneser); figure 6.12 (The Royal Society); figures 7.1 (*top*) and 10.9 (*right*) (Professor Silvan S. Schweber); figure 7.1 (*bottom right*) (Professor Franz Mandl); figure 7.5 (Professor Kip S. Thorne); figure 7.7 (Professor William Frazer); figure 8.1 (Professor Jerry Finkelstein); figure 8.2 (Professors Marvin Goldberger and J. D. Jackson); figure 8.8 (Professor John C. Taylor); figure 9.2 (Lawrence Berkeley National Laboratory); figures 9.3–4 (World Scientific); figure 10.1 (Dover Publications); figure 10.3 (Princeton University Press); figure 10.4 (Professor John C. Polkinghorne); figures 10.5 (*left*) and 10.6 (Professor Kenneth Ford); figure 10.7 (University of Chicago Press); and figure 10.11 (Cambridge University Press). I am indebted to all of these people and institutions for their kind permission to reproduce these figures.

My parents, Debra and Richard Kaiser, and my sister, Hilary, have provided many years of support, encouragement, and good humor, without which none of this would have come to fruition. And finally I want to thank my wife, Tracy Gleason. She has been sounding board, writing coach, therapist, and best friend all rolled into one. This book is dedicated to her with all my love.

Abbreviations

Lists of articles that make use of Feynman diagrams appear in appendixes A–F. References to these articles throughout the book will be made in an abbreviated form, citing the article's number within a given appendix. For example, Richard Feynman's paper "Space-time approach to quantum electrodynamics," *Physical Review* 76 (1949): 769–89, will be cited as "Feynman [A.4]," since it is the fourth entry in appendix A. The following abbreviations will also appear in the notes and appendixes. Where appropriate, items from archival collections will be cited as "Folder *x: y*," for box *x* and folder *y*.

AIP-EMD: American Institute of Physics, Education and Manpower Division, Records, 1951–73. Collection number AR15, Niels Bohr Library, American Institute of Physics, College Park, Maryland.

AJP: American Journal of Physics.

AMS: American Men of Science: A Biographical Dictionary. Edited by Jacques Cattell. 10th ed. 4 vols. Tempe, Arizona: Jacques Cattell Press, 1960.

BAS: Bulletin of the Atomic Scientists.

BDP: University of California, Berkeley, Department of Physics, Records, ca. 1920–62. Collection number CU-68, Bancroft Library, Unversity of California, Berkeley.

Birge, *History:* Raymond T. Birge, *History of the Physics Department, University of California, Berkeley.* 5 vols. Copies in the Bancroft and Physics Department Libraries, University of California, Berkeley.

CDP: Cornell University Department of Physics, Records. Ithaca, New York.

DiP: M. Louise Markworth, *Dissertations in Physics: An Indexed Bibliography of All Doctoral Theses Accepted by American Universities, 1861–1959.* Stanford: Stanford University Press, 1961.

FJD: Freeman J. Dyson, Papers. In Professor Dyson's possession, Institute for Advanced Study, Princeton, New Jersey.

FR: Fritz Rohrlich, Papers, 1948–94. Syracuse University Archives, Syracuse, New York.

HAB: Hans A. Bethe, Papers, ca. 1931–95. Collection number 14-22-976, Division of Rare and Manuscript Collections, Cornell University Library, Ithaca, New York.

HDP: Harvard University Department of Physics, Correspondence, 1953–57 and Some Earlier, and Correspondence, 1958–60. Call number UAV 691.10, Harvard University Archives, Pusey Library, Cambridge, Massachusetts.

HSPS: Historical Studies in the Physical Sciences; later issues entitled *Historical Studies in the Physical and Biological Sciences.*

IAS: Institute for Advanced Study, Archives, Princeton, New Jersey.

NBL: Niels Bohr Library, American Institute of Physics, College Park, Maryland.

NYT: New York Times.

Pauli, *WB:* Wolfgang Pauli, *Wissenschaftlicher Briefwechsel.* Edited by Karl von Meyenn. Vol. 3, *1940–1949.* New York: Springer, 1993. Vol. 4, pt. 1, *1950–1952.* New York: Springer, 1996. Vol. 4, pt. 2, *1953–1954.* New York: Springer, 1999.

PDP-AR: Princeton University Department of Physics, Annual Reports. Bound chronologically in Reports to the [University] President. Seeley G. Mudd Manuscript Library, Princeton, New Jersey.

PR: Physical Review.

PRL: Physical Review Letters.

PRSA: Proceedings of the Royal Society, Section A.

PT: Physics Today.

PTP: Progress of Theoretical Physics.

REM: Robert E. Marshak, Papers, 1947–88. Microfiche collection, call number M366, Physics-Optics-Astronomy Library, University of Rochester, Rochester, New York.

RPF: Richard Feynman, Papers, 1935–65. Collection number AR65, Niels Bohr Library, American Institute of Physics, College Park, Maryland.

RTB: Raymond Thayer Birge, Correspondence and Papers. Call number 73/79c, Bancroft Library, University of California, Berkeley. Letters written by Birge are filed chronologically. The items cited here are from

Boxes 39 and 40; explicit folder titles will not be cited. Letters written to Birge will be cited with box and folder titles.

SAG: Samuel A. Goudsmit, Papers, 1921–78. Collection number AR30260, Niels Bohr Library, American Institute of Physics, College Park, Maryland.

SHPS: Studies in History and Philosophy of Science.

SHPMP: Studies in History and Philosophy of Modern Physics.

Soken: Soryūshi-ron Kenkyū.

Sov Phys JETP: Soviet Physics JETP.

ZhETF: Zhurnal eksperimental'noi i teoreticheskoi fiziki.

· 1 ·

Introduction: Pedagogy and the Institutions of Theory

I would wish, therefore, that the man who claims to be scientific first tell me the method he employs for his scientific demonstrations and then show me how he has been trained in it.

GALEN, CA. AD 150[1]

RICHARD FEYNMAN AND HIS DIAGRAMS

Few scientists, living or dead, surpass Richard Feynman (1918–88) as a widely recognized scientific icon. His star rose early. Following undergraduate work at the Massachusetts Institute of Technology and graduate study at Princeton, Feynman served as the youngest subgroup leader in wartime Los Alamos—he was only twenty-five—leading a band of fellow physicists in the sprawling effort to build atomic bombs. As the war ground closer to a conclusion, physics departments throughout the United States jockeyed to hire Feynman, as word of his creativity and sheer calculating power began to spread. His boss at Los Alamos, Hans Bethe, managed to lure Feynman to Cornell, where he taught for five years before the California Institute of Technology enticed him away from Ithaca's winters to the golden groves of Southern California. At Caltech, Feynman's renown as an animated teacher grew to match that of his physics prowess. An invitation to teach Caltech's large introductory physics class during the late 1950s led to the famous *Feynman Lectures on Physics*—a three-volume set still known simply as "the red books" by admiring physicists the world over.[2] In later years, Feynman made a habit of giving informal lectures on

1. Galen, *On the Doctrines of Hippocrates and Plato* (1978), 113 (bk. 2, 3:17).

2. Feynman, Leighton, and Sands, *Feynman Lectures on Physics* (1965). The phrase "red books" refers to the original cover design, which featured a bold, red background. Never mind that the "experiment" of teaching the class was a flop: even Feynman came to admit within a few years that his approach was too difficult for the entering undergraduates, though it resonated with the grateful graduate students, postdoctoral students, and faculty who made a habit of sneaking into Feynman's lectures. Ibid., 3–5.

physics at neighboring industries and of teaching his "Physics X" class, a freewheeling class open to anyone with questions about science. All the while he made lasting contributions to physicists' understanding of electrodynamics, nuclear forces, solid-state physics, and gravitation, many of which still bear his name.

The Nobel Prize that Feynman shared with two fellow physicists in 1965 piqued the wider public's interest; by the mid-1980s, he had became a folk hero well beyond the world's clique of theoretical physicists. A visible role in the 1986 investigation of the explosion of the space shuttle *Challenger*—dunking a piece of O-ring rubber in a glass of ice water in the midst of a televised press conference—capped a decades-long process of becoming a household name. One year after his death, the American Physical Society, American Association of Physics Teachers, and American Association for the Advancement of Science jointly organized a daylong memorial. Since then, four biographical portraits and at least five collections of friends' and colleagues' reminiscences have been published, supplementing the two best-selling compilations of his own anecdotes and aphorisms that were published near the end of his life.[3] A book of his artwork and a compact disk of his celebrated bongo drumming were each on sale during the 1990s.[4] Television specials have supplemented his popular accounts of modern physics and of his own life story. During the late 1990s, the Apple Computer Company selected various portraits of Feynman, along with portraits of other iconic figures such as Albert Einstein, Pablo Picasso, and Charlie Chaplin, for their "Think different" advertising campaign. (See fig. 1.1.) More recently, no less an actor than Alan Alda depicted Feynman in a solo performance at the Vivian Beaumont Theater in New York City's Lincoln Center. The show sold out so quickly upon its opening in November 2001 that the producers brought it back for an even longer run the following spring. As

3. On the jointly sponsored memorial event on 18 Jan 1989 in San Francisco, see Lubkin, "Special issue: Richard Feynman" (1989), 23. Biographies include Gleick, *Genius* (1992); Schweber, *QED* (1994), esp. chap. 8; Mehra, *Beat of a Different Drum* (1994); and Gribbin and Gribbin, *Richard Feynman* (1997). Interviews and reminiscences include the Feb 1989 issue of *PT;* Ralph Leighton, *Tuva or Bust!* (1991); Brown and Rigden, *Most of the Good Stuff* (1993); Sykes, *No Ordinary Genius* (1994); and Mlodinow, *Feynman's Rainbow* (2003). Feynman's own anecdotes were published as Feynman with Leighton, *Surely You're Joking* (1985); and Feynman with Leighton, *What Do You Care* (1988). Both of these books were soon reissued, with wider circulation, in the New Age Books imprint series from Bantam. On the Feynman publishing industry, see also Goodstein, "Feynmaniacs should read this review" (1998).

4. M. Feynman, *Art of Richard P. Feynman* (1995); examples of his artwork appear in "Richard Feynman, Artist," *PT* 42 (Feb 1989): 86–87. Information on purchasing CDs of Feynman's bongo drumming is available in the back of Feynman with Leighton, *What Do You Care* (1988); and in Gribbin and Gribbin, *Richard Feynman* (1997), 287.

Figure 1.1. Richard Feynman featured in an Apple computer advertisement. (Source: *Scientific American* 280 [June 1999]: back cover.)

Alda explained, with more than a little Feynmanesque hyperbole, "Feynman's personality is so strong that if he was played by a three-foot-high dwarf of the opposite sex, you would still think it was Feynman up there."[5] Few academics answer to such depictions.

When the journal *Physics Today* dedicated a special memorial issue to Feynman in February 1989, the editor faced the difficult task of summing up Feynman's many contributions. She selected a simple graphic theme to unify the volume. "The diagram you see scattered throughout this issue is a reminder of the legacy Richard Feynman left us," the editor explained.[6] A simple line drawing, known to physicists as a "Feynman diagram," appeared atop each article in the issue, its white lines set against a dark field of mourning. (See fig. 1.2.) It was a fitting gesture. For all of Feynman's many contributions to modern physics, his

5. Alda as quoted in Overbye, "On stage" (2001), D5; *Playbill* 118 (Mar 2002): 15–23. The play, entitled *QED*, premiered in Los Angeles, with Alda playing Feynman, in March 2001.

6. Lubkin, "Special issue: Richard Feynman" (1989), 23.

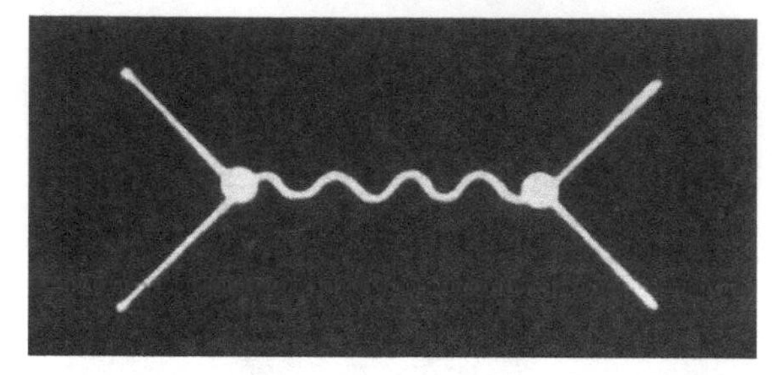

Figure 1.2. Feynman diagram from the special Feynman memorial issue of *Physics Today*. (Source: Lubkin, "Special issue: Richard Feynman," 22.)

diagrams have had the widest and longest-lasting influence. Feynman diagrams have revolutionized nearly every aspect of theoretical physics since the middle of the twentieth century. Feynman first introduced his diagrams in the late 1940s as a bookkeeping device for simplifying lengthy calculations in one area of physics—quantum electrodynamics, physicists' quantum-mechanical description of electromagnetic forces. Soon the diagrams gained adherents throughout the fields of nuclear and particle physics. Not long thereafter, other theorists adopted—and subtly adapted—Feynman diagrams for many-body applications in solid-state theory. By the end of the 1960s, some physicists even wielded the line drawings for calculations in gravitational physics. With the diagrams' aid, entire new calculational vistas opened for physicists; theorists learned to calculate things that many had barely dreamed possible before World War II. With the list of diagrammatic applications growing ever longer, Feynman diagrams helped to transform the way physicists saw the world, and their place within it.

Today physicists all over the world scribble down the diagrams when studying everything from the mundane to the bizarre. Whether calculating the properties of materials that will form electronic computer chips, the levitation of magnets above superconductors, the behavior of particles near black holes, or the origin of matter itself just fractions of a second after the big bang, physicists begin by drawing Feynman diagrams. Everything about using the diagrams has become routine. Particle physicists can now download computer programs that will both draw the standardized diagrams and evaluate the associated mathematical terms: the diagrams' lines and their mathematical content have become both algorithmic and automatic. As one physicist wrote in a recent preprint detailing the new automated diagram tools, "Explaining the necessity of [diagrammatic] one-loop calculations in the light of modern-day colliders is like carrying owls to Athens"—the centrality of Feynman diagrams to everyday practice, and the need to evaluate large numbers of them quickly, simply goes without saying.[7] The latest programs can generate and evaluate one thousand simple diagrams in about five minutes on an ordinary desktop computer.

Yet it hasn't always been this way. A generation of theoretical physicists earned doctorates for performing far less grandiose tasks just a few short decades ago. Evaluating a few Feynman diagrams by hand was a publishable feat in the early 1950s, and the stuff of which scores of dissertations were made. More important, both aspects of the diagrams' use and interpretation—how to draw them and how to calculate with them, that is, what they meant—became contested and shifting during the two decades after their introduction in the late 1940s. Today's automated computational utopia hides a rich history of competing appropriations and theoretical foundations for these bare line drawings. During the 1950s and 1960s, Feynman diagrams did not compel, by themselves, a unique meaning or interpretation. They were drawn differently and mustered in different fashions to varying calculational, and ultimately ontological, ends. Despite the diagrams' centrality today, and despite all the attention lavished on Feynman himself—his quirky genius, his lasting contributions to physics, his now-famous eccentricities—no attention has been paid to how his simple-looking diagrams actually came to be embraced by so many physicists for so many distinct applications. Indeed, more has been written about the hunt for Feynman's celebrated passenger van, bedecked in larger-than-life Feynman diagrams, than about how scores of physicists—and soon hundreds and thousands—came to traffic in the diagrams in the first place.[8]

Several physicists and historians have scrutinized various roots within Feynman's thinking for what became Feynman diagrams.[9] My project concerns instead what happened to the diagrams once they made the leap *out* of Feynman's head. How did the diagrams spread so quickly? What kinds of applications did physicists forge for the line drawings? Why did the diagrams remain central to physicists' work even as related methods of calculation came and

7. Hahn, "*FeynArts, FormCalc,* and *LoopTools*" (1999), 1. See also Tentyukov and Fleischer, "DIANA, a program for Feynman diagram evaluation" (1999); and Tentyukov and Fleischer, "Feynman diagram analyzer" (2000).

8. On the search for Feynman's van, see Gribbin and Gribbin, *Richard Feynman* (1997), 281–84; and Sykes, *No Ordinary Genius* (1994), 85–86.

9. Schweber, "Feynman" (1986); Mehra, *Beat of a Different Drum* (1994), chaps. 5, 6, 10–14; Schweber, *QED* (1994), chap. 8; and Galison, "Feynman's war" (1998). Feynman recounted his route to the diagrams in his Nobel Prize acceptance speech: Feynman, "Space-time view" (1966).

went? Feynman's protracted and largely private struggles to work out a consistent calculational scheme for quantum electrodynamics paled in comparison with the efforts required to equip other physicists with the new diagrammatic tool. Long after Feynman himself had grown accustomed to thinking and working with his diagrams, they remained neither obvious nor automatic for others. In fact, Feynman's particular ideas about the diagrams provided only one—and by no means the most important one—of several contrasting factors in determining how other physicists would treat them.

Rather than dwell on the isolated thoughts of a few Nobel laureates, therefore, I focus in this book on the *pedagogical* work involved in training large numbers of researchers to approach physical questions in similar ways. Unlike previous historical treatments of modern physics, this study follows neither the grand march of particular theories nor the lumbering progress of ever-growing experiments. Instead, I follow Feynman diagrams around, focusing on how physicists fashioned—and constantly refashioned—the diagrams into a calculational *tool*, a theoretical *practice*. My goal is to unpack the history of postwar theoretical physics from the ground up, as a story ultimately about crafting, deploying, and stabilizing the tools that undergird everyday calculations. Research tools such as Feynman diagrams never apply themselves; physicists have to be trained to use them, and to interpret and evaluate the results in certain ways. Stabilizing the new tool went hand in hand with training a new generation of theoretical physicists after World War II. The story of Feynman diagrams' spread thus illuminates larger transformations in what "theoretical physics" would be and how young physicists would become "theorists" after the war.

By following how physicists learned about and used Feynman diagrams from the late 1940s through the late 1960s, we bring broader changes in the infrastructure and intellectual development of postwar physics into focus. Everything about physicists' patterns of work came in for reevaluation after the war, from the methods of training young theorists, to the means of communicating new results and techniques, to decisions about what topics merited study and by what means. The diagrams likewise reveal the fissures and politics of becoming a young theorist after the war—from the international politics of the cold war, which shaped how physicists could communicate with colleagues in other countries; to the cold war's domestic doppelgänger, McCarthyism, and its effects on physicists' civil liberties and patterns of thought; to the generational politics between mentors and students, and the microsocial politics between competing research groups. In all these ways, following Feynman diagrams around helps us make sense of theoretical physicists' changing world.

This account of the dispersion of Feynman diagrams draws on several general themes, which I discuss in this introductory chapter and elaborate upon in the chapters that follow. Before turning to these issues in more detail, I should pause and discuss the book's title, *Drawing Theories Apart*. Two senses of the phrase are likely to be recognizable already: first, an emphasis upon drawing and similar pencil-and-paper work within theoretical physics; and second, the need to rethink the roles of theories versus tools in our accounts of modern physics. A third reference might not be as obvious for all readers: an inverted analogy with Bruno Latour's 1986 article "Drawing Things Together." As scholars in science studies will no doubt recognize in the chapters that follow, this book draws on several quintessentially Latourian themes: the building of networks, the importance of inscriptions, the work of translation and enrollment, and so on. Indeed, the very idea of following a nonhuman scientific object around as an organizing principle bears a certain Latourian signature.[10] Yet I follow a different line when it comes to the question of "immutable mobiles," a notion that Latour introduced in his 1986 article. Whereas Latour emphasizes "optical consistency" (even "immutability") as an essential feature of why diagrams and other scientific inscriptions carry so much force among scientists, I focus instead on unfolding variations within their work—on the production and magnification of local differences, and the work required to transcend these differences when comparing results from different places.[11] Hence the "apart" of my title, in place of Latour's "together."

PAPER TOOLS AND THE PRACTICE OF THEORY

Despite their centrality, the crafting and use of theoretical tools such as Feynman diagrams has not found an easy place within historians' and philosophers' traditional accounts of modern physics. Most studies have followed in the spirit of a joke that the wisecracking theorist George Gamow was fond of making. Gamow used to explain to his students what he liked most about

10. Latour, "Drawing things together" (1990 [1986]). See also Latour and Woolgar, *Laboratory Life* (1986 [1979]); Latour, *Pasteurization* (1988 [1984]); Latour, *Science in Action* (1987); and Latour, *Aramis* (1996).

11. This move parallels recent work by historians of early modern printing: rather than focus on the purported "fixity" of printed (rather than manuscript) texts, recent historians have highlighted the mutability of printed texts themselves and of the varied readings they could inspire. Cf. Eisenstein, *Printing Press* (1979); with Johns, *Nature of the Book* (1998); Eisenstein, "Unacknowledged revolution revisited" (2002); and Johns, "How to acknowledge a revolution" (2002). See also Darnton, *Great Cat Massacre* (1984); Chartier, *Order of Books* (1992); de Grazia and Stallybrass, "Materiality of the Shakespearean text" (1993); and Secord, *Victorian Sensation* (2000).

being a theoretical physicist: he could lie down on a couch, close his eyes, and no one would be able to tell whether or not he was working.[12] For too long, historians and philosophers have adopted Gamow's central (not to say sleepy) metaphor: research in theory, we have been told, concerns abstract thought, wholly separated from anything like labor, activity, or skill. Theories, worldviews, or paradigms seemed to be the appropriate units of analysis, and the challenge became charting the birth and conceptual development of particular ideas. In these traditional accounts, the skilled manipulation of tools played little role: theorists were assumed to write papers whose content other theorists could understand, at least in principle, anywhere in the world. Ideas, embodied in texts, traveled easily from theorist to theorist in these accounts, shorn of the material constraints that might make bubble chambers or electron microscopes (along with the skills required for their use) difficult to carry from place to place. The age-old trope of minds versus hands has been at play: a purely cognitive realm of ideas has been pitted against a manual realm of action. In short, more "night thoughts" than desk work, more *Weltbild* than *Fingerspitzengefühl.*

During the past decade, a rival vision of how to analyze work in theoretical sciences has begun to take shape.[13] Building upon these studies, this book begins with a simple premise: since at least the middle of the twentieth century—and, arguably, during earlier periods as well—most theorists have not spent their days (or, indeed, their nights) in some philosopher's dreamworld, weighing one cluster of disembodied concepts against another, picking and choosing among so many theories or paradigms. Rather, their main task has been to *calculate.* Theorists have tinkered with models and estimated effects, always trying to reduce the inchoate confusion of "out there"—an "out there" increasingly percolated through factory-sized apparatus and computer-triggered detectors—into tractable representations. They have accomplished these translations by fashioning theoretical tools and performing calculations. Theorists have used calculational tools, in other words, to mediate between various

12. Geoffrey Chew, who studied under Gamow as an undergraduate at George Washington University during the early 1940s, repeated Gamow's joke in a May 1997 interview with Stephen Gordon, as quoted in Gordon, *Strong Interactions* (1998), 15n10.

13. See esp. Olesko, *Physics as a Calling* (1991); Warwick, "Cambridge mathematics" (1992); Warwick, "Cambridge mathematics" (1993); Warwick, *Masters of Theory* (2003); Krieger, *Doing Physics* (1992); Pickering and Stephanides, "Constructing quaternions" (1992); Buchwald, *Scientific Practice* (1995); Galison and Warwick, *Cultures of Theory* (1998); Morgan and Morrison, *Models as Mediators* (1999); Kennefick, "Star crushing" (2000); and MacKenzie, "Equation and its worlds" (2003).

kinds of representations of the natural world. These tools have provided the currency of everyday work.

Focusing on theorists' tools cuts orthogonally across conceptual histories of theoretical physics, since physicists often improvise with calculational techniques across a wide range of distinct topics or fields of inquiry. Feynman himself explained near the end of his life that he chose various topics to work on—from the scattering of electrons to the superfluid behavior of liquid helium, from vortex rings and polarization waves in crystals to superconductivity, from the scattering of constituents inside protons to the scattering of gravity waves off of planets—because such problems all fell "in the range of my tools."[14] Concentrating on the separate theory domains with which such problems are usually associated—electrodynamics, solid-state physics, nuclear and particle physics, or gravitation—obscures deeper continuities in daily practice and calculational approach. Historians Andrew Warwick and Ursula Klein have given useful names to such calculational techniques, each drawing on an analogy to instruments: Warwick introduced the term "theoretical technology," and Klein the phrase "paper tools."[15] Since the late 1940s, generations of physicists have turned more and more often to Feynman diagrams as their paper tool of choice.

In order to build an account of theorists' uses of paper tools, we must understand how tools have functioned in historians' and sociologists' studies of the laboratory and field sciences. Central to many of these studies is the notion of skill. Students rarely look through a microscope the first time and see what they are supposed to see. The point becomes all the more obvious with the gargantuan apparatus of postwar "big science": there is nothing obvious or natural (at first) about pumping high-pressure liquid hydrogen into a bubble chamber, or searching for identifiable patterns in the meandering tracks of the millions of resulting photographs. With practice and over time, however, using scientific instruments eventually becomes obvious or even second nature for accomplished practitioners. As the eighteenth-century chemist Joseph

14. Richard Feynman, Jan 1988 interview with Jagdish Mehra, as quoted in Mehra, *Beat of a Different Drum* (1994), 429. The condensed-matter theorist David Pines drew a similar conclusion about Feynman's research trajectory: "Feynman became interested because he saw the polarons as an opportunity to test the power of his path integral approach." Pines, "Richard Feynman" (1989), 66.

15. Warwick, "Cambridge mathematics" (1992); Warwick, "Cambridge mathematics" (1993); Warwick, *Masters of Theory* (2003); U. Klein, "Techniques of modelling" (1999); U. Klein, "Paper tools" (2001); and U. Klein, *Experiments, Models, Paper Tools* (2003). For examples drawn from the life sciences, see Clarke and Fujimura, *Right Tools for the Job* (1992).

Priestley argued, scientists build up these skills, not (or not only) by reading elaborate written instructions for the instruments' use, but by practicing with the tools day in and day out:

> I would not have any person, who is altogether without experience, to imagine that he shall be able to select any of the following experiments, and immediately perform it, without difficulty or blundering. It is known to all persons who are conversant in experimental philosophy, that there are many little attentions and precautions necessary to be observed in the conducting of experiments, which cannot well be described in words, but which it is needless to describe, since practice will necessarily suggest them; though, like all other arts in which the hands and fingers are made use of, it is only *much practice* that can enable a person to go through complex experiments, of this or any other kind, with ease and readiness.[16]

As Priestley admonished, and as a number of historians and sociologists have concurred more recently, experimentalists must work hard to hone something like artisanal knowledge or craft skill in addition to an understanding of general principles. No amount of formal written instructions will suffice for producing this feel for the instrument—no student jumps from reading a manual to using an instrument correctly the first time. Michael Polanyi gave a name to those skills that Priestley argued "cannot well be described in words": *tacit knowledge.*[17]

Historians and sociologists have argued that tacit knowledge plays a central role when it comes to replicating someone else's instruments, even when the would-be replicator is already an expert experimentalist or instrument maker. In cases ranging from the design of modern-day lasers, to the use of early modern air pumps and glass prisms, to the establishment of electrical standards during the height of Britain's imperial rule, no amount of written instructions, supplied at a distance from the original site, proved sufficient for successful replication. Certain features of the instruments' design and use, according to these studies, remained literally ineffable—the rules required for actual use remained impossible to specify fully, even in principle, via textual instructions alone. The key to successful replication in each of these cases—the one way to transmit the tacit knowledge required to build and use these tools—was through extended personal contact. Only those scientists who worked face to

16. Priestley, *Experiments and Observations* (1775), 2:6–7, as quoted in Levere, "Measuring gases" (2000), 111; emphasis in original.

17. Polanyi, *Personal Knowledge* (1958); Polanyi, *Tacit Dimension* (1967). See also Lave, *Cognition in Practice* (1988).

face with those already "in the know" could develop the skills and master the practices necessary to build and use these instruments.[18]

Craftlike skill, local practices, and material culture might now seem appropriate categories for analyzing laboratory and field sciences, but what about theoretical sciences? Getting the "feel" in one's fingers for how to turn a dial or solder an electrical connection is one thing, but isn't theoretical work all about manipulating representations on paper—that is, isn't it *all* about texts? Once we shift from a view of theoretical work as selecting between preformed theories, however, to theoretical work as the crafting and use of paper tools, tacit knowledge and craft skill need not seem so foreign. Thomas Kuhn raised a similar point with his discussion of "exemplars." Kuhn wrote that science students must work to master exemplars, or model problems, before they can tackle research problems on their own. The rules for solving such model problems and generalizing their application are almost never adequately conveyed via appeals to overarching general principles, and rarely appear in sufficient form within published textbooks.[19] To put the matter in a more mundane way: theoretical physicists do not enter the field on the basis of correspondence courses, sending and receiving written instructions in the absence of any face-to-face training interactions.

If the point has been underappreciated by scholars in science studies, it was by no means lost on theoretical physicists themselves. Hans Bethe, for example, lobbied hard in 1950 with the American consul general in Genoa to gain a visitor's visa for an Italian physicist, Antonio Borsellino. "Professor Borsellino has made theoretical studies about matters which are directly connected with the experimental investigations" then underway at Cornell, Bethe explained. Despite Borsellino's publications, Bethe continued, he and his colleagues had been "unable . . . to obtain enough information on Professor Borsellino's theory to enable us to use it, and his personal presence here would therefore be of great value."[20] Nor was the matter restricted to theorists who worked an ocean

18. H. Collins, "TEA set" (1974); H. Collins, *Changing Order* (1992 [1985]); H. Collins, *Artificial Experts* (1990); Shapin and Schaffer, *Leviathan and the Air-Pump* (1985); Schaffer, "Glass works" (1989); and Schaffer, "Manufactory of ohms" (1992). See also Ravetz, *Scientific Knowledge* (1971); MacKenzie and Spinardi, "Tacit knowledge" (1995); Fujimura, *Crafting Science* (1996); Pinch, Collins, and Carbone, "Inside knowledge" (1996); Collins, de Vries, and Bijker, "Ways of going on" (1997); M. Jackson, *Spectrum of Belief* (2000); and Delamont and Atkinson, "Doctoring uncertainty" (2001).

19. Kuhn, *Structure* (1996 [1962]), 187–98.

20. Hans Bethe to the American consul general, Genoa, Italy, 28 Nov 1950, in *HAB*, Folder 9:60. See also the one-page memorandum by Guiseppe Coccioni, 19 Dec 1950, in the same folder. Bethe and Coccioni had difficulty gaining the visitor's visa for Borsellino because of the recently enacted McCarran Act, which blocked visas to former members of such groups as the Fascist Party

apart and who were not well acquainted personally. A few years earlier, in the summer of 1947, Enrico Fermi had complained that he was unable to make sense of one of Bethe's own recent papers, and hence could not reproduce and extend Bethe's calculation. Fermi and Bethe were both experts in the field in question, and they had worked closely together throughout the war years; they knew the territory and they knew each other quite well. All the same, however, only after Fermi and Bethe met together again that summer, where they could talk face to face about Bethe's work, could Fermi follow all the ins and outs of the calculation and carry it forward.[21] In both of these examples, senior theorists had difficulty making adequate sense of each other's work to build on it without informal personal communication. As we will see throughout part 1 (chaps. 2–4), the issue was magnified greatly when it came to younger theorists, still in the midst of their training. Something like tacit knowledge—or at least extended personal contact over and above the exchange of formal written instructions—proved crucial for spreading Feynman diagrams around.

Note that I wrote "something like tacit knowledge": Priestley and Polanyi notwithstanding, the category of tacit knowledge has not been without its detractors, and important cautions must be acknowledged before we import the category into studies of theoretical sciences. Myles Jackson, for example, raises the important point that items of scientific practice may remain "tacit" for many different reasons: trade secrecy and designs to protect priority are hardly the same as an in-principle inability to express all the nuances of experimental protocols. Stephen Turner asks how, if tacit knowledge is truly tacit, historians, philosophers, and sociologists can ever analyze it, let alone prove that it was the causal agent behind a scientist's work. Kathryn Olesko notes that much of what has formerly been classified as tacit knowledge—such as a scientist's judgments about how to analyze and reduce data—was in fact subject to explicit codification. And Peter Galison cautions against reifying skill: many aspects of laboratory life, Galison finds, were easily replicable at a distance, while others remained more stubbornly rooted to time and place.[22]

---

in Italy. Bethe explained that Borsellino had never been a party member and so was eager to head off potential State Department opposition from the start. We will examine similar problems that other physicists encountered regarding visas and passports during this period in chap. 9.

21. Enrico Fermi to Edwin Uehling, 26 Sep 1947, as quoted in Schweber, *QED* (1994), 232. The paper in question was Bethe's nonrelativistic treatment of the Lamb shift, which we will examine in chap. 2.

22. M. Jackson, *Spectrum of Belief* (2000), 10–13; Turner, *Social Theory of Practices* (1994), esp. chaps. 4, 6; Olesko, "Tacit knowledge" (1993); and Galison, *Image and Logic* (1997), 52–54.

Even with these well-placed cautions and caveats, I find tacit knowledge helpful as a phenomenological description of how Feynman diagrams spread. Exceedingly few physicists picked up the diagrams and applied them in research based only on reading articles about the new techniques. In almost every case, some form of informal personal communication—usually of an explicitly pedagogical kind, such as an adviser mentoring graduate students, or postdocs working closely together—can be traced behind the scenes of physicists' uses of Feynman diagrams. Looking at the work that physicists performed before and after they interacted with members of the growing diagram-using network quickly reveals a stark pattern: even years after the diagrams had been described at length in print, individuals—and even whole departments—that were not in personal contact with members of the diagrammatic network did not make any use of the diagrams, whereas those people who interacted regularly with other diagram users made frequent use of the diagrams. Without reifying "skill" or trafficking in inscrutable or mystical know-how, the concept of tacit knowledge remains useful because of its emphasis upon these nontextual means of transmission.[23]

Yet in one important example that we will consider in chapter 4—a small group of physicists working in the Soviet Union during the early years of the cold war—some theorists *did* pick up the diagrams and make use of them without any discernible personal contact with other diagram users. Clearly *something* about the diagrams could travel via texts alone. The question thus becomes not *whether* they could travel in principle, but rather *what exactly* traveled via these published sources? The handful of physicists who learned about the diagrams only from texts developed an adequate facility with one way to calculate with the diagrams. What seems not to have been transmitted via texts was a sense for how else the diagrams could be applied as useful tools. That is, some form of understanding could be packaged and transmitted via texts, but not the more improvisational uses developed by those groups of physicists who shared informal contacts. Infinite epistemic barriers—the "all or nothing" lurking behind most analyses of tacit knowledge—did not separate those who learned about the diagrams in person from the small minority who learned about them only from publications. Nor will we be aided by Kuhn's notion of "incommensurability": the gap between diagram users and the rest was not a hopeless conceptual mismatch so much as differences in physicists' preferences for one type of tool over another. A spectrum emerged, as physicists in various

23. A similar point is made in Trevor Pinch's review of Turner's *Social Theory of Practices:* Pinch, "Old habits die hard" (1997).

local settings made choices about what to work on, how best to deploy their hard-won tools, and what kinds of calculations their students should practice and master. Few aspects of these choices are captured by talk of "knowledge," "information," or "conceptual understanding." Rather, the idiom is one of skills and tools.

PEDAGOGY AND POSTWAR PHYSICS

Central to the argument of this book is that once we begin to examine the tools of theory, we must also consider the tool users—and thereby we enter the realm of pedagogy and training. Rather than construe "pedagogy" in the narrow sense of classroom teaching techniques—though these are certainly important —I interpret "pedagogy" more broadly. I am interested in the institutions of training by means of which young physicists became working theorists during the decades after World War II. Much of this training and apprenticeship took place outside the classroom. Thomas Kuhn raised similar themes with his discussion of exemplars in the theoretical sciences, though his prescient analysis left many questions open. In particular, he did not explore in any historical detail how such exemplars emerged, how students in various generations actually learned to solve exemplars and build upon them in their research, or whether students at different training centers learned about and leaned upon exemplars in distinct ways.[24] Kathryn Olesko and Andrew Warwick have begun to fill in Kuhn's picture. In their studies of nineteenth-century Königsberg and Cambridge (respectively), they have demonstrated that the transfer of theoretical skills was tied directly to new pedagogical instruments, such as new types of seminars, problem sets, paper-based examinations, and private coaching. As these new teaching techniques emerged over the course of the nineteenth century, so too did a distinct specialty called "theoretical physics," along with a critical mass of young physicists who now considered it their own.[25]

These detailed historical studies point to institutional and pedagogical reforms and their bearing on the form and content of knowledge. We learn from their accounts how new skills were inculcated in a particular setting and how these techniques became second nature to students in training there.[26] To understand physicists' incorporation of Feynman diagrams during the middle

24. Kuhn, *Structure* (1996 [1962]).

25. Olesko, *Physics as a Calling* (1991); Warwick, *Masters of Theory* (2003). Cf. Jungnickel and McCormmach, *Intellectual Mastery of Nature* (1986).

26. Cf. Traweek, *Beamtimes and Lifetimes* (1988); Smith and Wise, *Energy and Empire* (1989), esp. chaps. 1–5; Gooday, "Precision measurement" (1990); Gooday, "Teaching telegraphy" (1991);

decades of the twentieth century, we must remain similarly sensitive to changes in physicists' institutions and infrastructure. In particular, the diagrams' rapid dispersion raises the question of how theoretical skills—honed in local settings, subject to all the contingencies of time and place—became shared across so many distinct locations and research groups. That is, how did local practices take on the trappings of universal methods? Sustained pedagogical work proved crucial, connecting physicists at Cornell, Princeton, Berkeley, Cambridge, Tokyo, and a half dozen other sites in between. To understand these mechanisms of transfer, we must throw our historical net wide, tracing developments in each of these far-flung places; no single site will do.

During the 1940s and 1950s, the diagrams were adopted quickest and to the widest extent by physicists working in the United States—and thus it is crucial to understand the broader American setting within which Feynman diagrams began to disperse. The early postwar years presented unprecedented challenges to American physicists: following the dramatic (and well-publicized) wartime efforts of physicists on radar and the atomic bomb, and with the aid of such measures as the G.I. Bill, enrollments in physics graduate programs grew at nearly twice the rate of all other fields combined. Just four years after fighting had ended, the nation's physics departments were granting three times as many doctorates per year as the prewar highs—a number that would soon climb by another factor of three after the surprise launch of Sputnik. From the late 1940s through the end of the 1960s, the tremendous and rapid growth in American physics departments was felt viscerally in small, medium, and large departments alike. No aspect of daily life remained untouched, from the lack of office space and overcrowded laboratory conditions, to the proliferation of formal bureaucratic procedures where informal precedents had once held sway, to widespread feelings of "facelessness" and a perilous loss of "intimacy" amid the overflowing lecture halls. Even routine social events, such as the Berkeley physics department's annual picnics, became logistically exhausting affairs.[27] At the same time, the U.S. Atomic Energy Commission began to construct a coast-to-coast series of national laboratories, in part to keep the nation's physicists "on tap" and at the ready in case of the next outbreak of fighting.[28] The ever-growing cyclotrons at the new laboratories quickly flooded

Geisen and Holmes, *Research Schools* (1993); Kohler, *Lords of the Fly* (1994); Rudolph, *Scientists in the Classroom* (2002); and Kaiser, *Pedagogy and the Practice of Science* (2005).

27. Kaiser, "Cold war requisitions" (2002); Kaiser, "Suburbanization" (2004).

28. Seidel, "Home for big science" (1986); Forman, "Behind quantum electronics" (1987); and Westwick, *National Labs* (2003).

high-energy physicists with more data on particle scatterings and interactions than ever possible before.[29]

In the midst of skyrocketing enrollments, overflowing lecture halls, and unprecedented reams of experimental data, American physicists suddenly were faced with the twin questions: how do you train roomfuls of students instead of handfuls of disciples, and what do you train them to do? Institutionally, their responses centered on revamped postdoctoral training, refurbished summer schools, and new series of informal and inexpensive textbooks. Intellectually, most theorists interested in nuclear and high-energy phenomena redoubled their efforts to develop phenomenological models of use to their experimentalist colleagues. To many of these theorists, Feynman diagrams seemed to offer the ideal pedagogical resource for their newfound challenges: tailor made (or, rather, tailorable) for efficient, repeatable—and hence trainable—calculations, they gave theorists and their students a way to "get the numbers out" with greater efficiency than ever before. The oft-discussed pragmatic character of American science, and of postwar American theoretical physics in particular, must be understood in terms of these changes in physicists' infrastructure.[30] As we will see in chapters 4, 6, and 8, physicists working in other parts of the world faced different challenges after the war: not fables of abundance but rigors of reconstruction. Working within their own institutional constraints, when physicists outside the United States picked up Feynman diagrams and put them to work, they often did so in ways distinct from those of their American colleagues. Neither of the large-number pressures—student enrollments or new experimental data—affected physicists outside the United States in the same ways. These differences were inscribed in the calculations they undertook and the diagrams they drew.

OVERVIEW: THE TWO MEANINGS OF DISPERSION

Physicists' uses of Feynman diagrams from the late 1940s through the late 1960s present a series of puzzles. Practically no physicists could understand

29. Pickering, *Constructing Quarks* (1984), chap. 3; Galison, *How Experiments End* (1987), chap. 4; Galison, *Image and Logic* (1997), chap. 5–7; Brown, Dresden, and Hoddeson, *Pions to Quarks* (1989); and Polkinghorne, *Rochester Roundabout* (1989).

30. The phrase "get the numbers out" is borrowed from Schweber, *QED* (1994), chap. 3. See also Cini, "Dispersion relations" (1980); Schweber, "Empiricist temper regnant" (1986); Pickering, "From field theory to phenomenology" (1989); Assmus, "Americanization of molecular physics" (1992); and Galison, *Image and Logic* (1997), chap. 4. Cf. Kohler, "Ph.D. machine" (1990); and Oreskes, *Rejection of Continental Drift* (1999), chap. 5.

what Feynman was doing with his unusual doodles during his inaugural presentation in the spring of 1948, yet other physicists picked up the diagrams, often without comment, only a few months later. Seemingly standard rules for drawing and calculating with the diagrams were in place by early 1949, yet differences in the diagrams' pictorial form, calculational use, and theoretical interpretation quickly multiplied. In fact, the main purpose for which the diagrams had been fashioned seemed entirely hopeless for tackling the pressing problems of the early and mid-1950s, yet physicists clung to their diagrams, tweaking them here and there, even as they discarded other elements of their calculational instrumentarium. These puzzles point to three main questions, around which this book is organized. Chapters 2–4 address the question, *how* did the diagrams spread so quickly? Chapters 5–9 ask, for *what* did physicists actually use the diagrams during the two decades after their introduction? Given the great variety of uses and interpretations, chapter 10 examines the question, *why* did the diagrams "stick," even as related techniques came and went?

In order to make sense of the themes of transmission and differentiation, I draw upon the term "dispersion." One cluster of meanings is especially pertinent to the book's first question, regarding how the diagrams spread: "To distribute from a main source or centre . . . ; to put into circulation." A second meaning of "dispersion" points to the ever-expanding circle of competing uses forged for the diagrams: "To cause to separate in different directions . . . ; to spread in scattered order."[31] "Dispersion" captures at once the work required to make paper tools travel and the plasticity of those tools once they do travel; both meanings are essential. Here, in outline, are some of the main contours of the diagrams' dispersion:

How did the diagrams spread? If the tools of theory were not (or not only) conveyed as disembodied textual information, but rather involved something closer to craft skill and artisanal knowledge, then what mechanisms did physicists draw upon to put the tools into circulation? As we will see in chapters 2–4, Feynman diagrams spread by means of several new institutional arrangements, each taking form in the years after World War II. Most important within the United States was the rise of postdoctoral education for theoretical physicists. Although postdoctoral training had been introduced in the sciences during the early decades of the twentieth century, it became a standard element of

31. Simpson and Weiner, *Oxford English Dictionary* (1989), s.v. "disperse." For historiographical (rather than etymological) clarity, I have combined the *OED*'s definitions 4a and 4b for the first meaning of "disperse" quoted here and combined definitions 1a and 2b for the second meaning of "disperse." Cf. Jordan and Lynch, "Plasmid prep" (1992).

American theoretical physicists' training only after the war. Its champions custom-designed postdoctoral training for the cultivation of craftlike skill. Equally important, postdocs were seen as the best way to circulate new tools: after completing their graduate work, physicists almost always moved to a new site for their postdoctoral training before moving to a third site to begin teaching. Leading science policy makers engineered postdoctoral training to feature precisely these design specs: skill acquisition and circulation. Similar, though by no means identical, pedagogical mechanisms of circulation took form within other countries, such as Japan, just at the time that Feynman diagrams were introduced. Ancillary pedagogical materials rose up at the same time, further aiding the diagrams' spread: preprints (another prewar invention, put on a firmer foundation after the war), widely circulating (though unpublished) lecture notes, and soon new series of textbooks and reprint volumes. The new traffic in pedagogical texts reinforced, though rarely replaced, the primary circulation of students and postdocs.

What did physicists use the diagrams to do? As anthropologists such as Claude Lévi-Strauss have long emphasized, tools almost always prove malleable and multivalent. "Primitive peoples," explained Lévi-Strauss, constantly engage in bricolage—making do with the tools at hand. As he emphasized, a bricoleur's set of tools is not tied to a specific task or program; rather, the tools are accumulated over time and deployed on a case-by-case basis. The tool user's first move, in Lévi-Strauss's account, is thus retrospective: What tools do I already have? How have I used them in the past? What rearrangement might work now?[32] Lévi-Strauss clearly intended his account to apply to modern scientists and engineers as well; over the past twenty years, historians and sociologists have obliged, producing dozens of detailed descriptions of how experimental scientists tinker and improvise with tools, extending them piecewise for new investigations.[33] Theoretical physicists proved no less prolific at refashioning the tools at hand. Consider the examples in figure 1.3, each of which appeared with the label "Feynman diagram" between 1949 and 1954. As we will see throughout chapters 5–9, theorists tinkered with the new

32. Lévi-Strauss, *Savage Mind* (1966 [1962]), 16–22. See also Lynch, *Art and Artifact* (1985), chap. 1.

33. See esp. Gooding, Pinch, and Schaffer, *Uses of Experiment* (1989); Galison, *How Experiments End* (1987); Galison, *Image and Logic* (1997); Van Helden and Hankins, *Instruments* (1994); Pickering, *Mangle of Practice* (1995); Rasmussen, *Picture Control* (1997); and Holmes and Levere, *Instruments and Experimentation* (2000). Historians of life sciences have explored similar themes in terms of model organisms: Kohler, *Lords of the Fly* (1994); Rader, "Mouse people" (1998); Rader, "Of mice, medicine, and genetics" (1999); and Creager, *Life of a Virus* (2002).

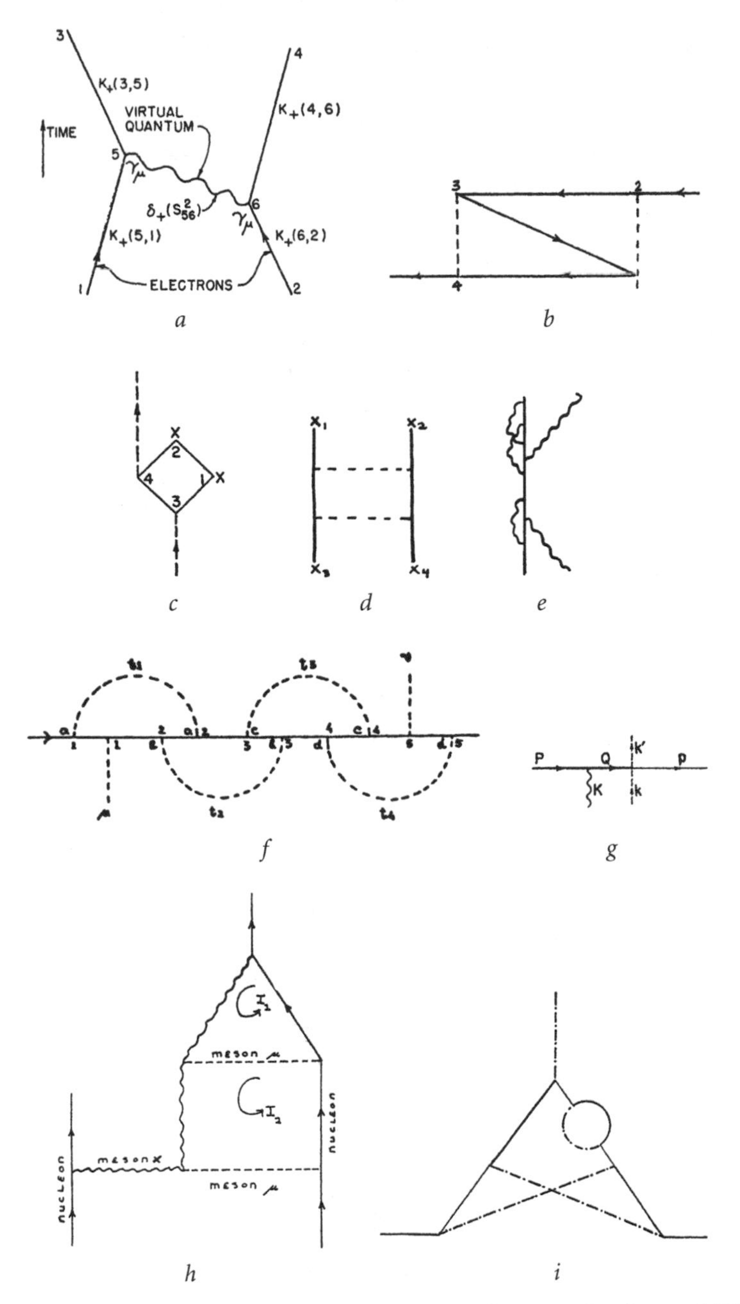

Figure 1.3. Feynman diagrams, 1949–54. (Sources: *a*, Feynman [A.4], 772; *b*, Villars, "Quantum electrodynamics" [1951], 65; *c*, Rohrlich and Gluckstern [A.54], 2; *d*, Gell-Mann and Low [A.43], 352; *e*, Low [A.65], 55; *f*, Salam [A.56], 735; *g*, Lenard [A.81], 971; *h*, Steinberger [A.6], 1182; *i*, Kroll and Ruderman [A.102], 235.)

diagrams, adapting them for use in the problems they deemed most relevant or pressing.[34]

These adaptations were neither random nor produced in vacuo. Instead, order may be brought to this scattershot display by focusing on theorists' training. Art historians have long been accustomed to classifying painters according to their training, using descriptive phrases such as "follower of van Eyk," or "from the school of Titian." Historians of science stand to make similar gains by considering the pedagogical links between young instructors and their students as these played out in various departments. In each local setting, physicists adapted features of the diagrams to better bring out aspects deemed most important for new kinds of calculations. More often than not, theorists trained their students to practice using the diagrams the way they used them. The diagrams drawn by young physicists at Cornell thus began to look different—and to be used in ways subtly distinct—from those drawn by their peers at Rochester, the University of Illinois, or Cambridge. (With practice, one can actually "predict" where a physicist was trained based on the diagrams he or she drew and the kinds of calculations in which the diagrams were enrolled.) Consider the examples in figure 1.4: in each pair, the diagram on the left comes from an adviser in one of the major training centers, and the example on the right from someone the adviser trained.[35] As we will see in chapters 5 and 6, mentors and students crafted the diagrams in locally varying ways, in pursuit of distinct types of calculations.

More than this, physicists in various places differed on what their diagrams purported to show. To some, the diagrams functioned as pictures of

34. Physicists developed several closely associated (and genealogically related) diagrams during the 1950s and 1960s, such as Lévy diagrams, diagrams for use with the Tamm-Dancoff approximation, Goldstone, Landau, and Cutkosky diagrams, and so on. Most of these stood in for classes of Feynman diagrams, such as all those with two particles in the intermediate state. These "Feynman-like" diagrams are also included in this study: where some physicists spoke of "Lévy diagrams," for example, others continued to label them simply "Feynman diagrams." Moreover, physicists routinely talked about relations between these types of diagrams. The label "Feynman diagram" did not extend infinitely, however: as we will see in chap. 10, there were other types of diagrams in play at the time, such as "dual diagrams," that physicists treated quite differently from Feynman diagrams and were often at pains to distinguish; they had little trouble telling them apart.

35. The examples from Cornell, Columbia, and Rochester all involve advisers and their graduate students. The cases of Chicago and the University of Illinois involve older theorists who learned about the diagrams from their colleagues. In the final example, from Oxford and Cambridge, the putative "advisee," John Ward, had already made use of Feynman diagrams for different types of analyses and talked extensively with his colleague, Abdus Salam, about how to use the diagrams for the types of calculations Salam had been working on; the diagrams Ward began to draw shifted accordingly.

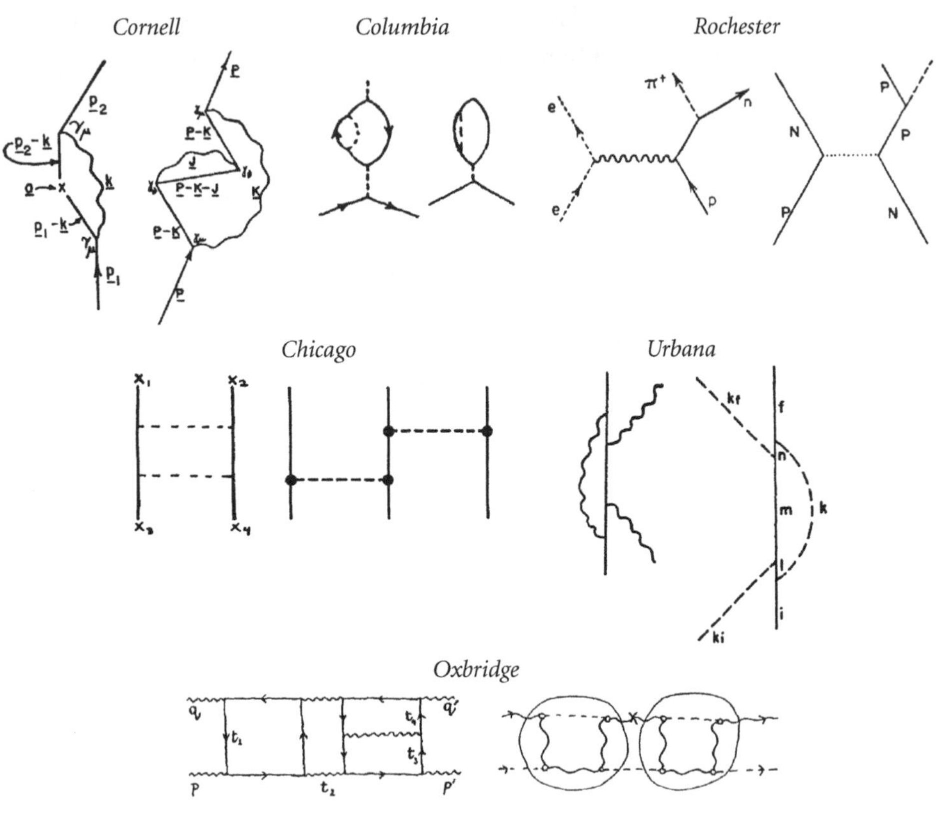

Figure 1.4. "Family Resemblances." Mentors and students crafted diagrams for different purposes. (Sources: *Cornell,* Feynman [A.4], 775; Frank [A.41], 1190; *Columbia,* Karplus and Kroll [A.11], 537; Weneser, Bersohn, and Kroll [A.90], 1258; *Rochester,* Marshak, *Meson Physics* [1952], 39; Simon [A.15], 574; *Chicago,* Gell-Mann and Low [A.43], 351; Wentzel [A.73], 684; *University of Illinois at Urbana,* Low [A.65], 55; Chew [A.117], 1749; *Oxford and Cambridge,* Salam [A.32], 223; Ward [A.46], 899.)

physical processes—they seemed to capture something essential about the mechanisms of the microworld. To others, the diagrams were no more than helpful mnemonic aids for wading through long strings of complicated mathematical expressions—they were not to be confused with the stuff of the real world. Still others developed the diagrams as tools for a new kind of diagrammatic reasoning—the diagrams' structural or topological features prompted and enabled the investigation of various symmetries that their associated mathematical expressions should obey. Most often, these distinct roles blurred together in practice. Feynman diagrams thus functioned as many other influential scientific diagrams have done—ranging from stratigraphic columns, to chemical formulas and atomic orbitals, to pictures of immunological antibodies—subject to the same slippage between realistic depiction, convenient calculational device, and heuristic guide to further research.[36]

Chapters 7–9 push the theme of the diagrams' many functions further, following physicists' efforts from the mid-1950s through the 1960s to pull together a new framework for studying nuclear particles' interactions. The Berkeley theorist Geoffrey Chew's "*S*-matrix program" grew out of the piecewise adaptation of Feynman diagrams, together with new rules (or, most often, rules of thumb) for their use. Chew and his students and collaborators came to read content in the diagrams' unadorned lines that ran counter to much of the diagrammatic machinery that had come before. Indeed, they argued that their new program, borne on the backs of reinterpreted Feynman diagrams—and which always seemed to be just around the corner, never quite at hand in final form—augured the death of quantum field theory itself, from which the diagrams' original rules had been derived. Chew's creative appropriation of the diagrams, which grew to dominate studies of the strong nuclear force during the 1960s, drew upon a mixed bag of conceptual resources. As he struggled to make sense of the strong force, Chew engaged in several other struggles as well, from political strife in the age of McCarthyism to unusual ideas about how to train large cadres of graduate students at a time when physics enrollments were booming. In pulling his particular package of diagrammatic techniques together—centered on what he called "nuclear democracy"—Chew drew metaphorically upon his earlier, intense efforts to forge "democracy" in other corners of his life. Physicists always appropriated Feynman's diagrams against a backdrop of local motivations and resources, forged by prior training and experience; Chew's fascinating work highlights the range of interpretations that physicists fashioned for Feynman diagrams.

Why did the diagrams stick? The challenge that Chew and his Berkeley group mounted to the prevailing framework for studying particles' interactions highlights a final question: why did physicists cling to Feynman diagrams, even as so many other elements of their toolkit came and went? Why, moreover, did physicists cling stubbornly to their Feynman diagrams when working on problems far beyond the domain for which the diagrams' effectiveness and efficiency had been demonstrated? In answering this final question, we must look beyond the myriad local differences and variations charted in earlier chapters and ask about more broadly shared features of physicists' training.

36. See esp. Rudwick, "Visual language" (1976); U. Klein, *Experiments, Models, Paper Tools* (2003); Park, "Quantum chemistry" (2005); and Cambrosio, Jacobi, and Keating, "Ehrlich's 'beautiful pictures'" (1993).

Art historians have long pondered such questions as why certain pictorial styles persist. Surveying the grand sweep of representational styles that have dominated at one time or another throughout world history, the resolution of this question can hardly be, "because that's just how the world looks." Likewise with Feynman diagrams, physicists always had choices about what tools to use and what problems to study. Many times during the postwar decades, physicists chose to work with Feynman diagrams even for those problems that—on the face of it—were *least* well-suited to diagrammatic treatment. No "natural fit" argument will do.

Feynman diagrams owed their persistence to broadly shared elements of physicists' training, which had become standard across the many local groups scattered throughout the world. Most physicists after the war came to draw and teach their Feynman diagrams according to much older pictorial conventions for depicting objects' paths through space and time. The visual connections with these older techniques, which had become standard fare for physics undergraduates by the early postwar years, help explain why Feynman diagrams stuck: they could be taught in ways that borrowed from more elementary skills that had already become second nature for most young physicists. On top of this, physicists often worked with their Feynman diagrams much the way they worked with other visual representations of particles' scattering, including the millions of bubble-chamber photographs that flooded the work of the postwar generation of nuclear and particle physicists. Issues of realism and reification, overlaid upon questions of pedagogical inculcation, help explain why physicists chose Feynman diagrams so consistently as their tool of choice.

* * *

Feynman diagrams are clearly artifactual. In this sense, we may speak of them as "constructions," or even as "social constructions." Yet to halt the investigation here would mean stopping short: it would do no more to clarify theorists' treatment of the diagrams than would equating all forms of written expression, from Romantic poetry to the federal tax code, as "social constructions." Our next task within postconstructivist science studies is to distinguish between kinds of constructions; the means by which constructions make sense for certain scientists in particular times and places; how they spread among large and heterogeneous communities; and how scientists put these constructions to work, generating new ideas or providing a heuristic scaffolding for others. Paper tools and pedagogy offer powerful means with which to tackle these questions.

PART I

Dispersing the Diagrams, 1948–54

· 2 ·

An Introduction in the Poconos

[Feynman's method] contains more arbitrary assumptions than Schwinger's which, however, do not affect the result; moreover, it is rather simpler than Schwinger's . . . and seems to lead to the same final results.

HANS BETHE, 1948[1]

Feynman introduced his diagrams in the spring of 1948 as an aid for tackling difficulties that had challenged theoretical physicists for decades. Following a brief review of these earlier efforts, this chapter turns to the diagrams' early dispersion, charting the diagrams' rising popularity between 1948 and 1954. During this brief period, the number of adherents of Feynman's new diagrammatic techniques grew from a lone user (Feynman himself) to include over one hundred diagram enthusiasts. All the more striking, this broad community grew without the aid of textbooks: the first English-language textbooks on the new methods were published only in 1955, after this first wave of conversion.[2] The diagrams' rapid dispersion depended much more strongly on personal contact and particular pedagogical relationships than on published texts. The task ahead (taken up in this chapter along with chaps. 3 and 4) thus turns on prosopography: who used the new diagrams, and how did they come to use them?[3]

1. Hans Bethe, "Electromagnetic shift" (1948), 13.

2. Two textbooks were published in 1953 that treated some of the new diagrammatic developments (one in Japanese and the other in Russian), although these books had no impact on the main community of diagram users centered in the United States until they were translated into English in 1956 and 1957, respectively. See Umezawa, *Quantum Field Theory* (1956); and Akhiezer and Berestetskii, "Quantum electrodynamics" (1957).

3. Cf. Shapin and Thackray, "Prosopography" (1974); Pyenson, "Who the guys were" (1977); and Geisen and Holmes, *Research Schools* (1993).

QUANTUM ELECTRODYNAMICS AND THE PROBLEM OF INFINITIES

As many Americans braced themselves for their entry into the Second World War, few bright features greeted their view of the recent past: the Great War, the failure of the League of Nations, the Great Depression, and the onset of fascism in Europe hardly bespoke ringing triumphs or accomplished glories. Yet as American and European-émigré physicists gathered for their wartime service in the early 1940s, they could recount with evident pride the accomplishments within physics from those same troubled decades: first Einstein's special and general relativity, and soon thereafter the Continental birth of quantum mechanics. Physicists, and especially the new specialists, theoretical physicists, had achieved unexpected insights into the worlds of the lightning fast, the cosmologically large, and the subatomically small. Even as the world around them seemed to lumber on with its troubles, physicists could recall heroic and revolutionary times.[4]

Still, as these same physicists knew well by the early 1940s, the riches of relativity and quantum mechanics resisted combination. Soon after Paul Dirac introduced his now-famous quantum-mechanical treatment of the electromagnetic field and his relativistic treatment of the electron in 1927–28, physicists in Europe, along with a string of young Americans on European postdoctoral visits, began to discover a sickness in quantum electrodynamics (QED), their new quantum-mechanical description of electromagnetic forces: straightforward calculation of nearly any process produced unphysical infinities, rather than finite numbers. The simplest approximation to the problem of electrons scattering off electrons, or of light scattering off electrons, produced results in good agreement with experimental data, but as soon as theorists tried to nudge their approximation another step or refine their calculation with additional correction terms, they invariably hit the troublesome infinities. When applied separately to problems of macroscopic motion or atomic dynamics, relativity and quantum mechanics fared exceedingly well. For questions involving both the very fast and the very small, however, theorists had more or less crawled to a standstill during the 1930s.[5]

4. For reviews, see esp. Galison, Gordin, and Kaiser, *History of Modern Physical Science* (2001), vols. 1, 2, and 4; Stachel, "History of relativity" (1995); Jammer, *Conceptual Development* (1966); Darrigol, *From c-Numbers to q-Numbers* (1992); and Beller, *Quantum Dialogue* (1999).

5. See esp. Schweber, *QED* (1994), chaps. 1, 2. Several of the original articles have been reprinted in Schwinger, *Selected Papers* (1958), and in Miller, *Early Quantum Electrodynamics* (1994). See also Weinberg, "Search for unity" (1977); Darrigol, "Origin of quantized matter waves" (1986); Pais,

Quantum Fields and Virtual Particles

In the midst of their efforts to cobble together a new quantum mechanics to treat the physics of atoms during the mid-1920s, several theorists—Werner Heisenberg, Pascual Jordan, Wolfgang Pauli, Paul Dirac, and others—began trying to quantize Maxwell's electromagnetic field as well. Jordan began the process in the midst of his work on Heisenberg's matrix mechanics, composing a long and difficult section of the famed *Dreimännerarbeit,* or "three-man paper" by Heisenberg, Jordan, and Max Born of 1926, on suggestive ways to quantize vibrating strings (seen as a first step toward treating waves and fields). The following year, Paul Dirac demonstrated that the electromagnetic field's infinite number of degrees of freedom could be decomposed as a sum over quasi-particulate oscillators, each corresponding to a specific frequency or energy—a representation, soon dubbed "second quantization," that Jordan quickly extended to matter fields as well. By 1927, Jordan had become convinced that all physical quantities—everything from the electrons and protons of ordinary matter to the electromagnetic fields that bound them together into atoms—arose ultimately from quantum fields. Although resisting some of Jordan's maneuvers at first, other theorists, including Heisenberg and Pauli, soon came to share Jordan's view.[6]

That same year, while still working out some consequences of his matrix mechanics, Heisenberg introduced his uncertainty principle. In its usual formulation, the uncertainty principle stipulates that a system's simultaneous position and momentum cannot be determined with arbitrarily high accuracy. A quickly derived variant held that the energy associated with a given process and the time over which that process takes place similarly cannot be determined with arbitrary accuracy: $\Delta E\, \Delta t \sim h$, where h is Planck's constant, and ΔE and Δt represent the uncertainty in energy and time, respectively. The science writer Isaac Asimov once described this energy-time uncertainty relation with a telling metaphor: a naughty schoolchild could break the classroom rules for short periods of time and avoid getting caught so long as the teacher's

Inward Bound (1986), chaps. 15–18; Kragh, *Dirac* (1990), chaps. 6–10; Miller, "Frame-setting essay" (1994); and Cao, *Conceptual Developments* (1997), pt. 2. An excellent concise summary may be found in Weinberg, *Quantum Theory of Fields* (1995), 1:1–48.

6. Dirac, "Quantum theory of radiation" (1959 [1927]); Born, Heisenberg, and Jordan, "Quantenmechanik" (1926); Jordan, "Quantenmechanik des Gasentartung" (1927); Heisenberg and Pauli, "Quantenelektrodynamik" (1929–30); Darrigol, "Origin of quantized matter waves" (1986); Pais, *Inward Bound* (1986), 334–40; Schweber, *QED* (1994), 23–56; and Miller, "Frame-setting essay" (1994), 18–28.

back was turned. A simple infraction (with a correspondingly small ΔE), such as sticking out his tongue, could be done over a relatively long period of time (large Δt), whereas a more outrageous stunt (a large ΔE), such as standing on his desk, would have to be wrapped up much more quickly (small Δt) so as to avoid getting caught before the teacher turned around again.[7] As with elementary schoolchildren, so with elementary particles: when Heisenberg and his colleagues began to ply his uncertainty principle in the context of quantized fields, they realized that the exact state of a quantum field could never be specified with absolute certainty at a given time. Like the schoolchildren's pranks, small fluctuations in the field would occur all the time, over small enough time intervals (as set by Heisenberg's energy-time relation).[8] Given Dirac's and Jordan's second-quantized representation of fields as collections of quanta, Heisenberg's uncertainty principle meant that quanta of a field could be created spontaneously (with some amount of energy, ΔE) as long as they were destroyed or reabsorbed within a correspondingly short time, Δt, so as to maintain $\Delta E \Delta t \sim h$. These short-lived excitations, soon dubbed "virtual particles," would pop into and out of existence all the time, "borrowing" energy and momentum and then paying them back again some time later.

By 1932 two young theorists, Hans Bethe and Enrico Fermi, had put virtual particles to work: in their scheme, virtual particles served as the carriers of forces between other particles. The electrostatic repulsion between like charges, for example, arose in their scenario by the exchange of virtual photons, the quanta of the electromagnetic field. One charged particle, such as an electron, could emit a virtual photon, which would carry some energy and momentum; the virtual photon, in turn, would be absorbed by a second electron, transferring its energy and momentum. Forces thus arose from direct, local interactions—the emission and absorption of virtual particles—rather than acting at a distance. Over the course of the 1930s, theorists elevated this simple mechanism into the heart of quantum field theories: all interactions arose from the exchange of virtual particles. The Japanese theorist Yukawa Hideki demonstrated that this scheme could work for forces beyond electrodynamics. As he showed in 1935, the attractive force between neutrons and protons, which kept

7. Asimov, *Asimov on Physics* (1976), 181. Heisenberg introduced his uncertainty principle in Heisenberg, "Über den anschaulichen Inhalt" (1927).

8. The question of how accurately various components of the quantized electromagnetic field may be measured at a given spacetime point, in the light of Heisenberg's uncertainty principle, became hotly contested during the early 1930s. See Landau and Peierls, "Erweiterung des Unbestimmtheitsprinzips" (1931); Bohr and Rosenfeld, "Measurability of electromagnetic field quantities" (1979 [1933]); Schweber, *QED* (1994), 111–12; and Hall, *Purely Practical Revolutionaries* (1999), 249–83.

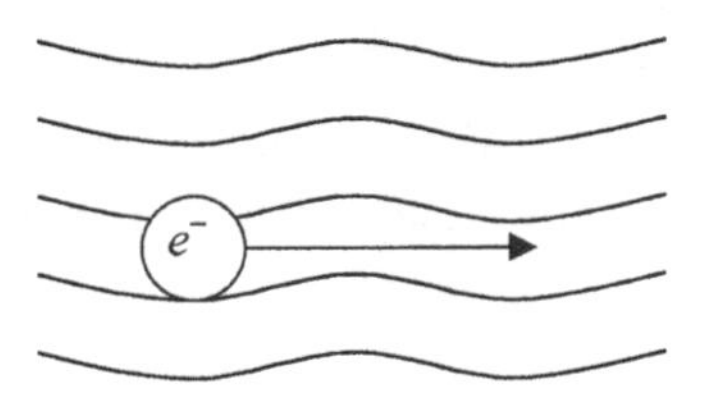

Figure 2.1. Self-energy. An electron traveling in its own electromagnetic field, much like a ball traveling in an incompressible fluid, will behave as if its total mass were different from its ordinary, mechanical mass outside the fluid.

them bound tightly within an atomic nucleus despite the protons' electrostatic repulsion, could be mediated by the exchange of virtual quanta from a new quantum field. If the mass of this new field's quanta were tuned just so, to lie between the mass of electrons and protons, then the range over which they could be exchanged while still obeying Heisenberg's energy-time uncertainty relation would correspond roughly to nuclear distances. Although no such "mesotrons"—named for their intermediate mass—had as yet been discovered, Yukawa's scheme at least showed that nuclear forces might be amenable to a quantum-field-theoretic treatment.[9]

Virtual particles thus offered theorists a new tool with which to model subatomic interactions, but not without a price. Nearly as soon as theorists began to study the behavior of virtual particles, they realized that the ghostlike states would give rise to corrections to electrodynamic quantities and processes. For example, an electron could spontaneously emit a virtual photon and reabsorb it a short time later; this would correspond to the electron's interacting with its own electromagnetic field. The energy and momentum carried by the virtual photon during its brief existence would affect the electron's overall energy and momentum; the "self-energy" interaction would change the effective mass that the electron would appear to have.[10] (See fig. 2.1.) Heisenberg and Pauli turned to the question of an electron's self-energy in 1929–30, calculating the effect of virtual photons on the electron's mass. They realized that the electron's effective mass ($m_{\rm eff}$)—that is, the mass that physicists would measure in experiments—would be a compound, made up of two parts. Being

9. Bethe and Fermi, "Wechselwirkung" (1932); Yukawa, "Elementary particles" (1935); Laurie Brown, "Yukawa's prediction of the meson" (1981); Darrigol, "Quantum electrodynamical analogy" (1988); Schweber, *QED* (1994), 78, 91; Carson, "Exchange forces," pt. 2 (1996); and Brown and Rechenberg, *Concept of Nuclear Forces* (1996), chap. 5.

10. Mathematical physicists had treated analogous problems—such as the motion of bodies in incompressible fluids and electric charges in electromagnetic fields—during the late nineteenth century. See Dresden, "Renormalization" (1993).

inherently quantum mechanical in origin, the virtual particles' contribution to the electron's mass (δm), the theorists assumed, would be relatively small compared with the electron's ordinary mechanical mass (m_0): $m_{\text{eff}} = m_0 + \delta m$, with $|\delta m| \ll m_0$. Instead, they immediately found that the self-energy contribution diverged to infinity: $\delta m \sim \infty$. Thanks to Heisenberg's uncertainty relation, the virtual photon could be emitted with any energy and momentum whatsoever—including infinite energy and momentum—so long as it "paid it back" sufficiently quickly. Theorists such as Heisenberg, Pauli, J. Robert Oppenheimer, and Victor Weisskopf knew they had to include all possibilities within their calculations—even the case of arbitrarily high energy and momentum—and thus, their self-energy calculations always returned infinity rather than a finite correction to the electron's effective mass.[11] The same problem occurred when they considered corrections to a photon's energy: a photon could spontaneously emit and then reabsorb a virtual electron-positron pair, and these, too, could carry any energy and momentum whatsoever.[12] Hence, the photon's self-energy diverged to infinity just as surely as the electron's did.

Theorists' first attempts to stanch the self-energy divergences during the late 1920s involved cutting off the relevant integrals at a high, but not infinite, value for the "borrowed" energy and momentum that the virtual particles could carry. By integrating to some large but finite constant, instead of to infinity, integrals for the electron's and photon's self-energies could be made to converge. This procedure, however, created problems of its own, the most important being that such arbitrarily cut-off integrals violated the symmetries of special relativity—the resulting expressions would no longer be relativistically invariant, rendering any finite answers ambiguous.[13]

Theorists faced similar problems during the 1930s when they tried to study other "radiative corrections," or corrections to basic electrodynamic quantities arising from virtual particles. Most important was vacuum polarization:

11. See Schweber, *QED* (1994), 86, 119–28; and Miller, "Frame-setting essay" (1994), 33–35, 58–69.

12. Dirac postulated that electrons should have antimatter cousins, with the same mass but opposite electric charge, based on the form of his relativistic equation for electrons. These antimatter partners were soon dubbed "positrons." See Dirac, "Quantum theory of the electron" (1928); Dirac, "Theory of electrons and protons" (1930); Dirac, "Theory of the positron" (1994 [1934]); Kragh, "Dirac's relativistic theory" (1981); Cassidy, "Cosmic ray showers" (1981); De Maria and Russo, "Discovery of the positron" (1985); Pais, *Inward Bound* (1986), 346–52; Galison, *How Experiments End* (1987), chap. 3; Kragh, *Dirac* (1990), 108–14; and Schweber, *QED* (1994), 56–69. Interestingly, the idea of pair production actually predated Dirac's equation: Bromberg, "Concept of particle creation" (1976).

13. Hans Bethe emphasized these difficulties with noncovariant formalisms in "Electromagnetic shift" (1948). See also Schweber, *QED* (1994), 112–19; and Miller, "Frame-setting essay" (1994), 66–67.

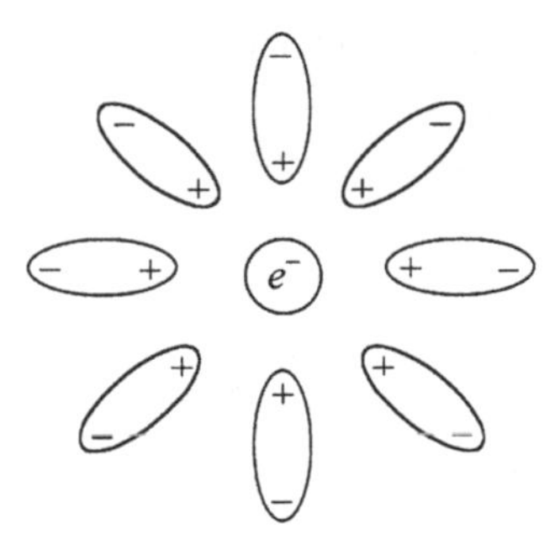

Figure 2.2. Vacuum polarization. A cloud of virtual electron-positron pairs surrounds an electron. The virtual pairs screen the original electron's charge, so that an observer would measure only the combined charge of the original electron plus the cloud.

the electric charge observed on an electron would be different from its "bare" charge. Thanks to the uncertainty principle, virtual electron-positron pairs would constantly be created out of the vacuum, so an electron would always be surrounded by a "cloud" of these virtual particles. The virtual pairs would align themselves so that the virtual positrons were closer to the original electron (opposite charges attract) and the virtual electrons further from the original electron (like charges repel). Therefore, any observer would measure the original electron's charge only as filtered or screened by this cloud of virtual pairs. (See fig. 2.2). Just as the electron had an effective mass different from its mechanical mass, therefore, an electron would carry an effective charge (e_{eff}) different from its bare charge (e_0): $e_{\text{eff}} = e_0 + \delta e$. Once again, theorists such as Edwin Uehling and Victor Weisskopf assumed that the change in the electron's charge stemming from vacuum polarization, δe, being inherently quantum mechanical in origin, would be a small correction, with $|\delta e| \ll e_0$; once again, they found instead that the vacuum polarization effects produced unphysical infinities instead of finite answers. The reason for the divergences was basically the same as for those in the self-energy calculations: the virtual particles could carry unlimited energy and momentum, and any effort to cut off the integrals by hand at some finite momentum would destroy the calculation's relativistic invariance.[14]

Beyond these troublesome infinites, a second problem plagued interwar efforts in QED: the formalism was notoriously cumbersome for performing calculations. Keeping track of the various ways that virtual particles could behave quickly became an algebraic nightmare. Electrons, for example, could

14. Uehling, "Polarization" (1935); Weisskopf, "Electrodynamics of the vacuum" (1994 [1936]); see also Schweber, *QED* (1994), 86–87, 112–19; and Miller, "Frame-setting essay" (1994), 64–65, 76–77.

interact with each other by exchanging virtual photons. In principle, they could shoot any number of virtual photons back and forth: they could exchange only one photon, two or three photons, sixty-seven thousand photons, three billion photons, and so on. The more photons in the fray, the more complicated the corresponding equations; and yet the full quantum-mechanical calculation depended on tracking each possible scenario and adding up all the contributions. All hope was not lost, at least at first: Heisenberg, Pauli, Jordan, and the rest knew they could approximate this infinitely complicated calculation because the measured charge of the electron, e, is so small: $e^2 \sim \frac{1}{137}$, in appropriate units. The electrons' charge governed how strong their interactions would be with the force-carrying photons: every time the electrons traded one more photon back and forth, the equations describing the exchange picked up one more factor of this small number, e^2. So a scenario in which the electrons traded only one photon would "weigh in" to the full calculation with the factor e^2, whereas electrons trading two photons would carry the much smaller number e^4—the latter term would contribute to the full calculation over one hundred times less than the former. The term corresponding to an exchange of three photons would carry the factor e^6—ten thousand times smaller than the simplest, one-photon-exchange term. Four photons exchanged and the corresponding term would be a mere one-millionth the size of the original. Although the full calculation extended in principle to include an infinite number of separate contributions—one for each way that the two electrons could exchange any number of virtual photons—in practice any given calculation could be truncated after only a few terms. This was known as a perturbative calculation: theorists could approximate the full answer by keeping only those few terms that made the largest contribution, since all the additional terms were expected to contribute numerically insignificant corrections.

This perturbative scheme appeared simple in the abstract, but was extraordinarily difficult to implement in practice. Hans Euler, a graduate student in Leipzig working under Heisenberg's direction, pushed the limits further than anyone else during the 1930s when he tried to calculate the scattering of light off of light. At the most basic level (ignoring virtual particles), light would never interact with itself. Heisenberg and his student realized, however, that virtual electrons and positrons could facilitate light-by-light scattering: one incoming photon could temporarily decay into a virtual electron-positron pair, and before the pair could be reabsorbed, a second photon might scatter off the virtual particles. Simple enough to formulate in words, this e^4 process proved ghastly to calculate quantitatively. Euler began working on the calculation with another of Heisenberg's students in the summer of 1934; eighteen months later, he had completed the calculation. First came the problem of infinities: Euler and his adviser decided to sidestep the divergences by restricting attention to very low energy photons, cutting off all integrals long before they ran off to infinity. Even then, the problem remained formidable. Euler's monumental effort (for which he received his doctorate) required tracking each minute variation in the ways the single pair of virtual particles could be emitted, scattered, and reabsorbed. Two photons came in and two came out, but six different possibilities could connect the various outcomes. Either the incoming photon on the left or the one on the right could first emit the virtual particles, off of which the other photon scattered. An incoming photon on the left could be exchanged within the equations for an outgoing photon on the right, and so on. Meanwhile, the virtual particles each carried spin and the photons each had a certain polarization, all permutations of which had to be summed over to arrive at the final answer. Each possibility made some quantitative contribution to the overall calculation; Euler quickly found himself swimming in hundreds of distinct terms. After setting up the calculation, he listed the many individual terms in large tables, spilling over six full pages of his published article. (See fig. 2.3.) Chock full of these and similar tables and equations, his article ran fully fifty pages in the *Annalen der Physik*.[15] All this just for the e^4 calculation (involving only one virtual electron-positron pair); no one dreamed of pursuing the next rounds of corrections, which would have involved additional electron-positron pairs in the mix, weighing in with e^6, e^8, and higher coefficients. Divergence difficulties, acute accounting woes—by the mid-1930s QED seemed an unholy mess, as calculationally intractable as it was conceptually muddled.

The great leaders of the interwar generation—Niels Bohr, Heisenberg, Pauli, Dirac—saw in these failures the seeds for yet another sweeping conceptual revolution. Just as the earlier successes with relativity and quantum mechanics had come only with great epistemological upheavals, these European theorists argued, so too would the latest challenges require an entire rethinking of issues such as measurement, observation, and the structure of space and time.[16] Fresh from their return from war work, however, a young generation

15. Euler, "Über die Streuung von Licht an Licht" (1936). The idea that virtual electron-positron pairs might mediate light-by-light scattering had first been introduced in Halpern, "Scattering processes" (1933); and a brief report on Euler's calculation appeared in Euler and Kockel, "Über die Streuung von Licht an Licht" (1935). See also Schweber, Bethe, and de Hoffmann, *Mesons and Fields* (1955), 1:323–24.

16. Rueger, "Attitudes towards infinities" (1992); Schweber, *QED* (1994), 82–92.

Reihen-folge	Reziproker Nenner	$\frac{1}{p_0^3}$	$\frac{1}{2}\frac{(\mathfrak{p}\mathfrak{g})}{p_0^5}$	$\frac{1}{4}\frac{\mathfrak{g}^2}{p_0^5}$	$\frac{1}{4}\frac{(\mathfrak{p}\mathfrak{g})^2}{p_0^7}$	$\frac{1}{8}\frac{(\mathfrak{p}\mathfrak{g})\mathfrak{g}^2}{p_0^7}$	$\frac{(\mathfrak{p}\mathfrak{g})^3}{8p_0^9}$	$\frac{\mathfrak{g}^4}{16p_0^7}$	$\frac{(\mathfrak{p}\mathfrak{g})^2\mathfrak{g}^2}{16p_0^9}$	$\frac{(\mathfrak{p}\mathfrak{g})^4}{16p_0^{11}}$
$g^1g^2g^3g^4$	$\frac{1}{N_1}=$	1	−2	9	5	−22	−14	30	56	42
	$\frac{1}{N_2}+\frac{1}{N_3}=$	2	−6	8	20	−10	−70	32	−28	252
	$\frac{1}{N_4}=$	1	−4	−3	15	24	−56	10	−140	210
	$\frac{1}{N_5}+\frac{1}{N_6}=$	1	−3	−1	9	8	−28	2	−44	90
$g^1g^2g^4g^3$	$\frac{1}{N_1}=$	1	0	9	3	0	0	30	28	22
	$\frac{1}{N_2}+\frac{1}{N_3}=$	2	0	8	8	0	0	32	0	64
	$\frac{1}{N_4}=$	1	−2	−3	7	10	−18	10	−60	66
	$\frac{1}{N_5}+\frac{1}{N_6}=$	1	−1	−1	5	4	−8	2	−28	42
$g^1g^3g^4g^2$	$\frac{1}{N_1}=$	1	−4	−5	17	48	−80	38	−372	414
	$\frac{1}{N_2}+\frac{1}{N_3}=$	2	−12	−4	68	72	−392	44	−840	2332
	$\frac{1}{N_4}=$	1	−6	3	33	−18	−182	−18	28	1030
	$\frac{1}{N_5}+\frac{1}{N_6}=$	1	−5	−5	25	56	−132	34	−484	738
$g^1g^4g^3g^2$	$\frac{1}{N_1}=$	1	−2	−1	5	6	−14	2	−28	42
	$\frac{1}{N_2}+\frac{1}{N_3}=$	2	−6	8	20	−50	−70	−8	252	252
	$\frac{1}{N_4}=$	1	−4	7	15	−48	−56	−6	252	210
	$\frac{1}{N_5}+\frac{1}{N_6}=$	1	−3	−1	9	8	−28	2	−44	90
$g^1g^4g^2g^3$	$\frac{1}{N_1}=$	1	0	1	3	0	0	−2	−4	22
	$\frac{1}{N_2}+\frac{1}{N_3}=$	2	0	−4	8	0	0	12	−60	64
	$\frac{1}{N_4}=$	1	−2	−1	7	2	−18	2	−28	66
	$\frac{1}{N_5}+\frac{1}{N_6}=$	1	−1	3	5	−4	−8	−6	12	42
$g^1g^3g^2g^4$	$\frac{1}{N_1}=$	1	−4	−3	17	36	−80	26	−312	414
	$\frac{1}{N_2}+\frac{1}{N_3}=$	2	−12	−16	68	192	−392	144	−1740	2332
	$\frac{1}{N_4}=$	1	−6	−5	33	62	−182	30	−564	1030
	$\frac{1}{N_5}+\frac{1}{N_6}=$	1	−5	−1	25	24	−132	2	−276	738

$$\begin{vmatrix} Z_1 \\ Z_2 = Z_3 \\ Z_4 \\ Z_5 = Z_6 \end{vmatrix} = -8\begin{vmatrix}0\\0\\0\\1\end{vmatrix} + \frac{8p_y^2}{p_0^2}\begin{vmatrix}1\\1\\1\\2\end{vmatrix} - 8\frac{p_y^4}{p_0^4}\begin{vmatrix}1\\1\\1\\1\end{vmatrix}$$

$$-\frac{4p_y^2}{p_0^4}\left[(\mathfrak{p}\mathfrak{g}^1)\begin{vmatrix}3\\2\\1\\4\end{vmatrix} + (\mathfrak{p}\mathfrak{g}^2)\begin{vmatrix}2\\1\\1\\2\end{vmatrix} + (\mathfrak{p}\mathfrak{g}^4)\begin{vmatrix}-1\\-1\\-2\\-2\end{vmatrix}\right]$$

$$-\frac{8p_y^4}{p_0^6}(\mathfrak{p}, -2\mathfrak{g}^1 - \mathfrak{g}^2 + \mathfrak{g}^4)\begin{vmatrix}1\\1\\1\\1\end{vmatrix}$$

$$-2\left[\frac{(\mathfrak{g}^1\mathfrak{g}^2)}{p_0^2} - \frac{(\mathfrak{p}\mathfrak{g}^1)(\mathfrak{p}\mathfrak{g}^2)}{p_0^4}\right]\begin{vmatrix}-1\\0\\0\\-1\end{vmatrix} + 2\left[\frac{(\mathfrak{g}^1\mathfrak{g}^4)}{p_0^2} - \frac{(\mathfrak{p}\mathfrak{g}^1)(\mathfrak{p}\mathfrak{g}^4)}{p_0^4}\right]\begin{vmatrix}0\\0\\1\\1\end{vmatrix}$$

$$+2\left[\frac{\mathfrak{g}^2\mathfrak{g}^4}{p_0^2} - \frac{(\mathfrak{p}\mathfrak{g}^2)(\mathfrak{p}\mathfrak{g}^4)}{p_0^4}\right]\begin{vmatrix}1\\1\\1\\1\end{vmatrix}$$

$$-2\frac{p_y^2}{p_0^4}\left[(\mathfrak{g}^1\mathfrak{g}^2)\begin{vmatrix}5\\3\\3\\5\end{vmatrix} + (\mathfrak{g}^4, \mathfrak{g}^1 + \mathfrak{g}^2)\begin{vmatrix}1\\1\\1\\1\end{vmatrix} - \frac{(\mathfrak{p}\mathfrak{g}^1)(\mathfrak{p}\mathfrak{g}^2)}{p_0^2}\begin{vmatrix}15\\7\\7\\15\end{vmatrix} + \frac{(\mathfrak{p}\mathfrak{g}^4)(\mathfrak{p}, \mathfrak{g}^1 + \mathfrak{g}^2)}{p_0^2}\begin{vmatrix}1\\1\\1\\1\end{vmatrix}\right]$$

$$-8\frac{p_y^4}{p_0^6}\left[-(\mathfrak{g}^1\mathfrak{g}^2) + \frac{4(\mathfrak{p}\mathfrak{g}^1)(\mathfrak{p}\mathfrak{g}^2)}{p_0^2} - 2\frac{(\mathfrak{p}\mathfrak{g}^1)(\mathfrak{p}\mathfrak{g}^4)}{p_0^2} - \frac{(\mathfrak{p}\mathfrak{g}^2)(\mathfrak{p}\mathfrak{g}^4)}{p_0^2}\right]\begin{vmatrix}1\\1\\1\\1\end{vmatrix}.$$

Figure 2.3. Hans Euler's calculation of light-by-light scattering. He parameterized the calculation as $H = C\int dp\, \Sigma_{\text{perm}} \Sigma_{\mu=1}^{6} Z_\mu / N_\mu$, summing over the six basic arrangements by which two photons could come in and two photons come out (with only one virtual electron-positron pair in between), as well as over the permutations of the virtual particles' spins and the photons' polarizations. (Here C is a constant and p is the virtual particles' energy.) His many tables listed the values for various numerator (Z_μ) and denominator (N_μ) terms within this overall sum. (Source: Euler, "Über die Streuung von Licht an Licht" [1936], 431, 441.)

of American physicists, along with many of their adopted European-émigré colleagues, began to approach the problem differently. As historians such as Sam Schweber, Paul Forman, and Peter Galison have emphasized, the American physicists' war projects during the 1940s helped to solidify an earlier tradition of practical calculation captured by the slogan "getting the numbers out."[17] Upon their release from Los Alamos, the Radiation Laboratories, and the Metallurgical Laboratory, physicists working in the United States treated the problems of quantum electrodynamics in this same numbers-oriented,

17. Schweber, "Empiricist temper regnant" (1986); Schweber, *QED* (1994), chap. 3; Forman, "Behind quantum electronics" (1987); Galison, *Image and Logic* (1997), chaps. 4, 9; and Galison, "Feynman's war" (1998).

practical-calculation way. Focusing on the evaluation of specific integrals rather than the grand epistemological edifice of modern physics, several young researchers produced schemes to remove the troublesome infinities from QED in relatively unambiguous ways, leaving only finite numbers behind. This program of removing the infinities was quickly dubbed "renormalization."[18]

Postwar Developments: Lamb Shift, Magnetic Moment, and Renormalization

Looking back—knowing how events unfolded after the war—several historians have pointed out that the essential ingredients for a successful renormalization program had been in hand by the late 1930s.[19] Yet this was certainly not how things appeared to the majority of theorists at the time. For these physicists—most of whose attention had been drawn away from the disheartening infinities of QED to more interesting developments in nuclear physics before the outbreak of World War II—the issue of tackling radiative corrections and extracting finite answers assumed center stage only in the late 1940s, thanks to three separate experimental investigations. Experimentalists at Columbia University's Pupin Laboratory, using surplus radiofrequency equipment left over from the wartime radar projects, made a series of high-precision measurements on atoms and electrons during 1947–48 that brought several theorists back to the territory of QED.[20]

In one of the Columbia experiments, Willis Lamb (a rare theorist turned experimentalist) and his student, Robert Retherford, measured a small but detectable shift in the energy levels of hydrogen atoms in the $2s$ and $2p$ states, even though Dirac's relativistic equation for the electron predicted that these two states should have exactly the same energy. In another series of experiments, I. I. Rabi and his students John Nafe and Edward Nelson measured a distinct splitting among hydrogenic energy levels when the atoms were placed in a magnetic field. Rabi and company examined the so-called hyperfine structure, which derived from whether the electron's spin was aligned in parallel or antiparallel with the spin of the nucleus's proton. Finally, Polykarp Kusch and Henry Foley, in still a different corner of Columbia's Pupin Lab, performed

18. Schweber, *QED* (1994); see also Laurie Brown, *Renormalization* (1993).

19. Schweber, *QED* (1994), 76, 89–92; Miller, "Frame-setting essay" (1994), 97–100.

20. Several physicists at the time emphasized the importance of the Columbia experiments for shifting their attention to QED and its problems, e.g., Freeman Dyson to his parents, 24 Jan 1948, in *FJD;* Bethe, "Electromagnetic shift" (1948); and Dyson, "Quantum electrodynamics" (1952). See also Pais, *Inward Bound* (1986), 451–52; and Schwinger, "Path" (1989), 46.

Figure 2.4. Physicists deep in discussion at the June 1947 Shelter Island conference. *Standing, left to right:* Willis Lamb, John Wheeler. *Seated, left to right:* Abraham Pais, Richard Feynman, Hermann Feshbach, and Julian Schwinger. (Source: Emilio Segrè Visual Archives, American Institute of Physics.)

similar experiments on the splitting between energy levels in gallium and sodium atoms placed in a magnetic field. In all three sets of experiments, the Columbia physicists found tiny deviations from the basic predictions of Dirac's equation—predictions made, that is, by neglecting the troublesome, infinity-producing radiative corrections.[21]

Lamb announced his experimental result for the $2s$-$2p$ energy-level shift at a small, private meeting organized by J. Robert Oppenheimer under the auspices of the National Academy of Sciences that was held on Shelter Island, off the north fork of Long Island, in June 1947. (See fig. 2.4.) Victor Weisskopf and Julian Schwinger immediately suggested that the measured effect might be due to interactions between electrons and virtual particles, incalculable though

21. See esp. Schweber, *QED* (1994), chap. 5. The Lamb shift was measured between the $j = 1/2$ $2s$ and $2p$ states of the hydrogen atom, where j is the combined orbital and spin angular momentum of the electron. Even before the war, spectroscopists had found evidence for a tiny splitting between the $j = 1/2$ and $j = 3/2$ $2p$ states.

they seemed to be. Bethe decided to check whether Lamb's result might be due to radiative corrections by performing a quick calculation on his train ride back from the meeting. Just to check the order of magnitude of the effects, Bethe calculated the self-energy of a bound electron within a nonrelativistic approximation (good enough for the low-speed electrons in question). To force his integrals to converge he cut them off by hand at the electron's rest energy, mc^2; since the results varied logarithmically with the cutoff, the order-of-magnitude calculation depended only weakly on this arbitrary value. His numerical result for the electron's energy-level shift came in remarkably close to Lamb's experimental value—approximately 1,040 megacycles, or 0.00012% of the $2s$ energy level itself—which helped to convince Bethe and many others that the new experimental results were indeed due to the physical effects of radiative corrections. Bethe immediately had copies of his manuscript mimeographed and sent to the other participants from the now-dispersed Shelter Island meeting. He wrote a special note to Lamb just five days after the meeting had ended: "Enclosed I am sending you a preliminary draft of a manuscript about the line shift. As you will see, the line shift merely explains your results very beautifully. . . . I think all this is very exciting and quite important." One week later, Weisskopf wrote to congratulate Bethe on his calculation. "It is a very nice way to estimate the effect and it is most encouraging that it comes out just right," Weisskopf wrote. "Your great and everlasting deed is your bright idea to treat the thing at first unrelativistically," since no one at the time knew how to complete a relativistic calculation self-consistently.[22]

Gregory Breit, Yale's senior theorist, suggested that a similar effect might explain the results from the Rabi-Nafe-Nelson and Kusch-Foley experiments. In these cases, Breit speculated, the observed deviations might stem from radiative corrections to the electron's magnetic moment. Because electrons carry an intrinsic angular momentum, or spin, of $\hbar/2$, Dirac's equation predicted that electrons should have an intrinsic magnetic moment of $\mu_0 = e\hbar/(2mc)$. Breit suggested that the effective magnetic moment of an electron plus its cloud of virtual particles would differ from this "bare" magnetic moment, much

22. Hans Bethe to Willis Lamb, 9 June 1947, in *HAB*, Folder 11:7; Victor Weisskopf to Hans Bethe, 17 June 1947, in *HAB*, Folder 12:47. Bethe's paper was published as Bethe, "Electromagnetic shift" (1947). Bethe's cutoff, the electron's rest energy, was the product of the electron's mass, m, times the square of the speed of light, c^2. On the Shelter Island conference and Bethe's nonrelativistic Lamb-shift calculation, see Schweber, *QED* (1994), 228–31 and chap. 4. Bethe's famous calculation is dissected in further detail in Schweber, *Relativistic Quantum Field Theory* (1961), 524–31. This was the paper that Enrico Fermi could only fully figure out after talking directly with Bethe during the summer of 1947.

the way an electron's charge would be affected by its virtual cloud.[23] Julian Schwinger, the wunderkind of Harvard's physics department, undertook a relativistic calculation of the electron's magnetic moment in the light of these radiative corrections and in November 1947—for the first time—managed to coax his new equations into finite form. When the dust settled, he found that the electron's effective magnetic moment (μ_{eff}) should be larger than Dirac's value (μ_0) by a tiny fraction: $\mu_{\mathrm{eff}} = \mu_0 + \delta\mu$, with $\delta\mu/\mu_0 = 0.001162$. Schwinger compared his calculated value with the Columbia experimentalists' recently measured values for the shift: 0.00126 ± 0.00019, 0.00131 ± 0.00025, and 0.00118 ± 0.00003.[24] Like Bethe's result for the Lamb shift just five months before, Schwinger's calculation convinced many physicists that the measured deviations from Dirac's predictions stemmed from radiative corrections.

Rumors of Schwinger's result spread quickly. Rabi wrote excitedly to Bethe in mid-November 1947 upon hearing incomplete reports of Schwinger's calculation. Three weeks later, Rabi's joy had turned to rapture: Schwinger's calculation "of our hyperfine structure anomaly is as correct as your theory of the Lamb-Retherford effect. God is great!" Bethe responded two days later in an equally buoyant mood: although Bethe still had not heard "a complete account" of Schwinger's work, "I am sure it is alright. . . . It is as exciting as in the early days of quantum mechanics."[25] The excitement was palpable the next month at the January meeting of the American Physical Society in New York. One of Bethe's graduate students who attended the conference wrote home that the sessions in general varied "from the profound to the puerile, though with a distinct preponderance of mediocrity." Yet one talk had clearly stood out: "The great event came on Saturday morning and was an hour's talk by Schwinger, in which he gave a masterly survey" of his recent results. "This talk was so brilliant that he was asked to repeat it in the afternoon session, various unfortunate lesser lights being displaced in his favour." (In fact, Schwinger delivered his talk a third time later that afternoon, in a still larger room, to meet popular demand.) "Tremendous cheers" greeted Schwinger's announcement that his calculated value of μ_{eff} matched the experimental results so closely. Bethe

23. Breit, "Relativistic corrections" (1947); Breit, "Intrinsic magnetic moment" (1947). Here $\hbar = h/(2\pi)$, where h is Planck's constant. Breit's suggestion was that in an external magnetic field, the spins of the virtual electrons and positrons would line up with that of the original electron, much the way compass needles line up with each other locally in the earth's magnetic field. The system of electron plus its virtual cloud would therefore carry greater angular momentum than the original electron alone, and hence, the electron's effective magnetic moment would be greater, too.

24. Schwinger, "On quantum-electrodynamics" (1948).

25. I. I. Rabi to Hans Bethe, 11 Nov 1947; Rabi to Bethe, 2 Dec 1947; and Bethe to Rabi, 4 Dec 1947, in *HAB,* Folder 12:8.

retained more composure when he summarized the results a few months later in a report for the Solvay conference: "The agreement between" Schwinger's theoretical value and the experimental values of the electron's magnetic moment "is too close to be ascribed to accident."[26]

Schwinger accomplished his calculation by returning to several prewar suggestions. Hendrik Kramers and Victor Weisskopf had each pointed out in the late 1930s that the roles of "effective" and "bare" quantities could be reversed. Physicists could never measure an electron's "bare" mass (m_0) or charge (e_0) and thus had no idea what these values were or should be; physicists could measure only the effective values, m_{eff} and e_{eff}. Kramers and Weisskopf suggested that one might be able to rearrange all calculations so that they were expressed only in terms of the observable quantities. Then the divergences of δm and δe would present much less of a problem: these quantities would always enter in combination with the unknown quantities, m_0 and e_0, which therefore could be "arranged" to have a finite difference, namely m_{eff} and e_{eff}. Without a relativistically covariant formalism for making actual calculations, however, ambiguities would always spoil these subtractions, which had hampered Kramers's and Weisskopf's program before the war.[27]

Kramers repeated his suggestion about reversing the roles of m_{eff} and m_0—a maneuver that came to be known as "mass renormalization"—at the 1947 Shelter Island meeting. His presentation impressed the young Schwinger, who was especially well prepared to receive Kramers's lesson. Schwinger had spent the war working on radar. Collaborating closely with electrical engineers, he had learned that many electrical calculations became tractable only if one "blackboxed" various circuit components, calculating in terms of effective input-output variables instead of the "fundamental" variables that appeared in the original equations. After hearing Kramers's suggestion at the Shelter Island conference, Schwinger began his magnetic-moment calculation using this effective-circuit approach, completing the work five months later. Spurred on by this success, Schwinger expanded his approach and within months had developed a means of manipulating his equations that satisfied the relevant

26. Freeman Dyson to his parents, 4 Feb 1948, in *FJD;* Bethe, "Electromagnetic shift" (1948), 34. On the need to repeat Schwinger's lecture, see Laurence, "New guide offered on atom research" (1948); and Mehra and Milton, *Climbing the Mountain* (2000), 26–27, 225. For more on the magnetic moment experiments, and on the improved versions carried out more recently, see Lautrup and Zinkernagel, "$g - 2$" (1999).

27. Kramers, "Interaction" (1994 [1938]); Weisskopf, "Electrodynamics of the vacuum" (1994 [1936]); and Weisskopf, "Self-energy" (1958 [1939]). See also Dresden, *H. A. Kramers* (1987), chaps. 4–5; Miller, "Frame-setting essay" (1994), 76–77, 82–83; and Schweber, *QED* (1994), 89–90, 103, 122–28.

symmetries of special relativity at each step of a given calculation—precisely what Kramers, Weisskopf, and all the others had been missing during the 1930s.[28] Although no one in the United States knew it at the time, the young Japanese theorist Tomonaga Sin-itiro was working out a relativistically covariant calculational scheme for QED between 1943 and 1948 that shared many essential features with Schwinger's formalism.[29] Working independently, Tomonaga and Schwinger thus introduced the first workable renormalization program: for the first time, theorists could calculate the effects of virtual particles reliably and compare their results with experiments.

Such calculations could inspire religious exhilaration in Rabi, but they were by no means simple or straightforward to conduct. Infinities tamed, theorists still faced the calculational morass through which Hans Euler and others had trod warily before the war. In October 1947, for example, Marvin "Murph" Goldberger, a graduate student working with Enrico Fermi in Chicago, wrote to Bethe to ask about a possible relativistic generalization of Bethe's treatment of the Lamb shift. Goldberger had already spent three weeks gearing up for the calculation and concluded (prematurely) that "the problem appears to be not too difficult." Bethe preached caution: "the calculation in the relativistic case is by no means simple. . . . There are approximately twenty different terms which have to be integrated." Bethe and one of his own graduate students had also begun work on the relativistic calculation, yet rather than scare off potential competition from Chicago, Bethe welcomed the newcomer's participation: "Because of the considerable complication of the calculation, I should find it desirable that the calculation be done at several places independently."[30] The situation did not change after Schwinger's and Tomonaga's introduction of their relativistically covariant formalism. When he reviewed the state of the field in the fall of 1948, Bethe noted that Schwinger's and Tomonaga's actual

28. See Schwinger's articles in Schwinger, *Selected Papers* (1958), along with his introduction on vii–xvii; and Schwinger, "Renormalization" (1983). See also Pais, *Inward Bound* (1986), 456–59; Mehra and Milton, *Climbing the Mountain* (2000), chaps. 7–8; and Schweber, *QED* (1994), chap. 7. On the relations between Schwinger's wartime radar work and his approach to renormalization, see Schwinger, "Two shakers of physics" (1983); Schwinger, "Greening of quantum field theory" (1996); and Galison, *Image and Logic* (1997), 820–27. Schwinger returned to a prewar suggestion by Paul Dirac: one could associate a unique time variable (t_i) with each quantum state, along with the unique set of spatial variables ($\mathbf{x}_i$) that one already associated with it. By expressing all the relevant quantities as functionals of a spacelike surface, $\sigma(t, \mathbf{x})$, Schwinger's equations displayed manifest relativistic covariance.

29. Darrigol, "Scientific biography of Tomonaga" (1988); Hayakawa, "Sin-itiro Tomonaga" (1988); and Schweber, *QED* (1994), chap. 6.

30. Marvin Goldberger to Hans Bethe, 13 Oct 1947; and Bethe to Goldberger, 20 Oct 1947, in *HAB*, Folder 10:38.

calculational methods were just as tedious and involved as the earlier methods had been. Even with Schwinger's and Tomonaga's efforts, it was no easier to conduct calculations within QED (indeed, to many people, it seemed even more difficult). The main difference was that now at least theorists could have greater confidence in the results of their labors.[31]

Schwinger and Tomonaga were not the only theorists working hard to fashion a consistent calculational scheme for QED after the war. Richard Feynman also worked on the problem during the late 1940s. In the course of his work, he began to draw simple pictures to help him keep track of his fast-multiplying equations. These line drawings—soon christened "Feynman diagrams"—provided what Schwinger's and Tomonaga's methods had not—namely, a streamlined method for making calculations in QED. Feynman's diagrams would eventually change the way physicists went about their daily business. Yet few would have predicted such a future based on Feynman's initial introduction of his new techniques.

INITIAL RECEPTION AND LINGERING CONFUSION

Feynman first introduced his novel diagrams at a special, by-invitation-only meeting at the Pocono Manor Inn in rural Pennsylvania during the spring of 1948. Under the auspices of the National Academy of Sciences, Oppenheimer helped to handpick a list of twenty-eight participants for this second annual retreat. The Pocono meeting, much like the previous year's Shelter Island meeting, had been convened "to get together a group consisting of the younger men, who would understand each other's jargon," one of the conference organizers explained.[32] Little could the organizers have foreseen how much work it would take to make the new "jargon" understandable.

Feynman developed his diagrams as a bookkeeping device for wading through complicated calculations. His ultimate goal was to find some way to tame the infinities that kept cropping up; he already had some ideas about how to cut off integrals before they diverged to infinity, in a way that would respect the relevant requirements of special relativity.[33] Now he wanted to find a reliable way of making perturbative calculations—to write down the algebraic

31. Bethe, "Electromagnetic shift" (1948), 23.

32. D. MacInnes to J. A. Wheeler, 15 Mar 1946, as quoted in Schweber, *QED* (1994), 163. On the organization of the Shelter Island, Pocono, and Oldstone conferences, see ibid., chap. 4.

33. Feynman, "A relativistic cut-off for classical electrodynamics" (1948); Feynman, "Relativistic cut-off for quantum electrodynamics" (1948). See also Schweber, "Feynman" (1986), 477–81; and Schweber, *QED* (1994), 414–22. We will examine Feynman's cutoff procedure in chap. 5.

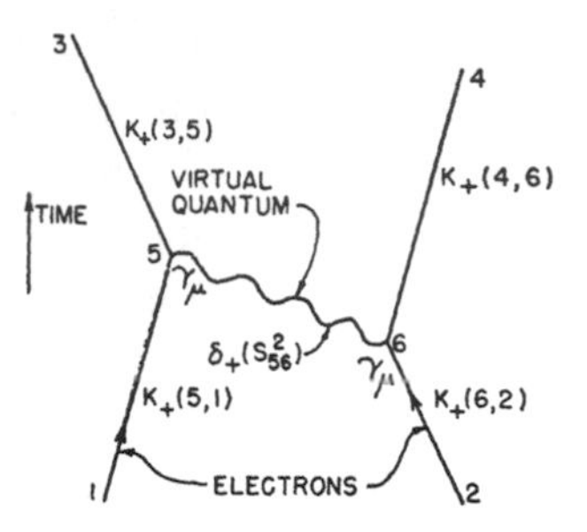

Figure 2.5. The simplest Feynman diagram for electron-electron scattering. (Source: Feynman [A.4], 772.)

form for these long series of integrals without confusing or omitting terms. He asked his listeners to consider again the simplest way that two electrons could scatter. In the simplest case, one electron could shoot out a force-carrying virtual photon, which would then be absorbed by the other electron. Feynman illustrated this process with the diagram in figure 2.5. This line drawing, Feynman explained, provided a shorthand to help calculate the one-photon contribution to the full calculation. The electron on the left had a certain likelihood of moving from the point x_1 to x_5, which Feynman abbreviated $K_+(5,1)$; the other electron similarly had a certain likelihood of moving from the point x_2 to x_6 and hence a factor of $K_+(6,2)$. This second electron could then emit a virtual photon at x_6. The photon had a certain likelihood of moving from the point x_6 to x_5, which Feynman labelled $\delta_+(s_{56}^2)$. Upon reaching the point x_5, the first electron could absorb the photon. The likelihood that an electron would emit or absorb a virtual photon also had a unique mathematical expression, derived from the interwar research, which could be written as $e\gamma_\mu$, where e was the electron's charge and γ_μ a vector of Dirac matrices. Having given up some of its energy and momentum, the electron on the right would then move from x_6 to x_4, much the way a hunter recoils after firing a rifle. The electron on the left, upon absorbing the photon and hence gaining additional energy and momentum, would scatter from x_5 to x_3. In Feynman's hands, then, this diagram stood in for the mathematical expression (itself written in terms of the abbreviations, K_+ and δ_+):

$$e^2 \iint d^4x_5 d^4x_6 K_+(3,5) K_+(4,6) \gamma_\mu \delta_+(s_{56}^2) \gamma_\mu K_+(5,1) K_+(6,2).$$

He integrated over both x_5 and x_6 because the emission and absorption events could occur anywhere in space and time. The integral thus also covered the

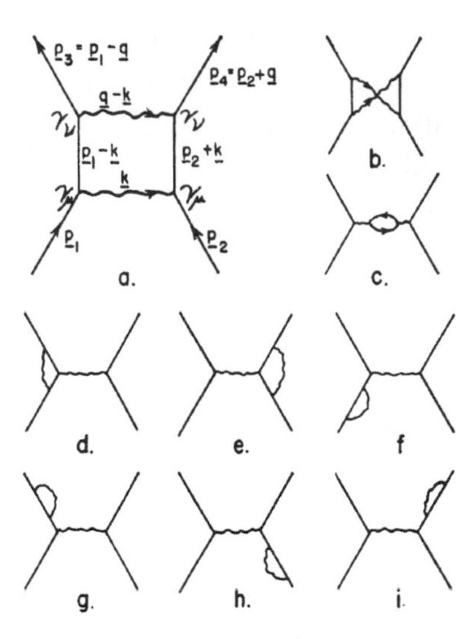

Figure 2.6. Feynman diagrams as "bookkeepers." (Source: Feynman [A.4], 787.)

case in which the electron on the left emitted a photon at x_5, which was later absorbed by the electron on the right at x_6.[34]

So much for calculating the simplest, one-photon term. Everyone at the Pocono meeting knew that the process in figure 2.5 was only the start of the calculation; the two electrons could scatter in all kinds of other ways, trading more and more photons in more and more complicated fashions, and both Feynman and his audience knew that these terms needed to be included as well. So Feynman pressed on. The next-simplest processes by which two electrons could scatter—that is, by trading two photons back and forth—corresponded to the diagrams in figure 2.6. As an example, consider figure 2.6*c*: the first electron could emit a virtual photon, which could disintegrate in midflight into a virtual electron-positron pair; these would in turn annihilate to form a new

34. Feynman [A.4], 771–73. As in any quantum-mechanical calculation, this quantity represented the probability amplitude, the absolute square of which yielded the probability that two electrons would scatter in this way. Actually, this was only half the calculation: because electrons are indistinguishable, Feynman next explained that one must include a similar integral describing the case in which the incoming electron on the left ended up, after scattering, as the outgoing electron on the right, and vice versa. I have simplified Feynman's notation somewhat, dropping factors of *i* and various subscripts (such as the electron labels *a* and *b*), but otherwise followed his notation. Feynman had already fixed upon these features of his approach before his presentation at the Pocono meeting, though other elements of his 1949 article included later developments. See Schweber, "Feynman" (1986); Mehra, *Beat of a Different Drum* (1994), chaps. 12–13; and Schweber, *QED* (1994), chap. 8.

photon, which would continue on its way and smack into the second electron. For each of these distinct diagrams Feynman could write down the associated mathematical contribution—a K_+ for each leg of an electron's motion, a δ_+ for each exchange of a photon, factors of $e\gamma_\mu$ at each vertex where electron and photon lines met, and so on.

For anyone working with the earlier methods, as theorists such as Hans Euler had done, the calculation of these e^4 correction terms had become a notoriously painstaking activity. It was all too easy—and frustratingly common—to confuse or, worse, omit some of the various terms that had to be included. The contributions that Feynman drew in figures 2.6*f*–*i*, for instance, could be maddeningly difficult to keep straight when manipulating the equations alone, especially considering that *each* of the corresponding equations stretched two or three lines long. Everyone in Feynman's audience knew how easy it was to get lost when using the standard methods.[35] (This was why Bethe had encouraged Goldberger in 1947 to work on his calculation in parallel with Bethe's own efforts, so they could provide a check on each other in the midst of so many different terms to keep track of and calculate.)

Standing at the blackboard in the Pocono Manor Inn that spring day, Feynman offered his colleagues an escape from the drudgery that had marred all their calculations for so long; no more cramped tables or equations without end, à la Euler. And yet, by all indications, his inaugural presentation of the diagrams was a flop. The odds were stacked against him: Feynman's presentation followed a virtuoso lecture by his young rival, Julian Schwinger, that stretched into an all-day affair, punctuated only by a break for lunch. Schwinger's magnetic-moment calculation had already brought widespread acclaim just a few months earlier. At the Pocono meeting, Schwinger demonstrated his new methods with a dazzling display of austere mathematical *sprezzatura*—a display that likely reminded several of his listeners of Oppenheimer's famous grandiloquence. Though few in the room could follow Schwinger's derivation line by line, all could appreciate that it took the form of a *derivation:* the performance in itself signaled to most that the problem of renormalization was in good hands.[36]

35. See, e.g., Dyson, *Disturbing the Universe* (1979), 54; and Pais, *Inward Bound* (1986), 452. For examples of the older style of computation, see Heitler, *Quantum Theory of Radiation* (1936); Wentzel, *Quantum Theory of Fields* (1949 [1943]); and the articles from the 1930s and 1940s reprinted in Schwinger, *Selected Papers* (1958); and in Miller, *Early Quantum Electrodynamics* (1994).

36. Wolfgang Pauli, who attended the Pocono meeting, later complained to Oppenheimer that Schwinger was "concealing behind a screen of mathematical virtuosity the assumptions which he silently uses." Pauli to Oppenheimer, 22 Feb 1949, in Pauli, *WB*, 3:637.

Coming late in the day, in contrast, Feynman's blackboard presentation was rushed and often interrupted. Rather than try to give a systematic introduction or justification of his new methods, Feynman chose to show off his diagrammatic tools with a series of worked examples. Yet no one seemed to be able to follow what Feynman was doing, or how his new doodles fitted in with the more established principles of quantum physics. Citing Heisenberg's uncertainty principle, for example, Niels Bohr raised repeated objections to the very notion that spacetime diagrams could be of any help for studying what Bohr insisted were inherently unvisualizable quantum phenomena. Dirac pressed Feynman on the question of "unitarity," that is, whether the probabilities for various interactions, when calculated with Feynman's new scheme, added up to one. Still others raised concerns about how to enforce the Pauli exclusion principle—yet one more pillar of prewar quantum theory—for electrons in intermediate states. At this point, as one of the auditors later recalled, Bohr again interrupted Feynman's scattered talk, "strode up to the blackboard and delivered himself of an encomium on the exclusion principle."[37] The official notes from the meeting, written up the next day and quickly mimeographed and distributed to participants, observe simply that "Bohr has raised the question as to whether this [Feynman's] point of view has not the same physical content as the theory of Dirac, but differs in a manner of speaking of things which are not well-defined physically"—hardly a resounding endorsement from the renowned Nobel laureate.[38] Most important, Feynman's weary auditors repeatedly wondered aloud what rules—if any—governed the diagrams' use. The official notes from the meeting dedicated relatively little attention to Feynman's presentation (certainly as compared to the lengthy review of Schwinger's work), and Feynman left the meeting disappointed, even depressed.[39]

37. Pais, *Inward Bound* (1986), 459. Edward Teller pressed Feynman in particular about the exclusion principle. See Schweber, *QED* (1994), 442.

38. No proceedings from the Pocono meeting were published, though at least two sets of notes are extant: the personal notes of Gregory Breit and the official notes by John Wheeler. Wheeler's notes, dated 2 Apr 1948, were duplicated and distributed to physicists throughout the country very soon after the meeting. Bohr's comment appears on p. 53. A copy of Wheeler's notes may be found in *NBL*, call number MP156. On the notes' distribution and use at the time, see Schweber, *QED* (1994), 631–32n143; Hans Bethe to Byron Cohn, 1 Nov 1948, in *HAB*, Folder 10:15; and Josef Jauch to J. Robert Oppenheimer, 6 May 1949, in *HAB*, Folder 11:2.

39. Wheeler's notes include thirty-seven pages on Schwinger's presentation, compared to eleven pages on Feynman's—and much of the space given to Feynman's presentation is taken up with large spacetime figures (similar to those in fig. 5.1, below), whereas the Schwinger material consists of tight, closely spaced algebraic arguments. See also Schweber, *QED* (1994), 196–97, 201, 436–45; Feynman, "Pocono conference" (1948); Pais, *Inward Bound* (1986), 452–58; Crease and Mann, *Second Creation* (1986), 137–38; and Gleick, *Genius* (1992), 255–61.

The story of Feynman's disappointing Pocono presentation has been told several times. Much more important—and completely overlooked—is the fact that the confusion surrounding Feynman's garbled presentation at the Pocono meeting did not dissipate overnight. Murph Goldberger remembers first meeting Feynman at an American Physical Society meeting in Washington, D.C., during the late 1940s. "I found myself," said Goldberger, "I don't remember exactly how—with a small group of people sitting under a tree in Rock Creek Park, and Feynman was talking about his new computational methods." Feynman "talked very fast, and drew lots of diagrams which I didn't understand or significantly comprehend." Meanwhile, back in Berkeley with his fellow postdocs, Goldberger and his peers continued to wonder how the new diagrams might be used: "The diagrams were fine, but you actually had to have some rules to calculate. And they had not been in their present highly developed form that you read about in the textbooks."[40] Under the shade of the trees in Rock Creek Park, much as in the quiet rooms of the Pocono Manor Inn, physicists struggled to make sense of Feynman's new diagrams. More often than not, they found them—and any purported rules for their use—thoroughly inscrutable.

Even physicists close to Feynman had trouble understanding how the new diagrams might be used. Hans Bethe provides one example. By this time, Bethe was one of Feynman's closest scientific allies. As Feynman's senior colleague at Cornell, and as an expert himself in QED and its problems, Bethe had discussed the new diagrammatic techniques with Feynman at length both before and after the Pocono meeting. Even so, Bethe had to send for help while he was visiting in England that summer. As Feynman recalled the exchange years later, Bethe had written, in effect, "I tried to do your thing, blah blah blah, and I got into the following difficulty." Feynman wrote back explaining the error, to which the grateful Bethe replied: "Well, the student [Bethe] flunked. Thank you very much." Nor did a single coaching session suffice: several months later, Bethe again wrote to Feynman, reporting that his new methods "work beautifully and I am completely converted. However, I don't get your result but instead . . . " And so the epistolary lessons continued.[41]

40. Goldberger interview (2002). Andrew Lenard, at the time a graduate student at the State University of Iowa, likewise recalls how frustratingly difficult it was to glean the rules governing the diagrams from Feynman's papers: Lenard, e-mail message to the author, 14 Dec 2000.

41. Feynman recalled the episode at a banquet in honor of Julian Schwinger's sixtieth birthday in 1978; his remarks are quoted in Mehra, *Beat of a Different Drum* (1994), 262–63. Feynman did not mention in these remarks what Bethe's "difficulty" had been. The correspondence continued in Hans Bethe to Richard Feynman, 24 Oct 1948, and Feynman to Bethe, n.d. (late Oct 1948), in *HAB*,

Other examples from the time (and less subject to Feynman's penchant for good storytelling) similarly point to physicists' difficulties when trying to use Feynman diagrams. In October 1948 Feynman wrote to Ted Welton, his old friend from undergraduate days, updating him on his progress. As a postdoc at MIT, Welton was working, like Feynman, on QED and its problems.[42] Feynman admitted to his friend that he still didn't have "the cold dope" when it came to renormalizing QED; in particular, Feynman continued to be troubled about how to treat closed loops within his diagrams. Yet even here he could not be entirely certain that his longtime correspondent and former study partner would be able to follow: "These terms come from closed loops, (in my way of talking, which I think you understand) in which two quanta are involved."[43] Feynman had been right to wonder whether Welton could follow: one year later, after Feynman's articles on his new techniques had appeared in the *Physical Review,* Welton asked his friend for reprints of the articles, explaining that he "would like to have them for souvenirs even though my chances of understanding them seem a little slim."[44]

Around the same time, a graduate student at Columbia wrote to Bethe about a talk Feynman had just given. Feynman had mentioned several unpublished diagrammatic calculations he had made and handed over to Bethe, and the student asked Bethe to send him the paperwork. Bethe obliged, closing, "I hope they will be understandable."[45] Robert Marshak, the senior theorist at Rochester who had attended the Pocono meeting, went straight to the source when stymied by the diagrams. In a 1949 article with a young theorist, Marshak explained that their calculation of pion production from collisions of nucleons neglected certain exchange terms, but that "Fortunately, it can be shown . . . that the error incurred is small." They supplied no justification for neglecting these terms, but a footnote revealed the source of their confidence: "This can be shown by writing out the different types of perturbation-theoretic terms or by means of a more direct method developed by R. P. Feynman; we are indebted to Professor Feynman for performing

Folder 3:42. In these letters, Bethe and Feynman discussed how to use Feynman's new diagrams and related mathematical techniques (such as how to combine denominators into a single function and perform integrals over the four-momenta) for calculating the Lamb shift to second order.

42. See, e.g., Welton, "Some observable effects" (1948).

43. Richard Feynman to Ted Welton, 30 Oct 1948, as quoted in Schweber, *QED* (1994), 450–51.

44. Ted Welton to Richard Feynman, 4 Nov 1949, in *RPF;* see also Feynman to Welton, 16 Nov 1949, in *RPF.*

45. Melvin Klein to Hans Bethe, 2 June 1949; and Bethe to Klein, 13 June 1949, in *HAB,* Folder 11:3.

the calculation at our request."[46] No master of the diagrams himself yet, Marshak had to watch silently as Feynman carried through the diagrammatic calculation.

Or consider the great Wolfgang Pauli, who directed an intense group in Zurich working on field theory in the late 1940s and throughout the 1950s. Pauli entered into correspondence with Paul Matthews, a graduate student at Cambridge, as Matthews was finishing his dissertation in 1950. Matthews exchanged letters with Pauli and with several others during that winter and spring on how to draw and calculate with the diagrams. In one of his earliest letters Matthews reported, "That great care must be taken" when calculating with "the different graphs is shown by the fact that we [Matthews and two postdocs, Fritz Rohrlich and David Feldman] have obtained between us three different answers." Some of them had failed to treat the photon contributions correctly, as Matthews had only just figured out: "Feldman only counts correctly when both photon lines are external. Thus his result for light-by-light is correct, but for Compton scattering is wrong. Rohrlich's equations are correct." There then followed a six-page enclosure, sent along to Rohrlich and Feldman as well as to Pauli, in which Matthews detailed everything from how to draw the various diagrams (which lines should be dashed or wavy; which lines should contain arrows), to where photon lines could and could not be inserted within a given diagram, to which kinds of diagrams contribute to which physical processes, to how—finally—to calculate with the pages and pages of tiny squiggles. Matthews wrote out and mailed the extended enclosure in the hopes that the four correspondents would be able to follow each other's calculations, since, at least until then, their separate calculations had failed to agree.[47]

The exchange of letters continued throughout that spring. Matthews dutifully mailed to Pauli copies of some of his brief letters to the editor published in the *Physical Review*, treating some of the same calculations about which the group had corresponded just a few months earlier. Yet even with all the correspondence, Pauli complained to Rohrlich that May, "Matthews has sent us several small notes [his letters to the editor], which are, however, very difficult to read for persons who are not experts in the higher approximations" of

46. Foldy and Marshak, "Production of π^- mesons" (1949), 1493.

47. Paul Matthews to Wolfgang Pauli, 25 Feb 1950, in Pauli, *WB*, 4.1:27–33, quotations from 27. Matthews indicated at the end of this letter that he sent copies to Rohrlich and Feldman. Matthews's correspondence with Rohrlich from May to December 1950 may be found in *FR*, Box 23, Folder "Correspondence, 1946–54."

calculating with Feynman diagrams.[48] Though Pauli's group in Zurich spent several more years in dedicated study of Schwinger's and Feynman's techniques, Pauli similarly had to admit to a younger colleague in the summer of 1952, "I myself am not enough of an expert in 'graphs' to be able to check the details" of a recent diagram-filled dissertation he had just received.[49]

In January 1952, Fritz Rohrlich asked Hans Bethe about the authors of a recent *Physical Review* article, chiding, "Don't you think they would save themselves quite a bit of trouble if they learned Feynman Electrodynamics?" Yet still, the gap between diagram cognoscenti and the great majority of other theoretical physicists had not closed. The following year—fully five years after Feynman's Pocono presentation—a senior Stanford theorist felt the need to explain in a letter of recommendation that his graduate student *had* learned how to use the diagrams, and had capitalized on the method in his thesis.[50] Such a statement would appear peculiarly unnecessary a few years later. Yet Leonard Schiff's need to spell this out so explicitly shows that even as late as May 1953, not all graduate students could be assumed to know how to use Feynman diagrams.

Feynman was fond of recalling in later years that when he was first developing his diagrammatic approach, he had been struck by just how "funny-looking" his new diagrams were; he remembered thinking how amusing it would be if one day the *Physical Review* were full of them.[51] Despite their familiarity today, it is crucial to remember that the "funny-looking" Feynman diagrams were neither automatic nor intuitive for physicists upon first sight in the late 1940s and into the 1950s. The new jargon which the Pocono conference organizers had hoped to foster did not spread easily or on its own.

48. Wolfgang Pauli to Fritz Rohrlich, 24 May 1950, in Pauli, *WB*, 4.1:99. See also Pauli to Rohrlich, 2 Mar 1950, in ibid., 46–47; Pauli to Rohrlich, 9 Apr 1950, in ibid., 72–73; Matthews to Rohrlich, 1 May 1950, in ibid., 4–5; and Matthews to Pauli, 18 May 1950, in ibid., 97–98. Matthews sent copies of his two letters to the editor to Pauli: Matthews [A.22] and [A.23].

49. Wolfgang Pauli to Gunnar Källén, 19 Aug 1952, in Pauli, *WB*, 4.1:708. My translation. The original reads, "Ich selbst bin nicht genug Experte in 'Graph's,' um die Details kontrollieren zu können." Pauli was referring to an unpublished dissertation by C. A. Hurst, completed in Cambridge, England, in Jan 1952. Hurst published details of his work in Hurst [A.52] and [B.9]. On Pauli's field-theory seminar in Zurich, see Schweber, *QED* (1994), 582–94.

50. Fritz Rohrlich to Hans Bethe, 22 Jan 1952, in *HAB*, Folder 12:13; Leonard Schiff to Edward Teller, 26 May 1953, in *RTB*, Box 25, Folder "Schiff, Leonard Isaac, 1915–1971."

51. Feynman discussed these early recollections with virtually identical wording in interviews conducted nearly twenty years apart. Cf. Feynman interview (1966); Feynman interview with Sam Schweber, Nov 1984, quoted in Schweber, *QED* (1994), 434; and Feynman with Robert Crease and Charles Mann, Feb 1985, quoted in Crease and Mann, *Second Creation* (1986), 137–38.

EVIDENCE OF DISPERSION

And yet the diagrams did spread—and quickly. Only one month after Feynman's pair of articles on the diagrammatic methods appeared in print, the *Physical Review* began to publish more articles that made use of the diagrams. Jack Steinberger submitted a diagrammatic article several months before Feynman's papers had been published. In what must have appeared bizarre to many of his readers, Steinberger began his diagram-laden calculations with the casual remark, "Since it is very convenient in these and other field theoretical problems to use the Feynman diagrams, the reader is assumed to be familiar with this mode of computation." This even though most physicists had received Feynman's initial, schematic description of his new techniques only a few short weeks earlier.[52] Somehow, within one year of Feynman's original introduction of his diagrams at a small, private meeting—an introduction, moreover, that had received merely fleeting, fraught attention—a growing band of young physicists began to use Feynman's brand-new calculational device. Indeed, most of these new converts had not even attended the Pocono meeting, and several, like Steinberger, picked up the diagrams without the benefit of Feynman's published descriptions. In the next chapter we will examine the principal means by which these early converts acquired their exposure to the diagrams; this will begin to clarify how a postdoc like Steinberger could assume—rightly or wrongly—that his relevant readers would already share his familiarity with the diagrams. But first consider the quickening pace of the diagrams' spread.

Following Steinberger's early article, Feynman diagrams appeared more and more frequently in the *Physical Review,* then as now the main American journal of physics. In fact, between 1949 and 1954 nearly as many diagrammatic articles appeared in the *Physical Review* as in the physics journals of Great Britain, France, Italy, West Germany, the Soviet Union, and Japan combined. Only the Japanese journal *Progress of Theoretical Physics* published a comparable number of diagrammatic articles during this period (roughly two-thirds the number in the *Physical Review*); the other journals carried only between three and nineteen diagrammatic articles each, that is, between 2% and 14% of the total in the *Physical Review.* (See appendixes A–F.) The *Physical Review* thus offers a crucial source for considering Feynman diagrams'

52. Steinberger [A.6], 1180. Feynman's first two articles on his diagrammatic techniques appeared in the 15 Sep 1949 issue of the *Physical Review:* Feynman, "Theory of positrons" (1949); and Feynman [A.4].

dispersion.[53] Anecdotal evidence, moreover, suggests that physicists approached the *Physical Review* differently from the way they approach it today. Most physicists in the United States at the time, including graduate students, bought personal subscriptions. In addition, the journal was not yet split into different subspecialties, and physicists often skimmed the two hundred or so pages of each biweekly issue cover to cover upon its arrival.[54]

Within the *Review*'s pages, only one month passed between the publication of Steinberger's article and the next diagrammatic installment. Cécile Morette, another young postdoc, submitted her diagram-filled article to the *Physical Review* in late June 1949. Like Steinberger's, Morette's piece was written months before Feynman's articles were published.[55] And so began a rush of articles incorporating Feynman diagrams. As figure 2.7 indicates, the number of diagrammatic articles within the *Physical Review* grew exponentially between 1949 and 1954. On average, every single issue of the journal (published biweekly) carried one diagrammatic article during 1952; by 1954, the number had nearly doubled. In fact, the average number of articles making use of Feynman diagrams within the journal doubled every 2.2 years. The rapid incursion of diagrammatic articles is all the more striking since the journal was still, at this time, devoted to research from the whole of physics, and physicists had begun to use Feynman diagrams only in a few subspecialties of the discipline.

At first, diagrammatic articles came from authors on the East Coast, from places like Cornell (where Feynman worked), the Institute for Advanced Study in Princeton, Rochester, and Columbia. Within a few years, however, the editors of the *Physical Review* received diagrammatic articles from physicists working throughout the Midwest—Chicago, Indiana, Iowa, and Illinois—as well as from the West Coast—Stanford, Berkeley, and UCLA. Joining these

53. Physicists often used the diagrams in print without citing Feynman's (or others') original articles on them, and, conversely, many articles that did cite Feynman's 1949 papers did not make any use of the diagrams. In other words, no simple citation analysis will do: one must read through each issue of the *Physical Review*, scanning for uses or discussions of Feynman's diagrams. For the same reason, electronic keyword searches, which can now be conducted for the entire *Physical Review* in its online archive, supplement but cannot replace the page-by-page hand searches. Similar page-by-page searches were conducted to produce appendixes B–F.

54. Samuel Goudsmit to Stephen White, 15 Apr 1949, in *SAG*, Folder 24:260; Leonard Loeb to Raymond Birge, 27 Oct 1954, in *RTB*, Box 19, Folder "Loeb, Leonard Benedict"; Loeb to S. A. Goudsmit, 19 April 1955, in ibid.; Raymond Birge to S. A. Goudsmit, 5 Apr 1955, in *RTB*, Box 40, Folder "Letters written by Birge, Jan–Apr 1955"; and Raymond Birge to Samuel Goudsmit, 22 Nov 1957, in *RTB*, Box 40, Folder "Letters written by Birge, June–Dec 1957."

55. Morette [A.8].

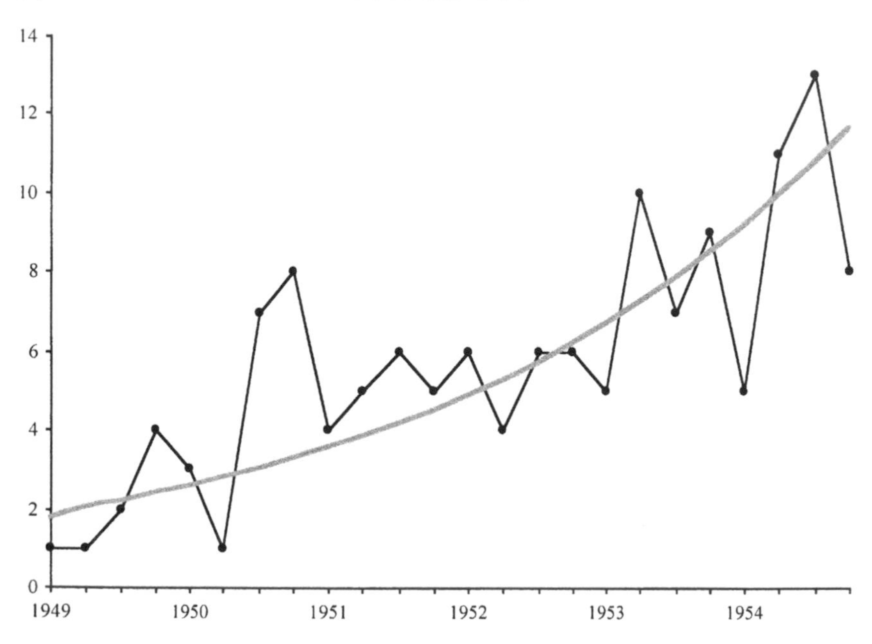

Figure 2.7. Number of articles using Feynman diagrams in the *Physical Review* during three-month intervals, based on the entries in appendix A. The solid gray line gives the best-fit curve, demonstrating that the number of articles rose exponentially during this period with a doubling period of 2.2 years.

domestic submissions were diagrammatic articles sent in to the *Physical Review* from Cambridge, Oxford, Birmingham, and Glasgow in Great Britain. From rather humble beginnings, Feynman diagrams fast became a staple tool for many physicists throughout the country and abroad—all before any English-language textbooks on the new techniques had been published.[56]

A quick glance at these many articles reveals that the diagrams during this early period were the near-exclusive property of *theoretical* physicists. Only 4 of the 139 articles in the *Physical Review* that included or discussed Feynman diagrams also included charts, tables, or figures of experimental data. Of these four, three were written by theorists working at national laboratories and the fourth by a theorist graduate student in contact with experimentalists

56. As noted above, one textbook treating the diagrams appeared in Japanese and one in Russian in 1953. The third edition of Walter Heitler's prewar textbook, *Quantum Theory of Radiation,* appeared in 1954, and included some brief sections on the Feynman-Dyson techniques. The first American textbooks—Schweber, Bethe, and de Hoffmann, *Mesons and Fields* (1955); and Jauch and Rohrlich, *Theory of Photons and Electrons* (1955)—appeared in 1955. We will examine these and other textbooks in chap. 7.

elsewhere.[57] Though by the 1960s Feynman's doodles would be commonplace on experimentalists' blackboards and in their publications, the diagrams initially spread only to theorists. Closer inspection of these articles shows, moreover, that it was not just any theoretical physicists who picked up the new diagrammatic techniques. Nearly all the published uses of Feynman diagrams in this period were by *young* theorists. The 139 articles listed in appendix A sprang from a total of 114 authors. Of these, four-fifths first began using Feynman diagrams while still in the midst of their training: thirty-nine as graduate students, and fifty-three as postdocs. Ten more began using Feynman diagrams while young instructors or assistant professors, fewer than seven years past their doctorates. Only ten authors first published diagrammatic calculations more than seven years after completing their dissertations.[58] The extreme youthfulness of the physicists in this sample provides one hint that something pedagogical was going on behind the diagrams' spread.

To be sure, not everyone was swayed by the diagrams at once. In addition to Julian Schwinger's famous avoidance of the diagrams, several other physicists continued to calculate field-theoretic quantities, such as the self-energy of an electron, without making any use of Feynman diagrams.[59] As the editors of the *Physical Review* received a rising stream of Feynman diagrams throughout the early 1950s, however, the few theorists who neglected—or actively shunned—Feynman's diagrammatic apparatus fell into a smaller and smaller minority.

The Role of Personal Contact

Though Feynman's published articles from 1949 surely helped introduce the new techniques to many physicists, these articles alone rarely proved sufficient. Nearly all physicists who used Feynman diagrams in print during the diagrams' first six years were in personal contact with other diagram users. Sometimes Feynman himself provided the needed personal communication. In part just

57. See the articles by Brueckner [A.17]; Kaplon [A.39]; Lindenfeld, Sachs, and Steinberger [A.72]; and Ross [A.108].

58. Information on authors' dissertations and appointments comes from *DiP; AMS;* Bitboul, *Retrospective Index* (1976); *Index to Theses* (1951–); *Nihon Butsuri Gakkai Meibo* (1956, 1963); *Nihon Hakushi Roku* (1985); Institute for Advanced Study, *Community of Scholars* (1980); and Pauli, *WB*, 4.2:287–89. No biographical information could be found for two of the 114 authors.

59. See, e.g., H. Snyder, "Quantum electrodynamics" (1950); H. Snyder, "Quantum field theory" (1950); Snow and Snyder, "Self-energies" (1950); Villars, "Energy-momentum tensor" (1950); Peaslee, "Boson current corrections" (1951); Peaslee, "Infinite integrals" (1951); and McConnell, "Diverging integrals" (1951), all of which more closely followed Schwinger's methods instead of Feynman's or Freeman Dyson's.

to get out of Princeton for a long weekend, for example, two postdocs at the Institute for Advanced Study, Freeman Dyson and Cécile Morette, ventured to Ithaca to visit with Feynman at the end of October 1948. As Dyson wrote soon after to his parents, the three proceeded to spend the weekend "in conclave discussing physics." Feynman's extended presentation of his work, spiced with his usual flippancy, "kept Cécile in fits of laughter" and impressed Dyson as "the most amazing piece of lightning calculation I have ever witnessed." A few months later, in February 1949, Feynman traveled to the Institute for Advanced Study for a brief visit. At the Institute, Feynman delivered eight hours of seminars on his diagrammatic approach and indulged in numerous other informal discussions with Institute members. As he had in his more private tutorials with Dyson and Morette, Feynman delivered his Institute presentations in his "enthusiastic mood, waving his arms about a lot and making everybody laugh," as Dyson reported to his parents. "This was a magnificent effort," Dyson continued, "and I believe all the people at the Institute began to understand what he is doing. I at least learnt a great deal."[60] In her article submitted that June, Morette thanked Feynman "for his illuminating lectures."[61]

A few other early diagram converts likewise interacted directly with Feynman. As noted above, for example, Robert Marshak, the head theorist at Rochester, turned to Feynman for help with the diagrams in 1949. That same spring, Bethe wrote to a Columbia graduate student who was struggling to apply Feynman's new diagrams to a calculation of the Lamb shift. Bethe suggested that the student visit Cornell rather than try to work out his questions through correspondence alone; such a trip would offer "some advantage . . . because you could then also see Feynman" as well as Bethe.[62] In October 1949, Fritz Rohrlich submitted his first diagrammatic article, which he wrote up at the Institute for Advanced Study, thanking "Professor R. P. Feynman for many helpful discussions."[63] A graduate student at Stanford, Marshall Rosenbluth, likewise

60. Freeman Dyson to his parents, 1 Nov 1948 and 28 Feb 1949, in *FJD*. Dyson further described his trip to Ithaca with Morette in a later interview, published in Sykes, *No Ordinary Genius* (1994), 77.

61. Morette [A.8], 1432n11.

62. Dyson described one of the trips from Ithaca to Rochester in letters to his parents, dated 19 Nov 1947 and 27 Nov 1947, in *FJD;* Hans Bethe to Melvin Klein, 8 Apr 1949, in *HAB,* Folder 11:3. Instead of visiting Cornell, the Columbia student (Melvin Klein) met with Bethe when Bethe visited his mother in New York City around Easter; see Bethe to Klein, 13 June 1949, in ibid.

63. Rohrlich [A.10], 360. By the time the article appeared, in Feb 1950, Rohrlich was at Cornell as a postdoc; it is not clear whether his discussions with Feynman took place while Feynman visited the Institute the previous winter or he and Feynman discussed their work while Rohrlich was at Cornell. Feynman left Cornell to take up a position at Caltech at the end of the 1949–50 academic year. See also

acknowledged Feynman's direct help in his diagrammatic calculation, published in 1950, adding that "The methods of calculation and the notation used in this paper are just those of Feynman unless otherwise indicated."[64] A year and a half later, Malvin Ruderman, a postdoc who had first encountered Feynman and his diagrams while a graduate student at Caltech, submitted a brief letter to the editor of the *Physical Review* using the diagrams for studying pion decay.[65]

Remaining outside the loop of personal communications could prove frustrating. A few weeks after Rohrlich submitted his first diagrammatic article, a graduate student at the University of Minnesota pleaded with him to send news of the latest developments: "If you have any ideas or projects to farm out, stemming from Feynman's work or your own or from anywhere, I should be very interested to hear of them. If one just reads the literature he is from a year to two behind.... Here in the Midwest we learn only of what we read in the newspapers, so to speak!" Hans Bethe similarly insisted on the need for personal contact. Endorsing an older theorist's proposal for a Guggenheim Fellowship, Bethe explained that given the very rapid changes in the field, one must "keep himself entirely abreast of current developments." Yet with "the very slow publication of the results, it is almost impossible to do so without close personal contact with the men leading in the field."[66] For Bethe, just as clearly as for the Minnesota graduate student, personal contact was the key. No one could afford to rely on published sources alone.

Beyond Feynman himself, the acknowledgments in the diagrammatic articles in the *Physical Review* reveal a remarkably closed set, as indicated in table 2.1: the names that recur most frequently in the acknowledgments—Freeman Dyson, Hans Bethe, Norman Kroll, Abraham Klein—were the same people who published the greatest number of diagrammatic articles during this period. Though comprising just one-seventh of all the authors who

---

the correspondence between Rohrlich, Paul Matthews, and Wolfgang Pauli cited above. Pauli wrote to Rohrlich throughout the spring of 1950 as if Rohrlich were Feynman's assistant, asking for news of Feynman's latest thinking as well as for copies of Feynman's reprints. See Pauli to Rohrlich, 2 Mar 1950, in Pauli, *WB*, 4.1:46–47; and Pauli to Rohrlich, 9 Apr 1950, in ibid., 72–73.

64. Rosenbluth [A.16], 616n4 and 615n1.

65. Ruderman [A.49]. Ruderman completed his Ph.D. at Caltech in 1951 (*DiP*); Feynman arrived at Caltech that same year, though he had given a series of lectures at Caltech during the winter of 1950. Ruderman helped to prepare mimeographed notes based on two of Feynman's courses at Caltech, as indicated on their cover sheets: Feynman, "Quantum electrodynamics" (1950); and Feynman, "High energy phenomena" (1951).

66. Leo Levitt to Fritz Rohrlich, 13 Nov 1949, in *FR*, Box 23, Folder "Correspondence, 1946–54"; ellipsis in original; Hans Bethe to Henry Moe, 2 Dec 1948, in *HAB*, Folder 11:17.

Table 2.1. Acknowledgments and authorship for articles in the *Physical Review* that used Feynman diagrams, 1949–54

Name	No. of Times Acknowledged by Other Authors	No. of Articles Contributed
Freeman Dyson	13	7
Hans Bethe	12	4
Norman Kroll	8	4
Richard Feynman	8	3
Abraham Klein	6	7
Abdus Salam	6	6
Abraham Pais	6	2
Paul Matthews	4	6
Fritz Rohrlich	4	4
Murray Gell-Mann	3	4
Robert Karplus	3	4
Francis Low	3	3
Kenneth Watson	3	3
Keith Brueckner	2	5
Malvin Ruderman	2	5
Marvin Goldberger	2	3
John Ward	1	4

used Feynman diagrams in the *Physical Review* between 1949 and 1954, this core group contributed more than half the total number of articles in the sample.[67] Often, the members of this core set thanked each other (Kroll thanked Dyson and vice versa; Bethe thanked Feynman and vice versa; Keith Brueckner thanked Goldberger and Kenneth Watson; Ruderman thanked Watson, Brueckner, and Kroll; and so on). Still more often, however, it was other young physicists—especially those making use of the diagrams for the first time—who thanked these authors. To a remarkable extent, diagram users were in personal contact with other diagram users.

* * *

67. Only five other authors contributed as many diagrammatic articles between 1949–54 as the physicists listed in table 2.1, though none of them was thanked in the acknowledgments to other peoples' diagrammatic papers: Edwin Salpeter (five papers), Michel Baranger (three papers), Maurice Lévy (three papers), Geoffrey Chew (three papers), and Stanley Deser (three papers). Similarly, both Rudolf Peierls and Victor Weisskopf were thanked three times each in acknowledgments but did not use Feynman diagrams in print in the *Physical Review* during this period.

Feynman faced an intensely skeptical audience when he first presented his stick-figure mnemonic aids at the Pocono meeting in the spring of 1948. Leaders like Niels Bohr, Wolfgang Pauli, and Paul Dirac worried that Feynman's calculational tool violated each of the major tenets of quantum theory—hardly a ringing recommendation for a method that purported to solve the persistent problems of quantum electrodynamics. Yet despite the poor reception and much lingering confusion, physicists—almost entirely young physicists—began to use Feynman's diagrams very soon after their initial introduction. The first diagram users deployed the technique months before Feynman had bothered to write up and publish his new methods. In a few short years, the twenty-seven doubters of the Pocono meeting had been overshadowed by scores of diagram enthusiasts. Behind these seemingly scattered, almost random personal contacts lay a particular pattern, which emerged by 1950. This pattern was centered primarily on one physicist—Freeman Dyson—and one institution—the Institute for Advanced Study. It is therefore to Dyson's efforts that we now turn.

· 3 ·
Freeman Dyson and the Postdoc Cascade

The best way to send information is to wrap it up in a person.

J. ROBERT OPPENHEIMER, NOVEMBER 1948[1]

Feynman diagrams first made the move from Feynman's disappointing Pocono presentation to physicists' lecture halls and research notebooks thanks to the efforts of his young colleague, Freeman Dyson. By means of his preprints and published articles, incessant lecturing, and informal recruitment among the younger generation of physicists, Dyson became the diagrams' earliest and most important ambassador. His efforts point to larger transformations then underway within American physics, affecting how physicists communicated new results and trained young members of the field. They also highlight the centrality of personal contact and tacit knowledge in making the new tools travel: although Dyson produced a series of highly influential articles and lecture notes, his written work alone rarely proved sufficient to get other people to use the new diagrams. To understand Dyson's role, therefore, we must first examine a major institutional change within the life course of young American physicists that unfolded during the middle decades of the twentieth century: the rise of postdoctoral training. Much like the nineteenth-century pedagogical reforms that introduced training seminars and problem sets in Germany and paper-based examinations and private coaching in Great Britain, postdoctoral training played a crucial role in establishing theoretical physics as a discipline in the United States.[2] It was a new institution, and just like the older pedagogical instruments, it helped to produce a new way of learning about and practicing physics. Without the rise of postdoctoral training and

1. J. Robert Oppenheimer, as quoted in Anon., "Eternal apprentice" (1948), 81.
2. Cf. Olesko, *Physics as a Calling* (1991); and Warwick, *Masters of Theory* (2003).

new institutions for fostering young postdocs in theoretical physics, Dyson's vigorous efforts—on paper and on blackboards—might well have come to nought.

THE RISE OF POSTDOCTORAL TRAINING

When the anthropologist Sharon Traweek embarked on her ethnographic fieldwork among American and Japanese high-energy physicists in the mid-1970s, postdoctoral study was a central, obvious, even taken-for-granted part of physicists' training. The "postdoc" stage—which often consisted of two or more separate postdoctoral appointments, each for one or more years—was a crucial phase for young physicists, during which they had to demonstrate independence and leadership ability, as well as prove themselves reliable and innovative. Here, among the postdocs, the next generation's leading physicists would begin to show themselves.[3] Over one-half of the people who received physics Ph.D.'s in the United States during 1975 anticipated postdoctoral study, whether their ultimate goals were to work in government laboratories, industrial research, or university departments.[4] Today, few physics-related positions are open to physicists who have not completed some form of postdoctoral training—a pattern hardly limited to physicists. Between 1981 and 1998, the number of university-based postdoctoral positions for young scientists and engineers in the United States doubled, reaching nearly forty thousand. The average time spent as a postdoc has also steadily increased since the 1970s: today's newest Ph.D.'s can look forward to three or more separate postdoctoral appointments of two to three years each before landing their first academic jobs.[5]

What appeared to Traweek's informants in the 1970s—or to physicists today—as the unquestionable, natural state of affairs was of rather recent vintage. Postdoctoral training for scientists in the United States began soon after World War I, thanks largely to the efforts of the influential scientists and policy makers George Ellery Hale, Robert Millikan, and Arthur Noyes, who worried about the scientific health of the nation during and after the war. They wondered how best the United States might achieve scientific parity with Europe, and with it, perhaps, stronger science- and technology-based defense

3. Traweek, *Beamtimes and Lifetimes* (1988), chap. 3. See also Calvert, Pitts, and Dorion, *Graduate School* (1972), 117–19.

4. National Research Council, *Postdoctoral Appointments* (1981), 130.

5. Travis, "Postdocs" (1992); Radetsky, "Modern postdoc" (1994); Lee, "Postdoc trail" (2001); Roman Czujko (director, Statistical Research Center, American Institute of Physics), personal communication, 10 May 2002.

capabilities. Some mechanism was needed, they concluded, to produce the kinds of weapons, from poison gas to submarines, that German scientists and engineers had recently provided their own government. Through the newly formed National Research Council (NRC), a branch of the National Academy of Sciences, Hale, Millikan, and Noyes petitioned private philanthropies, most importantly the Rockefeller Foundation, to help establish what they called "a new type of educational training."[6] Members of the NRC envisioned the new educational stage as the best way to improve the standards of research in the country—in which, after all, the doctorate itself had been introduced only a few short decades earlier. Rather than grant honors to already-established researchers or build centralized national laboratories, Hale, Millikan, and Noyes sought to "stimulate" younger scientists as a way of broadening the base of scientific competence within the country. The following year, the Rockefeller Foundation began to fund postdoctoral fellowships in physics and chemistry through the auspices of the NRC; fellowships in mathematics, astronomy, medicine, and biology followed within the next five years. The main idea behind the new fellowship program was to give recent recipients of the Ph.D. one or two years to hone skills and contribute to original research before assuming the burdens of teaching and administrative duties.[7]

By the early 1930s, approximately one in every six people who earned physics Ph.D.'s within the United States received postdoctoral felllowships.[8] More and more, a postdoc became a key element in a successful physicist's career. A survey of the newest 250 "starred" physicists in the 1937 edition of Jacques Cattell's *American Men of Science*—those singled out by their fellow specialists as being particularly distinguished in research—indicated that over one-third had completed some form of postdoctoral training.[9] Despite the early successes, however, the new program did not change higher education overnight. Although demand for postdoctoral fellowships peaked in the early 1930s with the onset of the Depression, by the late 1930s only 2% of all science Ph.D. recipients in the United States were awarded Rockefeller-funded NRC postdocs, which remained the predominant source of postdoctoral fellowships in

6. National Research Council, "Proposal for the endowment of research in physics and chemistry," unpublished proposal to the Rockefeller Foundation, 13 June 1918, as quoted in Assmus, "Postdoctoral fellowships" (1993), 161. See also Kevles, *Physicists* (1995 [1978]), chap. 8; and National Research Council, *Invisible University* (1969), 7–38.

7. Assmus, "Postdoctoral fellowships" (1993), 159–62; Rand, "National Research fellowships" (1951), 71–74.

8. Assmus, "Postdoctoral fellowships" (1993), 166–67.

9. Visher, "Younger starred scientists" (1939), 132.

the country.[10] A concerned education specialist proclaimed in the pages of the *Journal of Higher Education* in November 1944 that postdoctoral education remained "a field which needs more intensive cultivation." "Quite a few universities remain unaware of postdoctoral needs," he pressed, and often "no educational provisions beyond the doctorate" had been made.[11] Forged in the aftermath of World War I, postdoctoral fellowships had by no means become standard at the close of World War II.

The postdoc situation quickly garnered attention after the war. As with so many features of American science after World War II, the federal government stepped in where private philanthropy formerly had paid the bills. Upon their establishment in 1946, both the Atomic Energy Commission and the Department of Defense began funding postdoctoral positions as part of their research contracts. The National Science Foundation (NSF) began offering its own postdoctoral fellowships soon after its founding in 1950, leaping from 55 fellowships per year in 1952 to 180 in 1960. The National Aeronautics and Space Administration (NASA) likewise began funding postdocs during the late 1950s. By 1960, more than three-quarters of all postdocs in the United States were funded by some branch of the federal government, and the total number of fellowships had grown exponentially since the 1930s, just as the numbers of new Ph.D. recipients had continued to grow.[12] The result was nothing short of "unbridled expansion" in the number of postdoctoral positions, wrote the NSF's physics program director in the late 1950s.[13] National surveys at the time revealed that two-thirds of the faculty who taught at the graduate level in the physical and biological sciences believed that postdoctoral education was "becoming necessary or highly desirable for proper advancement"—even though only one-quarter of these faculty had been postdocs themselves. By the mid-1960s, the Physics Survey Committee of the National Academy of Sciences explained that within many subfields of the discipline, postdoctoral training for young physicists was "rapidly becoming a sine qua non."[14]

What were the reasons for postdoctoral training? During the early years, when scientists like Hale, Millikan, and Noyes labored to establish postdoctoral fellowships, the goal was to stimulate young scientists and thereby jump-start

10. Rand, "National Research fellowships" (1951), 74.

11. Israeli, "American postdoctoral education" (1944), 428, 430.

12. Berelson, "Postdoctoral work" (1962), 122–23. See also National Science Foundation, *Career Progression* (1988).

13. McMillen, "Research-associate positions" (1958). See also A. C. Helmholz to Clark Kerr, 5 Mar 1958, in *BDP*, Folder 1:28.

14. Berelson, *Graduate Education* (1960), 190–91; National Research Council, *Physics* (1966), 17.

the quality and quantity of American scientific research. Soon after postdoctoral appointments began, supporters could point to successes. Princeton's physics department chair, for example, noted in 1935 that his department had benefited from postdoctoral appointments for ten years, with the postdocs "working independently on their own problems and at the same time exchanging counsel with the other men around the laboratory. We have always welcomed these embryo physicists."[15] Soon after the war, "embryo physicists" assumed still greater responsibilities. By the 1950s and 1960s, large-scale research groups had become fixtures within the nation's research universities, and postdocs usually fulfilled important organizational, managerial, and instructional roles for teams of undergraduate- and graduate-level assistants. Postdocs after the war thus provided the glue that bound large research groups together. Citing these contributions in its 1966 report, the National Academy of Sciences Physics Survey Committee concluded simply that "postdoctoral personnel" were "essential to the present research effort in physics."[16]

Even as postdocs assumed greater leadership roles within large research groups, the educational function of postdoctoral fellowships began to change. After World War II, in the face of a tremendous upsurge in the rate of scientific publication and specialization, senior physicists declared that the Ph.D. degree no longer prepared young physicists adequately for careers in research. "The doctor's degree is no longer a terminal degree," explained an MIT physicist in the mid-1960s. "We have no other degree but we require our Ph.D. students to know more and more about less and less." Students and advisers alike thus came to believe that "further training and a broader experience are essential before they take up their life's work."[17] During the postwar period, the postdoctoral stage thereby acquired new importance: rather than focusing further on "book learning" or formal coursework—which many observers feared were filling more and more of graduate students' time—postdoctoral fellowships were

15. Henry DeWolf Smyth to Abraham Flexner, 15 Apr 1935, in *IAS,* Folder "General: Smyth, Henry D."

16. National Research Council, *Physics* (1966), 17–19. On the rapid establishment of "organized research units" within American universities after World War II, see Geiger, *Research and Relevant Knowledge* (1993), 47–61; on postdocs' postwar roles, see Assmus, "Postdoctoral fellowships" (1993), 182; Rand, "National Research fellowships" (1951), 76–77; McMillen, "Research-associate positions" (1958); Berelson, "Postdoctoral work" (1962), 128–29; and Traweek, *Beamtimes and Lifetimes* (1988), chap. 3.

17. S. Brown, "Post-doctoral training" (1966), 47. Similar observations had been made during the 1950s and early 1960s, as reported in Berelson, *Graduate Education* (1960), 194–96; Suits, "Postgraduate training" (1960), 88–95; and Douglass and Strandberg, "Education of a physicist" (1963).

supposed to enable young physicists to acquire those tacit skills, not codified within textbooks or quizzed on final examinations, that were essential for research. Postdoctoral training, more than any other stage in young physicists' careers, focused on cultivating tacit knowledge.

A second function of postdoctoral fellowships also became important. Given the nature of postdocs' peripatetic appointments—short term and nearly always at institutions different from the ones where they earned their Ph.D.'s—they *circulated* between different groups over time. As they traveled, they picked up new skills and helped to transfer them from one place to another. Both functions of postdoctoral training—providing glue within research groups and transferring skills among different groups—were new, and both played crucial roles in the development of modern physics.[18] As we will see, the increased circulation of postdocs after the war enabled Feynman diagrams to spread far and wide, thanks largely to the efforts of Freeman Dyson.

DYSON AS DIAGRAMMATIC AMBASSADOR

The First Apprentice

"Yesterday I came up here by train; a lovely trip over the mountains and up the spectacular Lehigh valley," Freeman Dyson wrote to his parents in September 1947, reporting his arrival at Cornell University. "The weather here is grey, damp and cool, but one can see already what a fine panorama of hills and lakes surrounds us." Following an enjoyable cruise on the *Queen Elizabeth* from England to New York, the young student began to get settled into his dormitory room, where he would spend the 1947–48 academic year as a graduate student in Cornell's physics department.[19] In many ways, it had been a longer journey still, from the realm of pure mathematics to theoretical physics.

18. For example, postdoc exchanges had proven crucial for bringing quantum mechanics to the United States from Europe during the late 1920s and 1930s: Coben, "Transmission of quantum mechanics" (1971); Kevles, *Physicists* (1995 [1978]), 213–21; Sopka, *Quantum Physics in America* (1988 [1980]); and Assmus, "Postdoctoral fellowships" (1993), 176–81.

19. Freeman Dyson to his parents, 20 Sep 1947, in *FJD;* see also Dyson to his parents, 15 Sep 1947, in *FJD.* Dyson regularly wrote letters to his parents during his studies in the United States between 1947 and 1949, sometimes writing more than one letter per week. Unbeknownst to him, his mother saved all his letters, which were rediscovered soon after her death in 1974. Dyson's letters to his parents are currently in his possession at the Institute for Advanced Study; his parents' replies are no longer extant. Eighty letters from Sep 1947 to Jul 1949 are in the collection; additional letters are extant from Dyson's visit to the Institute in autumn 1950, and regular correspondence resumed upon his return to the United States in autumn 1951.

His mathematical training had begun early, first at the elite and competitive Winchester College school for boys as a teenager, followed by two years at Cambridge University. As a second-year undergraduate at Cambridge, he had begun to publish short papers on number theory and statistics—several of them in *Eureka,* a journal run by the student mathematics club, others in the *Journal of the London Mathematical Society.* His brief stint at Cambridge, however, was interrupted by the war. Beginning in June 1943, at the age of nineteen, he plied his mathematical and statistical skills with the Operational Research Section of the Royal Air Force Bomber Command, busily computing analyses of bomber losses and the efficiency of various bombing strategies. He continued to pursue his mathematical interests on the side when he could, largely as an effort, he later wrote, "to keep me sane amid the insanities of the bombing campaign."[20]

During his time at the Bomber Command, Dyson also began to read Walter Heitler's textbook, *The Quantum Theory of Radiation* (1936). The textbook opened his eyes to "tantalizing hints of fundamental difficulties and unsolved problems" lurking within theoretical physics. He made up his mind that if he could succeed in proving a particular conjecture within number theory, he would become a mathematician, but if he could not produce the proof, then he would make his way into theoretical physics. Three months of intense work followed before Dyson decided he would become a physicist. On the strength of his previous work in mathematics he won a Trinity Fellowship to return to Cambridge soon after the war, where he quickly found in Nicholas Kemmer an accomplished theoretical physicist and a patient teacher who could help Dyson make the transition to physics. Dyson worked hard under Kemmer's tutelage, but Kemmer remained overburdened with teaching duties, and no one else at Cambridge appeared promising as a potential adviser. A chance encounter with Sir Geoffrey Taylor, a Cavendish physicist who specialized in fluid mechanics and who had worked at the wartime Los Alamos laboratory, convinced Dyson that he should pursue his studies at Cornell with Hans Bethe. "At the time I hardly knew that Cornell existed," Dyson later recalled. Yet with the help of a Commonwealth Fellowship, he was on his way to Ithaca less than a year later.[21]

20. Dyson, "Comments on selected papers" (1996), 6; a chronological list of his publications appears in Dyson, *Selected Papers* (1996), 587–95. Dyson described his experiences in the Bomber Command in more detail in Dyson, "Reflections" (1970). On the early development and use of "operations research" by physicists during World War II, see Fortun and Schweber, "Scientists and the legacy of World War II" (1993). For further biographical details on Dyson's childhood and mathematics education, see Schweber, *QED* (1994), 474–93.

21. Dyson, "Comments on selected papers" (1996), 8, 10–11. Although he did not prove the Siegel conjecture concerning approximations of numbers by rational fractions, he did manage to

Figure 3.1. Freeman Dyson, early 1950s. (Source: Sykes, *No Ordinary Genius* [1994], 73.)

Once Dyson arrived in Ithaca, he quickly settled into a routine. His dorm room struck him as "very comfortable and suitable for working in," he explained to his parents upon arrival. "It is not designed for social life, in fact the rules are 'No cooking, No alcohol, and No women.' . . . At present I am taking meals at the students' cafeteria, which is simply stacked with the most delicious food, and I shall have no difficulty in growing fat on $2 a day." (See fig. 3.1.) A few days later he wrote home to tell of his new peers within the physics department: "I have a table there in a room inhabited by 6 research students in theoretical physics; they are all serious students, and seem to know a lot about the subject"—just the sort of intensity that Dyson had found lacking in Cambridge. "I think I shall know a lot of things just by contact with these young people; they all keep quite quiet in working hours, in marked contrast to the Cavendish crowd, either because they have real problems to work on or because there are no girls around."[22] A month later, he expanded upon his daily routine in a long letter to his mother:

> You ask for more information about the pattern of my life. I am sorry not to have supplied this earlier; the reason why my letters are not very informative is partly

strengthen the original statement of the conjecture: Dyson, "Approximation to algebraic numbers" (1947). Dyson received his undergraduate degree in mathematics from Cambridge and completed several years of graduate study in physics but never formally received a Ph.D.

22. Freeman Dyson to his parents, 20 Sep 1947 and 25 Sep 1947, in *FJD*.

> that I haven't given much time to them, but also there is not a great deal to say; I am living a highly professionalized existence, without any private life to speak of, and wake up in the mornings thinking about mesons and photons, and there is not much one can say about that. The routine of the day is, roughly speaking, that I spend all the time at the Physics department, doing my own research and reading other people's. To relieve this programme, there are two courses of lectures which I go to, and two afternoons a week of lab work; also occasional seminars and a great deal more informal discussions; in the evenings I can stay here if nothing else is happening, but about one night in two there is either a party of some kind or a show; otherwise I can go for a walk or read in the browsing library. You will see from this that the life is in fact highly circumscribed; at present I have no desire for anything except this sort of life, and the only question is whether I shall be able to keep it up indefinitely.... At present I think the students I work with have practically no part in sports of any kind; most of them have wives and children which give them relaxation at week-ends, and others soon will. The most striking difference between this atmosphere and that of Cambridge is that here it is taken for granted that one works hard; and this strengthens the natural tendency I have to bury myself. For the first time in my life I can think about physics continually and without effort, and I want to confirm the habit before letting it drop.[23]

Dyson's description of his ascetic lifestyle did not always match other students' impressions of him. Michel Baranger, who arrived at Cornell as a graduate student in September 1949, recalls the stories that had already attained mythic status around Cornell's physics department. As Baranger heard it, Dyson used to come into the theoretical physics graduate students' office in the mornings, "read the *New York Times,* put his feet on the desk, and then fall asleep and take a nap. In the afternoon he'd go and talk with Bethe, but the other students assumed he wasn't very serious." Naturally, as the story got handed down, the joke was on Dyson's observers as soon as his publications began to appear.[24]

In addition to describing his fellow students, Dyson also began to write home about his new supervisor, Hans Bethe. Bethe often joined the graduate students for meals and worked closely with each of them as they struggled to make progress on their research. The German émigré gave a funny first impression; he was "the most unCambridgeish person" Dyson had ever met: "Bethe himself is an odd figure, very large and clumsy and with an exceptionally muddy old pair of shoes," Dyson explained to his parents. "He gives the impression of being very clever and friendly, but rather a caricature of a professor; however, he was second in command at Los Alamos, so he must be

23. Freeman Dyson to his mother, 29 Oct 1947, in *FJD.*

24. Baranger interview (2001).

a first-rate organiser as well."[25] The comical character quickly earned Dyson's respect: "I am seeing a lot of Bethe," Dyson wrote three weeks later, "far more in fact than I expected when I arranged this trip. I am more and more impressed with his intellect.... [He] thinks and argues in a most forceful way, and is never caught in a mistake."[26]

Bethe routinely assigned research problems to his students, handing out dissertation topics and postdoctoral projects from the long lists of open problems he continually drew up.[27] To Dyson he assigned the task of improving Bethe's own recent calculation of the Lamb shift, which had appeared in the *Physical Review* one month before Dyson arrived. Whereas Bethe had treated the problem nonrelativistically, Dyson was to try his hand at a fully relativistic calculation, but with the simplifying assumption that the electrons in the hydrogen atom carried no spin. In fact, Dyson's assignment was meant to be only an "interim programme," as he wrote home: having "had a good look at the frightfulness of the exact calculation," which would have included effects both from relativity and from the electron's spin, Bethe had "handed it over to one of the students in this room, by name [Richard] Scalettar; he has been struggling with it for the last two months and will take several more to finish it; he has not yet even managed to get it down to a finite result, and it is not obvious that this is possible." In the meantime, Dyson began to tackle his simplified calculation: "I am stimulated to work hard at it by frequent discussions with Bethe, and particularly by the feeling that I am on trial, and upon my success in this job will largely depend the amount of pull I shall have when new jobs come along," Dyson reported to his parents.[28]

In fact, he made quick work of the task, plunging in with the skills that he had already developed under Kemmer's direction in Cambridge. Kemmer had insisted that Dyson make a careful study—with Kemmer's guidance—of Gregor Wentzel's textbook, *Quantentheorie der Wellenfelder*, published in Vienna in 1943. (Kemmer had been a student of Wentzel's before the war.) Coming out in German in the midst of World War II, the book did not command

25. Freeman Dyson to his parents, 25 Sep 1947, in *FJD*. Dyson called Bethe "the most unCambridgeish person I know" in a letter to his parents, 9 July 1948, in *FJD*.

26. Freeman Dyson to his parents, 16 Oct 1947, in *FJD*.

27. See Freeman Dyson to his parents, 16 Oct 1947 and 29 Oct 1947, in *FJD;* Hans Bethe's handwritten notes, "Problems (1/28/48)," in *HAB*, Folder 3:42; Hans Bethe, memo to E. Baranger and S. Cohen, 1 Feb 1952, in *HAB*, Folder 10:1; and Baranger interview (2001). In a letter of recommendation for one of his students, Bethe explained that "He selected his thesis topic himself, in contrast to practically all our other students." Bethe to Harold Richards, 30 Jan 1950, in *HAB*, Folder 12:12.

28. Freeman Dyson to his parents, 16 Oct 1947, in *FJD*.

a wide audience among the English-speaking physicists in the Allied countries; Dyson recalls that Kemmer's personal copy was the only one available in Cambridge, and that there was perhaps only one more copy in all of England at the time.[29] Armed with this training in the rudiments of quantum field theory, Dyson completed his calculation for Bethe after two weeks of intensive work and after amassing a tall pile of scratch paper, several hundred pages high. When the dust had settled, he found results in good agreement both with Bethe's original calculation and with the experimental numbers being reported by the Columbia experimentalists. Moreover, he had demonstrated to Bethe (and soon to his fellow students) that he could calculate: whereas Bethe's own approximate, nonrelativistic calculation had taken up only two pages in the *Physical Review*, Dyson's calculation eventually filled ten densely packed pages in the journal. As Dyson remembers it, his calculation also demonstrated to Bethe the practical usefulness of quantum field theory: in Dyson's hands, Wentzel's elaborate formalism had actually led to a real number that could be compared with experiment.[30]

"Much to my surprise," Dyson wrote home, Bethe suggested that he write up the calculation and send it to the *Physical Review*. Having finished the calculation by late October, Dyson spent several weeks writing the paper—boiling down his tall stack of notes into the essential steps of the calculation—and submitted it in early December. On that same day, Bethe called him into his office and suggested that Dyson spend a second year in the United States; moreover, he should spend it with J. Robert Oppenheimer, who had just moved to the Institute for Advanced Study in Princeton, New Jersey.[31] Bethe explained to the Commonwealth Fellowship administrators why Dyson should receive an extension of his fellowship to cover the second year. After noting that Dyson "is

29. Dyson, "Comments on selected papers" (1996), 10.

30. Freeman Dyson to his parents, 16 Oct 1947 and 29 Oct 1947, in *FJD*. Dyson mentions the large pile of paper in Dyson, *Disturbing the Universe* (1979), 54, and discusses Bethe's reaction to his calculation in Dyson, "Comments on selected papers" (1996), 12. The results were published in Dyson, "Electromagnetic shift" (1948). When setting up his calculation, Dyson drew explicitly upon Wentzel's textbook, writing out the fields and their conjugate momenta, and then the Hamiltonian in terms of these field variables, before expanding the fields in terms of creation and annihilation operators. In contrast, Bethe's calculation made use of hole-theoretic perturbative methods as developed in Walter Heitler's textbook and eschewed the more formal field-theoretic approach. Bethe, "Electromagnetic shift" (1947); cf. Heitler, *Quantum Theory of Radiation* (1936). Like Bethe, Dyson forced his integrals to converge by putting in a cutoff by hand at momentum $k \sim mc$, where m is the mass of the electron and c is the speed of light. For more on Bethe's and Dyson's Lamb-shift calculations, see Schweber, *QED* (1994), 228–31, 497–500.

31. Freeman Dyson to his parents, 29 Oct 1947 and 7 Dec 1947, in *FJD*. Dyson's article was received at the *Physical Review* on 8 Dec 1947.

far superior, even to the best American graduate students we have at Cornell," Bethe continued: "There is a very compelling reason for the continuation of Dyson's study in this country. He has become interested in the new developments of Quantum Electrodynamics which have brought back to life the subject that was believed dead for almost 20 years.... All this work is going on in the United States at the present time, the main centers being Professor Oppenheimer's school at the Institute of Advanced Studies [*sic*], Harvard and M.I.T., and this department. If Dyson is enabled to spend a year with Oppenheimer, he will broaden his knowledge of this field to which he has already contributed so much, and will be able to spread his knowledge after his return to England."[32] Bethe highlighted the opportunity for Dyson to "broaden" his knowledge and then "spread" it—exactly the role that postdoctoral training had come to embody after the war.[33] The pace of it all left Dyson breathless. "All this shows how fundamentally right was the idea that made me change from mathematics to physics, in spite of many discouragements," he wrote to his parents. "I have done nothing in the last two months that you could call very clever or difficult; nothing one tenth as hard as my [Trinity] fellowship thesis; but because the problems I am now dealing with are public problems and all the theoretical physicists have been racking their brains over them for ten years with such negligible results, even the most modest contributions are at once publicised and applauded." Moreover, the prospects of spending the year at the Institute were heady, indeed: "I am not immune from the common disease of celebrity-hunting, and should like to make the acquaintance of Einstein and [Hermann] Weyl and [John] Von Neumann before I go home, which can only be done at Princeton."[34]

In the meantime, a new influence had entered Dyson's life. In the midst of his toils preparing his article on the scalar Lamb-shift calculation, Dyson's letters began to mention Cornell's other leading theorist, Richard Feynman. "Feynman is a man for whom I am developing a considerable admiration," Dyson wrote home in November 1947. "[H]e is the brightest of the young theoreticians here, and is the first example I have met of that rare species,

32. Hans Bethe to E. K. Wickman (director, Commonwealth Fund), 28 Feb 1948, in *HAB*, Folder 11:3.

33. The main exception in this case was that Dyson did not have a Ph.D.—indeed, he never received one. When asking the Cornell administration to offer Dyson a professorship a few years later, Bethe and his colleagues explained, "With all his varied activities Mr. Dyson has so far omitted to obtain his doctor's degree.... Obviously the degree in his case is a pure formality." Hans Bethe, Lloyd Smith, and Robert Wilson to Dean L. S. Cottrell, Jr., 27 Oct 1950, in *HAB*, Folder 10:20. See also Peierls, *Bird of Passage* (1985), 234–35; and Schweber, *QED* (1994), 556.

34. Freeman Dyson to his parents, 7 Dec 1947 and 14 Dec 1947, in *FJD*.

the native American scientist": "He has developed a private version of the quantum theory, which is generally agreed to be a good piece of work and may be more helpful than the orthodox version for some problems; in general he is always sizzling with new ideas, most of which are more spectacular than helpful, and hardly any of which get very far before some newer inspiration eclipses them. His most valuable contribution to physics is as a sustainer of morale; when he bursts into the room with his latest brain-wave and proceeds to expound it with the most lavish sound effects and waving about of the arms, life at least is not dull." Later in the same letter, Dyson described the scene at a party at the Bethes' house. Their son Henry, then five years old, "was not impressed" by the visitors, "in fact the only thing he would say was 'I want Dick. You told me Dick was coming,' and finally he had to be sent off to bed, since Dick (alias Feynman) did not materialise," Dyson recounted. "About half an hour later, Feynman burst into the room, just had time to say 'so sorry I'm late. Had a brilliant idea just as I was coming over,' and then dashed upstairs to console Henry. Conversation then ceased while the company listened to the joyful sounds above, sometimes taking the form of a duet and sometimes of a one-man percussion band."[35]

Soon it was not only Henry Bethe who delighted in Feynman's company. Dyson began to spend more and more time talking with Feynman, and they became fast friends. Describing Feynman as "half genius and half buffoon, who keeps all physicists and their children amused with his effervescent vitality," Dyson wrote home about his frequent chat sessions with Feynman—chat sessions that roamed from quantum electrodynamics to Feynman's love life and beyond. Feynman's new diagrammatic techniques initially struck Dyson—as they struck so many others—as inscrutable: "Feynman is a man whose ideas are as difficult to make contact with as Bethe's are easy," Dyson explained to his parents; he found Feynman's diagrams "completely baffling."[36] Yet Dyson had ample opportunity to learn more about Feynman's difficult ideas during the spring semester, around the time that Feynman presented his new diagrammatic methods at the Pocono meeting. A few months later, Dyson received an even more intense dose of Feynmaniana when they drove from Cleveland to Albuquerque together, Feynman to visit a girlfriend and Dyson

35. Freeman Dyson to his parents, 19 Nov 1947, in *FJD*. Other colleagues' children likewise delighted in playing with Feynman; see Wheeler, "Young Feynman" (1989), 25; and Tom Kinoshita, letter to the author, 19 Mar 2003. The "private version of the quantum theory" that Dyson referred to was Feynman's dissertation project, eventually published as Feynman, "Space-time approach" (1948).

36. Freeman Dyson to his parents, 8 Mar 1948 and 15 Mar 1948, in *FJD;* Dyson recalls finding Feynman's work "completely baffling" in Dyson, "Comments on selected papers" (1996), 12.

to do some sightseeing. The car trip gave Dyson the opportunity to see parts of the American Midwest and Southwest for the first time; he also got to share many intense hours deep in conversation with his new friend. Not all the talk focused narrowly on physics: in the midst of their "Odyssey," as Dyson called it, flooding closed parts of Route 66 in Oklahoma and they had to scramble for lodging in "what Feynman called a 'dive,' viz. a hotel of the cheapest and most disreputable character." (In Feynman's less sanitized version, it was a brothel.) After their cross-country road trip, Feynman spent several weeks in New Mexico, where he continued to work on his new diagrammatic calculations, while Dyson made his way by bus to Ann Arbor, Michigan, for the annual summer school in theoretical physics, held during July and August.[37]

The main attraction in Ann Arbor that summer was Julian Schwinger. Following his triumphant daylong presentation at the Pocono meeting on his own approach to rendering QED's infinities harmless, Schwinger strode into town with an entourage of eager acolytes in tow. "Yesterday the great Schwinger arrived, and for the first time I spoke to him," Dyson wrote home that summer. "His talks have been from the first minute excellent; there is no doubt that he has taken a lot of trouble to polish up his theory for presentation at this meeting. I think in a few months we shall have forgotten what pre-Schwinger physics was like."[38] Dyson listened to the lectures in the mornings and worked on his own research during the afternoons. His thoughts turned to ways of "tidying up" some details within Schwinger's new formalism, which paid immediate dividends: "I was very glad of this, as it enabled me to speak to Schwinger and the other 'big shots' and get some ideas out of them."[39]

Thus, by early August 1948, Dyson—and Dyson alone—had spent intense time working face-to-face with both Feynman and Schwinger, talking in detail

37. Freeman Dyson to his parents, 25 June 1948 and 2 July 1948, in *FJD*. For more on Dyson's work and friendship with Feynman, see Dyson, *Disturbing the Universe* (1979), 53–68; and Dyson's interview published in Sykes, *No Ordinary Genius* (1994), 73–74. Feynman's own, more bawdy recollection of the car trip appears in Feynman with Leighton, *What Do You Care* (1988), 65–66. Feynman's original motivation for the road trip had been to visit a girlfriend; he explained in a letter to Bethe upon arriving in Albuquerque, "See, I'm in New Mexico where love has drawn me, but on arrival love dispersed so I am now returned to work." In the meantime, Feynman's work was paying off, as he announced to Bethe: "I am the possessor of a swanky new scheme to do each problem in terms of one with one less energy denominator," that is, he had arrived at his trick for combining denominators. Feynman to Hans Bethe, 7 July 1948, in *HAB*, Folder 3:42; see also Feynman to Bethe, 22 July 1948, in the same folder; and Dyson to his parents, 8 Mar 1948, 11 June 1948, and 14 Nov 1948, in *FJD*.

38. Freeman Dyson to his parents, 22 July 1948, in *FJD*.

39. Freeman Dyson to his parents, 8 Aug 1948, in *FJD*. Dyson also gave a report on Schwinger's lectures to Bethe, who was in Europe at the time: Dyson to Bethe, 9 Aug 1948, in *HAB*, Folder 3:42.

about their rival methods. After the summer school ended, Dyson took a bus trip to Berkeley to spend a few weeks relaxing and then began his journey back across the country, again making his way by bus. "On the third day of the journey a remarkable thing happened," Dyson excitedly told his parents: "going into a sort of semi-stupor as one does after 48 hours of bus-riding, I began to think very hard about physics, and particularly about the rival radiation theories of Schwinger and Feynman."

> Gradually my thoughts grew more coherent, and before I knew where I was I had solved the problem that had been in the back of my mind all this year, which was to prove the equivalence of the two theories. Moreover, since each of the two theories is superior in certain features, the proof of the equivalence furnished incidentally a new form of the Schwinger theory which combines the advantages of both.
>
> This piece of work is neither difficult nor particularly clever, but it is undeniably important if nobody has done it in the meantime. So you can imagine that I became quite excited over it when I reached Chicago, and sent off a letter to Bethe announcing the triumph. I have not had time yet to write it down properly, but I am intending as soon as possible to write a formal paper and get it published.

As soon as he arrived at the Institute for Advanced Study that September, he set about writing his "magnum opus."[40]

Spreading the (Written) Word: Dyson's Textual Interventions

"After a very brief visit to Cornell, to collect my various belongings, I settled down to work at writing up the physical theories I mentioned in the last letter," Dyson reported to his parents near the end of September 1948, having just arrived in Princeton. "I found, as one often does when one comes to write, that the job was even bigger than I had imagined, and I was for about five days stuck in my rooms, writing and thinking with a concentration which nearly killed me. On the seventh day the paper was complete, and with immense satisfaction I wrote the number 52 at the bottom of the last page." He was clearly exhausted: "Having emerged from that, I feel I shall not do any more thinking

40. Freeman Dyson to his parents, 14 Sep 1948, in *FJD.* See also Dyson to his parents, 30 Sep 1948, in *FJD;* and Dyson to Bethe, 8 Sep 1948, in *HAB,* Folder 10:20. In his letter to Bethe, Dyson concluded that "Incidentally, the complete equivalence of Schwinger and Feynman is now demonstrated"—the bulk of his letter being occupied instead with how he had succeeded in reformulating Schwinger's methods so as to take advantage of the strengths of Feynman's approach, rather than on their equivalence per se. As Dyson saw it, Schwinger's approach had the advantage of maintaining manifest gauge invariance at each step (as he emphasized in his letter to Bethe of 9 Aug 1948), whereas Feynman's diagrams provided a highly efficient method for setting up the relevant integrals.

for the rest of the year."[41] Through his toil, he had managed to demonstrate the mathematical—though by no means conceptual—equivalence between Schwinger's and Feynman's formalisms; this was no mean feat, he explained to his parents, since "Feynman and Schwinger talk such completely different languages, that neither of them is able to understand properly what the other is doing." Worse still, no one else could understand what either of them was doing: neither had "published any moderately intelligible account of his work." Feynman had published nothing at all on his new diagrams, and Schwinger had published only a two-page letter to the editor; the first of Schwinger's long articles had only just been submitted from the Ann Arbor summer school, and would not appear in print until the middle of November. Thus, Dyson had to do more than simply demonstrate the equivalence of the two approaches in his own paper; he had to present the rudiments of both formalisms more or less from scratch—"in itself a valuable service to humanity," he mused to his parents.[42] Dyson submitted his paper to the *Physical Review* in October 1948.

Having recovered from the "ordeal" of writing his long paper within a few days, Dyson began to realize that in his effort to keep his paper as short as possible, he had ignored a series of elaborations and further developments to which he could now turn his attention; he anticipated writing "a whole string of papers." Sure enough, not quite two months later, he had begun to pull together a second long paper. By Christmas eve, he had finished nearly sixty pages of the manuscript, which "turned out to be even more formidable in length and difficulty than the first"; it eventually grew to eighty pages of typescript.[43] He finally submitted this second article to the *Physical Review* in February 1949. Although he had anticipated that the two articles would be long delayed in the reviewing stage, covering as they did difficult material about unknown methods and running far longer than the average submission for that journal, in fact they were both accepted almost immediately: each paper was accepted less than two weeks after submission. The acceptance notice for the second paper came so quickly, in fact, that Dyson suspected "some conspiracy" between the editor and his new boss, J. Robert Oppenheimer, especially since

41. Freeman Dyson to his parents, 26 Sep 1948, in *FJD*.

42. Freeman Dyson to his parents, 30 Sep 1948 and 4 Oct 1948, in *FJD*. Dyson's paper was published as Dyson [A.1]. In his paper, Dyson also briefly discussed Tomonaga's approach, which remained little known in the United States at the time. Dyson treated Tomonaga's and Schwinger's methods interchangeably, referring to the "Tomonaga-Schwinger" approach throughout his introduction and drawing a contrast between their work and Feynman's. In the body of his paper, Dyson focused only on Schwinger's and Feynman's work.

43. Freeman Dyson to his parents, 30 Sep 1948, 21 Nov 1948, and 24 Dec 1948, in *FJD*. Dyson's second article appeared as Dyson [A.2].

the second paper, like the first, was accepted without requiring any revisions.[44] The ultrafast turnaround time meant that both of Dyson's articles appeared in print months before anything by Feynman on the topic.[45]

In his long articles, Dyson did much more than merely demonstrate that Feynman's and Schwinger's unpublished methods could be reconciled. Still drawing on his knowledge of Wentzel's field-theoretic formalism, which had so impressed Hans Bethe the year before, Dyson demonstrated how to cast both Feynman's and Schwinger's approaches within a consistent field-theoretic framework. Moreover, Dyson's pair of articles became a "how-to" guide, enumerating carefully the step-by-step rules for drawing Feynman's new diagrams and explaining how their bare lines were to be translated uniquely into the mathematical elements of scattering amplitudes. It was thus Dyson, and not Feynman, who first codified the *rules* for the diagrams' use—precisely what Feynman's frustrated audience had hoped to hear at the Pocono meeting a few months earlier. Explicitly following Dyson's lead, the first textbook authors later included tables like that reproduced in figure 3.2 in their books. Thanks to Dyson's efforts (which we will examine in more detail in chap. 5), physicists eventually learned to build up their diagrammatic calculations one element at a time, as one would assemble toy building blocks, by snapping the appropriate mathematical function into place for each component of the line drawings.

In his second article Dyson went further still, surpassing even Schwinger's efforts to date, by generalizing Feynman's and Schwinger's distinct ways of removing the troubling infinities from QED to any order of a calculation. Until that time, Schwinger and Feynman had each removed the infinities only from the simplest type of divergent calculation, involving only one or two virtual particles; in other words, they had each found ways to extract finite

44. Freeman Dyson to his parents, 19 Oct 1948 and 11 Mar 1949, in *FJD*. Unfortunately, no referee reports have been preserved for articles that appeared in the *Physical Review* before 1964. (Martin Blume, editor in chief of the American Physical Society, personal communication, 16 Apr 2002.) Most likely, however, neither of Dyson's papers was thoroughly reviewed, given the standard practices of the day. Instead, the editor of the *Physical Review* at that time, Samuel Goudsmit, most likely asked one of Dyson's senior colleagues (in all likelihood, Bethe) to vouch for the papers in general terms, which would explain both the speed with which they were accepted and the fact that no revisions were required. The correspondence regarding other submissions to the journal at this time, revealing the relaxed reviewing practices, may be found in *SAG*, Folder 50:42, and in *HAB*, Folder 12:37.

45. Dyson's "Radiation theories" [A.1] was received at the *Physical Review* on 6 Oct 1948 and published in the 1 Feb 1949 issue; "*S* matrix" [A.2] was received on 24 Feb 1949 and published in the 1 June 1949 issue. Both of Feynman's articles, submitted in April and May 1949, appeared in the 15 Sep 1949 issue: Feynman, "Theory of positrons" (1949); and Feynman [A.4]. For more on Dyson's 1949 articles, see also Schweber, *QED* (1994), chap. 9.

The correspondence between diagrams and S-matrix elements in momentum space

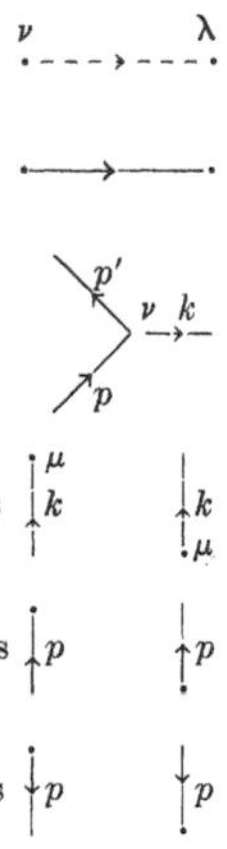

Component of Diagram	Factor in S-Matrix Element	
Internal photon line ν λ	$g_{\nu\lambda}\dfrac{1}{k^2 - i\mu}$	photon propagation function
Internal electron line	$\dfrac{i\not{p} - m}{p^2 + m^2 - i\mu}$	electron propagation function
Corner p' ν k p	$\gamma^\nu \delta(p - p' - k)$	
External photon lines μ k k μ	$\dfrac{1}{\sqrt{2\omega}}\, e_\mu(\mathbf{k}),\ \dfrac{1}{\sqrt{2\omega}}\, e_\mu(\mathbf{k})$	ingoing and outgoing photons
External negaton lines p p	$\sqrt{\dfrac{m}{\epsilon}}\, u(\mathbf{p}),\ \sqrt{\dfrac{m}{\epsilon}}\, \bar{u}(\mathbf{p})$	ingoing and outgoing negatons
External positon lines p p	$\sqrt{\dfrac{m}{\epsilon}}\, \bar{v}(\mathbf{p}),\ \sqrt{\dfrac{m}{\epsilon}}\, v(\mathbf{p})$	ingoing and outgoing positons

Figure 3.2. The "Feynman rules" in the momentum-space representation, following Dyson's prescriptions. (Source: Jauch and Rohrlich, *Theory of Photons and Electrons* [1955], 154.)

answers from the e^2 term within the infinite series, but neither had yet tackled the infinities lurking within the e^4, e^6, or more complicated terms. Dyson now demonstrated how the infinities could be removed systematically and self-consistently from any arbitrary calculation. In particular, Dyson showed that the ability to express results in terms of observable quantities, such as an electron's effective mass and charge, m_{eff} and e_{eff}, would not come undone as more complicated radiative corrections were included. Feynman diagrams in hand, Dyson thus showed how a renormalization program could be put together, stretching over and above the few worked examples that Feynman, Schwinger, and others had managed to cobble together. Although Dyson's use and interpretation of the diagrams remained different from Feynman's in important ways, as we will see in chapter 5, his pair of articles put the diagrams into circulation. As Dyson himself put it in a letter to his parents, his main achievement in these long papers was to make Feynman's diagrammatic methods "available to the public."[46]

A few days before mailing off his first article, Dyson had a long talk with Bethe, who was then visiting in New York City. Bethe and Dyson worried about

46. Freeman Dyson to his parents, 4 Dec 1948, in *FJD*.

offending Schwinger by publishing an account of his work before Schwinger's own detailed articles had appeared in print. They had few concerns where Feynman was concerned, "but with Schwinger it is different," as Dyson wrote home to his parents: "he certainly intends to present to the world a finished account of his work, but he wants it to be completely polished, and in fact the last word on the subject." Thus Bethe and Dyson agreed to step gingerly when it came to allotting credit within Dyson's manuscript: "the result of all this is that I am reversing the tactics of Marc Antony," Dyson explained, "and saying very loud at various points in my paper, 'I come to praise Schwinger, not to bury him.' I only hope he won't see through it." Dyson took similar care throughout his papers to give credit for the new diagrams to Feynman. Several months later, after Dyson had finished his second long article—while Feynman still had written nothing about his diagrammatic methods—Dyson began to "feel a little guilty for having cut in in front of him with his own ideas." Upon hearing that his second paper had been accepted, he wrote to Bethe that "I feel a little uneasy on the question of priorities, whether I have treated Feynman fairly in not waiting for him to publish his stuff first. Frankness on this subject would be very welcome." Bethe's response must have been reassuring, and Dyson consoled himself with the thought that if his efforts had encouraged Feynman to begin writing his own papers, then his prodding would constitute "a valuable service on my part."[47]

Despite Dyson's efforts to credit Feynman with the invention of the diagrams, many of the earliest diagram users spoke of the "Feynman-Dyson method," the "Feynman-Dyson formulation," and even of "Feynman-Dyson diagrams"—and, given Dyson's systematic derivation of the diagrams and their use, the attribution was not entirely unfair.[48] At first Dyson found the attribution surprising, even a bit embarrassing, as he described in a letter to his parents after the January 1949 meeting of the American Physical Society in New York: "On the first day the real fun began; I was sitting in the middle of the hall and in the front, with Feynman beside me, and there rose to the platform to speak a young man from Columbia whom I know dimly. The young man had done some calculations using methods of Feynman and me, and he did

47. Freeman Dyson to his parents, 4 Oct 1948 and 28 Feb 1949, in *FJD;* Freeman Dyson to Hans Bethe, 10 Mar 1949, in *HAB,* Folder 10:20. See also Dyson [A.1], 486; and Dyson [A.2], 1736–39.

48. The "Feynman-Dyson method" is referred to in Rohrlich [A.10], 357; and in Brueckner [A.17], 647; the "Feynman-Dyson formulation" is mentioned in Riddell [A.89], 1243. "Feynman-Dyson diagrams" or "graphs" appear in Brueckner [A.36], 600; and Roberts [A.37], 188–89. Wolfgang Pauli similarly referred to the "Feynman-Dyson-Kalkül" in correspondence with Gunnar Källén, 22 Dec 1949, in Pauli, *WB,* 3:724.

not confine himself to stating this fact, but referred again and again to 'the beautiful theory of Feynman-Dyson,' in gushing tones. After he said this the first time, Feynman turned to me and remarked in a loud voice, 'Well, Doc, you're in.'"[49] The effect on Dyson was immediate: "Since the New York meeting, I have felt some of the consequences of fame," Dyson wrote home excitedly a few weeks later. At the meeting it wasn't only the young Columbia physicist who "gushed" over Dyson's and Feynman's accomplishments. Oppenheimer delivered a keynote lecture in which he "spoke in enthusiastic terms of the work I have been doing," and upon Oppenheimer's public pronouncement, Dyson spent the following twenty-four hours with "one person after another pursuing me and asking to be told all about it." Oppenheimer's endorsement led immediately to a string of invitations for Dyson to lecture on his work: "beginning in March, a week in Chicago; middle of March, two days at Rochester and a week at Cornell; beginning of April, a week at Toronto; middle of April, the select conference organised by Oppy [the Oldstone conference, a follow-up to the Pocono meeting]; End of April, the annual meeting of the American Physical Society at Washington. With all this, I shall be at Princeton almost as little as Oppy. I have also invitations of a less formal kind from Harvard and Columbia." Karl Darrow, the perpetual secretary of the American Physical Society, invited Dyson to give a forty-minute invited lecture at the Washington meeting (rather than the usual ten-minute contributed talk), an honor Dyson accepted with some trepidation. In the meantime, after his first long paper appeared in print, the avalanche of correspondence from people "asking all kinds of questions, sensible and otherwise," led Dyson to daydream that "Soon I shall have to engage a secretary!"[50]

His hectic lecturing schedule certainly kept him busy during the winter and spring of 1949, long before his second article appeared in print. By the time he wrote his second article, however, Dyson had begun to take advantage of a new phase in physicists' communication habits: the circulation of preprints. As with so many major changes in the daily business of being a physicist, the preprint age did not dawn overnight. Several elements needed to be straightened out during the early postwar years: Should unpublished lecture notes

49. Dyson to his parents, 30 Jan 1949, in *FJD*. The speaker was most likely Francis Low, at the time a graduate student at Columbia who was working with Hans Bethe during Bethe's sabbatical year there in 1948. Kaiser, "Francis E. Low" (2001). Low probably learned about Dyson's work during an intense day Dyson spent at Columbia in early October, a few days before he submitted his first article on the diagrams. On Bethe's invitation, Dyson lectured to a group of Columbia students and physicists for three hours on the new material. See Dyson to his parents, 4 Oct 1948, in *FJD*.

50. Freeman Dyson to his parents, 30 Jan 1949, 15 Feb 1949, 28 Feb 1949, and 5 Apr 1949, in *FJD*.

that had been taken by graduate students and mimeographed be treated the same way as ordinary publications? Could government laboratories circulate unpublished reports without violating the regulations regarding government printing offices? Would preprints from national laboratories need to be kept in a safe, even if they treated nonclassified material?[51] These issues notwithstanding, more and more physicists began to demand preprints in order to keep up with the fast-moving research currents. An exasperated Hans Bethe pleaded with the Berkeley Radiation Laboratory's Robert Serber in October 1949, for example, that information about the Berkeley group's work "decreases more rapidly than the square of the distance from Berkeley." No longer should physicists across the country have to rely upon "long distance diffusion" to learn of physics results elsewhere, Bethe insisted; preprints should bring news of new work immediately.[52] Dyson quickly learned the lesson: whereas he had simply given the manuscript for his first paper to a typist at the Institute to prepare for submission to the journal, he "gently encouraged" the Institute staff to prepare one hundred copies of the manuscript of his second paper, to have them ready to distribute at the New York American Physical Society meeting. Oppenheimer likewise sent off preprints of Dyson's paper to a number of his colleagues. From that moment on, Dyson prepared preprints for all his articles, thus quickening the effectiveness of his textual interventions.[53]

He had also begun to learn about physicists' reading habits. While preparing his second article, he explained to his parents that he doubted whether most physicists would really work through the complicated arguments that made up the bulk of his paper. Instead, he would focus on the conclusions, "since this is the only part of the paper that the average reader will ever read." No matter—Dyson observed that the main function of his articles had been to "increase people's confidence" in the new methods, not necessarily to teach them all the minutiae of how to use them; "the details are not so important." If people felt confident about the new scheme in general, then he and his accomplices might be able to work with them to practice using the diagrams in actual calculations. His preprints and lectures, in the meantime, would "advertise"

51. See the correspondence between Hans Bethe, three of his graduate students, and representatives from John Wiley publishers and the General Electric Research Laboratory (where Bethe's lectures had been delivered), Jul–Sep 1946, in *HAB,* Folder 16:23; Richard Feynman to Cécile Morette, 5 June 1950, in *RPF;* Harold Agnew to Robert Bacher, 12 Mar 1952, in *HAB,* Folder 10:4; and Hans Bethe to Robert Serber, 5 Oct 1949, in *HAB,* Folder 12:25.

52. Bethe to Serber, 5 Oct 1949, in *HAB,* Folder 12:25.

53. Freeman Dyson to his parents, 22 Jan 1949 and 15 Jan 1949, in *FJD;* see also Dyson to his parents, 11 July 1950, 20 Nov 1950, and 29 Nov 1950, in *FJD.*

the new methods; they could function, in other words, more as a sales pitch than a user's manual.[54]

The advertisements clearly worked: after Dyson's preprints and articles came out, other physicists' presentations of the diagrams often became even more strongly Dysonian. Robert Karplus and Norman Kroll, for example, announced in their 1950 article on fourth-order corrections within QED that their purpose was "to demonstrate in a complete calculation of a particular example the feasibility of Dyson's program." In setting up and carrying out their extended calculations (which we will examine in chap. 6), Karplus and Kroll noted that "the methods of Dyson have been followed quite closely."[55] Paul Matthews wrote to Pauli and Rohrlich in the winter of 1950 about how to apply "Dyson's rules," a label that Matthews had already used in some of his brief letters to the editor of the *Physical Review.*[56] Three years later, a graduate student noted simply that "Dyson has shown" the relevant means of evaluating terms within QED.[57] Often Dyson's two articles from 1949 were cited without citations to Feynman's papers; this pattern persisted over the course of the 1950s, with Dyson's 1949 articles cited more frequently than Feynman's.[58] In preprint form and publications, Dyson worked hard to put Feynman diagrams in circulation.

Beyond these early papers, Dyson helped to put Feynman diagrams in circulation with a lecture course he delivered at Cornell during the autumn

54. Freeman Dyson to his parents, 15 Jan 1949 and 15 Feb 1949, in *FJD.* The articles clearly succeeded in building "confidence," especially when one considers the fact that Dyson's attempted proof of renormalizability in his second article was not watertight: he advanced a conjecture, rather than a proof, about how to handle overlapping divergences—that is, a special class of Feynman diagrams in which two separate divergences "line up" on top of each other in the accompanying integrals. Dyson sent a preprint of his second paper to Rudolf Peierls, writing that "It is very long, and not very well written, and some of the statements in it ought to be qualified with a 'perhaps' at the beginning and a 'maybe' at the end." Freeman Dyson to Rudolf Peierls, 23 Feb 1949, as quoted in Schweber, *QED* (1994), 545. Norman Kroll remembers finding Dyson's second article, which he received in preprint form from Oppenheimer, "very persuasive" and "relatively straightforward." Kroll interview (2002).

55. Karplus and Kroll [A.11], 536.

56. Matthews to Pauli, 25 Feb 1950, in Pauli, *WB,* 4.1:26–33, on 27. See also Matthews [A.3].

57. Anderson [A.112], 703. J. L. Anderson received his Ph.D. from Syracuse in 1954 (*DiP*).

58. Between 1949 and 1955, Dyson's "Radiation theories" [A.1] was cited 213 times, and his "*S* matrix" paper [A.2] was cited 249 times; during this same period, Feynman's "Theory of positrons" (1949) received 185 citations, and his "Space-time approach" [A.4] received 229 citations. Later the gap began to close: between 1949 and 1964, Dyson's two articles received 295 and 371 citations, respectively, while Feynman's two articles received 269 and 367 citations, respectively. *Science Citation Index,* s.v. "Dyson, Freeman," and "Feynman, Richard." Often Dyson's "*S* matrix" article was cited as the sole source for the so-called Feynman rules, as in the 1953 paper by Andrew Lenard, a graduate student of Josef Jauch's at Iowa. Lenard [A.81], p. 971.

1951 semester. Notes from the course, entitled Advanced Quantum Mechanics, developed Dyson's particular approach to QED, culminating in the use and interpretation of Feynman diagrams.[59] Something of Dyson's classroom style may be gleaned from a letter to Berkeley's department chair, Raymond Birge, from his son, who attended Dyson's course on nuclear theory at Cornell the following semester. "Dear Daddy," the letter begins,

> All I know about Dyson is what I've observed from taking a nuclear theory course from him. He is a pretty good lecturer, has a very lucid style but, at least in this course, rarely uses lecture notes, usually referring to a stack of *Physical Reviews* which he brings to class with him. He gets by with this most of the time because he has apparently given some thought to the lecture and knows the stuff so well any way, but sometimes he gets confused. He takes pains to make sure that everybody understands what he is saying and is patient about explaining things in detail.
>
> He is popular with the students and makes an effort to be so, I think.[60]

Dyson's patient explanations likewise suffused his Advanced Quantum Mechanics lecture notes, which provided far and away the most detailed and explicit instructions for using Feynman diagrams to date. Although both Bethe and Feynman had taught the course at Cornell before Dyson, Bethe remarked simply that Dyson's version had become the "model" for how the course should be taught from then on. Three graduate students from the course prepared hectographed copies of Dyson's notes for use by other members of the class; but soon word of the notes spread, and requests came in for more copies. Impressed by the response, the students proceeded to contact physics departments throughout the country to offer copies of the notes for $2.50; Berkeley's physics department, for example, ordered three copies.[61] The notes often served as a stand-in for more formal, published textbooks in graduate courses, even after several textbooks had been published by the late 1950s. Other times, graduate students used the notes for self-study, including, for example, the Berkeley theorist Henry Stapp, who still calls the yellowed notes, sitting on his office

59. Dyson, "Advanced quantum mechanics" (1951).

60. [Robert W. Birge?] to Raymond T. Birge, 18 May 1952, in *BDP*, Folder 5:33. See also Dyson to Raymond Birge, 5 June 1952, in the same folder. As Dyson wrote to his own parents, he found teaching "quite pleasant and I do not work very hard at it." Freeman Dyson to his parents, 29 Feb 1952, in *FJD*; see also Dyson to his parents, 2 Nov 1951, in *FJD*.

61. Hans Bethe to Aage Bohr, 13 Mar 1953, in *HAB*, Folder 10:1. Cf. Feynman, "Quantum electrodynamics" (1949); Bethe's notes from when he taught the course may be found in *HAB*, Folder 1:14. On the preparation and distribution of the notes from Dyson's course, see Stan Cohen, Don Edwards, and Carl Greifinger, form letter dated 9 Jan 1952; and Rebekah Young to S. Cohen, 15 Jan 1952, both in *BDP*, Folder 3:31.

shelf, "my bible."[62] The unpublished lecture notes, in turn, were cited dozens of times in the published literature, as well as in unpublished dissertations.[63] Nearly all the textbooks that were published on the diagrams in the 1950s and into the 1960s explicitly followed Dyson's treatment; at least three cited Dyson's 1951 notes in particular.[64]

Yet these textual interventions by Dyson—the two articles from 1949 and the lecture notes from 1951—proved to be only the start of Dyson's role as a diagrammatic ambassador. Much more important was his day-to-day work at the Institute for Advanced Study, where he worked with an active cohort of young postdocs. Dyson and his fellow postdocs arrived at that venerable institution at a time of great transition; it had just come under new management, and the consequences for the diagrams' spread proved to be enormous.

LIFE AND PHYSICS AT THE INSTITUTE FOR ADVANCED STUDY

Establishing Theoretical Physics at the Institute

In the early years of the Rockefeller-funded postdoctoral fellowships, physics postdocs could be most easily assimilated within experimental physics programs, where they could join many of the already-established laboratories throughout the United States. No such obvious stopping grounds existed for the minority of young scientists interested in theoretical physics, however, at the time the fellowships began. Instead, a small but steady stream of Americans, hopeful to learn about theoretical physics, traveled with Rockefeller funds to study with the European masters. Edwin Kemble, John Van Vleck, John Slater, Edward Condon, J. Robert Oppenheimer, and several others traveled to the great European centers of theoretical physics—Cambridge, Copenhagen, Göttingen, Munich, Zurich—during the 1920s and 1930s. Only in Europe could one "learn the music," as I. I. Rabi later put it, and not just

62. Wichmann interview (1998); Stapp interview (1998). Elliott Lieb similarly noted that copies of Dyson's 1951 lecture notes "still survive as treasured possessions in several private libraries." Elliott Lieb, "Foreword," in Dyson, *Selected Papers* (1996), xi–xii, on xi.

63. *Science Citation Index*, s.v. "Dyson, Freeman"; we will encounter several examples from Harvard students' dissertations below.

64. Schweber, Bethe, and de Hoffmann, *Mesons and Fields* (1955), 1:197, 203, 210, 422; Mandl, *Quantum Field Theory* (1959), 198; and Schweber, *Relativistic Quantum Field Theory* (1961), 435, 861. In addition, Josef Jauch and Fritz Rohrlich singled Dyson out as one of only two people who had worked particularly closely with them while they were writing their textbook. Jauch and Rohrlich, *Theory of Photons and Electrons* (1955), ix.

"the libretto" of research in theoretical physics.[65] Upon their return, many of these same physicists endeavored to build up domestic training grounds for young theorists. Soon after the war, one of the most important became the Institute for Advanced Study in Princeton, New Jersey.

The Institute for Advanced Study was established in 1930 with private funds from Louis Bamberger and his sister, Caroline Bamberger Fuld. The Bambergers sold their interests in their department-store chain to R. H. Macy in 1929 and proceeded to sail through the stock-market crash relatively unscathed, emerging as one of the nation's wealthiest families as the country slipped into the Depression. Having mastered the art of buying low and selling high, the Bambergers decided that the nation would benefit from sustained philanthropic donations, especially in such difficult economic times. The department-store magnates shared the same principles as Hale, Millikan, and Noyes: the country needed to build an infrastructure for high-quality independent research, and private philanthropy should help establish this infrastructure. Although at first the Bambergers thought that a new graduate school would benefit the country, Abraham Flexner, a longtime advocate of educational reform and an influential educational policy maker, convinced them that what the country really needed most was opportunities beyond the doctorate, both for young scholars fresh from their Ph.D.'s and for older professors to receive a temporary leave from teaching and administrative duties. Flexner envisioned a rural hideaway in which scholars could simply "sit and think"—an idea that was not without its critics. Vannevar Bush, soon to become the doyen of American science policy, quipped, "Well, I can see how you could tell whether they were sitting." Leopold Infeld, meanwhile, a Polish-born mathematical physicist who worked as Einstein's assistant at the Institute during the 1930s, wondered whether surrounding scholars in "isolation, comfort, and security" would leave their imaginations sterile. All the same, the Bambergers approved of Flexner's vision and endowed the new Institute with a gift of five million dollars, a tremendous sum in 1930.[66]

Flexner, who had just published a major comparative study of universities in Europe and the United States, became the Institute's first director. Nearly three years later, the Institute hired its first faculty member, Albert Einstein, when he decided to leave Nazi Germany and live permanently in the United States. Soon Flexner hired several other eminent mathematicians and mathematical

65. Rabi's famous remark is quoted in Rigden, *Rabi* (1987), 46. On the dearth of institutional support for theoretical physics within the United States before World War II, see in particular Schweber, "Empiricist temper regnant" (1986).

66. Bush's remark as reported in Anon., "Eternal apprentice" (1948), 70; Infeld, *Quest* (1941), 250.

physicists, including Oswald Veblen, John von Neumann, and Hermann Weyl—just those "celebrities" who made the Institute seem so enticing to the young Dyson several years later.[67] In short order, the Institute for Advanced Study developed a small, elite faculty of permanent members. Its School of Mathematics opened in 1933, followed by the School of Humanistic Studies and the School of Economics and Politics, both of which opened in 1935. Filling these schools with established, world-class scholars became a relatively easy task, given the intellectual migration out of fascist Europe.[68] But Flexner wanted to do more than hire a handful of famous intellectuals. Immediately he began to emphasize the Institute's unique role in bolstering postdoctoral training. In the Institute's earliest publications, Flexner announced that "the country has not hitherto possessed an institution in which young men and women could continue their independent training beyond" the doctorate. For just this reason, Flexner reminded Oswald Veblen in 1935, "the Institute should enroll only post-doctoral students." A week later, Flexner reiterated the message in a follow-up letter to Veblen: "Let us not lose sight of the fact that this Institute has no reason whatsoever for existence unless it offers opportunities beyond the Ph.D. degree, which are not obtainable in other institutions."[69]

Soon after its founding, Veblen, Flexner, and the Institute's next director, Frank Aydelotte, tried to improve the Institute's standing in theoretical physics, as part of its School of Mathematics. Throughout the 1930s and 1940s, Veblen's strategy focused on inviting senior physicists and mathematicians of worldwide reputation. Niels Bohr visited for several months in 1939, and Wolfgang Pauli spent most of the war years there. In 1940, after the Nazis occupied Denmark, Veblen floated a proposal to bring Bohr to the Institute permanently, so he could reestablish his Copenhagen Institute for Theoretical Physics. Bohr

67. Flexner, *Universities* (1930). Twenty years earlier, Flexner had achieved fame as a hard-nosed educational reformer, based on his scathing report on the nation's medical schools. On the founding of the IAS, see Institute for Advanced Study, *Bulletin* 10 (Oct 1941): 1–4; Institute for Advanced Study, *Bulletin* 11 (Mar 1945): 1–5; B. Stern, "History of the Institute for Advanced Study" (1964), chaps. 1–3; Regis, *Who Got Einstein's Office?* (1987), 8–16; and Porter, *Intellectual Sanctuary* (1988).

68. Institute for Advanced Study, *Bulletin* 11 (Mar 1945): 2–3. The School of Humanistic Studies was later changed to the School of Historical Studies, and the School of Economics closed soon after the war. During the late 1960s, under the new director, Carl Kaysen, the School of Social Sciences was added. See Anon., "Institute advances" (1970). On intellectuals' flight from fascist Europe, see Fermi, *Illustrious Immigrants* (1968); Weiner, "New site for the seminar" (1969); and the other essays in Fleming and Bailyn, *Intellectual Migration* (1969); Rider, "Alarm and opportunity" (1984); and Kevles, *Physicists* (1995 [1978]), chap. 14.

69. Flexner quoted from the IAS *Bulletin* in his letter to Veblen: Abraham Flexner to Oswald Veblen, 11 Dec 1935, in *IAS*, Folder "Faculty: Veblen, Oswald, 1934–5"; Flexner to Veblen, 20 Dec 1935, in the same folder.

did not relocate to Princeton, however, and budget problems hindered similar attempts to hire other senior theorists.[70] All the while, Veblen routinely underplayed Flexner's oft-stated goal of encouraging postdoctoral study, at least within the School of Mathematics. Few of Veblen's initiatives to improve theoretical physics at the Institute bore fruit. Hans Bethe advised Freeman Dyson, soon before Dyson was to move to the Institute for the fall of 1948, "not to expect to find too much going on" at the Institute.[71]

Despite Bethe's admonition, however, dramatic changes began to take place at the Institute for Advanced Study, sparked by the accession of its new director, J. Robert Oppenheimer. Having achieved worldwide fame for his role as the director of the wartime Los Alamos laboratory, Oppenheimer was in constant demand after the war. He left his Berkeley teaching post in 1947 to become the director of the Institute, in part to have a perch closer to his newfound consulting duties in Washington, D.C. Before accepting the directorship, Oppenheimer stipulated in April 1947 that he wanted to retain some time to work with young physicists. In October 1947, with Aydelotte still the director and Oppenheimer designated "director-elect," Oppenheimer pushed through a rapid-fire series of invitations to bring theorists to the Institute: first Pauli, Yukawa Hideki, and Abraham Pais, and then, a few weeks later, Bohr and George Uhlenbeck.[72] Nor were his initiatives restricted—as Veblen's had been—to inviting senior theorists. Oppenheimer reported to the Institute's board of trustees in April 1948 that there would be a sudden jump, by 60%, in the number of visiting fellows for the academic year 1948–49; all the increase would be limited to the School of Mathematics, and nearly all the incoming visitors would be postdocs in theoretical physics. Nearly as soon as he arrived, in other words, Oppenheimer began to flood the Institute with young theorist postdocs. As he reported to an Institute board of directors meeting in May 1951, he was doing nothing less than "attempting to create a school of very good physicists."[73]

70. Oswald Veblen to Abraham Flexner, 14 June 1937; Flexner to Veblen, 26 Jan 1938; and Flexner to Veblen, 14 Feb 1938, in *IAS*, Folder "Faculty: Veblen, Oswald, 1936–7"; Oswald Veblen to Frank Aydelotte, 15 Mar 1940; Veblen to Aydelotte, 23 Mar 1940; Veblen to Aydelotte, 10 Apr 1940; Veblen to Aydelotte, 28 May 1940; and Veblen to Aydelotte, 5 June 1940, in *IAS*, Folder "General: Theoretical Physics 1940 Proposals." See also the unsigned memorandum "Report on theoretical physics," 23 Oct 1937, in *IAS*, Folder "Physics, General." Veblen, Flexner, and Aydelotte had considered inviting Peter Debye, Paul Dirac, Enrico Fermi, and Erwin Schrödinger in addition to Bohr and Pauli.

71. Freeman Dyson to his parents, 2 June 1948, in *FJD*.

72. Minutes from the IAS board of trustees meetings of 1 Apr 1947, 2; 9 Oct 1947, 7–8; and 16 Dec 1947, in *IAS*, Box 10. See also Barnett, "Physicist Oppenheimer" (1951), 377–78.

73. "Director's Report," in the IAS board of trustees meeting minutes of 15 Apr 1948, 4, in *IAS*, Box 10; minutes from IAS board of directors meeting, 4 May 1951, 3, in *IAS*, Box 11.

The sudden changes engendered tensions between the new director and the lions of the School of Mathematics, several of whom, like Veblen, resented the incursion into their domain (not to mention their budget). All the same, Oppenheimer kept the young theorists coming. In 1953, twenty-five postdocs in theoretical physics resided at the Institute, bringing the total number who had visited between 1948 and 1953 up to nearly one hundred.[74] The Institute for Advanced Study rapidly became the only show in town for young theorists. Hans Bethe summarized the transition succinctly, writing in his long obiturary for Oppenheimer that on his arrival at the Institute, "it now became a centre for young physicists."[75]

Life at the Institute

For several years, the fledgling Institute had rented space in Fine Hall, the mathematics building at Princeton University. Finally it had acquired its own space in 1939, when it moved into a new building, Fuld Hall, built on a former farm plot a mile or so away from the university. The new setting gave the Institute a "little-vine-covered-cottage atmosphere," wrote a reporter in the *New Yorker* in 1949. A reporter from the *New York Times Magazine* expanded the sketch, describing Fuld Hall as a "simple Georgian building with cupola standing in serene, almost lonely, isolation against dogwood trees and woods."[76] (See fig. 3.3.) To Dyson, who arrived there for the second year of his Commonwealth Fellowship in September 1948, it was the building's interior rather than its exterior that deserved attention. "The most pleasant surprise" about the Institute, Dyson explained to his parents, "is that it is so small and intimate.... Inside, the Institute is beautifully decorated and furnished, there is a lounge with the [London] Times air edition and every other important newspaper and periodical, an excellent specialist library, a tea-room, and the various private work-rooms." Although years later a longtime resident, the art historian Erwin Panofsky, scoffed that Fuld Hall had been built "with isolated wings, like an insane asylum," and a later director of the Institute dismissed the

74. On the tensions between Oppenheimer and the IAS mathematicians over his new emphasis on theorist postdocs, see B. Stern, "History of the Institute for Advanced Study" (1964), 653–54, 664–66; and Regis, *Who Got Einstein's Office?* (1987), 140, 151–52. On the increases in numbers of young theorist postdocs, see J. Robert Oppenheimer, "Report of the director, 1948–1953," 10 Mar 1954, 6, in *IAS*.

75. Bethe, "Oppenheimer" (1968), 406; see also A. Klein, "Autobiographical notes" (1993), 14; and J. Bernstein, *Life It Brings* (1987), 87–88.

76. Institute for Advanced Study, *Bulletin* 10 (Oct 1941), 3; Anon., "Notes and comment" (1949), 23; Samuels, "Where Einstein surveys the cosmos" (1950), 14.

Figure 3.3. Fuld Hall at the Institute for Advanced Study. (Source: W. Davis, "Thinkers unlimited" [1962], 15.)

building as having "little architectural distinction," it nonetheless provided an institutional base within which the Institute could take root and grow.[77]

Oppenheimer's plan, upon arriving at the Institute in 1947, was to convert Fuld Hall into what he called "an intellectual hotel," as he told a reporter from *Time* magazine in the fall of 1948; the phrase stuck, and journalists repeated it for the next twenty years when reporting on the Institute.[78] More than ever before in its history, the Institute for Advanced Study under Oppenheimer became a "hotel" that catered to the youth. A *New Yorker* reporter seemed surprised during the spring of 1949 to find that "most of the Institute's geniuses" seemed "to be in their early twenties and still breathless from taking Ph.D.s on the run." Dyson, amused by the story, explained to his parents that in fact "some of the details are subject to a little poetic exaggeration; the girl who looked about eighteen and is an expert on non-associative algebras is actually Verena Haefeli, aged 26 and mother of a fat 2 1/2-year-old daughter named Katrina; the 22-year-old astrophysicist is actually 30 and his name is Dick Thomas." Even

77. Freeman Dyson to his parents, 14 Sep 1948, in *FJD;* Panofsky was quoted in Martin Mayer, "Leisure of the theory class" (1963), 20; Carl Kaysen penned an introduction to Robert Geddes's architectural study, Geddes, "Theory in practice" (1972), quotation on 53.

78. Anon., "Eternal apprentice" (1948), 81. Later reporters repeated Oppenheimer's phrase, "intellectual hotel," including Samuels, "Where Einstein surveys the cosmos" (1950), 15; Barnett, "Physicist Oppenheimer" (1951), 379; Bess, "Where the biggest ideas are born" (1955), 120; and Corry, "Visit to 'an intellectual hotel'" (1966).

given these exaggerations, however, Dyson thought that the whimsical piece was "a great deal more objective than the accounts in *Life* and *Time*" that had recently come out. Dyson himself had expressed surprise upon first arriving at the Institute in September 1948 at the large number of "good young" people around, far more than the "few eminent people" whose reputations he had already known. A reporter from the *New York Times Magazine* similarly remarked in November 1950 on "the seeming predominance of youthful people, and the American tempo" at the refurbished Institute.[79] Oppenheimer's move to inflate the ranks of young postdocs at the Institute had an immediate, palpable effect.

The new emphasis upon youth at the Institute also helped to shape the social atmosphere. Before the war, the Institute had been known in part for the high-tea snobbery of its members. During Oppenheimer's tenure there, some snobbish elements remained. Dyson, for one, wrote to his parents that "the general stratification of society at Princeton is much more developed than, for example, at Ithaca." Fritz Rohrlich, another young postdoc at the Institute, reported that the Institute—"this ivory tower of high-browed learning"—was "much more 'stuffed shirt'" than Cornell.[80] In place of snobbery, however, Oppenheimer emphasized the importance of "leisure" that the Institute could provide, especially "at a time when universities are so overcrowded and overwhelmed." While he was still only director-elect, *Time* magazine labeled the Institute simply "Oppy's Retreat," while other reporters routinely likened it to an English country day school. Indeed, Dyson took advantage of the Institute's rural surroundings, as he explained to his parents a month after arriving there. "In the afternoons I have managed to explore quite a lot of the country around here," he wrote. "It is excellent walking country, and I have met numbers of strange birds, insects and plants. The weather just now could not be better, and I hope to continue this form of exercise indefinitely." He had difficulty, however, finding young peers to join him on his jaunts: "Unfortunately my young colleagues are unwilling to join me, as they are obsessed with the American idea that you have to work from 9 to 5 even when the work is theoretical physics. To avoid appearing too superior, I have to say that it is because of bad eyes that I do not work in the afternoons; did you ever hear such nonsense." By the end of the academic year, Dyson reflected with amusement that he "should

79. Anon., "Notes and comments" (1949), 23; Freeman Dyson to his parents, 7 Apr 1949 and 26 Sep 1948, in *FJD;* Samuels, "Where Einstein surveys the cosmos" (1950), 14.

80. Freeman Dyson to his parents, 6 Feb 1949, in *FJD;* Fritz Rohrlich to Donald Menzel, 5 Nov 1949; and Fritz Rohrlich to Hans Bethe, 29 May 1950, in *FR*, Box 23, Folder "Correspondence, 1946–54." On life at the Institute before the war, see Infeld, *Quest* (1941), 246–338.

be living this life of idleness and leisured ease... in one of the most bourgeois societies that has ever existed; that is really something to chuckle about."[81]

But all was not quiet contemplation or workaholic ambition. Rambunctious postdocs had to be reminded to keep the noise from their touch football games down, so as not to disturb scholars in the Institute's library.[82] Beginning in February 1949, the young postdocs also held informal dances in the Institute cafeteria. The first one had been such a success, Dyson wrote to his family, that "it was agreed to have such a dance every Saturday from now on; it is a pity that nobody thought of doing it earlier." The idea had come from another of the young theoretical physics postdocs, Robert Karplus, who enjoyed dancing with his new bride. A *New Yorker* reporter attended one of these square dances a few weeks later, taking evident glee in the mixture of frivolity and intellectualism: "The participants were mainly Oppenheimer's young atomic physicists and mathematicians. It was like a square dance at any small college, except that more languages were in use. The boys were in shirtsleeves, slacks or jeans, and sneakers; the girls—some of them wives, some of them secretaries, some of them scientists—were mostly wearing peasant blouses, dirndls, and either saddle shoes or ballet slippers. The music was furnished by a phonograph, and the calling was routine, but every once in a while we overheard one of the young men in a square explaining '... then you form a rhomboid figure with discrete vectors at an angle of 2 pi over M.'" The reporter went on to describe a poker game among some young mathematicians later that night, who cheekily computed "the precise odds on filling any given bobtailed straight," but who had "wretched luck" all the same.[83] Postdocs frequently held other parties in the "Institute Housing Project" (later jokingly referred to as simply "the project"), at which the whisky flowed freely. Dyson eventually found that in the Institute housing area, there was "just so much social life that nobody has much chance to work or to sleep. However, it is all good fun."[84]

81. Anon., "Oppy's retreat" (1947): 82; Freeman Dyson to his parents, 16 Oct 1948 and 6 June 1949, in *FJD*.

82. J. Bernstein, *Life It Brings* (1987), 110. A senior economist at the Institute declared in 1950 that the postdocs in theoretical physics "are beyond all doubt the noisiest, rowdiest, most active, and most intellectually alert group we have here." Quoted in Barnett, "Physicist Oppenheimer" (1951), 378.

83. Freeman Dyson to his parents, 15 Feb 1949, in *FJD;* Anon., "Notes and comment" (1949), 23–24. Many physicists recalled the importance of informal square dances at wartime Los Alamos as an important element in fostering social cohesion and providing relaxation. See Brode, "Tales of Los Alamos" (1980), 153–55; and Serber with Crease, *Peace and War* (1998), 83.

84. Freeman Dyson to his parents, 19 Oct 1948 and 29 Sep 1950, in *FJD*. A decade later, a reporter from *Esquire* noted that "Each week end at the Institute is also likely to see at least one party that

So much for how the young postdocs filled their evenings; several journalists were equally surprised by how they filled their days. Reporters routinely noted that although the Institute had a "faculty," there were no regular classes, and no one in attendance worked toward a degree; in fact, there seemed to be no requirements at all. Oppenheimer explained to a *Time* reporter in 1948 that "None of the usual apparatus of education will be found at the Institute, nor will the people usually regarded as still capable of education."[85] He expanded on this in a letter to Wolfgang Pauli in February 1952. Oppenheimer hoped to lure Pauli, who had already spent several semesters over the previous years as a visitor, to become a permanent member of the Institute. "I need to remind you that our terms are short, and our summers long," Oppenheimer began. But Oppenheimer hoped that other aspects of life at the Institute would exert an even stronger appeal: "We now have, and will continue to have, I think, an interesting, varied and talented group of physicists here. It is not a school in the sense that even the younger people are not listening to lectures or working for doctor's degrees; but it is a school in the sense that almost everyone who comes learns of parts of physics, and sometimes of other science as well, which are new to him. It is a very fertile group, and the feeling of it is good; and for you it might be, as for me it is, a pleasant and exciting association."[86] Oppenheimer ran weekly "confessionals" in his office for the theorist postdocs, during which the group gathered to talk about how each person's research was proceeding. (See fig. 3.4.) Several additional seminar series ran throughout the year for the young postdocs to supplement their less structured, face-to-face discussions. Oppenheimer further emphasized in his director's report, filed in March 1954, that more often than not, theorist postdocs at the Institute learned more from each other than from the permanent faculty—a feature that would prove to be especially important in the case of Dyson and his cohort of postdocs.[87]

---

is a lalapalooza: mathematicians, most of them, drink." Martin Mayer, "Leisure of the theory class" (1963), 22. References to "the project" appear in Corry, "Visit to 'an intellectual hotel'" (1966), 50; and Anon., "Scholars: Paradise in Princeton" (1966): 70.

85. Anon., "Eternal apprentice" (1948), 70. See also Anon., "Oppy's retreat" (1947); Anon., "Thinkers" (1947); Anon., "Lighthouse keepers" (1948); Samuels, "Where Einstein surveys the cosmos" (1950), 14–15; Bess, "Where the biggest ideas are born" (1955), 25, 119–20; Hess, "Dr. Oppenheimer's Institute" (1956); W. Davis, "Thinkers unlimited" (1962); Martin Mayer, "Leisure of the theory class" (1963), 18; Anon., "Scholars: Paradise in Princeton" (1966), 70; and Corry, "Visit to 'an intellectual hotel'" (1966), 51.

86. J. Robert Oppenheimer to Wolfgang Pauli, 20 Feb 1952, in Pauli, *WB*, 4.1:553–54, on 553.

87. J. Bernstein, *Life It Brings* (1987), 90–91; Oppenheimer, "Report of the director, 1948–53," 7. Oppenheimer had taught his graduate students and postdocs at Berkeley during the 1930s in a similar way, with weekly group meetings or "confessionals." See Serber, "Early years" (1967).

Figure 3.4. Oppenheimer with several of his postdocs at the Institute. (Source: Anon., "Eternal apprentice" [1948], 71.)

Francis Low, who spent 1950–52 as a postdoc at the Institute, recently recalled happy times there. "It really was wonderful. I loved the Institute. I loved the Institute," he cooed. "It was a place where you went and you met your contemporaries, you saw where you were and who you were in the hierarchy. You were at the place where important things were going on." Low recalled an informal, social atmosphere among the postdocs—but also some self-applied tension. "It was a tense place, and I would go into my office every day. The tension was, you couldn't imagine how likely an important advance was. And the fear was not being there when dynamite was uncovered or discovered." Other theorists who had been postdocs at the Institute at this time describe their experiences in similar terms.[88]

As Low hinted, there could be a darker side to life as a postdoc amid all the excitement. Dyson himself felt very lonely, even several weeks after arriving in the fall of 1948. The pressure on young postdocs could be intense.[89] Some suffered from what one long-term member, Abraham Pais, called the "November Depression": having been the best prospects in their various Ph.D. programs, young physicists suddenly found themselves surrounded by peers who were every bit as good as—if not better than—they were. Feelings of loneliness, isolation, and even inadequacy could creep in by October or November. Pais thus took upon himself the task of "care and nursing of temporary junior physics members." He often felt the need to "keep a protective eye on the newcomers": "I would drop in, ask them what they were doing and how it was going, tell

88. Low interview (2001); Case interview (2002); and Kroll interview (2002). See also Kaiser, "Francis E. Low" (2001); J. Bernstein, *Life It Brings* (1987), chap. 5; Pais, *Tale of Two Continents* (1997), 179–84; and Johnson, *Strange Beauty* (1999), chap. 4.

89. Freeman Dyson to his parents, 16 Oct 1948, in *FJD;* see also Martin Mayer, "Leisure of the theory class" (1963), 26.

them it takes some time to adjust to new surroundings, stress that this is a very peculiar place, that it takes time to get going here, and that they would be just fine in a while. That was the kind of support Oppenheimer was congenitally incapable of providing; nor did he try." Dyson benefited from Pais's help—"he was in fact my shrink, that first year," Dyson recalled recently—and later adopted a similar role for others. "My most important function here," he said recently, has been that of "a psychiatric nurse." "People feel the terrible pressure to be all by yourself with a blank sheet of paper," Dyson explains, and they sometimes suffer from "the feeling that you have to do something great." The pressures led to occasional mental breakdowns and even suicides.[90]

In this environment—the wooded glen, the large mahagony desks, and the thrills and pressures of being among the best young theorists in the country—Dyson's campaign to put Feynman diagrams into circulation went into high gear.

THE POSTDOC CASCADE

In the midst of his bus trip after the 1948 Ann Arbor summer school session—soon after his epiphany about how to relate Feynman's and Schwinger's approaches—Dyson looked forward to his upcoming arrival at the Institute. His new result had come to him at just the right moment, he wrote to his parents. "Really this is a tremendous piece of luck for me, coming at the time it does," Dyson explained. "For I shall now encounter Oppenheimer with something to say which will interest him, and so I shall hope to gain at once some share of his attention." As he closed his long and happy letter, he remarked that "Tomorrow will be exactly a year since I landed [in the United States]. What a tremendous success the year has been! Who would have dreamed that I should be coming to Princeton with the thought not of learning but of teaching Oppenheimer about physics? I had better be careful."[91]

Much to Dyson's disappointment, Oppenheimer was in Europe when Dyson arrived at the Institute, and would not return until the middle of

90. Pais, *Tale of Two Continents* (1997), 259; Dyson interview (2001). Professor Tom Kinoshita likewise recalls that Pais invited Kinoshita and Nambu Yoichiro over to his apartment a month or two after they arrived at the Institute in the autumn of 1953, to make them feel more comfortable in their new surroundings. Kinoshita, letter to the author, 19 Mar 2003. Dyson further explained some of the difficulties at the IAS: "We put people into a vacuum here . . . which is a great help for some types of scholars, but which causes the mentally unstable to fall apart." Quoted in Stuckey, "Garden of lonely wise" (1975), 33.

91. Freeman Dyson to his parents, 14 Sep 1948, in *FJD*.

October. This turned out to be a blessing in disguise. Dyson's incoming cohort of postdocs was the largest yet at the Institute—1948–49 being the year that Oppenheimer inflated the usual enrollment by 60%—and the new building that was supposed to hold all the new postdocs had not been completed on time. The incoming postdocs in theoretical physics therefore all crowded around desks in Oppenheimer's large office while he was away. Instead of pairing off into tiny offices, as they would do once the new rooms were ready, the entire group sat around one large room, talking about physics and swapping ideas—an architectural arrangement perfectly suited to fostering informal exchanges and sharing tacit knowledge. "We have spent most of our time in argument, physical, political and otherwise," Dyson wrote home a few days before Oppenheimer's return. Dyson had been busy describing to his fellow postdocs the ins and outs of Feynman's still-unknown diagrams. "I had never understood anything I had heard Feynman say," Dyson's fellow postdoc, Norman Kroll, recalled recently. Upon talking with Dyson soon after they both arrived at the Institute that autumn, however, "I understood what Dyson did very well." Kenneth Case, another of the incoming postdocs, said recently that "the main thing I discovered when arriving at the Institute was there were these peculiar methods that Dick [Feynman] had developed. And nobody had the vaguest idea what they were, except Dyson," who promptly went about explaining it all to the group.[92]

Ten postdocs in theoretical physics entered the Institute with Dyson that autumn: Kenneth Case, Cécile Morette, Daniel Feer, Robert Karplus, Norman Kroll, Joseph Lepore, Sheila Power, Fritz Rohrlich, Jack Steinberger, and Kenneth Watson; David Feldman arrived during the second semester. Yukawa Hideki, a senior figure in Japanese physics and the theorist who had first suggested the idea of a meson during the mid-1930s, spent the academic year in residence at the Institute. Abraham Pais, a young theorist who arrived at the Institute in 1947—after having spent several months in a Gestapo jail in his native Holland during the war, followed by a brief postdoctoral stint with Bohr in Copenhagen—also worked with the young postdocs. Five of the postdocs had worked with Schwinger at Harvard (Case, Feer, Karplus, Lepore, and Rohrlich) and were well versed in his methods. Dyson quickly found that the informal instruction among his fellow postdocs ran both ways. Kenneth Case, for example, who shared an office with Dyson after the new building had been completed, "knows infinitely more physics than I do," Dyson wrote home,

92. Freeman Dyson to his parents, 4 Oct 1948 and 10 Oct 1948, in *FJD;* Dyson interview (2001); Kroll interview (2002); and Case interview (2002).

"having read and understood all the pre-1945 literature which I never learned: so he is a very useful person to have around as a walking encyclopedia." Norman Kroll, meanwhile, "has been doing during the last two months some work which I consider absolutely first-rate, on the Schwinger theory."[93] The group began to bond socially during the autumn semester: Steinberger joined Dyson on some of his walks around the Institute's grounds; Morette and Dyson took a weekend trip to visit Feynman in Ithaca; Dyson, Pais, Feer, Karplus, and Rohrlich enjoyed trips to the movies—Dyson always remarked with awe on the size of his American colleagues' cars—and, after a party with free-flowing whisky, Dyson wrote home that "my relations with these people have been on a much more friendly basis, so the whisky really did me a good service."[94]

Dyson's enthusiastic lessons about Feynman diagrams to his new friends did not attract Oppenheimer's allegiance upon his return that October. Instead, Oppenheimer expressed frequent opposition to Feynman's and Dyson's work, and criticized it bitterly.[95] Feeling "rather like Elijah in the wilderness," Dyson wrote to his parents of his need to "fight" Oppenheimer's "general attitude of hesitation." Oppenheimer allowed Dyson to give a few lectures at the Institute on the new techniques, but insisted, in his usual style, on interrupting the talks with cutting criticisms. (At one point, his fellow postdocs asked Dyson to repeat an entire two-hour presentation the following day, without Oppenheimer present.) Unable to best Oppenheimer in on-the-spot repartee, Dyson retreated to his typewriter one evening in October to write a memo to his new boss. Feeling "irritable" and stating in the opening line that he "disagree[d] rather strongly with the point of view" that Oppenheimer had been espousing, Dyson reiterated his faith that Feynman's diagrammatic methods were "considerably easier to use, understand, and teach" than either the prewar methods or Schwinger's and Tomonaga's recent approach to QED. Coming to the end of his one-page manifesto, he took a walk to clear his head, "and sat down on the grass to make up my mind whether I should send the letter off": "After some time I had finally decided to do it, and then suddenly the

93. Freeman Dyson to his parents, 14 Nov 1948, in *FJD*. Lists of members at the IAS may be found in Institute for Advanced Study, *Publications* (1955). On Pais's wartime and postwar experiences, see Pais, *Tale of Two Continents* (1997).

94. Quotation from Freeman Dyson to his parents, 19 Oct 1948, in *FJD;* see also Dyson to his parents, 10 Oct 1948, 1 Nov 1948, and 14 Nov 1948, in *FJD*.

95. Oppenheimer's criticisms extended beyond the Institute's walls; see in particular his report at the eighth Solvay conference, held in Brussels from 27 Sep to 2 Oct 1948. In marked contrast with Feynman's "algorithms," Oppenheimer concluded that the methods of Tomonaga and Schwinger had "the advantage of very great generality and a complete conceptual consistency." Oppenheimer, "Electron theory" (1950), 7.

sky was filled with the most brilliant northern lights I have ever seen. They lasted only about five minutes, but were a rich blood-red and filled half the sky. Whether the show really was staged for my benefit I doubt, but certainly it produced the same psychological effect as if it had been. I sent the letter off." The next day Oppenheimer saw Dyson, "said he was delighted with my letter," and offered to give Dyson another opportunity to lecture on the new diagrams. Yet once again Oppenheimer interrupted Dyson's presentation and adopted a dismissive tone throughout the proceedings. Only after Hans Bethe intervened directly with Oppenheimer a few weeks later did Dyson feel he received some earnest attention. After three more two-hour lectures at the Institute on the new diagrammatic techniques, Oppenheimer ended his criticisms, offering Dyson the simple handwritten note, "Nolo contendere," in November 1948.[96]

From that moment on, all remaining impediments to Dyson's efforts disappeared. With Oppenheimer on board, Dyson's recruitment work among the Institute's postdocs proceeded quickly, and he began to coordinate several collaborative calculations. By early December several of the postdocs were deeply engrossed in trying to figure out why Feynman, Schwinger, Bethe, Weisskopf, and others had gotten different answers in their second-order Lamb-shift calculations; this meant that the new recruits had to master the details of each approach in order to scour the previous calculations looking for mistakes.[97] Soon groups began undertaking their own diagrammatic calculations, all with Dyson's aid. Kenneth Watson and Joseph Lepore completed a long article that June on fourth-order radiative corrections to nuclear forces (rather than to electrodynamic ones), following Dyson's prescriptions to the letter and citing some of his unpublished work. Even more important became Robert Karplus and Norman Kroll's fourth-order calculation of the electron's anomalous magnetic moment, which they announced under the rubric of "Dyson's program." Karplus and Kroll's famous project had actually begun when Dyson handed them a piece of the overall calculation, instructed them in how to derive that piece, and worked with them as they continued their monumental diagrammatic calculation.[98]

96. Freeman Dyson to his parents, 10 Oct 1948, 16 Oct 1948, 19 Oct 1948, 1 Nov 1948, 4 Nov 1948, 21 Nov 1948, and 28 Nov 1948, in *FJD*. Dyson's memo to Oppenheimer, dated 17 Oct 1948, may be found in *FJD*. See also Schweber, *QED* (1994), 520–27; and Dyson, *Disturbing the Universe* (1979), 72–74.

97. Freeman Dyson to his parents, 1 Nov 1948 and 4 Dec 1948, in *FJD*.

98. Watson and Lepore [A.5]; Freeman Dyson to his parents, 15 Feb 1949, in *FJD;* Freeman Dyson to Hans Bethe, 10 Feb 1949, in *HAB*, Folder 10:20; Kroll interview (2002).

In this way, with Oppenheimer's blessing and Dyson's close guidance of his fellow postdocs, the Institute fast became a factory of Feynman diagrams. Besides the articles by Dyson and Feynman, the only articles in the *Physical Review* that made any use of the diagrams during 1949 all came from postdocs at the Institute (Jack Steinberger, Kenneth Watson, Joseph Lepore, David Feldman, and Cécile Morette).[99] The acknowledgments in Steinberger's early paper help to make sense of his otherwise surprising remarks, quoted in the previous chapter, that readers would be assumed to be familiar with the diagrammatic techniques. The Institute by this time could boast a sizable community dedicated to discussing the diagrams and their use; Steinberger concluded his article, "I am in debt to the physicists at the Institute for Advanced Study for their generous help, especially to Drs. K. M. Case, F. J. Dyson, N. Kroll, J. R. Oppenheimer, A. Pais, S. Power, F. Rohrlich, and H. Yukawa."[100] No wonder Steinberger could assume familiarity with the diagrams: all his *local* readers were already familiar with them.

Steinberger's experiences were some of the earliest of what rapidly became a familiar pattern. Over four-fifths of all the published uses of Feynman diagrams in the *Physical Review* during the period 1949–54 stemmed from the Institute's network of diagram users.[101] Very quickly, as the postdocs cycled through the Institute, they established a pedagogical cascade: they learned and practiced the diagrammatic techniques during their brief postdoctoral study, then fanned out across the country. When they took up teaching positions elsewhere, they began to train their own graduate students in how to use the diagrams; some of these students in turn got jobs and taught still more graduate students. The diagrams' initial, efficient dispersion thus arose from a productive coincidence: Dyson arrived at the Institute, diagrams in hand, just as Oppenheimer began to convert the Institute for Advanced Study into an "intellectual hotel" catering to large cohorts of postdocs in theoretical physics. The surprising pace and geographic scope of diagram use derived ultimately from the exigencies of training theoretical physicists in postwar America.

During 1950, six of the articles in the *Physical Review* that included or discussed Feynman diagrams were written by young physicists who had practiced

99. An additional brief letter to the editor, Matthews [A.3], came from a graduate student in Cambridge, England, with whom Dyson was in regular contact, as we will see in chap. 4.

100. Steinberger [A.6], 1185.

101. Of the remaining diagrammatic articles published in the *Physical Review* during this period, most came from physicists working at Cornell (with Feynman, Hans Bethe, and later Dyson), and a handful came from physicists elsewhere who were in personal contact with Feynman. Only two articles in appendix A—Belinfante [A.51]; and Macke [A.99]—came from authors with no discernible direct connections to Dyson, Feynman, or the IAS.

the techniques at the Institute.[102] Four more of the articles that year came from physicists who had worked with the now-dispersed Institute postdocs, such as Keith Brueckner and Gian Carlo Wick at Berkeley (with both Kenneth Case and Kenneth Watson). Robert Marshak's group at Rochester benefited both from Marshak's own discussions with Feynman and Dyson and from Julius Ashkin's continuing discussions with Feynman; Kenneth Case spent the summer of 1950 there working on diagrammatic calculations, and David Feldman's arrival in 1950 from his Institute postdoc solidified the pattern.[103] In fact, all the twenty-seven authors who used Feynman diagrams in the *Physical Review* during 1949–50 were in personal contact with either Feynman, Dyson, or a recent postdoc from Dyson's group at the Institute.[104] Pauli captured the system with his usual sarcasm when he asked Abraham Pais what Dyson and the rest of "the 'Feynman-school'" at the Institute were working on.[105]

102. Rohrlich [A.10], [A.26]; Karplus and Kroll [A.11]; Yang and Feldman [A.19]; Jost, Luttinger, and Slotnick [A.20]; and Karplus and Neuman [A.24]. Res Jost and Joaquin Luttinger had learned of Feynman's and Dyson's work while participating in Wolfgang Pauli's group in Zurich in 1948–49, Jost as a young instructor and Luttinger as a postdoc. Schweber, *QED* (1994), 582–94. Pauli was in residence at the IAS from 29 Nov 1949 until 30 Apr 1950, along with his "children" (Felix Villars, Jost, and Luttinger), as he called them in a letter to I. I. Rabi, 19 Dec 1949 (Pauli, *WB*, 3:722; see also Pauli, *WB*, 3:696–97, 915; and Pauli, *WB*, 4.2:560–62). Jost and Luttinger's first article in the *Physical Review* to incorporate Feynman diagrams was submitted in May 1950, when they were still in residence at the Institute. See Jost, Luttinger, and Slotnick [A.20]. Meanwhile, Jost and Luttinger, together with Pauli, had been corresponding with Dyson as they tried to read his and Feynman's articles, even before they arrived at the Institute from Zurich. See Pauli, *WB*, 3:574, 595–96; and Schweber, *QED* (1994), 670n38, 670n40.

103. See chap. 6.

104. Four of these authors (Matthews, Salam, C. B. van Wyk, and Ward) were students in England with whom Dyson worked closely, as we will see in chap. 4. Of the remaining twenty-three authors during 1949–50, two were Feynman and Dyson themselves, thirteen were postdocs at the Institute (Steinberger, Watson, Lepore, Feldman, Morette, Rohrlich, Karplus, Kroll, Jost, Luttinger, Murray Slotnick, Maurice Neuman, and C. N. Yang); one (Ning Hu) was a friend of Dyson's who spent 1948–49 at Cornell but saw Dyson regularly throughout the year; four were at Rochester (Ashkin, Theodore Auerbach, and Albert Simon all working with Marshak); two were at Berkeley (Brueckner and Wick, working with Watson and Case, recent postdocs from the Institute); and one was at Stanford (Rosenbluth, in contact with Feynman). Ning Hu thanked Feynman "for detailed exposition of his theory before publication" in Hu, "Quantum electrodynamics" (1949), 396; he also visited often with Dyson that year. Freeman Dyson to his parents, 28 Nov 1948, 4 Dec 1948, 11 Dec 1948, 24 Dec 1948, and 9 Jan 1949, in *FJD*. On Brueckner's work with Watson and Case, see Keith Brueckner, "Autobiographical notes," ca. 1986, in *NBL*, call number MB84; and Case interview (2002). Rosenbluth thanked Feynman for help in Rosenbluth [A.16], 616n4.

105. Wolfgang Pauli to Abraham Pais, 26 May 1949, in Pauli, *WB*, 3:655. In an earlier letter to Hans Bethe, Pauli referred simply to the "Princeton-group." Pauli to Bethe, 25 Jan 1949, in ibid., 621.

Soon the cascade of postdocs from the Institute picked up speed: Norman Kroll accepted an assistant professorship at Columbia in 1949 following his Institute postdoc; Fritz Rohrlich went to Cornell and Princeton after his post-doc, before joining another recent Institute postdoc, Josef Jauch, at the State University of Iowa; Kenneth Watson went to Berkeley, Indiana, and Wisconsin following his stay at the Institute; Robert Karplus went to Harvard and then Berkeley in the mid-1950s; Donald Yennie traveled from Princeton to Stanford; Murray Gell-Mann left the Institute for Chicago in 1952; Francis Low joined Gell-Mann in the Midwest, taking a job at Illinois following his postdoctoral studies at the Institute; and so on.[106] Where each of these young physicists went, younger graduate students began taking up the diagrams in their own dissertations and published articles. Kroll's students at Columbia, for example, followed his lead, quickly adopting the diagrams in their own work.[107] Lawrence Dresner, a graduate student at Princeton, thanked Rohrlich for helping him to master the technique.[108] Robert Lawson, a Stanford graduate student, thanked Donald Yennie for aid with diagrammatic calculations soon after Yennie arrived in Palo Alto.[109] Feynman diagrams spread to students in the Midwest, too, thanks to several Institute postdocs. Soon after Kenneth Watson moved to Wisconsin and Keith Brueckner to Indiana in 1952–53, their students began to use Feynman diagrams; no one from those institutions had used the diagrams before their arrival.[110]

Graduate students were not the only ones who picked up the diagrams; some of the Institute postdocs helped to convert peers and a few senior physicists as well. Gell-Mann's diagram converts at Chicago, for example, included both graduate students and older colleagues, including Gregor Wentzel. Wentzel's textbook on quantum field theory had introduced Dyson to the topic while he was still at Cambridge; though Wentzel continued to publish actively on field-theoretic topics after Feynman's and Dyson's work appeared, he never made any use of the diagrams in print until Gell-Mann reached town. Most

106. These physicists' changing institutional affiliations may be tracked by the bylines attached to their articles in the *Physical Review*, as well as by their listings in *AMS*.

107. Karplus and Kroll [A.11]; Kroll and Pollock [A.58]; Weneser, Bersohn, and Kroll [A.90]; Kroll and Ruderman [A.102]; and Yang and Mills [A.132]; see chap. 6.

108. Dresner [A.86], 202; see also Rohrlich [A.10], Rohrlich [A.26], Newcomb and Rohrlich [A.28], and Rohrlich and Gluckstern [A.54].

109. Lawson [A.100], 1279; see also Yennie [A.67].

110. Brueckner, Gell-Mann, and Goldberger [A.78]; Brueckner and Watson [A.96]; Kovacs [A.103]; Aitken et al. [A.106]. Julius Kovacs, Alfred Aitken, and Hormoz Mahmoud were all graduate students at Indiana (*DiP*).

likely, Gell-Mann had an opportunity to convince Wentzel of the diagrams' efficacy during the informal weekly "Quaker meetings" that the Chicago-area theorists held in the early 1950s.[111] Another of Gell-Mann's converts was Murph Goldberger, like Gell-Mann a young faculty member at the University of Chicago. Although Goldberger had heard something about the diagrams as a postdoc at the Berkeley Radiation Laboratory during 1948–49, he did not use the diagrams in print until he began working directly with Gell-Mann and other dispersed Institute postdocs a few years later.[112] In Urbana, Geoffrey Chew, a young associate professor, similarly thanked his new colleague, Francis Low, for "constant help" and "great patience in explaining to the author the intricacies of relativistic renormalization theory" in one of Chew's earliest diagrammatic articles. While Chew received help from Francis Low, his students likewise began to exploit the diagrams.[113] Similar examples link Institute alumni with other first-time Feynman-diagram users—students and colleagues alike—during this period.[114]

111. Gell-Mann and Low [A.43]; Wentzel [A.73]. Cf. Wentzel, "μ-pair theories" (1950), which did not use any Feynman diagrams, even when treating material for which many people at the IAS had begun using the diagrams. On the Chicago "Quaker meetings," see Johnson, *Strange Beauty* (1999), 115–16. Once at Chicago, Gell-Mann also helped several graduate students acquire the diagrammatic techniques: Fried [A.70]; Wyld [A.137].

112. Goldberger interview (2002). As Goldberger recalled recently, the Rad Lab's Robert Serber made a trip to Ithaca in the autumn of 1948 to learn about the diagrams firsthand from Feynman, with whom Serber had been friendly at Los Alamos during the war. Serber came back with some news of the diagrams, which the Berkeley Rad Lab postdocs, including Goldberger, tried to digest: Goldberger interview (2002). On Goldberger and Gell-Mann's collaboration at Chicago, see Pickering, "From field theory to phenomenology" (1989); and Johnson, *Strange Beauty* (1999), esp. chaps. 5 and 6.

113. Chew [A.117], 1754; Blair and Chew [A.83]; cf. Low [A.65]. Blair and Chew [A.83] was submitted in Mar 1953, whereas Chew wrote up [A.117], in which he thanked Low, over the summer of 1953 (while lecturing at the Les Houches summer school in France); presumably part of the "constant help" that Low provided Chew came during the winter and/or spring of 1953. Like Goldberger, Chew had been a postdoc at Berkeley's Radiation Laboratory during 1948–49, and so presumably heard something about Feynman diagrams from Serber following Serber's 1948 visit with Feynman; Chew did not use the diagrams, however, until he began working with Low at Illinois. For more on Chew's and Low's longtime collaboration, see Low, "Complete sets" (1985), 17; Pickering, "From field theory to phenomenology" (1989); and Kaiser, "Francis E. Low" (2001), 70–71.

114. For example, Felix Villars left his IAS postdoc in 1950 to begin teaching at MIT. As his unpublished lecture notes show (Villars, "Quantum electrodynamics" [1951]), Villars's approach to using the diagrams was extremely similar to Dyson's. Kerson Huang, then a graduate student at MIT, most likely learned of the diagrams from this source (Drell and Huang [A.92]). Gian Carlo Wick went to teach at the Carnegie Institute of Technology after working on the diagrammatic techniques in Berkeley; while in Pittsburgh, Wick advised Richard Cutkosky's diagram-filled dissertation (Cutkosky [A.123]).

Although the cascade of diagram-toting postdocs spread quickly, it did not reach everywhere at once. Residual geographic pockets remained in which students were unable to learn about the diagrammatic techniques. As late as October 1952—four and a half years after Feynman's Pocono presentation, and over three years after Dyson's and Feynman's articles on the diagrams had been published—Fritz Rohrlich advised an eager graduate student at the University of Pennsylvania that he should either choose a different thesis topic, better fitted to the Penn faculty's strengths, or "get a job and postpone" his thesis altogether; without any representatives from the Institute's network in town, it would be simply impossible to get up to speed with the diagrammatic methods.[115] Diagram use in two locations—Cornell and Harvard—frames the extent of the Institute's influence particularly clearly. In a sense, these two institutions cap opposite ends of the spectrum and help to clarify the importance of the Institute's pedagogical postdoc cascades.

Cornell

During the late 1940s and early 1950s, Cornell harbored a large number of diagram users. Feynman developed his doodles while teaching there in the late 1940s, and it was in Ithaca that he and Dyson first began discussing the novel techniques. Perhaps most important, Feynman and Dyson worked to get Cornell's leading theorist, Hans Bethe, up to speed with the diagrams. Bethe in turn managed a large number of theorist graduate students, many of whom published diagrammatic articles during or soon after their doctoral studies.[116]

In training this large group of students, Bethe received continuing help from Feynman and Dyson. Following Dyson's year at the Institute, he returned to England (as required by his Commonwealth Fellowship) to teach at the University of Birmingham, and he spent the autumn semester of 1950 back at the Institute in Princeton. In the meantime, Feynman left Cornell for a professorship at Caltech in 1950, although he continued to work closely with several Cornell graduate students from afar, including Michel Baranger, Laurie

115. Fritz Rohrlich to C. W. Ufford, 13 Oct 1952, in *FR*, Box 23, Folder "Correspondence, 1946–54." Rohrlich had been asked to supervise Robert Azendorf's thesis while Rohrlich was in Princeton and Azendorf a graduate student at the University of Pennsylvania; Princeton University policies forbade the arrangement, however, so Rohrlich wrote to Ufford to explain the situation.

116. Peshkin [A.29]; Beard and Bethe [A.40]; Frank [A.41]; Baranger [A.45]; Brown and Feynman [A.50]; Levinger [A.61]; and Mitra [A.131]. All the authors of these articles but Joseph Levinger were graduate students working with Bethe (*DiP*); Levinger completed his Ph.D. at Cornell in 1948 and remained as an instructor at Cornell until 1951 (*DiP, AMS*); he thanked both Bethe and Rohrlich in his acknowledgments (Levinger [A.61], 662).

Brown, David Beard, and Robert Frank.[117] Faced with Feynman's departure, Bethe arranged to have Dyson hired in his place. As he wrote to Dyson in the spring of 1950, "there was a unanimous opinion among the staff" of the Cornell department that "there was only one man in the world who could replace him [Feynman], namely you [Dyson]. Every one of the professors whom I asked for suggestions mentioned your name immediately and spontaneously." At the tender age of twenty-six, Dyson accepted Bethe's offer of a full professorship at Cornell, to begin in the autumn of 1951.[118] Upon his return to Cornell that autumn, he delivered his Advanced Quantum Mechanics course, the lecture notes from which soon traveled far and wide. While there, he was the principal adviser for at least two graduate students, William Visscher and M. K. Sundaresan, who capitalized on the diagrams in their theses. He also worked with other graduate students and postdocs, a collaboration that continued even after he left his Cornell professorship to become a permanent member of the Institute for Advanced Study in 1953.[119]

Several postdocs at Cornell likewise kept active with diagrammatic calculations. Fresh from his Institute postdoc, Fritz Rohrlich published two papers full of Feynman diagrams at Cornell before teaming up with a graduate student and another postdoc for two more diagrammatic papers.[120] Edwin Salpeter, a

117. Beard and Bethe [A.40]; Frank [A.41]; Baranger [A.45]; Brown and Feynman [A.50]; and Baranger, Bethe, and Feynman [A.93]. See also the correspondence between Feynman, Bethe, and several of these students, ca. 1951, in *HAB*, Folders 9:59, 10:29, 10:30, and 19:53; and between Feynman and Laurie Brown, ca. 1950–51, in Professor Brown's possession. See also Freeman Dyson to Wolfgang Pauli, 16 May 1951, in Pauli, *WB*, 4.1:304–5.

118. Hans Bethe to Freeman Dyson, 5 May 1950; and Dyson to Bethe, 23 May 1950, in *HAB*, Folder 10:20. Dyson deferred the offer for one year because the British fellowship with which he had first traveled to the United States for 1947–49 stipulated that recipients could not return to the United States to accept a permanent position for a period of two years. See the correspondence between Dyson and Bethe, ca. 1950–51, in *HAB*, Folder 10:20; Cornell Physics Department annual report, 1949–50, 19, in *CDP;* Cornell Laboratory of Nuclear Studies annual report, 1949–50, 4, in *CDP;* see also Schweber, *QED* (1994), 666n200; and Peierls, *Bird of Passage* (1985), 234–35.

119. Baranger, Dyson, and Salpeter [A.68]; Dyson et al. [A.127]; Visscher [A.133]; and M. K. Sundaresan, letter to the author, 3 Jan 2001. Dyson advised Visscher's dissertation, as noted in Visscher [A.133], 788, 793; Baranger interview (2001); Dyson interview (2001); and Dyson, "Comments on selected papers" (1996), 17–21. Oppenheimer made Dyson a permanent member of the IAS in 1953. Dyson opted to return to the Institute largely because he found catering to graduate students' dissertation needs too distracting from his own fast-changing interests. Freeman Dyson to Hans Bethe, 31 Aug 1952, in *HAB*, Folder 10:18; Cornell Physics Department annual report, 1952–53, 1–2, in *CDP;* K. T. Bainbridge to McGeorge Bundy, 24 Mary 1955, in *HDP,* Box 1, Folder "1955: Appointment, Bohr, A.N."; and Dyson interview (2001).

120. Rohrlich [A.10], [A.26]; Newcomb and Rohrlich [A.28]; and Rohrlich and Gluckstern [A.54]. See also Hans Bethe to Victor Weisskopf, 31 Oct 1949, in *HAB*, Folder 9:17; and Hans Bethe

Viennese physicist who earned his Ph.D. in 1948 with Rudolf Peierls in Birmingham, served as a research associate at Cornell from 1949 to 1953. During this time he worked closely with Bethe, applying Feynman's diagrams and kernel techniques to two-body bound-state problems; this collaboration led to the so-called Bethe-Salpeter equation, which we will examine in chapter 6.[121] Soon after completing his Ph.D. at Princeton, Sam Schweber filled in where Dyson had just left off, teaching the Advanced Quantum Mechanics course at Cornell in the autumn of 1953. While at Cornell, he worked closely with Bethe, Dyson, and several of Bethe's graduate students. Schweber went on to publish one of the first textbooks on the diagrammatic techniques in 1955, based on his version of the Cornell graduate course.[122] Andrew Sessler ventured to Ithaca straight from receiving his Columbia degree and continued to ply the techniques he had practiced with Norman Kroll's help during his doctoral studies.[123] A young Turkish physicist, Behram Kursunoglu, rounded out the group of diagram-wielding postdocs after earning his Ph.D. from Cambridge.[124]

At each level, then, from graduate students to the senior professor, Cornell physicists harnessed the diagrams and put them to constant use. Between 1949 and 1954, nearly as many articles in the *Physical Review* that made use of Feynman diagrams came from Cornell as from the Institute for Advanced Study.[125]

---

to G. Collins, 16 Jan 1953, in *HAB*, Folder 10:11. Joseph Levinger likewise recalls learning about the diagrams at Cornell with Rohrlich's help: Levinger, e-mail to the author, 13 Dec 2000.

121. Salpeter and Bethe [A.47]; Salpeter [A.59]; Baranger, Dyson, and Salpeter [A.68]; Salpeter [A.71], and Dyson et al. [A.127]. See also Salpeter interview (1978).

122. See Dyson et al. [A.127]; Schweber, Bethe, and de Hoffmann, *Mesons and Fields* (1955). Schweber taught the Advanced Quantum Mechanics course at Cornell during 1953–54 and wrote vol. 1 based on this course; Bethe and de Hoffmann contributed the second volume. *Mesons and Fields* [1955], 1:ix. Bethe noted that Schweber's Advanced Quantum Mechanics course at Cornell had been very successful and had been attended by several members of the faculty in addition to graduate students; his research, meanwhile, "has found the approval of Dyson himself." Hans Bethe to Saul Cohen, 8 Mar 1954, in *HAB*, Folder 10:12. M. K. Sundaresan recalls working closely with postdocs such as Schweber while learning to use the diagrams as a Cornell graduate student: Sundaresan, letter to the author, 3 Jan 2001.

123. Sessler [A.134]; and Andrew Sessler, e-mail to the author, 12 Dec 2000. Sessler earned his Ph.D. from Columbia in 1954, working on the hyperfine structure of helium-3 (*DiP*). See also Hans Bethe to H. H. Nielson, 5 Mar 1954, in *HAB*, Folder 12:29; and Norman Kroll to Fritz Rohrlich, 5 May 1954, in *FR*, Box 23, Folder "Correspondence, 1946–54."

124. Kursunoglu [A.139]. See also Hans Bethe to Abraham Pais, 12 Feb 1954, in *HAB*, Folder 11:5.

125. Of course, this comparison does not take into account the large number of postdocs who continued to publish diagram-laden calculations after they had left their positions at the Institute. The total number of diagrammatic articles in the *Physical Review* between 1949–54 contributed by IAS postdocs and postdoc alumni is fifty-five; twenty-four of these articles were written at the

Thanks to the constant two-way traffic, the group at Cornell functioned effectively as an extension of Dyson's Feynman-diagram factory at the Institute.

Harvard

Unlike the cases at the Institute and Cornell, Feynman diagrams did not exactly flourish at Harvard in the early 1950s. Julian Schwinger, Harvard's young prince of theoretical physics and Feynman's friendly rival in the race to renormalization, never used the diagrams; indeed, Schwinger scoffed years later that Feynman diagrams had brought "computation to the masses" and at best were a matter of "pedagogy, not physics."[126] Perhaps swayed by Schwinger's example, Wendell Furry, another of Harvard's key theorists, avoided the diagrams as well. Not surprisingly, neither Schwinger's nor Furry's students made much use of the diagrams in their work—published or unpublished—during the early 1950s.[127] The few times that diagrams showed up in the young Harvard physicists' work all may be traced fairly directly to interactions with Institute members.

The very first diagrammatic article from Harvard in the *Physical Review* came from Robert Karplus, who had learned how to use the diagrams for detailed perturbative studies while at the Institute during 1948–49 (and whose 1950 diagram-filled article with Norman Kroll on fourth-order electrodynamic corrections had quickly become a hallmark of what they termed "Dyson's program"). Karplus traveled to Harvard as an instructor immediately following his stay at the Institute. There he teamed up with a recent graduate student of Schwinger's, Abraham Klein, for a series of articles on the electrodynamic displacement of energy levels within simple atoms—essentially further scrutiny of the Lamb-shift effect. The great bulk of Karplus and Klein's three articles followed Schwinger's approach closely—the self-same approach that Klein had mastered for his long dissertation in 1950—and the second of the articles was even coauthored with Schwinger himself. Yet in the third and final article,

---

Institute itself. During the same period, twenty-one diagrammatic articles appeared from physicists working at Cornell.

126. Schwinger, "Renormalization" (1983), 343, 347. For recollections of graduate-student life under Schwinger's tutelage, see J. Bernstein, *Life It Brings* (1987), chap. 4; the essays collected in Deser et al., *Themes in Contemporary Physics* (1979); and Ng, *Julian Schwinger* (1996).

127. On Furry's work at this time, see, e.g., Furry, "Bound states" (1951); Katzenstein, "Radiative collisions" (1950). Jack Katzenstein completed his dissertation under Furry's direction; another of Furry's students, Maurice Neuman, picked up the diagrams as a postdoc at the Institute for Advanced Study: Karplus and Neuman [A.24]; and Neuman [A.48].

Karplus and Klein turned to Feynman diagrams in order to keep the various vertex-correction terms clear and distinct. This third article was also the only one in which the authors thanked "the members of the Institute for Advanced Study, Princeton, for an informative discussion." Only after Karplus's arrival, and bolstered by further discussion with Institute members, did Feynman diagrams arrive at Harvard.[128]

A second Harvard paper that discussed Feynman diagrams followed a few months later, though this hardly signaled a rising diagrammatic flood tide. Samuel Edwards enrolled as a graduate student under Schwinger in 1951; when he asked Schwinger what topic he should work on, Schwinger steered him in no uncertain terms to particle physics and quantum field theory, handing him a copy of his recent articles from the *Proceedings of the National Academy of Sciences.*[129] Two years later, Edwards—while temporarily visiting the Institute for Advanced Study—published a paper based closely on Schwinger's 1951 articles, in an attempt to develop a nonperturbative approach to QED. Halfway through the discussion, Edwards inserted a footnote, based on "some interesting discussions" with Jack Goldstein, a postdoc at the Institute. Goldstein had just completed his dissertation at Cornell under Bethe, working on the diagram-driven Bethe-Salpeter approach to two-body bound-state problems.[130] At the Institute, Goldstein showed Edwards that the particular set of terms that Edwards proposed to single out for further study (in the hopes that they could be resummed into a nonperturbative expression) were very similar to those that would be produced if one started with Feynman diagrams and applied the Bethe-Salpeter approximation. After noting the similarity, Edwards dropped all further mention of the diagrams for the remainder of the paper

128. Karplus and Klein, "Electrodynamic displacement," pt. 1 (1952); Karplus, Klein, and Schwinger, "Electrodynamic displacement," pt. 2 (1952); and Karplus and Klein [A.63], 855, 858. See also A. Klein, *Relativistic Theory* (1950), esp. 4, 6–19, which developed Schwinger's "interaction representation" formalism for use with meson field theories. Klein's dissertation, at 160 pages, is close to one and a half times the average length of Schwinger's other students' dissertations from this time. For more on Klein's collaboration with Karplus, see A. Klein, "Autobiographical notes" (1993), 16–18; and A. Klein, "Recollections of Julian Schwinger" (1996), 4–5. Klein remained at Harvard as a postdoc for three more years, during which time he published several more articles with increasingly idiosyncratic uses of Feynman diagrams (Klein [A.85], [A.87], [A.95], [A.114], [A.122], and [A.128]). These articles, together with the two articles written by Karplus during his Harvard stay (Karplus and Klein [A.63]; Fulton and Karplus [A.105]), make up eight of the ten articles from Harvard that used or discussed Feynman diagrams during the period 1949–54.

129. See Edwards, "Disordered systems" (1979), 212. The articles Schwinger gave Edwards were Schwinger's "Green's functions," pts. 1 and 2 (1951).

130. Goldstein, *Bound State Solutions* (1953). See also Salpeter and Goldstein, "Momentum space wave functions" (1953); and Hans Bethe to Polykarp Kusch, 23 Apr 1952, in *HAB*, Folder 11:4.

and continued the exposition in precisely the Schwinger-like manner in which he had begun—not exactly an auspicious beginning for Feynman diagrams at Harvard.[131]

Even so, there are a few hints of further diagram use at Harvard. Paul Martin, who completed his dissertation under Schwinger's direction in 1954, later recalled the hushed, underground quality of his and his fellow Harvard students' introduction to the diagrams: "In the dark recesses of the sub-basement of Lyman Laboratory, where the theoretical students retired to decipher their tablets, and where the ritual taboo on pagan pictures could be safely ignored, students scribbled drawings that disclosed profound identities between diagrams and sums of diagrams."[132] The basement dwellers had two key sources of aid for deciphering the "pagan pictures": Karplus himself, whom several of Schwinger's students thanked in their dissertations, and copies of Dyson's unpublished lecture notes from his 1951 Cornell course. More recently, Martin has written that his earlier reminiscences about the "pagan pictures" had been "seriously misconstrued": although Schwinger himself "might frown when one of us drew a Feynman diagram," the graduate students knew all about them, just the same; many simply "refrained" from using the diagrams around Schwinger.[133] The young Harvard theorists likewise refrained from making much use of the diagrams in their dissertations and published articles.

Stanley Deser, for example, while working with Schwinger, developed his adviser's nondiagrammatic Green's-function approach for formulating the two-body problem in his 1953 dissertation. Deser sprinkled throughout the discussion fleeting references to the diagrammatic language in which more and

131. Edwards [A.77], 286n4. Edwards never completed a Ph.D. from Harvard, or from any other U.S. institution: his name is not in *DiP*, nor does he have a listing in *AMS*. Another Schwinger student thanked Edwards for discussions in his 1952 dissertation: Arnowitt, *Hyperfine Structure* (1952), 86. Edwards spent at least part of 1953 in residence at the IAS, as his 1953 article in the *Physical Review* noted; by 1954 he was in Birmingham, England, working closely with Rudolf Peierls, as evidenced by his byline for his conference talk, Edwards, "Nucleon propagators" (1955). By the late 1970s, Edwards was working in the Cavendish Laboratory in Cambridge, England, as listed in Edwards, "Disordered systems" (1979).

132. Martin, "Schwinger and statistical physics" (1979), 71. Earlier in this essay, Martin likened the Schwinger "Green's functions" (1951) papers to a "cryptic testament," mastery of which "conferred on followers a high priest status" (70), which obviously contrasted with the "pagan" and "taboo" Feynman diagrams. Harvard's graduate students in theoretical physics were indeed housed in the basement at this time: see the summary of the Visiting Committee included in the 1959–60 annual report, in *HDP*, Box A–P.

133. Martin, "Julian Schwinger" (1996), 86. Abraham Klein reiterated this point: Klein, letter to the author, 1 Feb 2001.

more people outside Harvard were learning to converse: a handful of footnotes discussed "irreducible" and "crossed graphs" and such diagrammatic elements as "vertices." Moreover, though his dissertation did not contain any actual diagrams, the article that Deser published based on some of this work did. In his dissertation and his article, Deser thanked both Karplus and Klein, in addition to Schwinger.[134] Deser coauthored this article with his fellow Schwinger student, Paul Martin, who likewise thanked both Karplus and Klein in his own dissertation. Although no diagrams appeared in Martin's dissertation, just as none had appeared in Deser's, Martin drew explicitly upon Dyson's unpublished lecture notes.[135]

The only dissertation by one of Schwinger's graduate students that did include explicit Feynman diagrams provides the quintessential exception that proves the rule. Thomas Fulton included nine separate groups of Feynman diagrams throughout his 1954 dissertation; the early figures included "legends" labeling the photon and electron lines explicitly. Interestingly, although Fulton carefully noted in the opening section of his thesis the one-to-one correspondence between the diagrams and the formal expressions for their associated matrix elements, he made scant use of the diagrams in the real "meat" of his dissertation—the chapter entitled "Calculations"—in which he evaluated various energy-level shifts within positronium. Not only did Fulton cite Dyson's unpublished lecture notes; he also noted in his acknowledgments his particular intellectual genealogy: it was Karplus who suggested the problem to Fulton and who provided "much friendly advice and assistance . . . in the course of innumerable discussions and letters, especially in connection with the work represented in the earlier chapters of this thesis." These earlier chapters were the ones in which nearly all the explicit diagrams had appeared. Meanwhile, the acknowledgments continued, "Prof. J. Schwinger kindly consented to act as thesis advisor during the absence of Prof. Karplus in the fall term of 1953."[136] The effects of the Institute postdoc cascade are particularly prominent in this example, with diagrams appearing in that section of Fulton's dissertation that

134. Deser, *Relativistic Two-Body Interactions* (1953), I-4, II-10–11, IV-2, V-5–6, VI-1, and VI-7. See also Deser and Martin [A.84]. Abraham Klein, by this point a postdoc at Harvard, served on Deser's examination committee for the dissertation. As Klein has recalled recently, "At one point, Stan was having trouble explaining a point, an explanation that clearly cried out for the drawing of a diagram. At that point, I interjected, 'With all due respect . . .' I don't remember if I got the laugh my remark deserved, but certainly my relationship with Julian [Schwinger] survived." Klein, letter to the author, 1 Feb 2001.

135. Martin, *Bound State Problems* (1954).

136. Fulton, *Energy Levels* (1954). Feynman diagrams appear on 30, 35–36, 39–41, 61, and in the appendixes on 110 and 115; the acknowledgments are on 142. See also Fulton and Karplus [A.105].

Karplus had advised and quietly disappearing from those sections over which Schwinger had had the most sway.

No doubt aided by Karplus's discussions and Dyson's lecture notes, a few other Schwinger students produced diagrammatic articles soon after leaving Harvard.[137] Of the ten diagrammatic articles in the *Physical Review* that came from Harvard authors between 1949 and 1954, however, all but two had been written by the two physicists (Karplus and Klein) who had had the greatest contact with members of the Institute for Advanced Study. Only the Institute's postdoc cascade proved capable of piercing the "ritual taboo" against "pagan pictures" at Harvard.

A PEDAGOGICAL FIELD THEORY

Returning to table 2.1, we may now make sense of a pattern underlying the acknowledgment-and-authorship cycles among the core group of diagram users. Feynman tutored at least three of the entrants himself (Dyson, Bethe, and Ruderman). Bethe benefited from further discussions with Dyson, who, meanwhile, coached Salam and Matthews in person and via lengthy correspondence; they in turn worked closely with John Ward (as we will see in the next chapter). The bulk of the core group, however, learned and practiced the diagrammatic techniques as postdocs at the Institute for Advanced Study (Kroll, Pais, Rohrlich, Gell-Mann, Karplus, Low, and Watson). The three remaining members of table 2.1 each benefited from the Institute's postdoc cascades (Klein with Karplus, Brueckner with Case and Watson, and Goldberger with Gell-Mann and Low). Beyond the core group, the overwhelming majority of physicists who used Feynman diagrams in the *Physical Review* during the first six years of the diagrams' existence shared some connection with the Institute for Advanced Study. This connection usually took the form either of an extended postdoctoral stay at the Institute or of a close collaboration—as a peer or as a student—with someone who had recently been an Institute postdoc. During Dyson's first year in residence at the Institute, outside observers such as Pauli dubbed the place simply "the Feynman-school." Dyson's work dovetailed with larger institutional transformations: postdoctoral training in general was on the rise, with its built-in feature of circulation; and Oppenheimer had just

137. Stanley Deser went to the Institute for a postdoc after receiving his Harvard Ph.D. and wrote another diagrammatic article: Deser, Thirring, and Goldberger [A.113]. Richard Arnowitt completed his Ph.D. under Schwinger in 1952 and ventured to Berkeley for a postdoc. He thanked both Karplus and Klein in his dissertation (Arnowitt, *Hyperfine Structure* [1952], 86) and wrote a diagrammatic article from Berkeley: Arnowitt and Gasiorowicz [A.115].

begun to remake the Institute into the central training ground for postdocs in theoretical physics. These many factors converged, producing a rapid-fire spread of Feynman diagrams throughout the United States.

Of course, publications in the *Physical Review* alone cannot tell us how, when, or where each person first learned about the diagrams. Dyson's unpublished lecture notes, for example, were snatched up widely for self-study as well as for classroom instruction soon after they became available in January 1952. Certainly by the mid-1950s, if not the late 1940s, young physicists across the country could have learned about the diagrams without too much face-to-face contact with Institute alumni. But the fact remains that nearly the only physicists who actually used the diagrams did so after extended personal contact with members of the Institute's fast-growing guest list. Graduate students on several different campuses picked up the diagrams only after their young advisers arrived, fresh from their visits to Oppenheimer's "intellectual hotel." Instructors and even a few senior professors who had worked for years on QED and related topics first pressed the diagrams into service only after young colleagues arrived in town from the Institute.

Other young physicists, deprived of this kind of personal contact with the Institute's network of diagram users, directed their research efforts elsewhere. Rohrlich had to advise a graduate student at the University of Pennsylvania to choose a different topic or postpone his thesis altogether. In a similar way, Henry Stapp began graduate studies at Berkeley in 1952, intending to work on renormalization. He had gotten his first introduction to the diagrams by reading Feynman's and Dyson's 1949 articles; soon he got his hands on a copy of Dyson's 1951 lecture notes as well. But by the time Stapp arrived in Berkeley, the theorists with whom he had hoped to work (such as Gian Carlo Wick and Robert Serber) had left California because of the "loyalty oath" controversy, to which we turn in chapter 9. With no one to guide his research on diagrammatic topics, Stapp worked instead on analyzing the proton-proton scattering experiments on which his roommate, an experimentalist, worked.[138] Even dedicated study of Feynman's and Dyson's articles, without the benefit of direct contact with other diagram users, could prove insufficient for putting the diagrams to work. As Oppenheimer explained to a reporter from *Time* magazine in the autumn of 1948—just as Dyson began his recruitment and training work at the Institute—"the best way to send information is to wrap it up in a person."[139]

138. Stapp interview (1998). Stapp's experimentalist roommate was Tom Ypsilantis, who later worked on the Nobel Prize–winning team that discovered the antiproton in 1955.

139. As quoted in Anon., "Eternal apprentice" (1948), 81. The sociologist Randall Collins has recently attempted to retell the grand sweep of intellectual history—as it unfolded over the millennia

No uncrossable, epistemic barriers separated diagram users from nonusers. Physicists could certainly learn *something* about the diagrams from the published literature or from Dyson's unpublished lecture notes. Yet reading texts is not the same as using tools. Long after textual instructions became widely available, almost no one used the diagrams in actual calculations upon learning about them from texts alone; rather, they simply chose to work with different sets of tools.[140] Indeed, it is no accident that over one-third of the authors who first used Feynman diagrams in the *Physical Review* did so as graduate students, and nearly half as postdocs. Feynman diagrams are *practices,* and as such they must be *practiced.* Reading about the diagrams was one thing; practicing how to use them as research tools quite another. The distinction was not lost on physicists at the time: in recommending his graduate student for a position at Berkeley, for example, Leonard Schiff wrote that his student "understands the Feynman-Dyson techniques, *and has used them* in his thesis calculations."[141]

* * *

Between 1948 and 1954 the diagrams truly were "put into circulation" from a "main source or centre," recalling the first definition of "dispersion." One might be tempted, therefore, to equate the Institute for Advanced Study with what Bruno Latour has termed "centers of calculation." To Latour, such centers maintain their scientific prominence and authority much as imperial nations do, by having people and goods continually return from the colonized hinterlands to the imperial center. Latour further discusses how these "cycles of accumulation" enable the centers to "act at a distance," influencing remote

---

in Asia, the Middle East, Europe, and North America—exclusively in terms of personal networks linking mentors and students through the ages: R. Collins, *Sociology of Philosophies* (1998). See also the special issue dedicated to critical discussion of Collins's work, *Philosophy of the Social Sciences,* vol. 30 (June 2000). Standard accounts of network building within the science studies literature include H. Collins, *Changing Order* (1992 [1985]); Latour, *Science in Action* (1987); and Hughes, *Networks of Power* (1983).

140. Cf. Mario Biagioli's notion of the "anthropology of incommensurability," in contrast with epistemic or "meaning" incommensurability: Galileo and his Jesuit critics chose whether and when to work at understanding each other. Their differences, Biagioli argues, arose from differences in assumed methods, rather than from linguistic failures to translate one set of concepts into the "worldview" of the other. See Biagioli, *Galileo, Courtier* (1993), chap. 4. For a similar approach, see also Oreskes, *Rejection of Continental Drift* (1999).

141. Leonard Schiff to Edward Teller, 26 May 1953, in *RTB,* Box 25, Folder "Schiff, Leonard Isaac, 1915–1971"; emphasis added.

events.[142] Among other things, however, this model overlooks the crucial role played by pedagogy and training: the Institute for Advanced Study maintained its status as "the Feynman-school" precisely because its graduates did *not* continue returning to the Institute. Instead, the Institute's postdocs set up their own training centers, helping students and colleagues in new places practice using the diagrams. In place of Latour's "action at a distance," we are faced with a long string of local instructional interactions; the diagrams began to circulate thanks to an ever-widening pedagogical field, spreading through space and over time from its source at the Institute for Advanced Study. Much like the description of forces in the physicists' field theory, the diagrams' spread can be explained only in terms of causal, local interactions—a description built around personal contact and sustained, face-to-face training. Pace Latour, educational missionary work and imperialist exploitation need not be the same thing, past abuses notwithstanding.

142. Latour, *Science in Action* (1987), chap. 6. See also Latour, "Give me a laboratory" (1983); and Latour, *Pasteurization* (1988 [1984]).

· 4 ·
International Dispersion

[T]hey are all so much looking forward to having me in
Moscow that I must be careful not to be questioned to the point
of exhaustion by all the eager young theoreticians.

FREEMAN DYSON TO HIS PARENTS, APRIL 1956[1]

THE DIAGRAMS' DIASPORA

Personal contact, informal mentoring, and the postdoc cascade put Feynman diagrams into circulation within the United States. What about physicists in other countries, for whom face-to-face discussions with members of the diagrammatic network were more difficult, if not impossible? Were they able to learn how to use the new techniques based on written instructions alone? As we will see in this chapter, the American example proved to be the rule: personal contact and the institutions of training were crucial for spreading Feynman diagrams throughout the world. Transmission of the new techniques was not impossible without these mechanisms of transfer but was dramatically hampered by their absence.

The first thing to note is that physicists in other countries did not immediately rush to use the diagrams the way the American contingent did. (See table 4.1 and appendixes A–F.)[2] No other country's journals included nearly so many citations to Dyson's and Feynman's articles on the diagrams, nor did they publish as many diagrammatic articles themselves.

1. Freeman Dyson to his parents, 4 Apr 1956, in *FJD*.

2. Table 4.1 includes, for each country, the journals that included the highest number of citations to Dyson's and Feynman's early articles. Citation data comes from *Science Citation Index*, s.v. "Dyson, Freeman," and "Feynman, Richard." Data on numbers of articles comes from page-by-page searches of each journal. It is possible that some undercounting has occurred, since it is difficult to find articles that discussed the diagrams in words only. Every effort has been made to include such articles; electronic keyword searches of those journals that are now available online have been used to complement the manual searches. The number of authors listed for Great Britain includes those authors who contributed diagrammatic articles to the *Proceedings of the Royal Society* and/or to the *Physical Review*. In West Germany, the *Zeitschrift für Naturforschung* had been established at the

Table 4.1. Citations and publications of diagrammatic articles, 1949–54

	No. of Citations to Dyson [A.1], [A.2], and Feynman [A.4]	No. of Diagrammatic Articles	No. of Authors
United States (*PR*)	241	139	114
Japan (*PTP*)	100	97	92
Great Britain (*PRSA*)	35	23	26
Italy (*Il Nuovo Cimento*)	42	18	18
W. Germany (*Zeitschrift für Naturforschung*)	33	11	12
Soviet Union (*ZhETF*)	24	12	11
France (*Journal de Physique et le Radium*)	14	3	5

Physicists' experiences in three countries highlight the mechanisms behind the diagrams' international dispersion: Great Britain, Japan, and the Soviet Union. Both geopolitically and institutionally, these three countries' interactions with the United States covered a wide spectrum during the late 1940s and early 1950s. British physicists were in frequent, casual contact with their cold war allies (including Dyson's own travels back and forth across the Atlantic), and the diagrams spread much they way they did in the United States. In Japan, the diagrams spread thanks to a confluence of political, diplomatic, and institutional rearrangements that took form between 1948 and 1950—just as Feynman and Dyson introduced Feynman diagrams—and that led to an opening up of informal relations between physicists in Japan and the United States as well as among physicists throughout Japan itself. Meanwhile, Soviet physicists fell out of contact with their peers in the United States with the hardening of the cold war, which snapped off all informal communication just months before the diagrams were introduced. By following the fortunes of Feynman diagrams in these three countries, we learn how international relations and personal relations conspired to shape the dispersion of theoretical techniques.

end of the war as a self-conscious break from the legacy of the Third Reich. In the more established German journals, Feynman diagrams barely rated a passing glance: Dyson's and Feynman's articles received no citations at all during 1949–54 in the *Zeitschrift für Physik,* and only four citations in the celebrated *Annalen der Physik.* On the founding of the *Zeitschrift für Naturforschung,* see Carson, *Particle Physics* (1995), 410–13.

Given the contours of the cold war, Soviet physicists' engagement with Feynman diagrams provides the most stark contrast with the American case. Indeed, the Soviet example furnishes a natural experiment with which to test historians' and sociologists' ideas about tacit knowledge and the transfer of skills. The cold war's solidification produced so strict a separation between American and Soviet physicists during the period 1948–54 that any working knowledge of the diagrams gleaned by Soviet theorists had to come without the benefit of their American colleagues' hard-won tacit knowledge.[3] If new skills and techniques could be replicated *only* with the aid of personal contact and informal, face-to-face training, then we would expect to find no Feynman diagrams among the Soviet physicists' publications whatsoever. On the other hand, if theoretical skills and paper tools—being themselves "merely" explicit representations on paper, unlike the finger knowledge that experimentalists must develop with specific pieces of equipment—could travel with ease by means of written instructions alone, then we would expect to find roughly as many diagrammatic articles in the *Zhurnal eksperimental'noi i teoreticheskoi fiziki* (*Journal of Experimental and Theoretical Physics*) as in the *Physical Review.* Instead, we find an interesting pattern: a handful of diagrammatic articles appeared in the main Soviet journal, half of them by the same author. Aleksei Galanin's intensive struggle to learn how to use Feynman diagrams—in the high-pressure Soviet H-bomb program, under the watchful eye of Stalin's much-feared enforcer, Lavrentii Beria—proves at once the possibility, in principle, of learning new techniques from written instructions alone, as well as the extraordinary difficulty of doing so. The Soviet case likewise reveals how important informal training and pedagogical institutions were to distributing the newfound skills throughout the country, once the early working knowledge had been cultivated by a small cohort of theorists.

Galanin's early uses of Feynman diagrams point to a series of related questions: not only *whether* the new techniques could be learned at a distance, but *what* about the diagrams and their use could be transmitted by written texts alone. Who picked up Feynman's and Dyson's articles in the first place, under what conditions, and toward what ends? What struck Dyson's

3. Other historians and sociologists have highlighted the importance of Soviet examples for pursuing questions of how science and technology develop and spread, given the enforced isolation of Soviet and Western experts during much of the cold war: MacKenzie and Spinardi, "Tacit knowledge" (1995); Graham, *What Have We Learned* (1998); and H. Collins, "Tacit knowledge" (2001). Andrew Warwick has similarly charted an example of the interplay between geopolitics and the spread of theoretical techniques in his study of general relativity during World War I: Warwick, *Masters of Theory* (2003), chap. 9.

and Feynman's readers at the time as most salient, interesting, or useful? What did these readers consider to be the papers' essential content or meaning? And how were these readings conditioned by the pedagogical institutions in which they were embedded?[4] From today's vantagepoint, we have a good idea about what Dyson accomplished in his two landmark articles from 1949: not only did he derive the rules for using Feynman's diagrams, but he also established the mathematical equivalence between Feynman's, Schwinger's, and Tomonaga's different-looking approaches, and—most important of all—he proved that the divergences within QED could be removed systematically from any order of approximation. Yet our appreciation for Dyson's accomplishments must not cloud our view of how physicists first greeted his long and difficult articles upon their initial publication. The "obvious" importance and meaning of Dyson's work was not always obvious at the time, and his articles were subject to a wide range of interpretations—always interpretations within specified contexts. We must shift the burden of our analysis, then, from an attempt to plumb some essential residue of meaning inhering within Dyson's articles to the active appropriation of Dyson's and Feynman's work by physicists working at an increasing remove from the authors themselves. Much as physicists use test particles to probe the contours of underlying fields—think of the play of tiny iron filings near a magnet—we can use Dyson's famous pair of articles to probe the institutional arrangements and taken-for-granted structures within which physicists conducted research and trained their students in the early postwar years. Physicists' treatment of Dyson's papers reveals larger patterns in the underlying pedagogical field.

FEYNMAN DIAGRAMS IN GREAT BRITAIN

"Climbing Mt. Everest": Dyson's Preprints Arrive

As soon as he had pulled drafts together, Dyson sent copies of his papers to his former adviser, Nicholas Kemmer, at the University of Cambridge. Kemmer somewhat nonchalantly passed them along to one of his youngest graduate

4. The single best exemplar of this type of historical study, which has inspired my own work, is Warwick, "Cambridge mathematics" (1992); Warwick, "Cambridge mathematics" (1993). See also Secord, *Victorian Sensation* (2000); and Gingerich, *Book Nobody Read* (2004). Like this work, my interpretation draws on a long tradition of reader-response analyses in literary criticism. See esp. Barthes, "Death of the author" (1977); Jauss, *Aesthetic of Reception* (1982 [1974]); Iser, *Act of Reading* (1978 [1976]); Fish, *Is There a Text in This Class?* (1980); and Fish, *Doing What Comes Naturally* (1989).

students, Roy Chisholm—a few weeks into his graduate studies, just twenty-one years old—with the offhand remark, "these seem quite important." Kemmer assigned Chisholm the task of preparing a talk for Kemmer's group on Dyson's first paper within three weeks, to be followed by a talk on Dyson's second paper the following week. "Needless to say," Chisholm has recalled recently, "I did not understand my own seminars." Whether he understood the papers or not, Chisholm's talks were quite likely the first presentations anywhere outside the United States on Dyson's new techniques and Feynman's new diagrams.[5]

Dyson's dense papers were notoriously difficult to understand, especially when read in isolation; a few years later one of Dyson's earliest British converts would warn his own students that reading Dyson's two papers was "like climbing Mount Everest."[6] Despite their difficulty Chisholm pressed on, poring over the papers and deciding that he would like to work on the new material. In the absence of direct contact with Dyson, Chisholm developed an interesting interpretation of the papers. In addition to his work with Kemmer, Chisholm was also learning about general relativity and cosmology—a long-standing research specialty at Cambridge, but one that had largely fallen off most other physicists' agendas and out of their curricula. In fact, just months before Chisholm began to read Dyson's papers, Fred Hoyle at Cambridge, with Hermann Bondi and Thomas Gold in London, launched their ambitious new program for relativistic cosmology known as the "steady-state" model. Their model quickly became the major rival of the "big bang" model, also just then under development by George Gamow, Ralph Alpher, and Robert Herman in the United States.[7] As Hoyle had demonstrated recently, the universe in his model would assume the shape of de Sitter spacetime—an exact solution to Einstein's equations that had been worked out during the late 1910s and 1920s and that could be most easily studied by embedding the one temporal and three spatial dimensions we see around us within an extra, fifth dimension (subject to a specific relation between the five coordinates). Chisholm remembers several talks at Cambridge that autumn by the colorful steady-state group, sometimes with all three of them interrupting each other and talking simultaneously;

5. J. S. Roy Chisholm, e-mail to the author, 19 July 2000; Chisholm, letter to the author, 27 Sep 2000.

6. Quoted in Kibble, "Matthews" (1988), 564.

7. See Kragh, *Cosmology and Controversy* (1996). On general relativity in Cambridge, see Warwick, *Masters of Theory* (2003), chap. 9; on other physicists' neglect of the topic, see Eisenstadt, "Low water mark" (1989); and Kaiser, "ψ is just a ψ?" (1998).

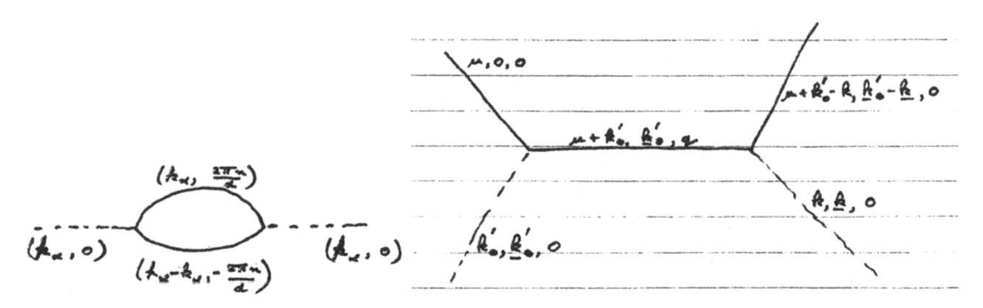

Figure 4.1. "Dyson graphs" in five-dimensional spacetimes. *Left,* diagram for a photon's self-energy. *Right,* diagram for Compton scattering. In each case, Chisholm carefully marked down the components of the five-dimensional energy-momentum vectors associated with each line, subject to the special boundary conditions on the fifth dimension. (Source: J. S. R. Chisholm's unpublished notes, n.d., ca. winter 1948–49.)

he also began to interact with Hoyle personally, and at one point considered working with him full-time on Hoyle's cosmological program.[8]

In the midst of all this activity and excitement about the new cosmological work, Chisholm decided that Dyson's techniques and Feynman's diagrams were just the thing for studying the behavior of matter in five-dimensional spacetimes. How would an electron behave, Chisholm wondered, in the interior of a thin, spherical, five-dimensional shell? He opened a new notebook and quickly began working with the new techniques, copying Dyson's examples as best he could to reproduce—and then generalize—the use of "Dyson graphs" for studying the self-energy of a photon and Compton scattering in such a five-dimensional space.[9] (See fig. 4.1.) What mattered most to Chisholm about Dyson's papers was that they contained convenient recipes for studying simple processes in space and time—and hence provided fodder for generalizing these results to more exotic spacetimes. In these early notes, Chisholm never worked beyond the lowest order in the radiative corrections—sometimes, as

8. Chisholm, letter to the author, 27 Sep 2000. Chisholm had earlier taken Bondi's course on general relativity and Hoyle's on statistical mechanics, while an undergraduate at Cambridge: Chisholm, e-mail to the author, 11 Mar 2003.

9. Chisholm began by writing down a set of coordinates in the full five-dimensional embedding space, subject to a constraint on the five coordinates, then worked out the various translations and rotations that would leave the spacetime invariant. He next took the limit of large radius for the embedded hyperboloid, such that along any observable patch the resulting four-dimensional spacetime would appear spatially flat. He then assumed that our universe ("the real world") had finite spatial extent, of width *d,* into the fifth dimension, and that all measurable quantities, such as electric charge, should be treated as averages over the five-dimensional shell. My thanks to Professor Chisholm for sharing copies of his unpublished notes, and for discussing his work: Chisholm, e-mail to the author, 8 Mar 2003.

on the right side of figure 4.1, neglecting radiative corrections altogether. Nor did Dyson's general proof of the renormalizability of QED to any order of approximation enter into Chisholm's purview at this early stage. Just weeks into graduate school, this eager young student read Dyson's papers and found meaning in them along specific directions—directions that dovetailed with the other new ideas he was just then encountering and with which he was learning to work. Needless to say, given the dearth of attention paid to general relativity on Dyson's adopted side of the Atlantic, no one in the United States seems to have found the same lessons in Dyson's papers at the time.[10]

So much for one eager student's appropriation of Dyson's work. A few weeks after Chisholm's initial presentations of Dyson's difficult papers, an older graduate student in Kemmer's orbit, Paul Matthews, gave a new round of talks on Dyson's manuscripts. Matthews read in Dyson's papers not tools to be applied to Fred Hoyle's pet project, but rather something closer to Kemmer's own interests. During the late 1930s, Kemmer had introduced many key ideas about how to represent mesons and their possible interactions with protons and neutrons—working with nuclear forces rather than electrodynamic ones. Matthews set himself the task of using Dyson's perturbative methods to study how virtual particles and radiative corrections would behave in a variety of Kemmer's meson models. Another key difference from Chisholm's first approach quickly surfaced: no doubt facilitated by Kemmer, Matthews also began corresponding directly with Dyson about his new work. Their interactions intensified, and soon Dyson was pointing out calculational errors in Matthews's notes and manuscripts.[11] Dyson's epistolary lessons in how to read his own papers helped to orient Matthews's developing approach. With this aid, Matthews began to make sense of Dyson's papers as a Dysonian.

In Dyson's second article, he not only demonstrated how to derive Feynman's diagrams and make perturbative calculations with their aid; he also demonstrated how to remove almost all the infinities from a given calculation. One loophole remained in Dyson's proof of QED's renormalizability. The loophole stemmed from certain high-order correction terms whose

10. In fact, after pressing on with his research for a year, Chisholm ultimately dropped his project without publishing any of the results from his tightly packed notes when his new supervisor, James Hamilton, showed little enthusiasm for the topic. By this time, after working closely with his fellow graduate students Paul Matthews, Richard Eden, Gordon Moorhouse, and Angas Hurst, Chisholm had mastered the intricacies of diagrammatic perturbative calculations; he went on to develop new, efficient means of evaluating Feynman diagrams in his dissertation. Chisholm, letter to the author, 27 Sep 2000; and e-mail to the author, 8 Mar 2003; see also Chisholm [A.34].

11. See, e.g., Matthews, "*S*-matrix" (1949), 1255n4. Kemmer's meson work appeared in Kemmer, "Field theory" (1937); and Kemmer, "Charge-dependence of nuclear forces" (1938).

divergences overlapped, leaving how to isolate and remove each divergence ambiguous. Dyson made some passing suggestions about how to handle these overlapping divergences near the conclusion of his long second paper, and that was where things stood when he began coaching Matthews on how to use the diagrams to study meson models. Soon after Matthews began this work, he, too, hit upon the problem of overlapping divergences—in fact, the problem in the meson cases was even worse than in QED.[12] The problem persisted for Matthews, who had begun working with another of Kemmer's new graduate students, Abdus Salam. In the spring of 1950, the problem still unresolved, Matthews publicly defended his dissertation. In addition to Kemmer, one of Matthews's examiners was Dyson himself, who was then in the midst of his first year teaching at the University of Birmingham (as required by his Commonwealth Fund fellowship). As Salam later recalled, the topic of overlapping divergences came up during the tense examination: "Dyson had asked Matthews about overlapping loop infinities: 'Have you come across these infinities? And if so, how do you resolve the problems posed by these?' Matthews had replied: 'You have claimed in your paper on quantum electrodynamics that these infinities, which occur in the self-energy graphs, can be properly taken care of. I am simply following you.' No further question on these infinities was asked; both Dyson and Matthews kept silent after this brief exchange."[13] With his examination over and his degree in hand, Matthews set off for vacation, bequeathing the problem of meson models' overlapping divergences to Salam.

Salam decided to seek help from the master. He reached Dyson by telephone on what turned out to be the night before Dyson was to return to the Institute for Advanced Study for the autumn 1950 semester.[14] Salam rushed from Cambridge to Birmingham that night, then took the train from Birmingham to London with Dyson the next day. As soon as Salam met up with him, he asked Dyson how he had solved the problem of overlapping divergences in QED.

12. Dyson [A.2], 1749–50; Matthews [A.3]; Matthews, "S-matrix" (1949); Matthews, "Application" (1949); Matthews "S-matrix" (1950); Matthews [A.22], [A.23]. See also Kibble, "Matthews" (1988), 559–61. The divergence problems were exacerbated in meson models in which protons and neutrons coupled to the derivative of the meson field: these derivatives added extra powers of momentum to the numerators of the relevant integrals, making them more strongly divergent.

13. Salam, "Excellences of life" (1989), 531.

14. Dyson received special permission from the Commonwealth Fellowship to spend the autumn 1950 semester back at the Institute for Advanced Study, on the strict condition that he return to Birmingham after three months. Dyson returned to Birmingham in Dec 1950, leaving England to accept his professorship at Cornell in July 1951. See Dyson to Hans Bethe, 10 May 1950, in *HAB*, Folder 10:20; and the other correspondence between Bethe and Dyson in this folder.

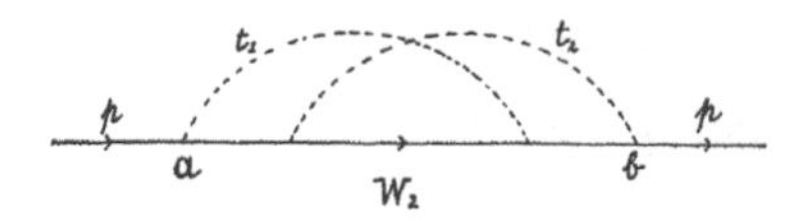

Figure 4.2. An overlapping divergence in an electron's self-energy diagram. (Source: Salam [A.32], 218.)

"But, I have no solution," came Dyson's reply. "I only made a conjecture." Shocked by the news (he had been certain that Dyson, his "hero," had already overcome the difficulty), Salam spent the rest of the day with Dyson learning about the background to Dyson's conjecture; Dyson also left some notes for Salam to study. They struck up an active overseas correspondence for the next several months, while Salam worked hard to turn Dyson's initial conjecture into a solid proof.[15]

These efforts culminated in a dense article that Salam submitted to the *Physical Review* in late September 1950. Working entirely with Dyson's notation, Salam introduced in detail what overlapping divergences were and how they could be treated. As an example, he included the diagram in figure 4.2. At first sight, this diagram seemed to be just another contribution to an electron's self-energy: a lone electron could emit virtual photons at various points in space and time, and then absorb the photons at different points. The example that Salam drew, however, came with a twist: the two dashed-line virtual photon loops overlapped for a stretch of the electron's solid straight line. Dyson's renormalization protocol worked unambiguously when one infinity could be dealt with at a time; but the overlapping loops in figure 4.2 allowed no such systematic subtractions, essentially because the variables involved in one divergent integral became entangled with the variables of a different divergent integral. Salam could no longer apply Dyson's recipe to fix one integral at a time. Instead, Salam developed a byzantine scheme of ordered subtractions and partial integrations. He went on to demonstrate that his scheme would work for any diagram with *n* vertices that contained overlapping divergences.[16] It wasn't

15. Salam, "Excellences of life" (1989), 532. In the article that grew out of their correspondence, Salam noted that "The author is deeply indebted to Mr. F. J. Dyson for indicating the considerations in Sec. II in an extremely helpful discussion, without which this work would not have been possible." Salam [A.32], 227. See also Kibble, "Salam" (1998), 388–89; and C. Angas Hurst, e-mail to the author, 25 June 2002.

16. Salam [A.32]. See also Ward [A.46]; Salam [A.56]; and Matthews and Salam, "Renormalization of meson theories" (1951). Ward concluded his article by "thank[ing] Mr. A. Salam for stimulating discussions and for reviving the author's interest in this subject" ([A.46], 901). On Salam's solution to the overlapping divergences problem, see also Schweber, *Relativistic Quantum Field Theory* (1961), 615–25.

pretty, but it did fill in a gap in Dyson's original proof. With initial coaching from Matthews followed by intense and extended discussions with Dyson, Salam had mastered Dyson's approach to Feynman's diagrams. Like Dyson, Salam saw the main purpose of Dyson's work to be the systematic removal of infinities from arbitrarily complex Feynman diagrams. Like Matthews, Salam became a Dysonian only with Dyson's direct intervention.

Moving beyond Cambridge

With these key interventions by Dyson—helping to get Matthews and Salam up to speed with the new techniques—Feynman diagrams began to spread rapidly within Kemmer's close-knit circle of students in Cambridge. The group was relatively large, having grown from five research students in 1947–48 to eight the following year. With Kemmer busy with his undergraduate teaching duties on top of other supervisory tasks, the students had grown accustomed to helping each other out. In 1947–48 two of Kemmer's students, Richard Dalitz and C. B. van Wyk, organized a seminar series among the graduate students in which they took turns making presentations for each other, either based on recent publications in the literature or drawn from their own emerging research. As one veteran of the group has recalled recently, this informal forum served for several years as "the main teaching vehicle" for the group: they "taught each other by only semi-formal communication."[17] Dyson visited Kemmer's group as soon as he arrived in England from the Institute for Advanced Study in the summer of 1949, further strengthening the ties he had already begun to forge with Paul Matthews; Matthews, in turn, quickly became a local authority on the new techniques, freely mentoring younger members of Kemmer's group and helping them practice using the diagrams. Keith Roberts, Richard Eden, and C. Angas Hurst all submitted diagrammatic articles as they were completing their dissertations in Cambridge during 1950–51; Cambridge degree in hand, C. B. van Wyk sent off his first diagrammatic letter to the editor of the *Physical Review* in 1950 as soon as he arrived at his new position at the University of Natal in South Africa. John Ward, at the time a graduate student at Oxford University, became a de facto member of the Cambridge group, making frequent visits to the Kemmer students' informal seminar during 1949–50. During that autumn and winter, he submitted several brief letters to the editor of the *Physical Review* containing the early fruits of his diagrammatic

17. Quotation from Gordon Moorhouse, e-mail to the author, 16 July 2000; see also Roy Chisholm, letter to the author, 27 Sep 2000, and e-mail to the author, 8 Mar 2003; C. Angas Hurst, e-mails to the author, 13 Dec 2000 and 25 June 2002.

labors. Included within his deceptively dimunitive-looking offerings was a one-page derivation of what would soon come to be known as the "Ward identity," showing that the conservation of electric charge led necessarily to a specific relation between various renormalization constants in Dyson's framework.[18]

Kemmer's group at Cambridge quickly became the functional equivalent in Britain of Dyson's postdoc cohort at the Institute for Advanced Study: young theorists practiced using the diagrams in informal congress, then fanned out throughout the country, bringing their new skills with them. Gordon Moorhouse spent two years with Kemmer's group during 1947–49, before finishing his dissertation the next year at the University of Birmingham, where Dyson had just taken up residence. Then Moorhouse set off for a three-year postdoctoral fellowship at the University of Glasgow. Not quite a year into his Glaswegian stay, a graduate student there, E. A. Powers, picked up the diagrams and began to use them in print. Soon Moorhouse was joined in Glasgow by Roy Chisholm, then completing his thesis work. After his earliest encounters with Dyson's manuscripts several years earlier, Chisholm had abandoned his original interest in five-dimensional spacetimes and—with much help from Matthews in the meantime—had begun to use Feynman diagrams in a more recognizably Dysonian fashion. Upon arriving in Glasgow, Chisholm teamed up with an émigré physicist from Vienna, Bruno Touschek, to cowrite an article that would become Touschek's first use of the diagrams.[19] Meanwhile Leon Castillejo, a graduate student at University College London, became an adopted member of Kemmer's students' circle, traveling to Cambridge nearly every week during 1949–50 to meet with the diagram-wielding students. He brought his new working knowledge of the diagrams with him back to London, where he coached a fellow graduate student, Michael Redhead, on the new techniques. After spending a year at the Institute for Advanced Study, John Ward visited Australia's University of Adelaide for a year; while he was there,

18. C. A. Hurst, e-mails to the author, 13 Dec 2000 and 25 June 2002. See also Roberts [B.1], [B.4], and [A.37]; Eden [B.7], [B.10], [B.13], and [B.16]; Hurst [A.52], [B.9]; and van Wyk [A.25]. After he had completed his dissertation, Richard Eden worked with a younger graduate student at Cambridge, G. Rickayzen, helping him learn the new techniques; see Eden and Rickayzen [B.15]. Ward's articles appeared as Ward [A.9], [A.12], and [A.14]. Today a generalized form of the identity introduced in [A.12] is known by the twin moniker the "Ward-Takahashi identity"; see, e.g., Peskin and Schroeder, *Quantum Field Theory* (1995), 238–44.

19. Moorhouse, e-mail to the author, 16 July 2000; Chisholm, letter to the author, 27 Sep 2000, and e-mail to the author, 8 Mar 2003; Moorhouse [A.74]; Power [B.6], [B.14]; Chisholm [A.34]; Chisholm and Touschek [A.80]. On Touschek, see Pauli, *WB*, 4.2:287–89.

a young lecturer at Adelaide, H. S. Green, submitted his first diagrammatic article.[20]

As the Cambridge circle expanded, British theoretical physicists gained a second diagrammatic stronghold at the University of Birmingham. Dyson taught there during 1949–51, before returning to the United States to teach at Cornell and (ultimately) to take up his permanent post at the Institute for Advanced Study. While in Birmingham, Dyson introduced the senior theoretical physicist, Rudolf Peierls, to the new diagrams; Dyson also worked closely with several students. Richard Dalitz undertook postdoctoral studies in Birmingham, where he benefited from Dyson's tutelage and began publishing his first diagrammatic articles. H. Pierre Noyes, who had earned his Ph.D. at Berkeley and spent 1950–51 as a postdoc in Birmingham, likewise thanked Dyson for introducing him to the new techniques, even citing Dyson's well-attended Birmingham lectures in the footnotes of his earliest diagrammatic contribution.[21] When Dyson accepted his call to Cornell, Peierls hired Paul Matthews to take his place; Matthews taught at Birmingham for the next five years. Among the students with whom he worked were G. E. Brown and Gordon Feldman, two more young theorists who began to make regular use of the diagrams. After spending nearly two years in Pakistan, Matthews's collaborator, Abdus Salam, returned to Cambridge as a lecturer in 1954. Salam's return not only revived interest in the diagrams among the new crop of Cambridge graduate students—John C. Taylor, John Polkinghorne, and Richard Phillips soon began submitting diagrammatic articles—but also strengthened ties between Cambridge and Birmingham, as Matthews and Salam visited each other frequently and continued their active and fruitful collaboration.[22]

Perhaps the most striking illustration of the power of face-to-face mentoring to spread the diagrams comes from the experiences of a postdoctoral researcher, J. G. Valatin. During the late 1940s, Valatin had been studying at the Institute of Experimental Physics in his native Budapest; in the early 1950s,

20. Chisholm mentions Castillejo's visits in his letter to the author, 27 Sep 2000; Redhead thanked Castillejo for "many lively discussions" in [B.17], 235. On Ward's visit to the University of Adelaide for the 1953–54 academic year, see Dalitz and Duarte, "Ward" (2000): 99; Green's article appeared as Green [A.121]. See also Hurst, "Green" (2001).

21. Dalitz [B.2], [B.3]; Noyes [A.60], 346n12 and 348.

22. Kibble, "Matthews" (1988), 563–65; Kibble, "Salam" (1998), 389–91; Gordon Feldman, e-mail to the author, 7 Dec 2000; John C. Taylor, e-mail to the author, 24 June 2002; Brown [B.11]; Feldman [B.20], [A.130]; Taylor [A.125], [A.136]; Polkinghorne [B.23]; Phillips [A.138]. Matthews's and Salam's active collaboration produced several articles from this period: see Salam and Matthews [A.79]; and Matthews and Salam [B.18] and [A.107].

he began to tackle QED from his new home as a postdoc at Niels Bohr's Institute for Theoretical Physics in Copenhagen. Then, in August 1953, he arrived in Birmingham, where he quickly began discussing physics with Peierls and Matthews. (Peierls himself had just returned from a yearlong visit to the Institute for Advanced Study.) Valatin had come from Copenhagen with a long manuscript already completed on the divergences within QED, along with most of a second manuscript—all without the aid of Feynman diagrams or Dyson's techniques. Peierls communicated Valatin's already-completed paper to the *Proceedings of the Royal Society* and also began working with Valatin on how Dyson's and Feynman's techniques could aid in Valatin's continuing research. The marks of Peierls's discussions are clear: near the end of Valatin's second paper, which had been almost but not entirely finished before his arrival in Birmingham, he began to mention Dyson's work and the ways that Feynman diagrams could be used in unpacking the effects of vacuum polarization; he also thanked Peierls for aid "in the course of [the paper's] later development." The discussions continued, and that same month Valatin dashed off a brief letter to the editor of the *Physical Review* that made more conspicuous mention of Feynman diagrams. Over the following year, he submitted two more lengthy articles from Birmingham, now with Feynman diagrams more central to his arguments.[23] Personal contact and informal tutelage with members of the diagrammatic network had made the difference.

With Dyson's direct aid, nurtured by the active study group at Cambridge, twenty-six authors in Great Britain published fifty-two diagrammatic articles (in both the *Physical Review* and the *Proceedings of the Royal Society*) between 1949 and 1954. Just as in the United States, nearly all these authors were graduate students (62%) or postdocs (19%) when they first learned to use Feynman diagrams.[24] These physicists, such as Valatin, picked up the diagrams only after direct intervention from physicists already "in the know." Dyson's ocean crossings quickly led to many others, reinforcing the ties between the American and British networks of diagram users: Paul Matthews, Abdus Salam, John Ward, Keith Roberts, and Rudolf Peierls all spent time in residence at the Institute for Advanced Study between 1950 and 1953.

23. Quotation from Valatin [B.19], 239. See also Valatin, "Molecular coupling effects" (1947), from Budapest; Valatin, "Quantum electrodynamics" (1951), from Copenhagen; and Valatin, "Singularities" (1954), submitted by Peierls from Birmingham, which lists Valatin's address as Bohr's institute in Copenhagen. His later publications, which made use of Feynman diagrams, all listed Birmingham as his instituitonal address: Valatin [B.19], [A.94], [B.21], and [B.22]. See also Institute for Advanced Study, *Community of Scholars* (1980), 384.

24. Information on British degrees comes from Bitboul, *Retrospective Index* (1976); and from *Index to Theses* (1951–).

FEYNMAN DIAGRAMS IN JAPAN

Setting the Stage: Rebuilding Physics under Occupation

In the months and years after the end of World War II, life in Japan and life in the United States could hardly have seemed more different.[25] While many Americans greeted the end of the war with dreams of large cars and suburban ranch houses, the Japanese suddenly had to adjust to life under U.S. military occupation. Whole cities had been destroyed—some by firebombing with conventional explosives, others by the new flames of atomic weapons—and a general cultural malaise seemed to settle over much of the population.[26] As Americans feted their new heroes—the physicists behind radar, the proximity fuse, and the atomic bomb—Japan's physicists labored under newfound hardships. Shortages of housing, food, funding, and even paper marked their new existence. Nor was their scientific equipment safe: in November 1945, U.S. military authorities seized and destroyed all known cyclotrons in Japan—despite protests from several American scientists—thinking that no instruments of nuclear physics could be left safely in the hands of the former enemy.[27] Many young Japanese physicists slept in their offices for months after the war, having no other homes to which they could return.[28]

Despite the shortages and hardships, however, new signs of hope began to emerge. One group, surrounding Tomonaga Sin-itiro in Tokyo, gathered weekly in a rickety Quonset hut beginning in 1946, hoping that their esoteric talk of integrals and electrons would help to rebuild their scientific community. That same year, Yukawa Hideki founded a new journal, *Progress of Theoretical Physics.* Although the Japanese Physical Society had maintained a European-language journal before the war, it had featured articles in German at least as often as in English. Now, with the postwar balance of scientific power viscerally obvious, Yukawa's new journal—financed with his family savings and independent of any official scientific body—would publish almost exclusively

25. On theoretical physics in interwar Japan, see esp. Ito, *Making Sense of* Ryōshiron (2002). Unless otherwise noted, all translations from Japanese in this section were done by Kenji Ito, to whom I am grateful. Japanese names appear in traditional order within the text (with family name first); citations to published sources follow whichever convention the authors used, which was most often Western name ordering.

26. Dower, *Embracing Defeat* (1999).

27. Weiner, "Cyclotrons and internationalism" (1975); Weiner, "Retroactive saber rattling?" (1978); Nakayama, "Destruction of cyclotrons" (2001).

28. Nambu, "Personal recollections" (1988); Hayakawa, "Sin-itiro Tomonaga" (1988); Brown and Nambu, "Physicists in wartime Japan" (1998).

in English. Its earliest issues carried reports of research that had been conducted during the war, together with translations of papers that had been published in Japanese during these same years. One of the very first articles to appear in the new journal was Tomonaga's piece, originally published in Japanese in 1943, in which he derived a relativistically invariant formalism for making QED calculations.[29] Although paper shortages and other delays marked the early years of *Progress of Theoretical Physics*—today, the fragile and yellowed pages from the first volumes look even more weathered and worn than those of the *Physical Review* from the same period—it quickly became the major organ in which Japanese physicists could publish their work and aim for an international audience.

As Yukawa began publishing his new journal, Japanese physicists could also look once again to the *Physical Review* to learn about recent developments in their field. Although shipments of the American journal had been interrupted by the war, the new Civil Information and Education (CIE) libraries, under the auspices of the U.S. occupation forces, began to stock copies of the *Physical Review* alongside other American publications, such as *Time* and *Newsweek*. The first CIE library opened in Tokyo in November 1945; in August 1947, the U.S. authorities set up a series of branch libraries throughout Japan. The precious copies of the *Physical Review* did not circulate outside the CIE libraries; several Japanese theorists recall writing out entire articles, longhand, so they could ponder the results after they left the libraries' readings rooms.[30] As soon as the American forces lifted their ban on international mail—no Japanese could send letters overseas until the spring of 1948—Japanese physicists wrote to their colleagues in the United States, hoping to find more direct ways of getting copies of the *Physical Review*. Kiuchi Masazō, the president of the Physical Society of Japan and a professor at the University of Tokyo, wrote immediately to his counterparts at the American Physical Society. "For our part," he explained, "we are very anxious to learn of the research conducted in your country." Perhaps the Americans would like to return to "the same friendship and favor as before the war," by agreeing to exchange copies of their journals. "Now there are very few files of the 'Physical Review,' the 'Reviews of Modern Physics,' the 'Journal of Chemical Physics,' etc. in Japan, which keeps the Japanese physicists from being informed of the world-wide trend of modern physics. This is, indeed, the most serious obstacle in our research."

29. Tomonaga, "Relativistically invariant formulation" (1946); cf. Schweber, *QED* (1994), chap. 6. See also Yukawa's autobiography: Yukawa, *Traveller* (1982 [1957]).

30. Kinoshita, "Personal recollections" (1988), 8; Ōneda, "Personal recollections" (1988), 16. See also Nakayama, "International exchange" (2001), 238–39.

The American Physical Society responded by donating complete sets of the *Physical Review* and *Reviews of Modern Physics,* covering the period from 1941 to 1948, to ten different universities and institutions throughout Japan.[31]

Within a year, physicists joined their journals in the back-and-forth traffic between Japan and the United States. Yukawa became the first Japanese physicist to visit the United States after the war when he ventured, on Oppenheimer's invitation, to the Institute for Advanced Study in September 1948. Yukawa's visit had not been easy to arrange: the Supreme Command of Allied Powers (SCAP) had enacted a ban on all international travel by Japanese scientists; only after Oppenheimer interceded directly with General Douglas MacArthur, SCAP's commanding officer, did Yukawa obtain permission to leave Japan.[32] Travel restrictions were eased near the end of 1949, and Japanese scientists began arranging for foreign visits immediately. By the end of 1951, nearly nine hundred Japanese scientists had traveled abroad; over eight hundred had gone to the United States, usually with financial assistance from SCAP's General Headquarters. Between 1949 and 1952, over one thousand Japanese students also studied abroad with scholarships from General Headquarters and the Japanese government; again, the largest portion ventured to the United States for their training.[33]

Beginning in 1949, General Headquarters also oversaw a tremendous expansion of the Japanese university system. The American authorities stipulated that there should be at least one national university in every prefecture of Japan, in imitation of the state-university system within the United States—a decree aimed at weakening the monopoly of the older imperial universities. By the early 1950s, the number of university students and faculty had leaped tenfold over the prewar numbers. American-style graduate schools also began to open in Japan with the 1949 expansion of the university system, although they remained rare. Until then, and for a long time afterward, most Japanese physicists completed undergraduate degrees (roughly equivalent to a master's degree from most American universities at the time), and then joined the

31. M. Kiuchi to American Physical Society, 3 Aug 1948; K. K. Darrow to M. Kiuchi, 28 Sep 1948; K. K. Darrow to H. A. Barton, J. R. Oppenheimer, J. T. Tate, and G. B. Pegram, 28 Oct 1948; see also the form letter from Komei Hashikawa (chief of the Publication Section of the Scientific Research Institute, Tokyo), 1 July 1948; and W. V. Houston to John Tate, 22 Jan 1949, all in *SAG,* Folder 51:51. On the ban on international mail before 1948, see Nakayama, "International exchange" (2001), 238. Before the APS sent back issues of the *Physical Review,* a few individual physicists had donated their own copies directly to colleagues in Japan; Berkeley's Luis Alvarez sent copies to Ryōkichi Sagane at the Tokyo Imperial University. Nakayama, "International exchange" (2001), 238.

32. Nakayama, "Sending scientists overseas" (2001), 250–51.

33. Ibid., 252–56.

research groups of senior professors. The next formal degree that Japanese physicists might obtain would be the Doctor of Science degree (D.Sc.), which would be granted on the basis of research already completed; few if any obtained Ph.D. degrees. In terms of physicists' life cycles, the D.Sc. functioned more like the culmination of a postdoc than like a Ph.D.; physicists usually spent ten years in active research and teaching before obtaining their D.Sc.'s. Before the expansion of the universities, explained a Japanese journalist in 1950, a young researcher toiling toward a D.Sc. degree would be "bound to a supervisory professor by the iron-like chains of feudalistic master-apprentice bondage."[34] With the quick upspringing of new universities, however, came a new pattern in young physicists' careers: now physicists began to *circulate,* bound far less tightly to any single "feudalistic master." Often they would complete their undergraduate studies at one institution and then work in several different groups, scattered throughout the new university system, before applying to obtain their D.Sc.'s.[35] Although not quite the same as the newly expanded postdoctoral system within the United States—with the postdoc cohort at the Institute for Advanced Study nearly doubling at just this time—the suddenly expanded Japanese university system played a similar role in dispersing Feynman diagrams throughout the country.

One other institutional element became remarkably important for the dispersion of Feynman diagrams in Japan. During the war, Yukawa, Tomonaga, Sakata Shōichi, and others had formed what they called the Meson Club, an informal, semiannual gathering of physicists interested in mesons and other aspects of nuclear and particle physics. The Meson Club had met between 1941 and 1943, before wartime duties and shortages had forced the group to halt its meetings. By 1948, the group had been reconfigured as the Soryūshiron Group (Elementary Particle Theory Group). The numbers of participants quickly outstripped the old Meson Club's—by 1950 there were more than 130 members, most of whom had just obtained their undergraduate degrees and were years away from their D.Sc.'s. The rapid growth continued: the Soryūshiron Group counted 200 members in 1952, and nearly 300 by 1955. The group began to sponsor informal meetings throughout Japan, affording members the opportunity to work together even though they might have belonged to different universities. As Ōneda Sadao recalled many years later, the Soryūshi-ron

34. Kaneseki, "Elementary particle theory group" (1974 [1950]), 231.

35. Nakayama, "Occupation period" (2001), 37–39; Kaneseki, "Elementary particle theory group" (1974 [1950]), 230–32.

Group "played a crucial role for people like myself who belonged to a small group," away from the main centers of physics research.[36]

In October 1948, the group also began distributing its own informal, mimeographed newsletter, *Soryūshi-ron kenkyū* (*Studies in Elementary Particle Theory*, hereafter *Soken*). Paper remained scarce; envisioning an ephemeral newsletter rather than a formal scholarly journal, the editors scraped by with even lower-quality paper than that used for *Progress of Theoretical Physics*, and today extant copies of the early *Soken* newsletters are correspondingly even more yellowed and fragile. Frustrated by the continuing delays with *Progress*, caused in large part by paper shortages, the editors of *Soken* included more and more informal, unrefereed preprints in their newsletter alongside brief summaries of talks that had been given at recent meetings, job notices, and other announcements from the Soryūshi-ron Group. More than this, *Soken* quickly became a vital connection between Japanese theorists and their international colleagues. From its earliest issues, *Soken* printed correspondence between Japanese physicists and colleagues abroad—the newsletter having been founded a few months after the ban on international mail had been lifted. Most of these letters had been addressed to individual Japanese theorists and were reprinted with all the ordinary apparatus of personal correspondence still included: return addresses and even postmarks were simply recopied by hand onto the mimeographed masters for distribution. The earliest issues included letters from Oppenheimer, Abraham Pais, and Wolfgang Pauli (in English) and from Werner Heisenberg (in German), among others. One early correspondent, Frederick Belinfante at Purdue University in Indiana, decided as "a matter of politeness" to correspond in "a neutral language," choosing to write lengthy updates on his research in Esperanto rather than English. When Japanese physicists were allowed to travel abroad, beginning in 1949, they regularly sent updates back to be printed in *Soken*, describing their travels and the research they had learned about from their international colleagues. In 1949 alone, *Soken* printed over fifty letters from abroad; that same year, *Soken* had soared to more than 1,200 pages in total, reaching nearly 2,000 pages by 1951.[37]

36. Ōneda, "Personal recollections" (1988), 15. See also Konuma, "Social aspects" (1989), 542–43. On the Meson Club, see Hayakawa, "Development of meson physics" (1983).

37. Frederick Belinfante to Koba Zirō, 11 Dec 1948, as reprinted in *Soken* 2, no. 3 (1949): 115–20; and Belinfante to Miyamoto Yoneji, 6 Mar 1949, as reprinted in *Soken* 2, no. 3 (1949): 128–36; Belinfante, "Supermultatempa Teorio" (1948); Belinfante, "Pri la Kalkulado de Elektromagnetaj Fenomenoj" (1949). See also Konuma, "Social aspects" (1989), 542–43; and Kaneseki, "Elementary particle theory group" (1974 [1950]), 230.

The new epistolary outlet obviously brought news of new developments to the Japanese community much more quickly than published articles could. It also played an important socializing role for the new generation: Nambu Yoichiro recalls that the newsletter "served as a classroom for social etiquette." Through *Soken*, the young physicists learned "what is fit to print and what is not." As Nambu noted, he and his colleagues sometimes learned the lessons "the hard way"; the new form of interaction was not without its problems.[38] For example, Kenneth Greisen, an experimentalist at Cornell, wrote to Hayakawa Satio in November 1948—in a letter quickly reproduced in *Soken*—explaining in detail the status of several competing experimental efforts to study meson interactions in cosmic ray showers. "An important consideration in these matters," Greisen explained, "is the reliability of the experiments and the interpretation of the experiments. We who are closer to them may hear or see some things which do not appear in print. For your help, I list some of [these] things." There followed seven detailed paragraphs, giving Greisen's behind-the-scenes evaluation of how things stood. "Most of the work with high pressure chambers (Lapp, etc.) has had an error in the interpretation due to use of a wrong value (too low) for the specific ionization," he explained; "I do not believe the experiment of Schein, Jesse and Wollan," he continued.[39] Theorists in Japan, cut off from the informal discussions surrounding such experiments, were not able to judge such things as easily for themselves.

The difficulties of making up for such isolation by means of foreign letters quickly became clear. Four months later, Greisen wrote back to Hayakawa (also reprinted in *Soken*), complaining that "I was somewhat embarrassed by the enthusiasm with which you accepted and repeated my steatments [*sic*]. I intended primarily to advise you not to accept uncritically the experimental results published in the journals because many of them contain some inaccuracy. Similarly, you should not have complete faith in my steatments [*sic*] either." There followed another six dense paragraphs in which Greisen tried to explain in more detail his own doubts about several published results, and the bases upon which he drew such conclusions.[40] Clearly, the new correspondence opportunities represented an improvement over the immediate postwar silence; yet as Greisen, Hayakawa, and scores of other *Soken* readers learned

38. Nambu Yoichiro, e-mail to the author, 13 Apr 2003; see also Kaneseki, "Elementary particle theory group" (1974 [1950]), 251–52; Ōneda, "Personal recollections" (1988), 15–16; and Konuma, "Social aspects" (1989), 542–44.

39. Kenneth Greisen to Hayakawa Satio, 16 Nov 1948, as reprinted in *Soken* 2, no. 3 (1949): 227–30.

40. Greisen to Hayakawa, n.d. (ca. Mar 1949), as reprinted in *Soken* 2, no. 4 (1949): 136–39.

quickly, even informal correspondence could not always replace personal visits and face-to-face discussions.

Tomonaga's Tokyo Group

When word of the new Feynman diagram techniques reached Japan early in 1949, an active community surrounding Tomonaga Sin-itiro was already well primed to receive the news and to act on it. In the spring of 1946, Tomonaga began to rally an active group of young theorists. Breaking with the long-standing tradition of mutual isolation between imperial universities and all other institutions, the group drew members from several different schools in the Tokyo area—students and researchers from Tokyo University began to converse freely with members from the Tokyo Bunrika Daigaku (Tokyo Education University, where Tomonaga worked) and elsewhere. During the war, Tomonaga had worked out many elements of his relativistically invariant formalism for QED. Now his ragtag group of young followers set to work extending his formalism and applying it to problems in both QED and in studies of meson interactions. Some weeks, the group gathered in his Quonset hut—which doubled as Tomonaga's makeshift office and residence—to talk about some members' recent calculations or to arrange further collaborations on new projects; other times, they entered into intense discussions of a recently published article. Still other weeks, the students worked together to make sense of sections of Walter Heitler's prewar textbook, *Quantum Theory of Radiation* (1936), with Tomonaga's aid, to keep calculational skills sharp. By all accounts, the Tokyo group was informal yet spirited: they threw themselves into their physics, despite—or because of—the lasting material deprivations all around them.[41] (See fig. 4.3.) Several members of Tomonaga's circle went on to distinguished international careers: Nambu Yoichiro, Kinoshita Tōichirō, and Nishijima Kazuhiko, among others.

Tomonaga's circle logged many hours in the Tokyo CIE library. During the summer of 1947, they noticed a brief announcement in *Newsweek* of Willis Lamb's high-precision measurement of an energy-level splitting within hydrogen atoms, and they realized—just as surely as the American physicists who heard Lamb's news directly—that the effect might be due to radiative corrections within QED. Their suspicions were confirmed two months later when they read (again in the CIE library) Hans Bethe's *Physical Review* article, in

41. Nambu, "Personal recollections" (1988); Kinoshita, "Personal recollections" (1988); Hayakawa, "Sin-itiro Tomonaga" (1988); Brown and Nambu, "Physicists in wartime Japan" (1998); Schweber, *QED* (1994), chap. 6; and Tōichirō (Tom) Kinoshita, letter to the author, 19 Mar 2003.

Figure 4.3. Tomonaga Sin-itiro (*center, pointing*) with members of his study group, ca. 1948. (Source: University of Tsukuba, Tomonaga Memorial Room, courtesy AIP Emilio Segré Visual Archives.)

which he studied the Lamb shift within a nonrelativistic approximation. The news pushed Tomonaga's circle even more fervently into the hunt for renormalization. The group worked collectively, seeking ways to apply Tomonaga's relativistically invariant formalism to tackle the infinities of QED. By the spring of 1948, the group had collected several results, all showing that—at least in the first approximation—the infinities within QED could be handled by making charge and mass renormalizations, that is, by consistently trading in the e_0 and m_0 of the starting equations for the e_{eff} and m_{eff} that an experiment would be able to measure. Although neither side knew it yet, Tomonaga's entire approach, right down to his notation and formalism, showed remarkable resemblance to Julian Schwinger's emerging work.[42] In one of the earliest packages

42. Tomonaga's and Schwinger's similarity in approach and notation is not without explanation: although they worked in mutual isolation during the war and afterward, they both drew strongly upon the interwar work by people like Heisenberg and Pauli. Tomonaga had studied with Heisenberg in Leipzig between 1937 and 1939 before returning to Japan with the outbreak of World War II; he and several other Japanese theorists continued to study Heisenberg's papers intensely. Although Heisenberg had not found a solution to the problem of QED's infinities, both Tomonaga's and

sent overseas after the ban on international mail had been lifted, Tomonaga sent a series of papers and manuscripts to Oppenheimer, who received them as soon as he returned from the Pocono conference (at which Schwinger had dazzled his listeners with his daylong lecture on renormalization, after which Feynman had haltingly introduced his diagrams). Oppenheimer immediately cabled Tomonaga with a gracious reply (soon reprinted in one of the earliest issues of *Soken*), and arranged for a brief report on Tomonaga's work to appear in the *Physical Review.* He also sent copies of Tomonaga's letter to the participants of the Pocono conference, highlighting the significance of the Tokyo group's achievements and their similarity to Schwinger's approach. Acknowledging receipt of Tomonaga's follow-up manuscript, Oppenheimer replied (in another letter reprinted in *Soken*): "I am looking forward to a rapid improvement in the possibilities of travel for you and your colleagues, and hope that before long you will spend some time with us at the Institute where we shall all welcome you so warmly."[43] Although the travel ban kept Tomonaga from visiting the Institute for another year and a half, by the summer of 1948 physicists in both the United States and in Japan began to speak of a "Tomonaga-Schwinger" approach to QED and renormalization.

In the course of these intense and fast-moving developments, the Tokyo group focused on how to make better, more efficient, and more reliable perturbative calculations. As the Tokyo group learned quickly, there was always a danger that theorists might overlook, or otherwise confuse, terms that should contribute to a given order in their perturbative expansion. These issues took on a certain immediacy thanks to competition from other groups. Sakata Shōichi, the senior physicist at Nagoya, had suggested in 1946 that the infinities of QED could be removed if one assumed that electrons interacted with a new, hypothetical field—the so-called *C*-meson—in addition to photons.[44] Tomonaga's group took immediate interest in the compensating effects of the

---

Schwinger's programs grew out of careful study of the same starting point. Moreover, both Tomonaga and Schwinger worked on radar during the war, and both became impressed with the effective-circuit approach. See Schwinger, "Two shakers of physics" (1983); Darrigol, "Scientific biography of Tomonaga" (1988), 3, 25–26; Maki, "Tomonaga" (1988); and Schweber, *QED* (1994), chaps. 6–7.

43. J. Robert Oppenheimer, telegram to Tomonaga Sin-itiro, 14 Apr 1948, as reprinted in *Soken* 1, no. 2 (1948): 61; Oppenheimer to Tomonaga, 28 May 1948, as reprinted in *Soken* 1, no. 2 (1948): 147. Oppenheimer's letter to the Pocono participants is reprinted in Schweber, *QED* (1994), 198–200; Tomonaga's brief paper was published as Tomonaga, "Infinite reactions" (1948).

44. Abraham Pais had come up with the same basic idea in 1945. See Pais to Sakata Shōichi, 8 Feb 1949, as reprinted in *Soken* 2, no. 3 (1949): 242–44; Pais, "Electron and nucleon" (1945); Pais, "Elementary particles" (1946). Wolfgang Pauli, whose regularization method (developed with

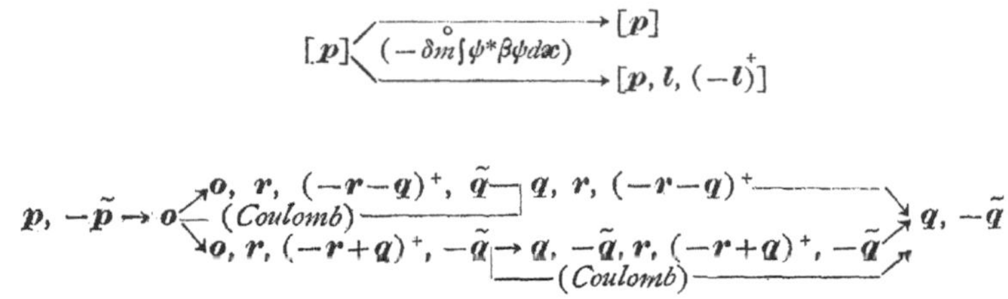

$$[\boldsymbol{p}] \longrightarrow [\boldsymbol{p}]$$
$$[\boldsymbol{p}] \xrightarrow{(-\delta \mathring{m} \int \psi^* \beta \psi d\boldsymbol{x})} [\boldsymbol{p}, \boldsymbol{l}, (-\overset{+}{\boldsymbol{l}})]$$

$$\boldsymbol{p}, -\tilde{\boldsymbol{p}} \rightarrow \boldsymbol{o} \nearrow \boldsymbol{o}, \boldsymbol{r}, (-\boldsymbol{r}-\boldsymbol{q})^+, \tilde{\boldsymbol{q}} \rightarrow \boldsymbol{q}, \boldsymbol{r}, (-\boldsymbol{r}-\boldsymbol{q})^+ \searrow \boldsymbol{q}, -\tilde{\boldsymbol{q}}$$
$$\boldsymbol{o} \text{ — } (Coulomb) \text{ — } \boldsymbol{q}, \boldsymbol{r}, (-\boldsymbol{r}-\boldsymbol{q})^+$$
$$\boldsymbol{o} \searrow \boldsymbol{o}, \boldsymbol{r}, (-\boldsymbol{r}+\boldsymbol{q})^+, -\tilde{\boldsymbol{q}} \rightarrow \boldsymbol{q}, -\tilde{\boldsymbol{q}}, \boldsymbol{r}, (-\boldsymbol{r}+\boldsymbol{q})^+, -\tilde{\boldsymbol{q}} \nearrow \boldsymbol{q}, -\tilde{\boldsymbol{q}}$$
$$-\tilde{\boldsymbol{q}} \text{ — } (Coulomb) \text{ — } \nearrow \boldsymbol{q}, -\tilde{\boldsymbol{q}}$$

Figure 4.4. Arrow notation for perturbative calculations. (Sources: *top,* Koba and Tomonaga, "On radiation reactions, I" [1948], 294; *bottom:* Koba and Takeda, "Radiation reaction, II" [1948], 414.)

hypothetical meson but quickly disagreed with Sakata over the value of the *C*-meson's charge needed to cancel the offending infinities. The source of their disagreement, as eventually became clear, was that the Tokyo physicists had mistakenly dropped a few key terms from their perturbative expansion.[45] Nor was the problem limited to intergroup rivalries within Japan. An early paper by one of Sakata's students included a note added in proof arguing that Bethe and Oppenheimer had unwittingly dropped crucial terms in their own recent perturbative calculation.[46] Both between Tokyo and Nagoya and between Japan and the United States, questions frequently turned on whether all possible contributions to a given order in a perturbative calculation had been properly taken into account.

Given the importance of identifying and including all possible contributions to a given perturbative order, Tomonaga's group began tinkering with its own graphic means of keeping track of the necessary terms. In the spring of 1948, Tomonaga and Koba Zirō initiated a text-based notation, using arrows to denote various possible processes. For example, they explained that the notation in figure 4.4 (*top*) indicated that an electron with momentum $\boldsymbol{p}$ could remain in that state, or it could split into an electron plus a virtual electron-positron pair (with momenta $\boldsymbol{p}$, $\boldsymbol{1}$, and $-\boldsymbol{1}$, respectively). Both possibilities needed to be included, since the divergence arising from the virtual

---

Felix Villars) also shared similarities with the Pais-Sakata method, likewise came to admire Sakata's approach, especially as it was further developed by Sakata's student Umezawa Hiroomi: Pauli to Umezawa Hiroomi, 3 May 1949, as reprinted in *Soken* 2, no. 1 (5 Aug 1949): 253; and Kamefuchi, "Quantum field theory" (2001), 8, 10.

45. Darrigol, "Scientific biography of Tomonaga" (1988), 21–23; Hayakawa, "Sin-itiro Tomonaga" (1988), 56–57; Itō, "My positive and negative contribution" (1988); Kamefuchi, "Quantum field theory" (2001); and Schweber, *QED* (1994), 267–68.

46. Tanikawa, "Radiation damping" (1947), 53. The paper in question was Bethe and Oppenheimer, "Reaction of radiation" (1946).

electron-positron pair would precisely cancel a second divergence, appearing in a different part of the overall calculation. For the next installment of this research, Koba worked with Takeda Gyō, another young member of Tomonaga's circle. Submitted just three weeks after the first paper, it made clear that the Tokyo group had fast become adept at using its new arrow formalism: whereas only two simple examples had appeared in the first paper, a dozen more complicated versions, such as figure 4.4 (*bottom*), filled the second paper.

Yet even these arrows proved insufficient. Six months later, Koba and Takeda submitted the third and final paper in this series, itself published in two separate installments. The perturbative terms they now sought to calculate had become so intricate that they turned to a new graphic means to try to keep them all straight. They explained the need for their new "transition diagrams": "In order to dispose of all possible radiative corrections in a general case, it is essential to analyze systematically the complicated chains connecting the initial and final states not only through the least necessary number of intermediate states, but also through certain detours including the emission and reabsorption of a virtual photon or the creation and annihilation of a virtual electron-positron pair, which just account for the radiation reaction. For this purpose we have introduced a 'transition diagram method,' which turns out [to be] an effective tool for the discussion of higher order processes."[47] The new diagrams allowed one to "command a view of the whole connection between initial and final states which appears in the perturbation calculus of a certain complicated process," they continued. Their transition diagrams had a momentum axis running vertically and a time axis running horizontally. An electron with momentum $\boldsymbol{p}$ would therefore be represented by a horizontal line at a height $\boldsymbol{p}$ with an arrow moving to the right; a positron with momentum $\boldsymbol{p}$ would be represented by a horizontal line with a height $-\boldsymbol{p}$ with an arrow moving to the left.[48] (See fig. 4.5.) The emission of virtual photons would appear as wavy vertical lines connecting the two momentum states of an electron; the absorption of virtual photons would appear as dashed vertical lines. They illustrated their technique by considering the four distinct ways in which Compton scattering—the scattering between an electron and a photon—could occur in the lowest approximation. (See fig. 4.6.) Armed with these line drawings to delineate all the processes that contributed to a

47. Koba and Takeda, "Radiation reaction, III" pt. 1 (1949), 61.

48. Ibid. Note that Koba and Takeda worked in terms of spatial-momentum three-vectors, rather than relativistic energy-momentum four-vectors.

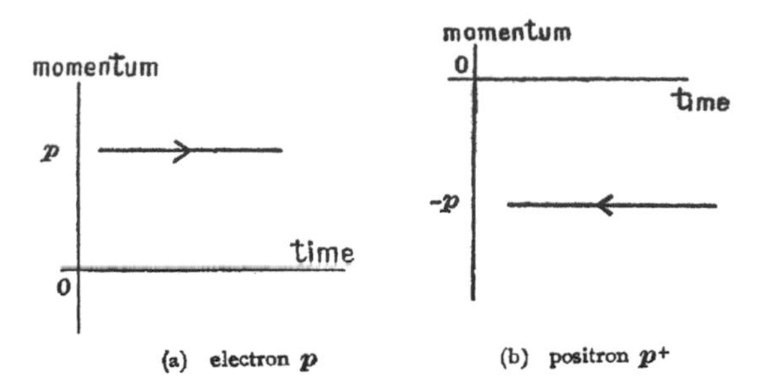

Figure 4.5. Transition diagrams. (Source: Koba and Takeda, “Radiation reaction, III,” pt. 1 [1949], 62.)

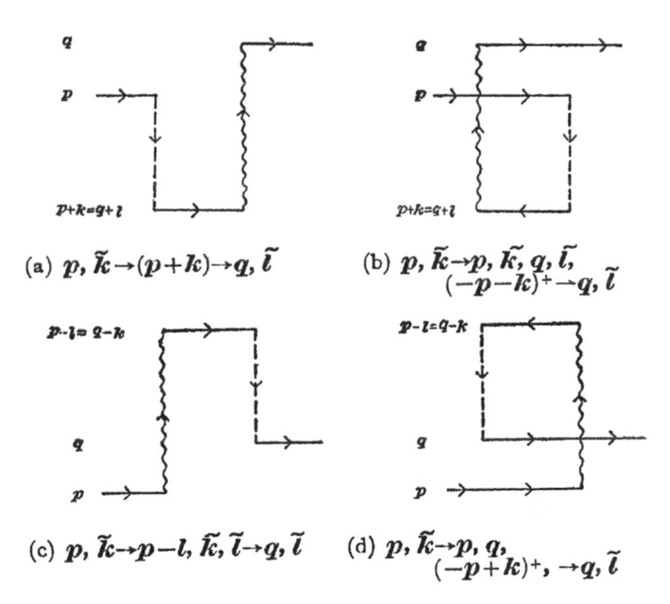

Figure 4.6. Compton scattering, as illustrated by transition diagrams. (Source: Koba and Takeda, “Radiation reaction, III,” pt. 2 [1949], 64.)

given order, Tomonaga’s two apprentices could then begin to study radiative corrections within QED, such as the self-energy of an electron, with more confidence that they would include all the appropriate terms.[49] Many young members of Tomonaga’s group quickly recognized the usefulness of these line

49. Koba and Takeda, “Radiation reaction, III” pt. 2 (1949). One possible source for the form of Koba and Takeda’s transition diagrams was prewar energy-level diagrams from atomic physics, which had played a central role in the early years of quantum mechanics. These depicted atomic transitions as vertical lines between the horizontal lines of distinct energy levels.

drawings. Nambu Yoichiro, who shared an office (and hence makeshift living quarters) with Koba at the time, submitted a paper just three weeks after Koba and Takeda wrote up their transition-diagrams work, making use of a similar graphic scheme for his own perturbative calculations.[50] At last, by October 1948, members of the Tokyo group seemed to have a reliable method for pursuing perturbative calculations.

Enter Feynman Diagrams

Yet Koba and Takeda's transition diagrams were not to last. In a note added in proof soon before their final installment appeared in the April–June 1949 issue of *Progress of Theoretical Physics,* they announced that they had just read an article in the *Physical Review* by F. J. Dyson. Dyson's article introduced "diagrams in ordinary space-time, and we hope our momentum-diagram will act as an intermediary between the formalism" of Dyson's paper "and the conventional perturbation calculation. Judging from Dyson's work, Feynman's radiation theory, of which we know very little, seems also to employ some diagram method."[51] The paper by Dyson to which they referred was his first article on the diagrams—indeed, the first article by anyone on Feynman diagrams—which was submitted to the *Physical Review* in October 1948 and published in February 1949. (In fact, Dyson's article was received at the *Physical Review* on 6 October 1948, while Koba and Takeda's article on transition diagrams was received at *Progress of Theoretical Physics* on 4 October 1948.) The similarity between Koba and Takeda's transition diagrams and Dyson's reports of Feynman diagrams had not been lost on the Tokyo theorists, nor was it lost on Dyson himself. Dyson added a brief footnote to his own paper just before it went to press, after he read a short letter to the editor in *Progress of Theoretical Physics* by Koba and Takeda that had described (in words only) some of the rudimentary elements of their transition-diagram scheme. The editors of *Soken* reprinted Dyson's reaction upon learning of the Koba-Takeda work, underlining his final sentence for emphasis: "After this paper [Dyson's first article on Feynman diagrams] was written, the author was shown a letter, published

50. Nambu, "Level shift" (1949). On Nambu's close working relationship with Koba, see Nambu, "Personal recollections" (1988), 3–5. Sakata's students Umezawa Hiroomi and Kawabe Rokuo developed a similar graphic scheme for making perturbative calculations, also working in momentum space and using vertical lines to denote transitions between various states. See Umezawa and Kawabe, "Vacuum polarization" (1949).

51. Koba and Takeda, "Radiation reaction, III," pt. 2 (1949), 141.

in *Progress of Theoretical Physics*... by Z. Koba and G. Takeda, [which] briefly describes a method of treatment of radiative problems, similar to the method of this paper.... *The isolation of these Japanese workers has undoubtedly constituted a serious loss to theoretical physics.*"[52]

Only a little news of Dyson's work had trickled into Japan before his first paper was published. Yukawa Hideki arrived at the Institute for Advanced Study at the same time as Dyson, in September 1948; he immediately began sending reports to his colleagues in Japan, and many of the reports were quickly published in *Soken.* A month after he arrived in Princeton, for example, Yukawa wrote that "Here, Feyman's [*sic*] theory is popular among young people." Translating what he had heard about Feynman's work—by way of Dyson's informal lectures at the Institute—into the familiar formalism of his Japanese colleagues, Yukawa explained that the new work "is a well-organized way to calculate $\Phi(+\infty)$ to $\Phi(-\infty)$ in Tomonaga's formula. In other words, if you follow the changes in the temporal order... various virtual terms disappear automatically, which makes calculation very simple."[53] Yet Dyson did not prepare any formal preprints of his first diagrammatic article, written up so hastily upon his arrival at the Institute. Nor did I. I. Rabi talk about Dyson's results when he visited Tomonaga's group in Tokyo in November 1948, talking excitedly about Schwinger's recent work instead.[54] Only once Dyson's first article appeared in print (and after copies of the 1 February 1949 issue of the *Physical Review* arrived in Japan) did Tomonaga's group learn more of what Dyson had done. If Yukawa's early reports had not sufficiently stirred interest, the title of Dyson's paper probably would have caught the Tokyo group's attention: he announced his work under the rubric "The Radiation Theories of Tomonaga, Schwinger, and Feynman" and demonstrated the formal equivalence of Tomonaga's and Schwinger's approach to Feynman's still-unpublished methods. Around the same time, Yukawa sent a preprint of Dyson's second paper,

52. Quoted in *Soken* 2, no. 4 (1949): 145; emphasis in original. Dyson's footnote originally appeared in Dyson [A.1], 486–87, and referred to Koba and Takeda, "Radiative corrections" (1948). Quite likely, Yukawa Hideki (then visiting at the Institute for Advanced Study) had shown Dyson Koba and Takeda's letter to the editor.

53. Yukawa Hideki to Sakata Shōichi, 30 Oct 1948, as reprinted in *Soken* 2, no. 3 (1949): 61–62. By discussing the means of calculating the evolution of the quantum state Φ from the far past to the far future, Yukawa was clearly referring to Dyson's work. Dyson's first paper, [A.1], introduced his S-matrix formalism explicitly in relation to Tomonaga's and Schwinger's formalisms, as providing one means of relating $\Phi(+\infty)$ to $\Phi(-\infty)$.

54. On Rabi's visit, see Hayakawa, "Sin-itiro Tomonaga" (1988), 58; and Miyamoto, "Personal recollections" (1988), 65.

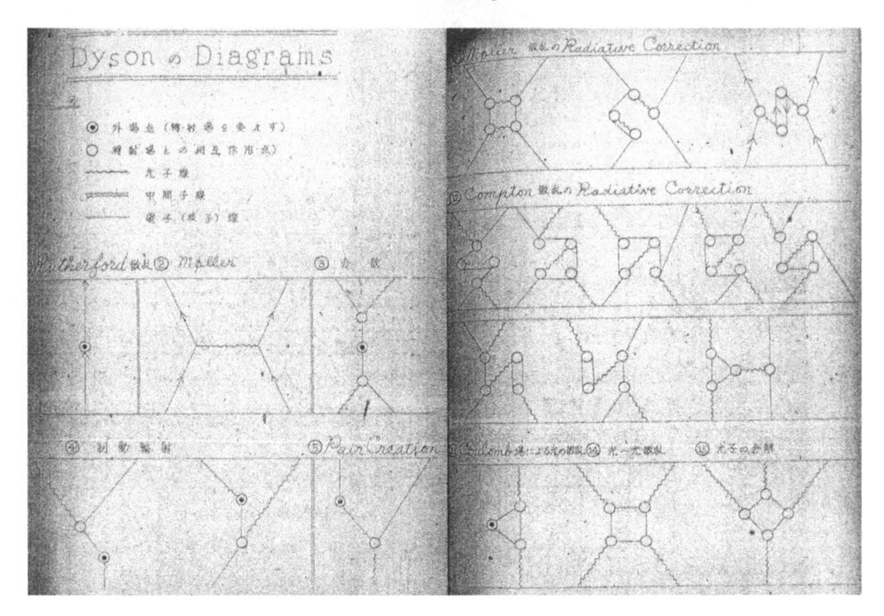

Figure 4.7. "Dyson diagrams." (Source: Fukuda and Itō [D.3], 165, 167.)

which had become available in late January. Immediately Tomonaga's group threw itself into concerted study of Dyson's work.[55]

Within months, the members of Tomonaga's group began making use of Feynman's diagrams, based on what they had learned from Dyson's papers. Already primed by Koba and Takeda's transition diagrams, they scrutinized Dyson's work with special care. The August 1949 issue of *Soken* carried a lengthy "homegrown" user's manual to Dyson's first paper prepared by Fukuda Nobuyuki and Itō Daisuke, both young researchers at the Tokyo Education University, where Tomonaga taught. Fukuda and Itō explained that Dyson had "proposed very skillful and simple methods" to replace the "very tedious and pains taking" means of performing perturbative calculations.[56] Three pages followed in which Fukuda and Itō spelled out, step by step, the means of relating Dyson's formalism to Tomonaga's more familiar approach. Next came four full pages filled with "Dyson diagrams," such as the examples in figure 4.7. Two months later, Takeda Gyō himself submitted a brief letter to the editor of *Progress of Theoretical Physics* making use of Dyson's diagrammatic approach; his former coauthor, Koba Zirō, followed three months after that,

55. Nambu, "Personal recollections" (1988), 5; Kinoshita, "Personal recollections" (1988), 9; and Kinoshita, letter to the author, 19 Mar 2003.

56. Fukuda and Itō [D.3], 162.

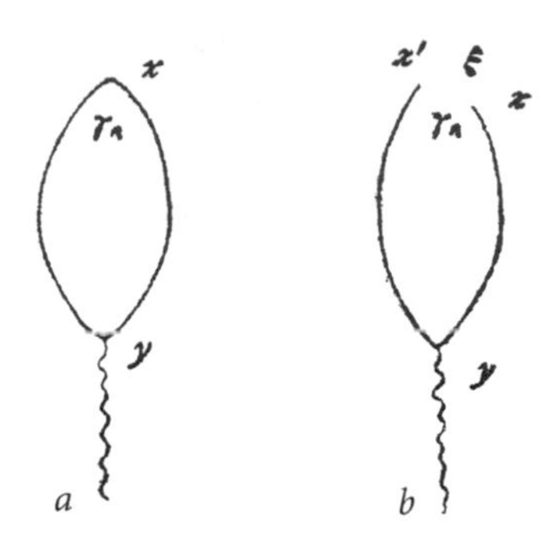

Figure 4.8. *a*, Feynman diagram for the lowest-order contribution to vacuum polarization. *b*, Takeda's modified diagram for the same process. (Source: Takeda [C.1], 573.)

including Feynman diagrams in an article published in the January 1950 issue of *Soken*.[57]

Even with all the close scrutiny of Dyson's papers, however—intense discussions in Tomonaga's seminar, plus Fukuda and Itō's helpful summary and extended translation into the familiar Tomonaga formalism—the Tokyo theorists' initial appropriation of Feynman diagrams revealed several subtle differences from Dyson's own work. Consider, for example, Takeda's 1949 letter to the editor, which became the first published use of Feynman diagrams in *Progress of Theoretical Physics.* He returned to a familiar problem within QED, the divergences associated with vacuum polarization—that is, the infinities that arose from the incessant popping into and out of existence of virtual electron-positron pairs. "According to Feynman-Dyson's theory," Takeda briskly explained, one could write an expression for the vacuum polarization in the lowest order of approximation, based on a single simple Feynman diagram. (See fig. 4.8*a*.) Next to this diagram, Takeda included a second diagram (fig. 4.8*b*), explaining that "This just means to adopt an open loop having a mouth with small breadth Z_μ instead of a closed one." Takeda's aim was to combine Dyson's work with Heisenberg's prewar suggestion that QED's infinities could be evaded by introducing a universal "fundamental length"—some finite, smallest distance closer than which no objects could pass. Like Roy Chisholm's initial efforts with Feynman diagrams in Cambridge (see fig. 4.1), Takeda thus appropriated Dyson's work in a rather non-Dysonian way. He duly wrote down a modified mathematical expression for the vacuum polarization, taking into account the "open mouth" of his new diagram.[58]

57. Takeda [C.1]; Koba, Kotani, and Nakai [D.9].

58. Takeda [C.1], 573–74. In his alternate calculation, Takeda assigned the argument of the electron's propagator to be $(x - y + Z/2)$, while the positron's propagator carried the argument

Takeda's diagrammatic maneuver is striking and no doubt would have appeared bizarre to Dyson or the other Institute postdocs. To Dyson, there was no such thing as an "opened" closed loop. The diagram in figure 4.8*b* no longer depicted virtual particles, precisely because the solid electron and positron lines no longer met at a single point. In Dyson's scheme, if the two lines never rejoined at a single point, that meant that the two virtual particles never repaid the energy thcy had borrowed, according to the Heisenberg uncertainty principle. To one steeped in Dyson's methods, in other words, figure 4.8*b* depicted the interaction between a photon (*wavy line*) and two *real* electrons (*solid lines*), rather than a photon encountering a virtual electron-positron pair. As such, figure 4.8*b* had nothing at all to do with vacuum polarization. Within Dyson's formalism, Takeda's accompanying mathematical expression for figure 4.8*b* was therefore neither correct nor even relevant, since the diagram contributed to a wholly separate physical process.

The next article in *Progress of Theoretical Physics* to make use of Feynman diagrams signaled a similarly non-Dysonian adoption of the new techniques. Fukuda Hiroshi (no relation to Nobuyuki) and Miyamoto Yoneji, both working at Tokyo University, pressed Feynman diagrams into service for their study of various meson-decay processes. They worked beyond the lowest-order approximation, including the effects of the first round of radiative corrections. Although they worked beyond the lowest order, however, they did not begin their calculation with Feynman diagrams and use Dyson's rules to translate each diagram step by step into a mathematical expression. Instead, the young Tokyo theorists wrote out all the relevant integrals in full—using Tomonaga's formalism, without the aid of any diagrams—and only later "illustrated" the various complicated integrals with associated Feynman diagrams. They ignored Dyson's formal rules for translating Feynman diagrams into equations altogether (and thereby avoided some of Takeda's complications), choosing instead to write down the expressions the best way they knew how.[59] Feynman

---

$-(x - y - Z/2)$. He then expanded his integral to lowest order in Z. Heisenberg had quietly dismissed his "fundamental length" idea long before Takeda's paper; no other physicists in Japan or elsewhere seem to have pursued this line further. On Heisenberg's fundamental length, see Schweber, *QED* (1994), 100–101; and Miller, "Frame-setting essay" (1994), 43–44, 79–81.

59. Fukuda and Miyamoto [C.2]. This brief letter to the editor announced results of the full calculations, presented in Fukuda, Hayakawa, and Miyamoto [C.9]. The letter did not contain any explicit details of the calculation; in the longer paper, the means of calculation were laid out in full, and in them it becomes clear that the theorists wrote down their matrix elements first, using Tomonaga's formalism, and only later appended the Feynman diagrams as illustrations. For example, after

diagrams had entered the Tokyo theorists' purview but had hardly edged out their native routine.

The first articles published in *Progress of Theoretical Physics* that put Feynman diagrams to use in a recognizably Dysonian fashion came from Kinoshita Tōichirō. Kinoshita was already well versed in how to calculate using Tomonaga's formalism, as well as the pitfalls of attempting to perform complicated perturbative calculations without additional bookkeeping aids. In his earliest published work, submitted to *Progress of Theoretical Physics* in July 1948, he had worked as an "apprentice" with Koba Zirō and Endō Shinji on a study of radiative corrections to an electron's scattering in an external electromagnetic potential. Eight months later, he and his collaborators had to retract several of their printed conclusions upon realizing that they had overlooked an important term in their perturbative calculation. By April 1949, Kinoshita and his collaborators had thus redone their full calculation with the aid of Koba and Tomonaga's arrow notation (as in fig. 4.4).[60] Just as the corrected calculation was sent off to the journal, Dyson's first paper arrived in Japan. As Kinoshita later recalled, he immediately made a careful study of it and "volunteered to present Dyson's paper at [Tomonaga's Tokyo] seminar, which quickly turned into a very lively free-for-all."[61] In preparing for his presentation before such an alert and vigorous group—and having himself only recently been chastened by the difficulties of perturbative calculations—Kinoshita probably scrutinized Dyson's paper even more closely than most of his peers had yet done.

Nine solid months of work later, Kinoshita's diligence began to pay off, as he applied Feynman diagrams and Dyson's calculational techniques to a series of problems. *Soken* carried his first diagrammatic preprint in December 1949. A few months later, he submitted a brief letter to the editor of *Progress of Theoretical Physics* with the results of his diagrammatic investigation of Sakata's *C*-meson hypothesis. Armed with Dyson's diagrammatic machinery, Kinoshita set out to check the consistency of the *C*-meson hypothesis at fourth order. (Previous calculations by Sakata's and Tomonaga's groups had considered only second-order contributions.) Of the many Feynman diagrams

---

writing down a particular matrix element, they wrote that "The decay process $\tau^+ \rightarrow \pi^+ + \gamma$ is visualized by Feynman's diagram in Fig. 1" (Fukuda, Hayakawa, and Miyamoto [C.9], 353).

60. Endō, Kinoshita, and Koba, "Reactive corrections" (1948); Endō, Kinoshita, and Koba, "Errata" (1949); and Endō, Kinoshita, and Koba, "Reactive corrections" (1949). Kinoshita's reference to himself as Koba's apprentice for this work was made in Kinoshita, letter to the author, 19 Mar 2003.

61. Kinoshita, "Personal recollections" (1988), 9.

that contributed to the fourth-order calculation, Kinoshita explained, one gave rise to a divergence that the C-meson remained powerless to compensate. No relation between the coupling constants of the electron and C-meson could make the fourth-order calculation converge, and thus, Sakata's trick could not be the whole answer to the riddles of QED and its infinities.[62] Around the same time, Kinoshita submitted a long paper on radiative corrections within meson theories, in strict analogy to Tomonaga's and Dyson's approach to QED. He studied the electromagnetic interactions between mesons and photons, showing how "Dyson's program" could be extended to cover the new interactions. Once again, Dyson-style Feynman diagrams filled his analysis. Kinoshita teamed up with another young theorist from Tomonaga's seminar, Nambu Yoichiro, to push Dyson's methods and Feynman's diagrams into further studies of particles' interactions.[63] Kinoshita worked hard to master the intricacies of Dyson's calculational methods—immediately tackling Dyson's paper in order to present it to Tomonaga's seminar—and after several months of devoted attention, he became an early and prolific champion of the new techniques.

Soon theorists in Tokyo had more than just Kinoshita's example to guide them in their incorporation of Feynman diagrams and Dyson's associated techniques. On Oppenheimer's personal invitation, Tomonaga spent the academic year 1949–50 in residence at the Institute for Advanced Study. While in the United States, Tomonaga had the opportunity to learn firsthand about the fast-breaking developments. He heard Schwinger lecture several times, listened to Dyson, and learned in greater detail, month by month, some of the ins and outs of making calculations with Feynman diagrams—all in much greater detail than had been conveyed by Dyson's publications alone. While he was in Princeton, Tomonaga sent regular updates to his students and colleagues back in Japan, many of which were published in *Soken*.[64]

62. Kinoshita [C.7], 336.

63. Kinoshita [C.11], 480. Early reports on this research appeared as Kinoshita and Nambu [D.7]; and as Kinoshita and Nambu, "Electromagnetic properties" (1950). The two authors extended this research in Kinoshita and Nambu [C.14]. To this day, Kinoshita holds the record for having calculated the highest-order perturbative corrections to electromagnetic quantities: over the decade of the 1980s, he and his collaborator, W. B. Lindquist, calculated *eighth-order* corrections to an electron's magnetic moment, a calculation involving 891 distinct Feynman diagrams. For a recent review and update, see Hughes and Kinoshita, "Anomalous g values" (1999).

64. Although by the time Tomonaga arrived Dyson was no longer in residence at the Institute, making only occasional visits, several other members of Dyson's original postdoc cohort remained at the Institute during the year Tomonaga visited, including such devoted diagram users as Kenneth

Dispersing the Diagrams throughout Japan

Over time, Feynman diagrams became a steady fixture of the Tokyo theorists' calculations and publications. Between 1949 and 1954, ninety-two physicists working in Japan published a total of ninety-seven diagrammatic articles in *Progress of Theoretical Physics,* with an additional sixty-six diagrammatic preprints appearing in *Soken* between 1949 and 1952. Half these authors were young theorists trained in Tomonaga's Tokyo seminar; their contributions accounted for more than half of all the diagrammatic preprints and articles.[65] The shared research topic, motivation, and tight-knit study group—bolstered by Tomonaga's extended visit to the Institute for Advanced Study and his return to Tokyo in the summer of 1950—help explain why Tomonaga's group eventually mastered the new diagrammatic techniques and made such regular use of them. What about the scores of Japanese theorists working outside Tomonaga's circle? By 1954, articles that incorporated Feynman diagrams had been published by physicists working throughout Japan: all the way from Hokkaido, in the north, to Osaka, Kyoto, Nagoya, Kanazawa, Hiroshima, Wakayama, and beyond. Hence, we are faced with the same question as for physicists working in the United States and Great Britain: how did an esoteric technique, which had proved neither obvious nor easy to master upon first viewing, spread so quickly?

Like their counterparts in the United States and Great Britain, the physicists in Japan who began to use Feynman diagrams were almost exclusively young theorists. Their average time between obtaining undergraduate degrees and D.Sc. degrees was 10.6 years; this period corresponds roughly to American physicists' graduate school and postdoctoral stages. On average, the Japanese theorists began to use Feynman diagrams right in the middle of

Watson, Joseph Lepore, Jack Steinberger, Cécile Morette, and Robert Karplus. See also Tomonaga Sin-itiro to Koba Zirō, 18 Jan 1950, as reprinted in *Soken* 2, no. 1 (1950): 205–7; Tomonaga to Miyazima Tatsuoki, 11 Feb 1950, as reprinted in *Soken* 2, no. 2 (1950): 11–13; Tomonaga to Koba, 12 Feb 1950, as reprinted in *Soken* 2, no. 2 (1950): 3–4; Tomonaga to Kotani Masao, 10 Mar 1950, as reprinted in *Soken* 2, no. 2 (1950): 10–11; and Tomonaga to Kotani, 14 May 1950, as reprinted in *Soken* 2, no. 3 (1950): 205–6.

65. Many of the authors who published diagrammatic articles in *PTP* also wrote preprints in *Soken;* an additional twenty-five authors contributed diagrammatic preprints to *Soken* between 1949 and 1952 without publishing accompanying articles in *PTP. Soken* preprints have been collected only for the period 1949–52. By 1953, the earlier complaints about *PTP* that had driven many Japanese theorists to publish preprints in *Soken*—paper shortages and frustrating publication delays—had largely subsided.

their postundergraduate training: 4.4 years after obtaining their undergraduate degrees, just over 6 years before obtaining their D.Sc. Only three authors in Japan published their first diagrammatic articles after obtaining their D.Sc.'s (Miyazima Tatsuoki, Husimi Kōdi, and Nakabayasi Kugao), and two more wrote their first diagrammatic articles in the same year they obtained their D.Sc.'s (Taketani Mitsuo and Utiyama Ryōyū). All five of these physicists began using Feynman diagrams while collaborating with younger diagram users from Tokyo, or with colleagues who had been tutored by the Tokyo group. All the rest first picked up the diagrams as young theorists, still in the midst of their training. As in the United States and Great Britain, the new techniques required extended practice, and older physicists simply did not retool.[66]

Consider next where the diagrams spread, and by what means. Nearly every diagrammatic article published in Japan between 1949 and 1954 can be traced back to extended personal communication with members of Tomonaga's circle. Three novel institutional rearrangements, all taking shape between 1948 and 1950—just when Feynman diagrams were introduced—helped put the diagrams into broad circulation throughout Japan. First was the founding of the Elementary Particle Theory Group in 1948, together with the group's newsletter, *Soken*. The meetings facilitated collaborations extending between members of different universities. Tokyo theorists began to coauthor diagrammatic articles with physicists at several other universities, including Osaka, Hokkaido, Shiga, Kyushu, and Wakayama universities; in each case, these articles were the first from physicists in any of these universities to make use of the diagrams. Acknowledgments in many other diagrammatic articles, written by theorists not working in Tokyo, further noted help received from the Tokyo group.[67]

The earliest diagrammatic article from beyond Tomonaga's orbit, for example, came from a trio of young theorists working at Tohoku University in

66. Information on undergraduate and D.Sc. degrees comes from several sources: Kaneseki, "Elementary particle theory group" (1974 [1950]), 222–29; *Nihon Hakushi Roku* (1985); *Gakushikai Shimeiroku* (1940–); and *Nihon Butsuri Gakkai Meibo* (1956, 1963). S.B. information has been collected for 86% of the authors in Japan, and D.Sc. information for 79% of them. I am grateful to Kenji Ito for helping to collect this information.

67. See, e.g., Nakano et al. [C.18]; Matsumoto, Fujinaga, and Watari [D.40]; and Tokuoka and Tanaka [C.60]. One year after Watari Wataro collaborated with Matsumoto Masahiko, Matsumoto's young colleague at Shiga University, Shintomi Taro, submitted a diagrammatic article: Shintomi [D.65]. See also the acknowledgments in Ozaki [C.10]; Ogawa, Yamada, and Nagahara [C.22]; Koba, Mugibayashi, and Nakai [C.26]; Ōneda [D.36], [C.32]; Koba, Kotani, and Nakai [D.9], [D.15], [C.35]; Utiyama, Sunakawa, and Imamura [C.46]; Takahashi and Umezawa [C.56]; Minami et al. [C.58]; and Iso [D.61]. Beginning in 1952 these interuniversity exchanges became even more solidly institutionalized, with the foundation of several new centers. See Konuma, "Interuniversity institutes" (1988).

Figure 4.9. Feynman diagrams for meson scattering. (Sources: *left,* Fukuda and Miyamoto [C.2], 149; *right,* Ōneda, Sasaki, and Ozaki [C.4], 169.)

the city of Sendai, two hundred miles north of Tokyo. They had been working in relative isolation on the same problem as the Tokyo theorists Fukuda Hiroshi, Miyamoto Yoneji, and Hayakawa Satio and learned of the Tokyo efforts only through the Elementary Particle Theory Group. The two groups immediately began to talk regularly. Only at this point, having already published four diagram-free installments of their project in *Soken* and another two in *Progress of Theoretical Physics,* did the Tohoku group turn to Feynman diagrams. When it did, it used them in precisely the same way that Fukuda, Hayakawa, and Miyamoto had done: thanking the Tokyo theorists "for their discussions and advices," the Tohoku group included the same five Feynman diagrams as after-the-fact illustrations of quantities it had already calculated without the aid of any diagrams. Indeed, comparing the pictorial form of their diagrams with those of the Tokyo group reveals the extent of Tomonaga's group's influence. (See fig. 4.9.) Two months later, one of the Tohoku theorists, Ozaki Shoji, followed up with a longer article on the topic, reproducing the same five Feynman diagrams (again included as post hoc illustrations of calculations he had already written out in full). As before, he thanked Fukuda, Hayakawa, and Miyamoto, "who gave me frequently advices and promoted the work." By August 1950, two more Tohoku theorists had begun making some use of the diagrams.[68]

A second new arrangement helped to circulate the diagrams. The Elementary Particle Theory Group, the Japanese Ministry of Education, and some private companies introduced new fellowships to allow young physicists to spend time at different universities within Japan.[69] Umezawa Hiroomi, for example, completed his undergraduate degree at Nagoya in 1947, working with Sakata Shōichi. He remained in Nagoya, where he quickly assumed prominence in Sakata's group, mastering the details of Sakata's *C*-meson hypothesis.

68. Ōneda, "Personal recollections" (1988), 15–17; Fukuda and Miyamoto [C.2]; Fukuda, Hayakawa, and Miyamoto [C.9]; Ōneda, Sasaki, and Ozaki [C.4], 176; Ozaki [C.10], 394; Nakabayasi and Sato [D.24], [C.23], and [D.33].

69. Ōneda, "Personal recollections" (1988), 17; Kamefuchi, "Divergence problem" (1988), 128.

He became a prolific author and advised several younger members of Sakata's group, obtaining his D.Sc. from Nagoya in 1952. Yet he began to make use of Feynman diagrams only in December 1950: he won a traveling fellowship from the Chubu-Nippon Press (based in Nagoya), which allowed him to spend the autumn 1950 semester in Tokyo. Since his parents lived in Tokyo, he was able to keep his costs low, so he used his fellowship to bring a younger Nagoya student, Kamefuchi Susumu, with him. Together they attended "the famous Tomonaga Seminar that was taking place at a shabby building" in the city, as Kamefuchi later recalled. They worked with Tomonaga's group for several months, in other words, immediately after Tomonaga had returned from the Institute for Advanced Study. At the end of the semester, Umezawa and Kamefuchi returned home to Nagoya. By December 1950, Umezawa had added Feynman diagrams to his arsenal of calculating techniques, making steady use of the diagrams for the remainder of his career. Kamefuchi likewise turned to the diagrams once he began to publish some of his early research in *Soken* and *Progress of Theoretical Physics.* Only after Umezawa returned to Nagoya and resumed his advising duties for many of Sakata's students did Feynman diagrams become a regular fixture of the Nagoya group's publications. By May 1952, Umezawa's influence had extended to neighboring universities: Goto Shigeo, working at nearby Gifu University, thanked Umezawa for "valuable discussions" in Goto's first diagrammatic article, the first such article to appear from anyone at Gifu.[70]

The third institutional feature driving the diagrams' dispersion was the 1949 decree from the occupation authority's General Headquarters to expand the Japanese university system. With the creation of so many new universities came new jobs for young physicists. The Tohoku theorists Ozaki Shoji and Ōneda Sadao, who had first learned about the diagrams in 1949 from Tokyo's Fukuda, Miyamoto, and Hayakawa, moved to the newly created Kanazawa University during the summer of 1950 and there began to train new recruits;

70. Kamefuchi, "Quantum field theory" (2001), 10; Kamefuchi, "Divergence problem" (1988), 128. Diagrammatic articles from Nagoya theorists that were submitted after Umezawa's return include [C.20], [C.21], [C.22], [C.28], [C.55], [C.56], [C.58], [C.59], [D.29], [D.32], [D.44], [D.53], and [D.60]. Kamefuchi thanked Umezawa for help in his early diagrammatic articles: Kamefuchi [D.44], [C.55]; Goto Shigeo thanked Umezawa in his first diagrammatic article: Goto [C.50]. Sakata's colleague Taketani Mitsuo maintained contact with the Nagoya group in 1949–51, when he first began to use Feynman diagrams. Although he lived mostly in Tokyo at the time, his official institutional affiliation remained Nagoya until the summer of 1951, and he traveled to visit Sakata's group fairly regularly during this period. Most likely, Taketani helped another young Nagoya theorist, Yamada Eiji, to learn about the new diagrammatic techniques. Yamada published an early diagrammatic report in *Soken* in Jan 1950 [D.10], but no other papers from Nagoya made any use of Feynman diagrams until Umezawa returned from his study in Tokyo.

Takeda Gyō moved from Tokyo to Kobe University, where he continued to publish diagrammatic works.[71] But by far the most important shift came when young theorists moved from Tomonaga's Tokyo group to Osaka. Osaka City University became one of the earliest of the new universities built upon the General Headquarters plan, and it quickly became home to several of the Tokyo group's most prominent and prolific diagram enthusiasts, including Nambu Yoichiro, Koba Zirō, Hayakawa Satio, Nishijima Kazuhiko, and Yamaguchi Yoshio. Fresh from their work with Tomonaga, these theorists submitted twelve diagrammatic articles to *Progress of Theoretical Physics* and fourteen additional reports to *Soken* by the end of 1952. With their help, eleven other young theorists at Osaka University and Osaka City University began to publish diagrammatic articles, beginning in May 1951.[72] Thanks to the Tokyo transplants, nearly 30% of all the articles in *Progress of Theoretical Physics* and *Soken* making use of Feynman diagrams between 1949 and 1954 appeared with Osaka bylines. The Tokyo-turned-Osaka theorists likewise began to help theorists in neighboring Kyoto get up to speed with the new techniques.[73]

In the early 1950s, the circulation of young theorists within Japan was supplemented by a series of exchanges between Japan and the United States. Younger theorists began to make the trip that only their elders Yukawa and Tomonaga had made before. Hayakawa Satio traveled to MIT and Cornell during 1950–51, sending regular updates to *Soken;* Nambu and Kinoshita both

71. Ōneda, "Personal recollections" (1988), 17; Hori [D.37], [D.49], [C.49]. Several months before Hori Shoichi submitted his first diagrammatic paper, Ōneda acknowledged working with him (in both [C.32] and in [D.36]); Hori thanked both Ozaki (in [C.49]) and Ōneda (in [D.49]). On the founding of Kanazawa University, see http://www.kanazawa-u.ac.jp/history/. Takeda's diagrammatic papers from Kobe include Takeda [D.45], [D.46], [C.47], and [D.51].

72. See [C.6], [C.9], [C.12], [C.14], [C.18], [C.26], [C.27], [C.29], [C.30], [C.31], [C.34], [C.35], [C.39], [C.41], [C.42], [C.46], [C.53], [C.54], [C.58], [D.7], [D.9], [D.11], [D.12], [D.15], [D.19], [D.20], [D.21], [D.26], [D.30], [D.31], [D.39], [D.42], [D.47], [D.50], and [D.54].

73. As early as Dec 1949, several theorists at Tokyo Bunrika Daigaku were in touch with Kyoto's Araki Gentaro, as they put the finishing touches on their diagrammatic article; two months later, Nambu Yoichiro, recently arrived in Osaka from Tokyo, also exchanged counsel with Araki as Nambu completed his third diagrammatic article. That spring, Araki helped two younger Kyoto theorists, Enatsu Hiroshi and Ataka Yasushi, use the diagrams in a short report. Baba et al. [C.3]; Nambu [C.6]; Enatsu, Ataka, and Takano [D.23]. During spring 1950, another Kyoto theorist, Katayama Yasuhisa, turned to Koba Zirō, one of the most active of the Tokyo theorists to arrive in Osaka, for help applying the new diagrammatic techniques; Katayama also discussed the new types of calculation with Tohoku's Ozaki Shoji, who had begun learning about Feynman diagrams from the Tokyo group several months earlier. Katayama [D.16]; Ozaki [C.10]. Kita Hideji and Munakata Yasuo, both in Kyoto, learned about the diagrams from Nambu. Kita and Munakata [D.25].

took up residence at the Institute for Advanced Study in 1952. The trips were revealing for both travelers and hosts: writing soon after his arrival in Boston, Hayakawa observed to a Japanese colleague that "Schwinger is just Tomonaga's copy. Only he is strong. After all, Japan is the best in field theory, and in the US, only Feynman can match Japan."[74] The following year, physicists from both countries had a chance to make such evaluations on their own, when Kyoto hosted Japan's first international physics conference since the end of the war. Feynman, Murray Gell-Mann, Robert Marshak, Abraham Pais, and dozens of their diagram-wielding colleagues from the United States traveled to Kyoto for the meeting, further solidifying what had already become a robust community of diagram users.[75]

FEYNMAN DIAGRAMS IN THE SOVIET UNION

Timing Is Everything

Just as in Japan, geopolitics shaped the dispersion of Feynman diagrams in the Soviet Union.[76] With the Soviet Union, the exigencies of world affairs cut at the same time, but the other way: immediately before Feynman diagrams were introduced, the cold war made it impossible for physicists in the United States and the Soviet Union to exchange ideas in person or to discuss techniques informally. The separation of physicists in the United States and the Soviet Union was not new with the coming of the cold war. Soviet physicists'

74. Hayakawa Satio to Kobayashi Minoru, 2 Sep 1950, as reprinted in *Soken* 2 (Oct 1950): 165–67. Hayakawa wrote many other letters to Japanese colleagues from both MIT and Cornell, which were reprinted in *Soken* 2 (29 Aug 1950): 178–87; *Soken* 2 (Oct 1950): 158–60, 165–70; and *Soken* 3 (5 Feb 1950): 211–19. See also Nambu, "Personal recollections" (1988); Kinoshita, "Personal recollections" (1988); and the panel discussion reprinted in Brown et al., *Elementary Particle Theory in Japan* (1988), 41. Returning from his stay in the United States, Hayakawa spent time at the University of Delhi in India before settling back in Japan; there he helped R. C. Majumdar and several students learn how to use the new diagrams. The following year Majumdar and a student submitted a diagrammatic article to *PTP:* Majumdar and Mitra [C.57]. See also M. K. Sundaresan, letter to the author, 3 Jan 2001.

75. Konuma, "Interuniversity institutes" (1988), 25. See also the correspondence in *HAB,* Folder 10:49; and the eleven-page memorandum circulated to American physicists before their departure for the Kyoto meeting entitled "What will be expected by the Japanese of their American visitors?" circulated by Harry Kelly and John Wheeler on 28 July 1953, a copy of which may be found in *REM,* "IUPAP," microfiche 122.

76. Unless otherwise noted, all translations and transliterations from Russian in this section were done by Karl Hall, to whom I am deeply indebted.

isolation, combined with occasional suspicion by Party functionaries of Western scientists and their theories, had been in force since the mid-1930s. Signs of the new isolation were clear as early as 1933, when the state imposed travel bans on the influential physicist Abram Ioffe, and reached fever pitch by 1937 with the Stalinist purges: "bourgeois" or "idealist" influences in physics were to be shunned, the authorities made clear, and genuinely "Soviet" theory was to be put in its place.[77] Soviet physicists' hopes that a new era of international scientific exchanges might be opening had been raised at the end of World War II. An international meeting was held in 1945 to celebrate the 220th anniversary of the Soviet (formerly Imperial) Academy of Sciences, with wide, positive news coverage throughout the West, and a similar meeting was planned for 1947. Yet the latter meeting was canceled by Stalin at the last minute: by that time the cold war had hardened, and renewed calls went out from Party bureaucrats and Academy philosophers to purge physics of its overdependence on Western sources. Several leading Soviet theorists were publicly upbraided for "groveling before the West," for "uncritically receiving Western physical theories and propagandizing them in our country," and for producing insufficiently Soviet textbooks. A meeting had been planned for mid-March 1949—just one month after Dyson's first article on Feynman diagrams had appeared in the *Physical Review*—to carry these denunciations further and to "clean house" in Soviet physics.[78]

The isolation that was so harshly reimposed upon Soviet physicists by their government in the late 1940s deprived them of any chance for informal, face-to-face exchanges with physicists on the other side of the iron curtain. Because of this postwar isolation, Soviet theorists did not pick up and use Feynman diagrams in anything like the ways their colleagues in the United States, Great Britain, and Japan did. Whereas young theorists (beyond Feynman and Dyson themselves) in these other countries had begun using the new techniques in print as early as autumn 1949, the first published uses of Feynman diagrams in the Soviet Union did not appear until 1952. Only eleven authors put the diagrams to use in the main Soviet journal between 1952 and 1954, publishing a total of twelve articles—a far cry from the exponentially rising flood of diagrammatic articles filling such journals as the *Physical Review.*

77. Holloway, *Stalin and the Bomb* (1994), 26–8; Hall, *Purely Practical Revolutionaries* (1999), esp. chaps. 4, 9, and 10.

78. Holloway, *Stalin and the Bomb* (1994), 206–13; Hall, *Purely Practical Revolutionaries* (1999), 705–14; and Kojevnikov, "Stalin's Academy" (1996), 43–48. Soviet biologists similarly suffered from the new restrictions on foreign travel and exchange with Western colleagues: Krementsov, *Cure* (2002), chaps. 4–5.

One might suspect that difficulties in receiving copies of the *Physical Review* led to Feynman diagrams' delayed appearance within Soviet publications.[79] Yet difficulties in receiving the *Physical Review* cannot be the whole story. In fact, officers of the American Institute of Physics made special efforts to distribute copies of the *Physical Review* "behind the iron curtain," as their internal memoranda reveal, thinking that their flagship journal would "have important propaganda value." Official copies mingled with cheap pirated editions: before the Soviet government signed international copyright statutes, they regularly made photo-offset copies of American journals, producing what young physicists called the "grey *Phys. Rev.*" Soviet articles in the *Zhurnal eksperimental'noi i teoreticheskoi fiziki* (*Journal of Experimental and Theoretical Physics*) often cited the most up-to-date publications from the *Physical Review*.[80] What was lacking in the Soviet case was not written texts or formal publications, but the ancillary institutional support for informal, pedagogical exchange between physicists already "in the know" about the new tools and those who had not yet learned how to use them.

H-Bomb Assignment

Several physicists and historians have suggested that Stalin canceled the March 1949 meeting, planned to denounce "foreign influences" in Soviet physics, out of fears that it would disrupt full-scale work on the atomic bomb project. The incident, according to this interpretation, was but one aspect of the more general relationship between physicists and the Soviet state at this time—namely, that nuclear weapons work spared Soviet physics from the kinds of Party meddling that so famously wrecked Soviet biology at the same time, in the form of Lysenkoism.[81] Only when it came to weapons work did the Party

79. Boris Ioffe discusses such delays in "Landau's theoretical minimum" (2002). My thanks to Professor James Bjorken for bringing Ioffe's paper to my attention.

80. Samuel Goudsmit to Karl K. Darrow, 3 Jan 1951, in *SAG*, Folder 51:51; see also E. M. Webster to Eilen Neuberger (publications manager at the American Institute of Physics), 12 Mar 1953, in the same folder. On the "grey *Phys. Rev.*," see Meinhard Mayer, "Reminiscences" (1990), 34. An article submitted to *ZhETF* on 5 June 1950 included citations to several articles in the *Physical Review* from 1949 and 1950, including one article that had been published in the 15 May 1950 issue; a note added in proof to this same *ZhETF* article cited an article from the 15 August 1950 issue of the *Physical Review*. See Baldin and Mikhailov, "Obrazovanie" (1950), 1063. Similarly, a *ZhETF* article that was submitted on 29 Aug 1951 cited a *Physical Review* article that had been published in the 1 June 1951 issue: Galanin [E.2], 470.

81. Linde, "Physics fostered freedom" (1992); Holloway, *Stalin and the Bomb* (1994), 210–11. Alexei Kojevnikov cites additional Russian sources that draw the same conclusion in Kojevnikov, "Stalin's

functionaries tolerate (indeed, encourage) a heavy reliance upon "Western bourgeois cosmopolitan" physics: with Stalin's approval, Lavrentii Beria, the official Party overseer of the atomic bomb work and the much-feared former director of the secret police, spurred the scientific director, Igor Kurchatov, and others to rely upon espionage when building their first nuclear weapons. "Copying" American physics in this particular realm was beyond rebuke, even by the most shrill Party ideologues.[82] Physicists working at the secret weapons laboratory, code-named Arzamas-16 (and nicknamed "Los Arzamas"), had assured access to the latest Western journals, including the *Physical Review* as well as nontechnical magazines such as the *Bulletin of the Atomic Scientists.*[83]

Deeply ensconced in the secret weapons program, Feynman diagrams made their initial, if halting, entry into Soviet physics. A small team of young theorists, including Vladimir Berestetskii, Aleksei Galanin, Boris Ioffe, and Alexei Rudik—all protégés of Isaak Pomeranchuk in Moscow—were assigned to work on a specific task for the H-bomb project in the summer of 1950. In order to predict whether certain H-bomb designs could work in principle, the Soviet theorists needed to know how to calculate the flow of energy that would be carried away from the fusion reaction region by out-flying radiation. The escaping radiation would carry energy away from the interaction region, thereby cooling it down; the question was whether the radiation would carry energy away too quickly, robbing the thermonuclear reaction of its required heat and causing the bomb to fizzle. Unlike details of the fusion reaction itself, this part of the calculation depended only on knowing how to calculate the cross section for Compton scattering, that is, the scattering between high-energy photons (the emitted radiation) and electrons (in the material surrounding the

Academy" (1996), 46n73. Both Kojevnikov and Karl Hall caution against framing the analysis too narrowly around a monolithic Party versus a homogenous, passive physics community. In particular, Kojevnikov notes that no documentary evidence exists linking the cancellation of the Mar 1949 conference with concerns about the nuclear weapons project: Kojevnikov, "Stalin's Academy" (1996), 46–47; and Hall, *Purely Practical Revolutionaries* (1999), 697–714.

82. Norris and Arkin, "Russian/Soviet weapons" (1993); Khariton and Smirnov, "Khariton version" (1993); Sagdeev, "Russian scientists" (1993); Leskov, "Dividing the glory" (1993); Holloway, *Stalin and the Bomb* (1994), 90–97, 102–8, 296–97, 310–12; Khariton, Adamskii, and Smirnov, "Way it was" (1996); Holloway, "New light" (1996); Reed and Kramish, "Trinity at Dubna" (1996); Goncharov, "Thermonuclear milestones" (1996); and Ioffe, "Top secret" (2001). Yuli Khariton, one of the leading scientists in the Soviet nuclear weapons project, insists that espionage was used only to help build the Soviets' first fission weapon—detonated in August 1949 and code-named "Joe I" in the West—but that Soviet work on later fission and fusion weapons was conducted without any significant help from Western sources.

83. Gorelik, "Sakharov" (1999), 99.

fusion-reaction region). This was a problem in electrodynamics, not nuclear physics. Yakov Zel'dovich, famous for his rough-and-ready, back-of-the-envelope approach to calculation, had given a rough estimate for this radiation scattering, which was trustworthy only to within a factor of two or so. Yet the bomb designers knew that this crucial quantity had to be calculated to a much higher accuracy—with an uncertainty of a few percent, at most—since the entire question of whether the H-bomb design would work hung on this delicate balancing of the energy. So Pomeranchuk's young charges were instructed to find some way to make this calculation in a more precise way.[84]

Pomeranchuk had noticed the recently published articles on the new approach to QED by Schwinger, Feynman, and Dyson but had not yet mastered the new techniques. He pointed his disciples to these works, and they scoured the articles intensely. One of the worked examples that Feynman included as an appendix to his long article involved using the diagrams to calculate radiative corrections to Compton scattering, corrections that would enter at the percent level. This was precisely the calculation and the level of accuracy that Berestetskii, Galanin, Ioffe, and Rudik now needed to master, and in a hurry. A few weeks into their work, they learned to their surprise that Rudik had not been cleared for the top-secret work. The remaining three theorists threw themselves into trying to make sense of the new calculating techniques, based on Feynman's and Dyson's articles alone. After more than a year of full-time effort, they succeeded. They worked so hard to try to understand the ins and outs of the articles, in fact, that Galanin and Ioffe even made their own translations of the articles into Russian, hoping that the line-by-line scrutiny of the articles required for translation would pay dividends in their mastery of the diagrammatic tools. Berestetskii's close study of the articles similarly bore fruit in a lengthy review article published in 1952.[85] With time and intense effort, in

84. Ioffe, "Top secret" (2001), 25–28. On the importance of understanding the radiation reaction in the early Soviet H-bomb work, see also Gorelik, "Sakharov" (1999), 98–99. On Pomeranchuk, see Josephson, *Red Atom* (2000), 216–17.

85. Ioffe, "Top secret" (2001), 25–28. Ioffe remembers that he had already begun trying to understand Schwinger's, Feynman's, and Dyson's papers on his own during the spring of 1950, but that it had been difficult, since "At that time nobody in Moscow was proficient in these new QED methods." Ioffe, "Landau's theoretical minimum" [2002], 9. Feynman worked through lowest-order corrections to Compton scattering in Feynman [A.4], 787–89. Galanin's translation of Feynman's "Theory of positrons" (1949) and [A.4] appeared in a 1951 pamphlet along with Ioffe's translation of Dyson [A.2], in *Problemy sovremennoi fiziki: Sborniki sokrashchennykh perevodov i referatov inostrannoi literatury* [Problems of Modern Physics: Collections of Abbreviated Translations and Abstracts of Foreign Literature], series 3, issue 11 (1951); Berestetskii's lengthy review appeared as Berestetskii, "Teoriia" (1952). In a 1955 pamphlet in the *Problemy sovremennoi fiziki* series, Galanin,

isolation, these three theorists learned how to put Feynman diagrams to work in the context of their assigned task. Just when leading Soviet theorists were being lambasted in public for their narrow dependence upon Western physics, Pomeranchuk's students, under cover of the high-priority H-bomb program, smuggled in the new techniques and put them to work.

Cool Reception in Landau's Seminar

The details of Ioffe's, Galanin's, and Berestetskii's calculations, along with the specific motivation for undertaking them, were "born secret," hardly the stuff they could discuss openly with colleagues in the Soviet Union, let alone with colleagues outside the country. Pomeranchuk suggested that Ioffe use his calculation of the radiative corrections to Compton scattering from his H-bomb work for his Ph.D. dissertation. His defense was held at Igor Kurchatov's institute at the end of 1953, since that was the only institution in which such secret materials could be discussed. Ioffe recently recalled a "curious episode" that occurred during his thesis defense: "The dissertation was 4-letter secret ('Top Secret, Special Folder, t.s.s.f.'), although it was about a purely theoretical problem, solved a model problem, and never mentioned the 'Tube,'" the H-bomb design for which the calculations had been carried out. At the end of the defense, one of the Scientific Council members proclaimed, "'I have no doubt that the dissertation is all right, but I didn't understand just one point: Why is it so secret?' The chairman L. A. Artsimovich answered, 'It's very good that you didn't understand it.'"[86]

His dissertation might have been labeled "top secret," but Ioffe, along with Galanin and Berestetskii, knew that the new diagrammatic techniques they had worked so hard to acquire had utility and interest far beyond this specific miliary application. As they were wrapping up their H-bomb calculations, Galanin published three long articles that made use of the new techniques (without the original motivation for practicing how to use them), which he submitted to the *Zhurnal* beginning in late August 1951. His first three installments—the earliest publications in the country to make any use of the diagrams—presented many of the fruits of his group's close readings of Feynman's and (especially) Dyson's articles. Galanin's first paper—the longest of the three—tackled the exact same

Ioffe, and Berestetskii published translations of Feynman and Schwinger articles from 1951, including Schwinger's famous "Green's functions" (1951) papers. My thanks to Karl Hall for bringing these publications to my attention, and for providing a translation of the Berestetskii review article.

86. Ioffe, "Top secret" (2001), 28–29.

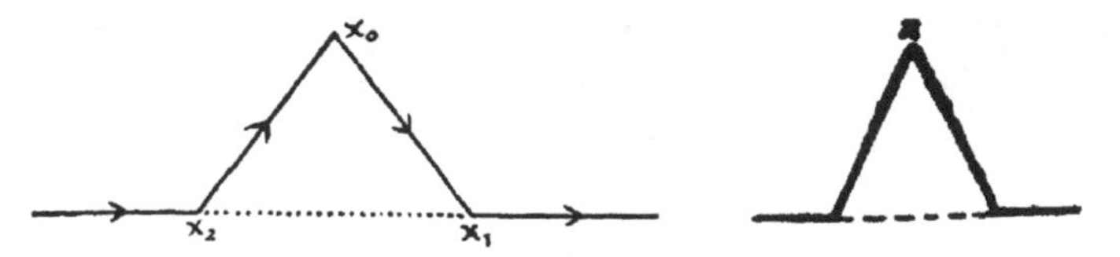

Figure 4.10. Dyson's and Galanin's first Feynman diagrams. (Sources: *left,* Dyson [A.1], 501; *right,* Galanin [E.1], 458.)

calculation that Dyson himself had used for introducing the diagrams in his first article: corrections to an electron's behavior in an external electromagnetic field. Galanin followed nearly all of Dyson's prescriptions carefully, including how to draw the relevant Feynman diagrams.[87] (See fig. 4.10.) Galanin's second paper applied the new formalism to the very problem that he, Ioffe, and Berestetskii had calculated for their H-bomb work: radiative corrections to Compton scattering. His third paper in this early series, submitted one year later, drew heavily upon recent work published in the *Physical Review* by both Murray Gell-Mann and Francis Low and by Hans Bethe and Edwin Salpeter, applying Dyson's formalism and Feynman's diagrams to problems involving bound states.[88]

Note Galanin's selective reading of Dyson's difficult papers: in none of his three articles did Galanin discuss Dyson's generalization of how to remove the infinities from any given order of approximation (the subject that occupied nearly all of Dyson's second article and that Dyson considered his most important contribution). Instead, during his months-long scrutiny of Dyson's and Feynman's published work, Galanin homed in on a select few of their worked examples, building his diagrammatic arsenal by focusing on those particular calculations that were most crucial to the secret H-bomb work. Galanin and his compatriots needed to calculate real numbers to high accuracy for specific interactions between electrons and photons; from the welter of details in Feynman's and Dyson's publications, Galanin picked out and developed those elements of most immediate concern.

87. Galanin [E.1]. Galanin made one alteration to Dyson's prescriptions: whereas Dyson, following Schwinger and Tomonaga, first used a contact transformation on the quantum fields to go into the interaction representation, Galanin remained in the Heisenberg representation. Galanin made some fleeting remarks in his article that by skipping the transformation to the interaction representation, his perturbative expansion could be trusted even if the full, infinite series of perturbative terms did not converge, although this was not demonstrated. Other than this one change, he adopted Dyson's approach very closely. Galanin published a brief summary of his work in Galanin, "Radiatsionnye popravki" (1951).

88. Galanin [E.2], [E.4]. Cf. Gell-Mann and Low [A.43]; and Salpeter and Bethe [A.47].

Galanin, just as surely as Dyson, knew that publishing formal research articles was only one way to spread new ideas and methods, and by no means the most efficient way. At the time, there was one prominent place where Soviet theorists (young and old) gathered to learn the latest developments in their field: Lev Landau's weekly seminar, held every Thursday afternoon in the Moscow Institute for Physical Problems. Landau's seminar, much like Kemmer's group in Cambridge and Tomonaga's in Tokyo, featured intense discussions in which young theorists took turns presenting the latest research from the international physics journals or, occasionally, presenting some of their own work. Landau (or simply "Dau" to his friends and students) was famous for the breadth of his knowledge within physics—a breadth he sought to encourage in his students as well, with a rigorous series of examinations on various topics that became known as Landau's "Theoretical Minimum." Landau coauthored a world-renowned series of textbooks, known today as simply the "Landau-Lifshitz" series, built in the same spirit as his "Theoretical Minimum": students of his multivolume *Course of Theoretical Physics* were to practice deploying certain calculational methods, such as Lagrangians and order parameters, across the wide spectrum of topics throughout physics. In the late 1940s and early 1950s, with the ranks of his graduate students growing and his textbooks entering into wider and wider circulation, Landau was at the height of his influence.[89]

Landau kept tight control over what material would be fit for presentation in his weekly seminar. He also developed a reputation, rather similar to Oppenheimer's, for interrupting speakers with cutting criticism whenever he thought that their presentations had wandered into "pathology"—Landau's favorite term for physics that was not necessarily incorrect, but rather stale or overly pedantic. One of the longtime members of Landau's seminar compared it to a "Cossack army": "There was a sense of battle between [the speaker] and Landau, naturally of much interest to all those present, who were always very numerous, including staff of the Moscow and Dubna institutes as well as visitors from Leningrad, Kharkov, Kiev, and Novosibirsk."[90] The loyalty

89. See especially Hall, *Purely Practical Revolutionaries* (1999), chaps. 6–13; and Hall, "Short course" (2005).

90. Akhiezer, "Teacher and friend" (1989), 51–52. Landau's Moscow seminar became legendary even in its day and has been described by many of his former students and colleagues. See the essays in Khalatnikov, *Landau* (1989), some of which also appeared in abridged form in *Physics Today:* Khalatnikov, "Reminiscences" (1989); Ginzburg, "Landau's attitude" (1989); and Akhiezer, "Recollections" (1994). See also Ioffe, "Landau's theoretical minimum" (2002); and Hall, *Purely Practical Revolutionaries* (1999), chap. 13.

Figure 4.11. "Dau said...," by A. A. Yuzefovich. (Source: Khalatnikov, *Landau* [1989], following 208.)

of Landau's students to their fabled leader and his dominance of the Moscow community of theoretical physicists were captured in the cartoon in figure 4.11.

One might have expected Landau to welcome the new diagrams and feature them in his influential seminar upon hearing about them from Galanin, Ioffe, and Berestetskii. Landau was no stranger to QED, having cut his teeth on a notoriously difficult problem—how accurately electric and magnetic fields can be measured, subject to the uncertainty principle—with Rudolf Peierls in 1930. Not only had he worked on QED himself, but also for over a decade he had directed his students to master the means of making perturbative calculations as well.[91] Reinforcing the interest in these assignments, Landau's group, and many Soviet theorists in addition, learned about the new experimental developments in QED, such as the Lamb shift and measurements of the electron's anomalous magnetic moment, from a lengthy review article by Iakov Smorodinskii published in 1949. Although Smorodinskii's article had been written too early to include any news of Feynman's or Dyson's diagrammatic work, it did highlight Hans Bethe's nonrelativistic, approximate calculation of the Lamb shift, as well as Schwinger's and Tomonaga's early work on renormalization. Beyond Smorodinskii's review, forty-two articles on QED appeared in the Soviet *Zhurnal* between 1946 and 1954 (accounting for nearly 30% of the journal's output in theoretical high-energy physics), while over one hundred other articles appeared on closely allied topics such as relativistic quantum

91. Landau and Peierls, "Erweiterung des Unbestimmtheitsprinzips" (1931); Hall, *Purely Practical Revolutionaries* (1999), 249–83. See also Berestetskii and Landau, "O vzaimodeistvii mezhdu elektronom i pozitronom" (1949); Akhiezer, "Recollections" (1994), 36–37; Akhiezer, "Teacher and friend" (1989), 44–46; and Ioffe, "Landau's theoretical minimum" (2002), 7–8.

mechanics, quantum field theory, and field-theoretic studies of nuclear forces.[92] QED was definitely on Soviet theorists' agendas at the time.

Yet instead of embracing the new diagrams and renormalization methods, Landau shunned them. Ioffe, Galanin, and Berestetskii, along with their teacher Pomeranchuk, tried several times to interest Landau in their newly discovered diagrammatic techniques. Each time they were rebuffed by Landau, who found nothing sufficiently of interest in their hard-won squiggles. The little he heard about Feynman's and Dyson's techniques for removing the divergences from QED struck Landau—as they struck many other members of the European theoretical physics elite at the time—as little more than temporary trickery. A deeper conceptual overhaul was needed to place quantum field theory on a solid footing, Landau maintained; there was little sense fiddling with Feynman diagrams in the meantime.[93] As Ioffe recalled, "Two attempts to present Feynman's papers at Landau's seminar failed: the speakers were thrown off the podium after 20 or 30 minutes of talking. Only the third attempt succeeded (if I remember correctly, this was in 1951 or even in 1952). But still he had no interest in these problems."[94] Landau's lack of interest in the diagrams did not dissipate quickly. As late as autumn 1954—over three years after Galanin submitted his first diagrammatic articles to the *Zhurnal*—Landau barked at a young graduate student that it would be "immoral" to chase such "fashions" as Feynman diagrams.[95]

After continuing to fail in their efforts to interest Landau in the new techniques, Pomeranchuk and his students decided to start their own weekly seminar. It was held in the same place and on the same day as Landau's seminar, starting two hours before Landau's larger seminar began. Ioffe became the secretary of the group and gave the first presentation in the group's inaugural meeting on 1 October 1951, lecturing on Dyson's articles. Landau avoided the new seminar, teasing its young organizers for wasting their time on such matters. Despite Landau's disapproval, Pomeranchuk's seminar began to attract more and more participants, and slowly news of the diagrammatic techniques began to spread.[96]

92. Smorodinskii, "Smeshchenie" (1949). Although Smorodinskii's review had been written too early to include any discussion of Feynman diagrams, it did review Feynman's covariant cutoff procedure, first published in 1948. My thanks to Karl Hall for tallying the contents of *ZhETF* during this period.

93. See Brown and Rechenberg, "Landau's work" (1990), 67–68, 73–74. Cf. Kragh, *Dirac* (1990), 183–88; Schweber, *QED* (1994), 595–605; and Cao, *Conceptual Developments* (1997), 203–4, 214–17.

94. Ioffe, "Landau's theoretical minimum" (2002), 10–11.

95. Dzyaloshinskii, "Landau through a pupil's eyes" (1989), 90.

96. Ioffe, "Landau's theoretical minimum" (2002), 11–12.

Among the first to pick up the diagrams were two of Landau's young protégés, Alexei Abrikosov and Isaak Khalatnikov. Most likely, they learned about the diagrams in Pomeranchuk's seminar; the only people thanked in their first diagrammatic article, published in 1954, were Pomeranchuk, Galanin, and Ioffe. Whereas Ioffe and Galanin had met only stubborn resistance from their elder, Landau, it was perhaps easier to talk freely with their peers among the younger guard.[97] Abrikosov, in turn, was more experienced than many young theorists in the strategies of how to present new ideas to the master, and he became one of the first people to convince Landau to pay more attention to Feynman diagrams. Landau finally relented and began to learn something about the diagrams during the winter of 1953–54, with the aid of Abrikosov, Khalatnikov, Galanin, and Ioffe. Landau attended the weekly seminars by the experimentalists at the Institute for Theoretical and Experimental Physics (ITEP) in Moscow, after which he began to meet for several hours with Galanin and Ioffe in their ITEP office. They patiently explained how to calculate radiative corrections within QED with the aid of the diagrams. Then Landau would join his students Abrikosov and Khalatnikov back at his home institution, the Moscow Institute for Physical Problems, to continue working on the new diagrammatic techniques.[98]

Beginning in February 1954, Landau published a series of brief papers with Abrikosov and Khalatnikov that made use of the diagrams. They were studying the asymptotic behavior of Green's functions within QED: taking into account radiative corrections, how did the amplitudes behave at high momenta (or, correspondingly, short distances)?[99] Even though Landau threw himself into the research, giving some lectures on the project even before they had completed the calculations, Khalatnikov recalls that a division of labor

97. Landau, Abrikosov, and Khalatnikov, "Removal of infinities" (1954), 610. Abrikosov recalled that it took practice to learn how to talk about physics with Landau, whose sharp interruptions could overwhelm inexperienced students. Abrikosov witnessed this happening in discussions with Landau about Feynman diagrams during the mid-1950s. See Abrikosov, "Recollections" (1989), 30.

98. Ioffe, "Landau's theoretical minimum" (2002), 12–13; Ioffe, "If Landau were alive now" (1989), 153–54; and Khalatnikov, "Reminiscences" (1989), 37, 39. The ITEP, founded in 1945, was originally called simply "Laboratory 3": it was the third laboratory established especially for work on the nuclear program (both weapons and reactors). See Josephson, *Physics and Politics* (1991), 338; and Josephson, *Red Atom* (2000), 216–17.

99. Similar work was published around the same time in Stueckelberg and Petermann, "Normalisation des constantes" (1953); and in Gell-Mann and Low [A.124]. These papers were later seen as laying crucial groundwork for the idea of "running coupling constants" and the "renormalization group," i.e., that radiative corrections to an electron's charge would lead to different measured values of the coupling constants at different momenta or distance scales. See Shirkov, "Historical remarks" (1993); and Gross, "Chasing the Landau ghost" (1990).

developed: Abrikosov and Khalatnikov focused on the details of how to calculate with the diagrams, while Landau urged them on with predictions about the ultimate physical conclusions that the overall calculation might point toward. After the papers had been submitted, Landau admitted to a friend, "This is the first work where I could not carry out the calculations myself."[100] For the first time in his illustrious career, Landau had failed to master a technique within theoretical physics. Feynman diagrams remained beyond the pale.[101]

First Contact

With the aid of Pomeranchuk's new seminar, plus a textbook published in 1953 by Berestetskii and Alexander Akhiezer—and despite Landau's active antipathy—a slow trickle of diagrammatic papers by Soviet physicists began to appear. By the end of 1954, the twelve research articles in *Zhurnal eksperimental'noi i teoreticheskoi fiziki* that made use of Feynman diagrams had been supplemented by a handful of additional brief notices in *Doklady Akademii Nauk SSSR* (*Proceedings of the Soviet Academy of Sciences*), many of them by Galanin, Ioffe, and Pomeranchuk themselves.[102] The diagrams really took off, however, becoming taken-for-granted tools among the Soviet theorists, only after physicists in the Soviet Union and in the United States began to meet

100. Khalatnikov, "Reminiscences" (1989), quotation on 39. In addition to Landau, Abrikosov, and Khalatnikov, "Removal of infinities" (1954), see also Landau, Abrikosov, and Khalatnikov, "Electron Green function" (1954); Landau, Abrikosov, and Khalatnikov, "Photon Green function" (1954); and Landau, Abrikosov, and Khalatnikov, "Electron mass" (1954).

101. This work from 1954 led to Landau's better-known, radical conclusions about the ultimate fate of quantum field theory, which he announced with increasing bravado beginning in 1955. He became convinced that the physical charge of an electron should actually *vanish:* a "bare," ingoing charge on an electron would not only be shielded or screened by radiative corrections, but would in fact be entirely compensated—in effect, canceled—by the effects of virtual particles. If the physical charge was truly zero, then QED was at best a phenomenological theory that could be applied only in certain energy ranges; at worst it was a deeply flawed shell that captured little of how the submicroscopic world really worked. As we will see in chap. 8, Landau's charge against quantum field theory broadened over the course of the mid- and late 1950s. See also Brown and Rechenberg, "Landau's work" (1990), 67–76; and Gross, "Chasing the Landau Ghost" (1990), 97–100.

102. In addition to Galanin, "Radiatsionnye popravki" (1951); and the Landau-Abrikosov-Khalatnikov papers from 1954; see also Skorniakov, "Izluchenie" (1953); Akhiezer and Polovin, "Radiatsionnye popravki" (1953); Ioffe, "O raskhodimosti" (1954); Ioffe, "Sistemy kovariantnykh uravnenii" (1954); Zel'dovich, "K teorii π-mezonov" (1954); Zel'dovich, "O raspade" (1954); Galanin, Ioffe, and Pomeranchuk, "Perenormirovka" (1954); and Klepikov, "K teorii vakumnogo funktsionala" (1954). See also Akhiezer and Berestetskii, *Kvantovaia elekrodinamika* (1953). Akhiezer and Berestetskii's textbook was translated into English by the U.S. Atomic Energy Commission: Akhiezer and Berestetskii, "Quantum electrodynamics" (1957).

again at conferences and workshops. Only then could they exchange news more freely and discuss techniques face to face. Just as in the United States, Great Britain, and Japan, personal contact was the key.

No Soviet physicists were allowed to travel beyond the Warsaw Pact countries to participate in international conferences until after Stalin died in 1953; indeed, it took another two years before the first tentative exchanges between East and West began to take place. As early as July 1954, Igor Kurchatov, scientific director of the Soviet nuclear weapons program, complained to the authorities that Soviet physicists were suffering because of their isolation from Western scientists. The Soviet Academy of Sciences echoed Kurchatov's assessment and began preparing international meetings.[103] One of the first was a small workshop on theoretical physics, focusing in particular on QED, held in Moscow in early April 1955. So eager were the organizers to learn about the new techniques directly from the source that Feynman and Dyson were the only physicists from the United States whom the Soviet Academy invited. Arriving in January 1955, the invitation came as something of a surprise to Dyson; he reported to his family that "It seems to be a genuine and serious affair," and that he hoped to be able to attend, out of curiosity for the chance to peer behind the fabled curtain, if for nothing else. Several weeks later, he had to report the sad news: he would not be able to attend the meeting after all because the American authorities, citing the 1950 McCarran Internal Security Act, would not guarantee that he would be able to return to the United States following the meeting. Dyson, after all, still held a British passport, and the authorities did not like the idea of the young theorist consorting with Communists. Feynman had also hoped to attend the meeting and had initially accepted the invitation. Like Dyson, however, he was barred from making the trip: the Atomic Energy Commission refused to let the Manhattan Project veteran and Defense Department consultant visit the Soviet Union. Feynman's and Dyson's dilemma became front-page news when a *New York Times* reporter picked up the story. As the journalist reported, Feynman explained that "the conference would 'have been valuable on both sides,' American and Russian. He said that he had received several reprints and publications by Mr. Landau and had hoped he would have been able to induce him to explain in detail the nature of his recent work."[104] Physicists in the Soviet Union and the United States

103. Holloway, *Stalin and the Bomb* (1994), 352.

104. Freeman Dyson to his parents, 17 Jan 1955, in *FJD;* see also Dyson to his parents, 1 Mar 1955, 18 Mar 1955, 23 Mar 1955, and 14 Apr 1955; and Dyson to Abdus Salam, 15 Mar 1955, in *FJD;* Dyson to Hans Bethe, 24 Mar 1955, in *HAB,* Folder 10:19; Schwartz, "U.S. barred visits" (1955). A friend of Dyson's, the Swedish theorist Gunnar Källén, was allowed to attend the Moscow workshop, and he

had learned the same lesson: reprints and articles were not enough. Informal, personal exchange was needed to really understand each other's work.

A few months after Dyson's and Feynman's foiled trip, the United States did participate with the Soviet Union in an international scientific conference. In August 1955, physicists from around the world gathered in Geneva for the "Atoms for Peace" conference, at which they talked about possibilities for civilian nuclear power production. Immediately after the success of the Geneva meeting, calls rang out in both the United States and the Soviet Union to follow up on the "good feelings" of the Geneva experiment by holding more international meetings.[105] Most significant for theoretical physicists was a series of exchanges that began in 1956. Coincidentally, on the very same day in February 1956 invitations arrived for several American physicists to visit Moscow while Americans sent invitations to several of their Soviet colleagues to attend a meeting in Rochester, New York. Emboldened by the public-relations success of the Geneva conference, the United States government acquiesced in both sets of meetings, and excited chatter about the Moscow trip occupied much of the hallway discussion at the American Physical Society meeting that March.[106] Three Soviet physicists attended the sixth annual "Rochester conference" at the University of Rochester in April 1956, and Dyson was pleased that the Russian he had studied could be put to good use. One evening during the conference Dyson had dinner with Vladimir Veksler, the leading Soviet designer of particle accelerators. Veksler told Dyson (as he reported excitedly to his parents in a letter that night) that "they are all so much looking forward to having me in Moscow that I must be careful not to be questioned to the point of exhaustion by all the eager young theoreticians."[107]

Fourteen physicists from the United States visited Moscow the next month. Included among the American delegation to the May 1956 meeting were several of the world's earliest and most enthusiastic users of Feynman diagrams: not only Dyson himself, but also Jack Steinberger, Keith Brueckner, Robert

---

reported that the meeting "was quite an ordinary conference," at which "most of the papers were perhaps not so very interesting" after all: Källén to Dyson, 19 Apr 1955, a copy of which is in *HAB*, Folder 10:19.

105. Anon., "Atom parley rules" (1955); Anon., "Wider exchange in atomics asked" (1955). See also Fermi, *Atoms for the World* (1957); and Holloway, *Stalin and the Bomb* (1994), 352–54.

106. Freeman Dyson to his parents, 29 Feb 1956 and 17 Mar 1956, in *FJD*. See also the correspondence in *REM*, "Moscow conference, May 1956," microfiche 46.

107. Freeman Dyson to his parents, 4 Apr 1956, in *FJD;* Anon., "Russian hails parley" (1956); Rosenbaum with Marshak and Wilson, "Physics in the U.S.S.R." (1956); Polkinghorne, *Rochester Roundabout* (1989), 60.

Marshak, and Murray Gell-Mann. Making the arrangements embroiled the physicists in complex four-way negotiations between the U.S. State Department, the U.S. Atomic Energy Commission, the Soviet consulate in Washington, D.C., and the Soviet Academy of Sciences in Moscow.[108] But at long last all visas were obtained and permissions granted, and the physicists spent two weeks in the Soviet Union. According to Luis Alvarez's diary of the trip, after the first week or so, "both Russians and visitors have gotten quite used to being with each other, so it is no longer the case that everyone is consciously trying to be on his best behavior. We just acted naturally and had a wonderful time. There was lots of laughter and some hot arguments about physics between the theoreticians—just what goes on at any gathering of American physicists."[109] Taking a break from the laughter and debate, Dyson wrote to his parents that "In general I am amazed to find what an exaggerated reputation I have in Russia. . . . And they have evidently read what I write, not only in the Physical Review but also in the Scientific American." Dyson, like Alvarez, noted the easy informality of his hosts: "The meetings themselves, apart from being too long, are lively and interesting. The Russians are informal, more like Americans than West Europeans, and they contradict and argue and make jokes freely in the meetings." (See fig. 4.12.) Jack Steinberger similarly told a newspaper reporter that "the Americans and other foreigners mixed freely with Russian scientists" during the two-week visit.[110]

Dyson had hoped that the May 1956 meeting would be the start of many more exchanges, and he did not have to wait long before his next visit with Soviet theorists.[111] In September 1956 an international conference on theoretical physics was held in Seattle, and a delegation of Soviet theorists received clearance to attend (including Nikolai Bogoliubov, a senior mathematical physicist from Moscow, and Karen Ter-Martirosyan, one of Landau's younger associates). As with the May 1956 trip to Moscow, many of the earliest diagram afficionados attended the meeting, including Robert Karplus, Joseph Lepore, and Malvin Ruderman. The conference featured a session on QED led

108. See the correspondence in *REM,* "Moscow conference, May 1956," microfiche 46.

109. Alvarez, "Further excerpts" (1957), 24–25; see also Alvarez, "Excerpts from a Russian diary" (1957).

110. Freeman Dyson to his parents, 15 May 1956, in *FJD;* Steinberger quoted in Salisbury, "Curtains are parted" (1956), 12. The May 1956 Moscow visit received much attention from the press: Anon., "U.S., Soviet atom men meet" (1956); Raymond, "Americans in Soviet" (1956); Rosenbaum with Marshak and Wilson, "Physics in the U.S.S.R." (1956); Alvarez, "Excerpts from a Russian diary" (1957); Alvarez, "Further excerpts" (1957); and the many press clippings in *REM,* "Moscow conference, May 1956," microfiche 47.

111. Dyson to his parents, 17 Mar 1956, in *FJD.*

Figure 4.12. Physicists conversing during the May 1956 Moscow physics conference. *Left,* Freeman Dyson (*center*) talking with Nikolai Bogoliubov (*left*) and Dmitrii Shirkov (*right*). *Right,* Jack Steinberger (*left*) talking with Lev Landau (*right*), while the British physicist L. Riddiford (*center*) listens. (Source: Rosenbaum with Marshak and Wilson, "Physics in the U.S.S.R." [1956], 29, 31.)

by current members of the Institute for Advanced Study. During the trip, the Soviet physicists also visited the Institute for Advanced Study, where they met with Dyson for two days.[112] The following year, another delegation of Soviet physicists visited the United States (including the theorists Dmitrii Blokhintsev and Lev Okun'), this time to attend a special conference on nuclear physics at Stanford University chaired by Hans Bethe. The Soviets also attended the American Physical Society meeting that was held at Stanford immediately after the special conference.[113]

Soviet Diagrams

After the reestablishment of informal, personal contact between theorists in the two countries, Soviet theorists began to take up Feynman diagrams at a fast clip. The editors of the *Zhurnal* received more diagrammatic submissions dur-

112. See William Phillips to Eyvind Wichmann, 9 Aug 1956; and Verna Hobson (assistant to J. Robert Oppenheimer), "Memorandum to file re: visit of Russian physicists," 27 Sep 1956, in *IAS,* Folder "General: Russian Physicists' Visit to Institute, Sept. 1956"; and previous correspondence in the same folder; and Freeman Dyson to his parents, 17 Oct 1956, in *FJD.* The Soviet delegation to the Seattle conference was also invited to visit Berkeley: Clark Kerr to Robert Brode, 1 Aug 1956; and A. C. Helmholz to N. N. Bogoliubov, 14 Sep 1956, in *BDP,* Folder 6:22. It appears, however, that they did not visit Berkeley: a copy of their itinerary in the *IAS* files shows that they visited Columbia University and Brookhaven National Laboratory for three days immediately upon leaving Seattle, and then visited with Dyson at the Institute for Advanced Study for two days before returning to the Soviet Union.

113. Anon., "Russians at Monterey" (1957); Davies, "Soviet physicists" (1957); Davies, "Russian gives U.S. pure science lead" (1957).

ing July 1956—just weeks after the lengthy Moscow conference—than during any previous month. The papers continued to pour in at more than twice the rate at which diagrammatic articles had been submitted before the first visits between Soviet and American physicists. The number of authors contributing the papers likewise grew at over twice the earlier rate. Perhaps more important, the number of first-time diagrammatic authors also shot up soon after the first visits, nearly doubling, from nine (between July 1955 and June 1956) to seventeen (between July 1956 and June 1957)—the single largest increase between 1951 and 1959.[114] Reinforcing the new trend, many Soviet physicists welcomed the arrival of a new textbook by the Moscow-based mathematical physicists Nikolai Bogoliubov and Dmitrii Shirkov, published just over a year after the Americans' first visit.[115]

More changed than just the rate at which Soviet theorists published diagrammatic articles. During the early 1950s, before Soviet physicists were allowed to fraternize with colleagues in the West—and before Feynman diagrams became a staple tool on Soviet soil—Soviet theorists had hardly been idle or uncreative. Instead, when they wanted to study various problems in, for example, meson scattering, they developed their own diagrammatic methods. Whereas several groups within the United States had already begun to use Feynman diagrams for these kinds of calculations—cavarlierly dissociating Feynman diagrams from the specific set of rules that Dyson had worked so hard to pull together for the diagrams' "proper" use, as we will see in chapters 5 and 6—Soviet theorists turned to their own doodles instead. Consider, for example, the line drawings in figure 4.13, introduced in a 1950 article to keep track of the various ways in which photons, nucleons, and mesons could interact. Just as the Japanese theorists Koba Zirō and Takeda Gyō had invented their own line drawings to keep track of perturbative terms prior to learning about

114. Between July 1951 and June 1956, the number of diagrammatic articles submitted to *ZhETF* rose at an average rate of 4.7 articles per year, whereas the number of submitted articles rose at a rate of 10.0 articles per year between July 1956 and June 1959. Similarly, the total number of authors contributing diagrammatic articles to *ZhETF* between July 1951 and June 1956 rose at an average rate of 4.6 authors per year, whereas the total number of authors rose at a rate of 10.5 per year between July 1956 and June 1959. Based on data in appendix E.

115. Bogoliubov and Shirkov's textbook was first cited in *ZhETF* in an article submitted in July 1957 (*Science Citation Index*, s.v. "Bogoliubov, N. N.") The book was later translated as Bogoliubov and Shirkov, *Theory of Quantized Fields* (1959 [1957]). Bogoliubov and Shirkov, both based at the Mathematics Institute of the Soviet Academy of Sciences in Moscow, had earlier written a long review article making use of Feynman diagrams, published prior to the first visits between American and Soviet physicists: Bogoljubow and Schirkow, "Quantentheorie der Felder" (1955). The first American textbooks to treat Feynman diagrams were published in 1955 and received only a handful of citations within *ZhETF*, beginning in 1957.

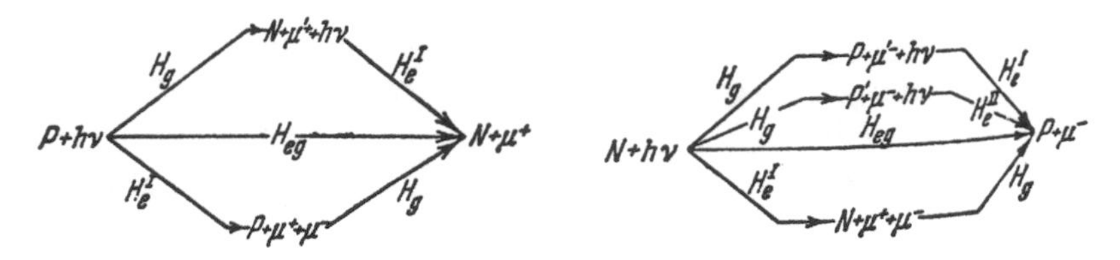

Figure 4.13. Reaction diagrams for keeping track of various particle interactions. (Source: Baldin and Mikhailov, "Obrazovanie" [1950], 1058.)

Feynman diagrams (see figs. 4.5 and 4.6), so too did these Soviet theorists create their own helpful pictorial mnemonic aids.

Unlike the Japanese drawings, however, the homegrown Soviet diagrams were not edged out overnight upon the arrival of Feynman diagrams. In fact, they coexisted for some time with Feynman's diagrams, sometimes appearing in the same articles. In his very first publication to make any use of Feynman diagrams, for example, Yakov Zel'dovich included a variant of the Soviet reaction diagrams to establish which virtual states of pions and nucleons he planned to study. He turned to Feynman diagrams only later in the paper, when undertaking a specific calculation of the pions' self-energy. Only when completing a perturbative calculation according to Dyson's step-by-step rules did Zel'dovich turn to Feynman diagrams; when picturing various interaction possibilities, he chose to stick with his countrymen's drawings instead. Examples of the Soviet reaction diagrams, akin to figure 4.13, could be found in the *Zhurnal* as late as 1957.[116]

Soviet physicists began to use Feynman diagrams in place of these kinds of reaction diagrams—in a more casual way, not tied to specific perturbative calculations—only after they had begun to meet with their American counterparts. As we will see in chapter 6, physicists such as Robert Marshak at the University of Rochester had become accustomed to using Feynman diagrams as pictures of allowed processes, independent of any details of Dysonian calculations, as early as the spring of 1950. Soon after Marshak and his colleagues visited with Soviet physicists, they, too, began to pick up Feynman diagrams for these kinds of picturing—that is, to fill the role that their own reaction diagrams had been invented to fulfill. One telling example comes from the work of Lev Okun', a young theorist working at the Moscow Laboratory of Thermotechnical Studies. He had been quite active during the mid-1950s, publishing a long series of papers analyzing the flood of new particles that

116. Zel'dovich, "K teorii π-mezonov" (1954); see also Pontecorvo, "Processes of production" (1956), 136–37; Kulakov, "Inelastic proton-proton scattering" (1957), 478; and Nelipa, "Problem of excited states" (1958), 983.

Soviet and American experimentalists kept finding with their new accelerators. Simple text-and-arrow notations sufficed for his purposes, allowing him to represent which short-lived mesons interacted with which other particles. He had worked closely with several of the earliest Soviet adopters of Feynman diagrams—publishing papers jointly with Pomeranchuk, Ioffe, and others—but had never seen the need to include Feynman diagrams in his work, since he was not using perturbative techniques to study the strongly interacting particles. Only after the arrival of people like Marshak did Okun' replace his reaction diagrams and arrow notations with simple Feynman diagrams. (He was an active participant in the May 1956 Moscow conference, delivering three separate talks.)[117] Details of his calculational approach had not changed significantly—he still was not completing perturbative calculations. Yet he adopted Feynman diagrams where formerly other notations had satisfied him.[118] In ways that no amount of written texts had done before, talking with his American colleagues and watching them deploy Feynman diagrams for tasks similar to his own convinced him to put Feynman diagrams to work.

TACIT AND EXPLICIT KNOWLEDGES

Feynman diagrams infiltrated the rising generation of British theorists much the way they had done in the United States: thanks to Dyson's direct, personal tutelage, accompanied by a cascade of newly trained recruits circulating out from Cambridge and Birmingham. In Japan, meanwhile, Tomonaga's Tokyo group had been primed to tackle Feynman diagrams more than any other group of physicists outside the United States. They were already world-class experts on the general topic of QED and its problems, and they had logged several years practicing the finer points of making complicated perturbative calculations. They had even invented their own notations and line-drawing schemes

117. A copy of the meeting's program may be found in *REM*, "Moscow conference, May 1956," microfiche 46.

118. Most of Okun's work focused on selection rules among possible decay processes in the simplest approximation, independent of radiative corrections. Cf. Okun' [E.82] and Okun' and Pomeranchuk [E.105] with Okun's prediagrammatic work: Okun', "K-meson charge exchange" (1956); Okun' and Pomeranchuk, "Conservation of isotopic spin" (1956); Okun' and Shmushkevich, "Capture of K^--mesons" (1956); Kobzarev and Okun', "Spin of the Λ-particle" (1957); Okun', "Isotopic invariance" (1957); Okun', "Probabilities of Σ-particle decay" (1957); Ioffe, Okun', and Rudik, "Problem of parity" (1957); Okun', "μ-decay" (1957); Ivanter and Okun', "Theory of scattering" (1957); Okun' and Rudik, "Possible correlations" (1957); Kobzarev and Okun', "Simultaneous creation" (1957); Okun' and Pontecorvo, "Slow processes of transformation" (1957); Kobzarev and Okun', "Decay probabilities" (1958); and Okun', "On K_{e3} decay" (1958).

to keep track of the many terms involved in such calculations. Prompted in part by Yukawa's early letters from the Institute for Advanced Study about Dyson's work, and by the manifestly shared purpose between the Koba-Takeda transition diagrams and the Feynman diagrams that Dyson reported, Tomonaga's group worked hard to master Dyson's papers as soon as they arrived.[119] Learning to wield the tools from printed recipes alone, however, proved remarkably difficult, as witnessed by Takeda's "opened" closed loops and Koba's "illustrative" early examples of Feynman diagrams. Only after nearly a year of dedicated study did some young theorists (such as Kinoshita) get up to speed in making Dysonian calculations. Tomonaga's intense, yearlong visit to the Institute, followed by his return to Tokyo in the summer of 1950, solidified the transfer. The new techniques and skills began to travel beyond Tomonaga's tight-knit circle only thanks to several well-timed contingencies: the appearance of a new organization and mimeographed newsletter, the inauguration of new fellowships, and the sudden expansion of the national university system. Politics and institutions, in other words, aligned just in time to foster effective pedagogy.

Politics likewise shaped pedagogy in the Soviet Union, where more than a year of high-pressure work on a top-secret military project was required before three young theorists had learned how to complete diagrammatic calculations much the way Feynman and Dyson did. Lacking a robust pedagogical environment in which to spread the new techniques, however, Berestetskii's, Galanin's, and Ioffe's extended efforts to master the diagrammatic approach nearly came to nought. Only when they took matters into their own hands, establishing their own training center to rival Landau's, did they begin, slowly, to put the diagrams into circulation. Only with the return of informal contact with their American colleagues did Soviet physicists pick up the diagrams at anything resembling theorists' pace in other countries.

Using Feynman's and Dyson's articles as test particles, we may thus probe the pedagogical infrastructures in each country. Dyson had written his early articles in part to serve as recipe books, making as explicit as he could the

119. As several sociologists have noted recently, one can often aid in the transfer of tacit knowledge and craft skills by deploying "second-order measures of skill": explicit instructions about how the skills are relevant, for what they are to be used, and how long it often takes to learn and/or use the new techniques. In their analysis, such explicit discussion rarely suffices to supplant the informal, tacit knowledge built up from personal communication and embodied practice, but the second-order measures can nonetheless help reduce the time needed to pick up the new skills. Yukawa's letters might well have functioned in this way, drawing the Japanese community's attention to Dyson's work and emphasizing that it was relevant to their own program of research—and hence deserving of extra-special scrutiny. Cf. Pinch, Collins, and Carbone, "Inside knowledge" (1996).

step-by-step rules needed for undertaking diagrammatic calculations. Feynman similarly had included several lengthy appendixes in his first diagrammatic article, working through five or six calculations from start to finish, hoping that the explicit examples would help readers get up to speed with the new tools. With the aid of these textual instructions, *something* could be learned about the diagrams, even without the intense face-to-face training sessions and the postdoc cascade that put the diagrams into circulation within the United States. But what exactly spread via texts alone? Not Dyson's own vision of how the diagrams should be used: Matthews and Salam began to calculate like Dyson, focusing on how to remove divergences from arbitrarily complex Feynman diagrams, only after interacting with Dyson directly. When they read the papers in isolation from these focused conversations, young theorists like Chisholm, Takeda, and Galanin found other elements in the papers to focus on, practice, and develop.

The Soviet case throws the issue into greatest relief: what the Soviet theorists did not gain until personal contact was reestablished with American physicists was a feel for how *else* the diagrams might come in handy. Only after several extended visits did some Soviet theorists pick up various "rules of thumb" for deploying the diagrams—far beyond the cookbook, rule-driven instructions that Dyson had managed to bundle into print. Lacking the institutions of personal contact and pedagogical inculcation, Soviet theorists did not see Feynman diagrams as their own calculational panacea, often preferring to stick with their own reaction diagrams rather than extend Feynman diagrams. And this leads us to the second major theme of the book: for *what* were theorists actually using the diagrams?

PART II

Dispersion in Form, Use, and Meaning

· 5 ·
The Seeds of Dispersion

[Feynman diagrams are] simultaneously a picture of one kind of process ... and an abbreviation for the amplitude contributed by this process.

FELIX VILLARS, 1951[1]

In pursuing the question of how the diagrams spread in part 1, the postdoc cascades emanating from the Institute for Advanced Study, along with parallel circulation mechanisms in other countries, assumed central place. In these descriptions, Feynman diagrams appeared almost like batons in a relay race—stable objects that retained their meaning and form as they were passed from one user to another in a growing network. This initial appearance of stability, however, is misleading: the cascade model misses the rich heterogeneity of ways that physicists used the diagrams during this period. Feynman diagrams were not stable vessels, carrying by themselves an essential or inherent meaning. Rather, they were actively appropriated and deployed in a fast-expanding assortment of ways. The diagrams—like any scientific text or research practice—required interpretation and did not impose any inner meaning upon the world by themselves.

In physicists' hands, Feynman diagrams proved tremendously plastic, highlighting the second meaning of "dispersion": "To cause to separate in different directions.... To spread in scattered order." The diagrams' pictorial forms, relations with other elements of formalism, and calculational roles varied widely. It will do us little good to approach this variety by assuming that theorists after Feynman and Dyson simply "misunderstood" or "misinterpreted" the true meaning of the diagrams, or to appeal to some preinterpretive residue that later physicists failed to comprehend. Young theoretical physicists fashioned new roles for the diagrams, tweaking their pictorial form and reworking their

1. Villars, "Quantum electrodynamics" (1951), 66.

relations with other calculating tools to better serve an array of tasks—many of which had not even been recognized as problems, let alone solutions, when Feynman and Dyson introduced the diagrams. None of these new uses or interpretations was predetermined by something in the diagrams themselves.[2] As we continue to follow the diagrams, our gaze will remain tightly focused on the theorists' hands—not on some purported "inner truth" of the diagrams themselves—as young physicists fashioned and exploited their diagrammatic handiwork.

Whereas part 1 addressed *how* the diagrams spread, therefore, part 2 focuses on *what* was spreading. The diagrams' dispersion in pictorial form, with their attendant dispersion in calculational role, derived from at least two sources. In the first place, the diagrams' original progenitors—Feynman and Dyson—disagreed about the diagrams' proper forms, uses, and meanings. The split between Feynman and Dyson provided one key ingredient for the diagrams' later "scattered order." The other crucial factor came from the failure of Feynman's and Dyson's original diagrammatic techniques when physicists tried to tackle the behavior of particles beyond the familiar electron and photon. Most physicists did not use the diagrams for the same kinds of tasks that Feynman and Dyson did; in particular, few continued to ply them for problems in electrodynamics. When physicists fashioned the diagrams for investigations of new particles and interactions, Feynman's and Dyson's work offered only limited guidance, and no single set of rules held sway. Young theorists improvised and tinkered with the diagrams, crafting new features to better fit a given problem—all of which dispersed the diagrams across many types of calculations.

Signs of the rapid dispersion were not lost on physicists at the time. In mid-June of 1951, Paul Matthews and Abdus Salam—having quit Cambridge for postdoctoral study at the Institute for Advanced Study—delivered an invited talk at the American Physical Society meeting in Schenectady, New York. When their turn came to speak, the two postdocs greeted their listeners with row after row of Feynman diagrams. Building on Dyson's work in the realm of electrodynamics, Matthews and Salam unleashed Feynman's stick figures for problems in nuclear physics. Though they could report much progress in their research, they cautioned in their conclusions that "The difficulty, as in all this work, is to find a notation which is both concise and intelligible to at least two people

2. Cf. Pickering, *Mangle of Practice* (1995); and Barnes, Bloor, and Henry, *Scientific Knowledge* (1996), chap. 3.

of whom one may be the author."[3] More than just a droll conclusion to a difficult paper, the postdocs' quip pointed to a growing problem: finding concise and intelligible notation remained a challenge during the diagrammatic age. The fast-multiplying array of symbols and notation surrounding Feynman diagrams caused the editors of the *Physical Review* no small amount of frustration. Beginning in 1951, lamenting what the assistant editor decried as "varying and conflicting notation" within the sea of diagrammatic submissions, they tried to enforce uniformity, circulating memos with notational instructions to authors who submitted papers with the diagrams. The effort was a flop—"everyone hated it," recalled one physicist recently—and less than two years later the editors admitted defeat. "We were unsuccessful in our attempt to obtain general acceptance of our suggestion (e.g., Feynman demurred)," the assistant editor explained in March 1953. "We have therefore given up trying to be reformers, and now let the authors do as they please," restricting their efforts to "point[ing] out obvious inconsistencies" within a given paper.[4] Laissez-faire had won, and the diversity of diagrammatic notations—springing from Feynman's and Dyson's ingoing split and exacerbated by physicists' shifting research topics—continued to grow.

THE FEYNMAN-DYSON SPLIT

From the very beginning, Feynman and Dyson held different ideas about how the diagrams should be drawn, interpreted, and used. For Feynman, doodling simple spacetime pictures preceded any attempts to derive or justify his new calculational scheme. To Dyson, the diagrams could be of any help only if they were first derived rigorously from a specific field-theoretic basis. To Feynman, his new diagrams provided pictures of actual physical processes, and hence added an intuitive dimension beyond furnishing a simple mnemonic calculational device. To Dyson, the line drawings were never more than "graphs

3. Matthews and Salam, "Renormalization of meson theories" (1951), 314. As noted on 311, their talk was originally delivered at the American Physical Society meeting on 16 June 1951.

4. The original memo is no longer extant in either Hans Bethe's or Samuel Goudsmit's papers; Michel Baranger discussed the memo during Baranger interview (2001). Samuel Pasternack, assistant editor for the *Physical Review*, explained the abolition of the editors' system in a letter to Hans Bethe, 2 Mar 1953, in *HAB*, Folder 12:8. Cf. Williams, "Coherent systems" (1954), 8; IUPAP SUN Commission, "Symbols and units," (1956), 23; John Van Vleck to Hans Bethe, 26 Mar 1954, in *HAB*, Folder 9:35. Similar correspondence regarding the APS SUN (Symbols, Units, and Nomenclature) Committee, ca. 1954–55, may be found in *HAB*, Folder 9:34; and in *REM*, microfiche 122, "International Union of Pure and Applied Physics." See also Alborn, "Negotiating notation" (1989).

on paper," handy for manipulating long strings of equations but not to be confused with the stuff of the real world. Feynman worked almost entirely in terms of particles, and had striven since his undergraduate days to remove fields from physicists' descriptions altogether. Dyson domesticated Feynman's unconventional approaches, fashioning the diagrams into markers for strictly field-theoretic applications. And whereas Feynman experimented with his new theoretical tool, working out makeshift rules for their use by comparing his work with more familiar results, Dyson first demonstrated how the diagrams could be used in a systematic, rule-bound way for perturbative calculations. Given each of these original contrasts over how to treat the diagrams, it is small wonder that later uses of Feynman diagrams displayed such dizzying variety.

Feynman's "Intuitive Pictures"

Part of what had stumped Feynman's listeners at the Pocono meeting was that he had provided few clues as to where his diagrams actually came from. Though it was lost on his circle of auditors at the spring 1948 meeting, Feynman's diagrammatic approach to renormalization had been based on treating particles, rather than fields, as the primary ingredients. Indeed, throughout Feynman's undergraduate and graduate studies, he had struggled to eliminate fields altogether, replacing them with direct interactions between various particles.[5] From this ingoing stance, Feynman had developed his path-integral approach to quantum mechanics for his dissertation, based on integrating particles' behavior over all possible paths through space and time; along the way, while tinkering with his eccentric language of path integrals, he had resurrected John Wheeler's suggestion of treating positrons as electrons moving backward in time.[6] To Feynman, his diagrams, haltingly introduced at the Pocono meeting

5. Before graduating from college, Feynman had begun trying to formulate a classical electrodynamics in which there was no electromagnetic field: by removing the field, he could remove its infinite number of degrees of freedom, leaving only the particles' finite degrees of freedom behind. Soon after entering graduate school, however, he realized that this solution failed to account for radiation resistance; hence, he and his adviser, John Wheeler, pursued their now-famous particle-based, action-at-a-distance approach to electrodynamics, which included both advanced and retarded potentials. See Feynman, "Space-time view" (1966); Wheeler and Feynman, "Interaction with the absorber" (1945); Wheeler and Feynman, "Classical electrodynamics" (1949); Schweber, "Feynman" (1986), 455–61; Wheeler with Ford, *Geons* (1998), 164–67; Mehra, *Beat of a Different Drum* (1994), chap. 5; and Schweber, *QED* (1994), 379–89.

6. Feynman, "Space-time approach" (1948); see also Feynman, "Space-time view" (1966); Mehra, *Beat of a Different Drum* (1994), chaps. 6, 10; Wheeler with Ford, *Geons* (1998), 117–18; and Schweber, *QED* (1994), 389–97.

for treating the renormalization of QED, had flowed seamlessly from this larger set of unprecedented—and virtually unknown—ruminations on classical and quantum physics. Though he finally published much of his reformulation of nonrelativistic quantum mechanics in a lengthy article in 1948, these strange, unorthodox path-integral methods could claim almost no dedicated followers for the better part of two decades. It was no exaggeration when he commented years later on the reasons for his failure to convince his listeners at the Pocono meeting: "My machines came from too far away."[7]

Despite this distance in methods, or perhaps because of it, Feynman believed fervently that the diagrams were more primary and more important than any derivation they might be given. In fact, Feynman continued to avoid the question of derivation in his articles, lecture courses, and correspondence. While in New Mexico during the summer of 1948, after driving cross-country with Dyson, Feynman gave an update to Bethe on his progress. Using his new diagrams, he had derived a result for vacuum polarization, quickly adding a string of parenthetical caveats: "(by the same hocus pocus I spoke to you often about) (i.e.: dropping obviously screwy stuff) (Not that I believe it) (etc.)." His result, moreover, could be made gauge invariant only "by trickery."[8] He kept something of this playful, cavalier attitude in his published work as well. Nowhere in Feynman's 1949 article on the diagrams, for example, were the diagrams' specific features or their strict one-to-one correlations with specific mathematical expressions derived or justified from first principles.[9] Instead, Feynman avowed unapologetically that "Since the result was easier to understand than the derivation, it was thought best to publish the results first in this paper." Feynman relied on the "physical plausibility" of the diagrammatic approach, and "in the interest of keeping simple things simple the derivation will appear in a separate paper." He finally submitted this "separate paper," containing the mathematical derivation and justification of his renormalization results (though not of the diagrams themselves), a full thirteen months later, well after other physicists had begun to use the diagrams.[10]

7. Feynman interview with Sam Schweber, Nov 1984, as quoted in Schweber, *QED* (1994), 436.

8. Richard Feynman to Hans Bethe, 7 July 1948, in *HAB*, Folder 3:42.

9. Feynman [A.4].

10. Quotations from Feynman [A.4], 770. The later paper was published as Feynman, "Mathematical formulation" (1950), and contained neither explicit diagrams nor even discussion of the diagrammatic approach. Instead, it extended Feynman's Lagrangian method, first developed in his dissertation for the case of nonrelativistic quantum mechanics, for use in relativistic electrodynamics problems. This "derivation" paper was submitted in early June 1950, whereas Feynman's two 1949 articles were submitted in early April and early May 1949.

A year after that, Feynman submitted another long paper to the *Physical Review*, introducing still further calculating tricks to simplify lengthy calculations. He began with the pronouncement "The mathematics is not completely satisfactory. No attempt has been made to maintain mathematical rigor. The excuse is not that it is expected that rigorous demonstrations can be easily supplied. Quite the contrary, it is believed that to put the present methods on a rigorous basis may be quite a difficult task, beyond the abilities of the author." Instead, as in his 1949 articles, Feynman "illustrated [the new techniques] with simple applications."[11] When his good friend Ted Welton, himself an expert on QED, asked for reprints of his 1949 articles as "souvenirs, even though my chances of understanding them seem a little slim," Feynman responded with copies of the papers, along with a lesson in how to read them: "I am enclosing reprints of my papers. I gather from your letter that you did not try to read them because if you had I assure you you would find them very simple, at least if you don't try to prove that all the things I say are correct. You know how I work so most of it is just a good guess. All the mathematical proofs were later discoveries that I don't thoroughly understand but the physical ideas I think are very simple. Start with the one about positrons. I wish you luck."[12] For Feynman, the diagrams, coupled with basic ideas about particles' propagation and scattering, came first. The diagrams held such an immediacy and obviousness—at least for Feynman, if not for everyone else—that details of their derivation could be safely postponed.

A closer look at Feynman's famous pair of articles from 1949 bears this out. His first paper, "The Theory of Positrons," developed the idea that positrons, the positively charged antimatter cousins to the familiar electrons, could be treated as ordinary electrons traveling backward in time.[13] From the very beginning, Feynman introduced spacetime pictures to track the propagation of these wandering electrons forward and backward through time. (See fig. 5.1.) The treatment in this first article was entirely divorced from QED: the electrons and positrons either propagated freely, subject to no interactions, or traveled in a fixed, classical, external field. Using these spacetime pictures, Feynman

11. Feynman, "Operator calculus" (1951), 108.

12. Richard Feynman to Ted Welton, 16 Nov 1949, in *RPF;* see also Welton to Feynman, 4 Nov 1949, in *RPF.*

13. The idea of treating positrons as electrons traveling backward in time had also been introduced by Stueckelberg, "Mécanique du point matériel" (1942). See also Feynman, "Space-time view" (1966); Schweber, "Feynman" (1986), 460–61; Feynman, "Reason for antiparticles" (1987); and Schweber, *QED* (1994), 428–34, 576–82.

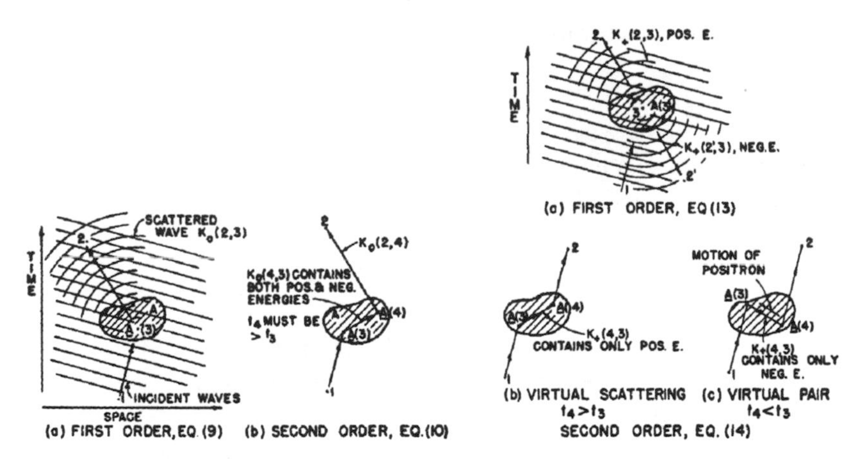

Figure 5.1. Spacetime pictures for electrons' and positrons' motion in an external potential. (Source: Feynman, "Theory of positrons" [1949], 751–52.)

developed his "kernel" methods with a series of semiclassical examples of electrons' motion—semiclassical because he treated the electrons and positrons quantum-mechanically, but the electromagnetic potential classically. The central ingredient of this approach, the kernel, $K(2, 1)$, gave the probability amplitude for an electron to travel from spacetime point x_1 to spacetime point x_2. In other words, given the electron's wave function at x_1, $\psi(1)$, the wave function at some other spacetime point x_2 (with $t_2 > t_1$) could always be obtained by

$$\psi(2) = \int d^3\mathbf{x}_1 K(2, 1)\psi(1).$$

Feynman treated $\psi(x)$ as a one-particle wave function, rather than as a quantized field: it was simply a function whose absolute square gave the probability for finding an electron at position $\mathbf{x}$ at time t. Unlike Kemmer's students in Cambridge, such as Dyson, in other words, who had been drilled to treat quantized fields, ψ, as operators, decomposed (via "second quantization") into sums over creation and annihilation operators, Feynman clung tenaciously to his particle-based, semiclassical approach.

Feynman could relate earlier and later wave functions by means of this integral formula because the kernel, K, was a Green's-function solution of the electron's equation of motion. For both algebraic and conceptual simplicity, Feynman began with the convenient fiction of treating the electrons as spinless particles subject to the Schrödinger equation of nonrelativistic quantum mechanics, written in the form

$$i\frac{\partial\psi}{\partial t} = H\psi,$$

where H was the Hamiltonian for the system (single electron plus external field). The kernel satisfied

$$\left(i\frac{\partial}{\partial t_2} - H_2\right)K(2,1) = i\delta(2,1),$$

where $\delta(2,1)$ stood for the four-dimensional Dirac delta function, $\delta^4(x_2 - x_1)$. (The subscript 2 meant that these operators acted upon the x_2 variables of K.) This expression allowed Feynman to work either in terms of the system's Hamiltonian, H, or in terms of the *solutions* of the equation of motion, K. As Feynman explained, "For some purposes the specification in terms of K is easier to *use* and *visualize.* We desire eventually to discuss quantum electrodynamics from this point of view."[14]

Long before he could tackle QED, however, Feynman first showed—by example, rather than general prescription—how to approximate K in a perturbative series if the electron was acted upon by a weak external field, how to rewrite his kernels as solutions to Dirac's relativistic equation for spinor electrons, and how to keep the bookkeeping straight when treating positrons as electrons moving backward in time. The kernel for a free electron could be obtained from the (relativistic) Dirac equation much the way he had derived the original kernel from the (nonrelativistic) Schrödinger equation. Using a common integral representation of the Dirac delta function, and the usual quantum-mechanical substitution of momenta for derivatives, the kernel could be written easily in the form

$$K_+(2,1) = \int \frac{d^4p}{4\pi^2}\frac{ie^{-ip\cdot x}}{(\boldsymbol{p} - m)},$$

where Feynman used the notation $\boldsymbol{p} = \gamma_\mu p_\mu$, with a sum over the index μ implied.[15] He explained this result even more briefly in the midst of his second

14. Feynman, "Theory of positrons" (1949), 750; emphasis added. As in all quantum-mechanical calculations, the relevant quantity is an amplitude whose absolute square yields a probability; in this case, $|K(x,y)|^2$ yields the probability for an electron to move from spacetime point x to y. I have followed Feynman's notation, using italic letters to denote four-vectors and boldface letters to denote spatial three-vectors, i.e., $x = (t, \mathbf{x})$; the numbers "1" and "2" thus stand for x_1 and x_2, respectively.

15. Feynman, "Theory of positrons" (1949), 752, 757. Here γ_μ is a four-vector, each component of which is a 4×4 matrix. These matrices are Dirac's famous generalization of Pauli's 2×2 spin

article, writing simply that "The reason an electron of momentum $\boldsymbol{p}$ propagates as $1/(\boldsymbol{p} - m)$ is that this operator is the reciprocal of the Dirac equation operator, and we are simply solving this equation." With a similar sleight of hand, Feynman introduced the Fourier transform of the photon propagator, δ_+, writing, "Likewise light goes as $1/k^2$, for this is the reciprocal D'Alembertian operator of the wave equation of light."[16]

Feynman's formulation throughout his "Theory of Positrons" paper thus continued to sidestep questions of QED altogether: rather than treat the force of electrodynamics quantum-mechanically, as arising from the exchange of virtual photons, Feynman built up the semiclassical motion of electrons diagrammatically. Without exchanged quanta, the electrons and positrons could not interact with themselves or with each other; and without these interactions, there were neither unphysical infinities nor a theory of QED to give rise to them.[17] His unpublished presentations followed the same pattern: in lectures he delivered at Cornell in 1949 and at Caltech during 1950 and 1953, spacetime diagrams and proto-Feynman diagrams appeared long before the problems of QED had even been broached. Diagrams were simply everywhere.[18] In fact, diagrams similar to the one in figure 5.2 (below) appeared on the *second page* of the 1951 lecture notes, where they were used to introduce a general distinction between "past, present, [and] future."[19]

Armed with these semiclassical descriptions of particles' motion, along with his developing diagrammatic schemes, Feynman proceeded to implement

---

matrices, familiar from nonrelativistic quantum mechanics. The "+" subscript indicates that the kernel applies only to positive-energy states. Indices on four-vectors during the late 1940s were rarely written explicitly in raised and lowered form, as they are today; the scalar product of two four-vectors, a_μ and b_μ, was written $a \cdot b = a_\mu b_\mu$, rather than $a \cdot b = a_\mu b^\mu$.

16. Feynman [A.4], 775. His treatment of the photon propagator was thus particularly schematic, since it had long been known that a proper treatment of the photon was more tricky: given the gauge freedom, quantizing a massless vector field was still an unsolved problem in 1949. Feynman ignored this entire issue and continued to work with his simple Fourier-transform solution for δ_+. S. N. Gupta and Konrad Bleuler independently demonstrated in 1950 how to handle these subtle gauge issues when quantizing a massless vector field (such as the electromagnetic potential). See Gupta, "Longitudinal photons" (1950); K. M. Bleuler, "Eine neue Methode" (1950); and Schweber, *Relativistic Quantum Field Theory* (1961), 240–53.

17. Feynman, "Theory of positrons" (1949).

18. See the diagrams within Feynman, "Quantum electrodynamics" (1949), 30, 32, 34, 42, 45, 49–53, 77, 84, 86, 88, 90–91, 96–97; Feynman, "Quantum electrodynamics" (1950), 2–3, 5–6, 10–11, 18–21, 28, 34, 38–39, 66–67, 72, 79–80; and Feynman, "Quantum mechanics III" (1953), 7, 9, 15–16, 57, 59, 62, 76, 79, 82, 86, 90, 93–94, 97, 102–3, 105, 107, 109, 114–18, 124, 126–28, 133–34, 137, 139–40. The lecture notes from 1953 became the basis for Feynman, *Quantum Electrodynamics* (1961).

19. See Feynman, "High energy phenomena" (1951), 2–3.

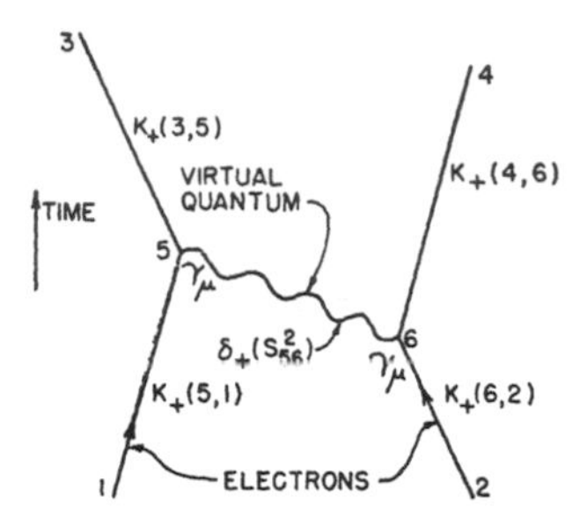

Figure 5.2. The simplest Feynman diagram for electron-electron scattering. (Source: Feynman [A.4], 772.)

these tools in a full assault on QED in his second article. This second article contained the diagrammatic prescriptions, but not the diagrams' derivation, for curing QED of its ills. Rather than demonstrate the general principles involved, Feynman again presented his methods by working out several examples. As in the "Theory of Positrons" paper, Feynman began with diagrams. He explained how the first diagram (reproduced in fig. 5.2) could be used to study electron-electron scattering. The diagram provided a shorthand for an associated mathematical description. Plugging in the kernels for electrons and photons to travel from one point to another—K_+ and δ_+—and for electrons and photons to scatter—$e\gamma_\mu$—Feynman could write down a simple expression corresponding to his diagram. The associated mathematical term gave a fairly good estimate for the probability that two electrons would scatter. This first estimate could be improved, however, by considering more complicated ways in which the incoming electrons could interact to produce the two outgoing electrons. The diagrams for the simplest correction terms, Feynman continued, would each involve four vertices, as he illustrated with the nine diagrams reproduced in figure 2.6. As before, Feynman could walk through the associated mathematical contribution from each of these diagrams, starting with free-particle propagators K_+ and δ_+ and connecting them at the various vertices with the factors of $e\gamma_\mu$. For Feynman, the entire analysis was driven by the diagrams: hapless electrons zigzagged from point to point, scattering from an external field here, emitting a virtual photon there.

After a few more worked examples, stemming from his growing collection of particle-based diagrams, Feynman cautioned his readers that his diagrams and their associated mathematical amplitudes were "no more than a re-expression of conventional quantum electrodynamics. As a consequence, many of them are meaningless."[20] They were "meaningless" because, just like the integrals

20. Feynman [A.4], 776.

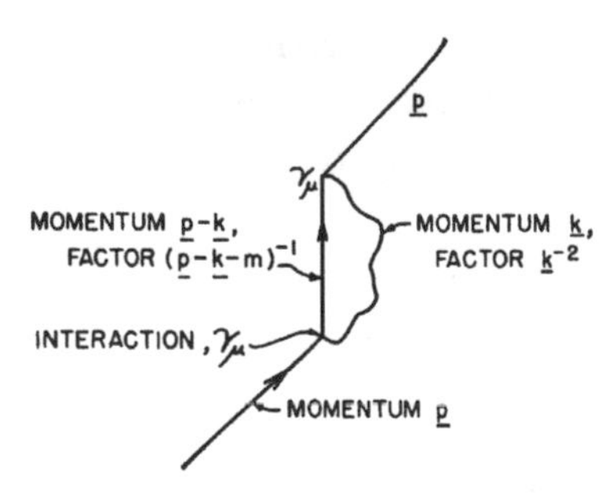

Figure 5.3. The lowest-order contribution to an electron's self-energy. (Source: Feynman [A.4], 774.)

within QED that had been studied since the 1930s, Feynman's diagram-derived integrals still diverged: the integrals associated with most of the diagrams within figure 2.6 ran off to infinity, rather than providing finite, numerical answers. Divergences appeared in even simpler diagrams, in fact, including the diagram for an electron's self-energy (see fig. 5.3). The amplitude associated with this self-energy diagram could be written easily, based on his previous examples:

$$A = \left(\frac{e^2}{i\pi}\right) \int d^4k \gamma_\mu \frac{1}{(\boldsymbol{p} - \boldsymbol{k} - m)} \gamma_\mu \frac{1}{k^2}.$$

The virtual photon could be emitted with any momentum k, and each of these possibilities had to be included—hence the integral over d^4k. Simple power-counting arguments, however, showed that this integral diverged linearly as $k \to \infty$: the factor d^4k contributed as $k^3 dk$, and the integrand went as k^{-3} for large k, leaving an integral of the form $\int dkF(p) \sim kF(p)$, where $F(p)$ was some function independent of k. This quantity clearly ran off to infinity, given the limits on the k integration. Physically, this divergence arose because the virtual photon could carry any momentum whatsoever, including infinite momentum.

In order to tame the persistent infinities, Feynman turned to a complicated combination of calculating maneuvers, some of which he had invented and others borrowed from the likes of Wolfgang Pauli.[21] These steps amounted

21. Feynman had introduced a relativistically invariant cutoff method in 1948; a few months later, he learned about Wolfgang Pauli's and Felix Villars's gauge-invariant regulator method. Pauli had first spelled out his approach in a lengthy letter to Julian Schwinger in Jan 1949 and sent a copy to Hans Bethe; Bethe in turn showed the letter to Feynman, who incorporated the regulator scheme in his own calculations. See Feynman, "A relativistic cut-off for classical electrodynamics" (1948); Feynman, "Relativistic cut-off for quantum electrodynamics" (1948); Wolfgang Pauli to Julian Schwinger, 24 Jan 1949, in Pauli, *WB*, 3:609–19; Wolfgang Pauli to Hans Bethe, 25 Jan 1949, in Pauli,

to *changing* the governing laws of electrodynamics for very small distances (or, equivalently, very large momenta). In particular, Feynman adopted a new form for the photon kernel. In place of

$$\delta_+(s_{12}^2) = \frac{-1}{\pi}\int d^4k\frac{e^{-ik\cdot x_{12}}}{k^2},$$

Feynman now posited

$$f_+(s_{12}^2) = \frac{-1}{\pi}\int d^4k\frac{e^{-ik\cdot x_{12}}}{k^2}C(k^2),$$

$$C(k^2) = \int_0^\infty d\lambda\frac{-\lambda^2}{k^2-\lambda^2}G(\lambda),$$

where s_{12} was the spacetime distance between x_1 and x_2, and the new "convergence factor," $C(k^2)$, went as $1/k^2$ for large k. This added power of k^2 in the denominator guaranteed that the integral over d^4k would converge even in the $k \to \infty$ limit. The price paid for this k convergence was the presence of the arbitrary constant λ in the intervening integrations. Rather than evaluate the final integral over $d\lambda$, Feynman explained that "for practical purposes we shall suppose hereafter that $C(k^2)$ is simply $-\lambda^2/(k^2-\lambda^2)$ implying that some average (with weight $G(\lambda)d\lambda$) over values of λ may be taken afterwards."[22] For "practical purposes," then, many of Feynman's answers depended explicitly on the arbitrary parameter λ. The ad hoc nature of this substitution, $\delta_+ \to f_+$, was hardly lost on him; he noted that this particular form for f_+ was "not determined by the analogy with the classical problem," and that his choice "is arbitrary and almost certainly incorrect." The left-over dependence on the arbitrary parameter λ clearly required some as-yet-unknown remedy. Yet at least Feynman could use the convergence factor $C(k^2)$ to demonstrate how actual calculations *could* be made. As Feynman put it, "The desire to make the methods of simplifying the calculation of quantum electrodynamic processes more widely available has prompted this publication before an analysis of the correct form for f_+ is complete."[23]

Feynman proceeded to introduce a series of handy calculational techniques for evaluating his diagram-derived integrals, the most important of which was

---

WB, 3:619–21 (the original is in *HAB*, Folder 10:25); Pauli and Villars, "Invariant regularization" (1949); Schweber, "Feynman" (1986), 477–81; Mehra, *Beat of a Different Drum* (1994), chaps. 13, 14; and Schweber, *QED* (1994), 414–22, 451–52.

22. Feynman [A.4], 776.

23. Feynman [A.4], 778.

his "new swanky scheme" for combining denominators, as he called it in a letter to Bethe during July 1948.[24] Combining his denominator-identity with several algebraic steps and changes of integration variables, Feynman arrived at a new expression for the electron's self-energy, corresponding to figure 5.3:

$$A = \left(\frac{e^2}{2\pi}\right)\left[4m\left(\ln\left(\frac{\lambda}{m}\right)+\frac{1}{2}\right) - \boldsymbol{p}\left(\ln\left(\frac{\lambda}{m}\right)+\frac{5}{4}\right)\right].$$

Unlike the original divergent integral, this expression was finite for any finite parameter λ and varied relatively slowly as λ varied. Better still, as Feynman next demonstrated, the λ terms exactly canceled when he added in the other radiative corrections that entered at the same order of approximation, leaving a final answer that was independent of λ.[25] With this long series of steps, Feynman had shown how to remove the infinity from the lowest-order contribution to an electron's self-energy. Diagrams in hand, Feynman had conquered at least one of the infinities of QED. For Feynman, calculating in QED meant rehearsing each of these many steps, one diagram and integral at a time.

One obvious source for Feynman's doodles, made all the more clear by this reference to "past, present, and future" when introducing his diagrams in unpublished lecture notes, was Minkowski diagrams. Physicists used Minkowski diagrams (also called "spacetime diagrams"), long a staple element in American undergraduates' training in physics, to track classical objects' motion in space and time. Hermann Minkowski had introduced spacetime diagrams for use with Einstein's special relativity in 1909. Against a time axis drawn vertically and a single spatial axis drawn horizontally, objects' "worldlines," or paths through space and time, could be charted.[26] An object at rest in a given frame of reference would trace out a worldline moving vertically straight up the page: its spatial position would not change as time went by. An object traveling with a constant speed in a particular direction, on the other hand, would trace out a worldline inclined away from the vertical at some angle. (See fig. 5.4.)

24. Richard Feynman to Hans Bethe, 7 July 1948, in *HAB*, Folder 3:42. Feynman's trick for combining denominators, which had been derived independently by Julian Schwinger, was based on the integral identity $1/(ab) = \int_0^1 dx[ax + b(1-x)]^{-1}$. Using this integral (as well as further identities obtained by differentiating both sides with respect to a or b), Feynman could combine several denominators within his lengthy integrals into one function raised to a given power, plus an integral over the new auxilliary parameters, x.

25. See Feynman [A.4], 777–78, referring to the three diagrams within his figure 4 on 775.

26. See Galison, "Minkowski's space-time" (1979). Stanley Goldberg charts the incorporation of special relativity into the curricula for undergraduate and graduate physics instruction in the United States between 1920–1980 in Goldberg, *Understanding Relativity* (1984), pt. 3, esp. 277–93.

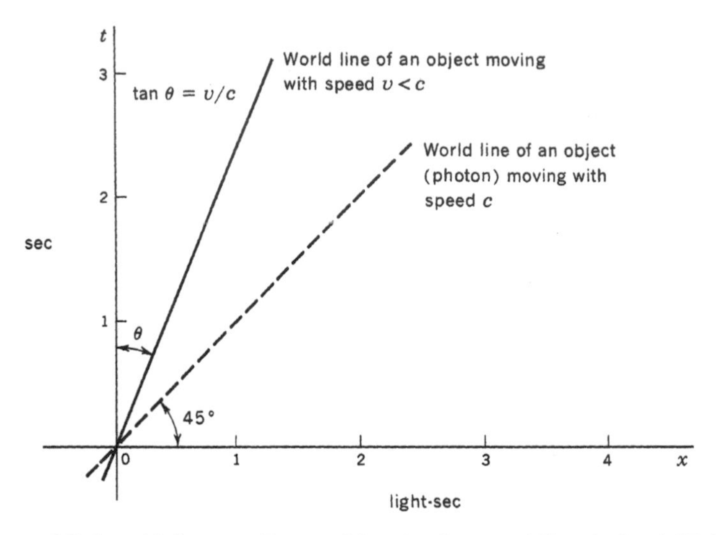

Figure 5.4. Minkowski diagram. (Source: Mermin, *Space and Time in Special Relativity* [1968], 160.)

According to custom, students learned to scale the speed of light to one, so that light rays would travel along 45° diagonals in Minkowski diagrams, as demonstrated in figure 5.4. Furthermore, because Einstein's special relativity elevated the speed of light to an absolute speed limit obeyed by all objects, any other object's worldline would necessarily have a slope greater than 45°; that is, all objects, from airplanes to automobiles to atoms, would dance around at speeds less than the speed of light. A worldline that extended straight horizontally would picture the unphysical situation of an object traveling with infinite speed, taking no time to traverse greater and greater spatial distances.

In Feynman's published as well as unpublished presentations, he spoke repeatedly about electrons' and positrons' "worldlines"—vocabulary clearly borrowed from the Minkowski-diagram tradition—and labeled several of his diagrams "space-time pictures."[27] The diagrams within his "Theory of Positrons" article had explicit space and time axes (see fig. 5.1), and a time axis appeared in the first Feynman diagram within his "Space-Time Approach" article (fig. 5.2). Feynman continued to use many of the pictorial conventions

27. E.g., Feynman, "Theory of positrons" (1949), 749–51; and Feynman [A.4], 775. Feynman announced in his 1949 lecture notes, after introducing a simple Minkowski diagram depicting a static, external potential, that "In fact, we shall frequently make use of such space-time diagrams in the future to represent particles in motion and waves in space." Feynman, "Quantum electrodynamics" (1949), 45. Dyson later explained that Feynman "had this wonderful vision of the world as a woven texture of world lines in space and time, with everything moving freely." Dyson, *Disturbing the Universe* (1979), 62.

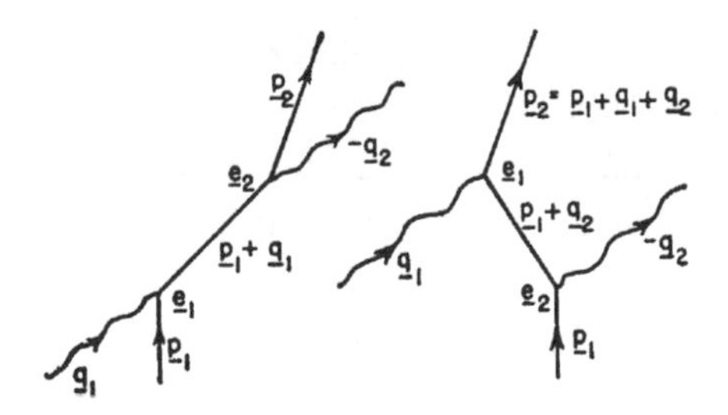

Figure 5.5. Compton scattering, to lowest order. (Source: Feynman [A.4], 775.)

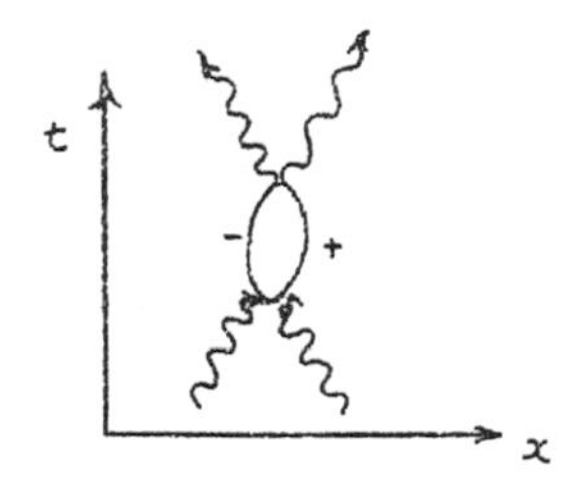

Figure 5.6. Spacetime diagram for light-light scattering. (Source: Feynman, "Quantum electrodynamics" [1949], 49.)

of Minkowski diagrams—such as tilted propagation lines to show the temporal development of physical processes, inclined according to the customary scaling of the speed of light—even after he had switched from the spacetime representation to the momentum-space representation, for which such tilted lines carried no obvious meaning. (See figs. 5.3 and 5.5.)

In a particularly interesting example, Feynman drew a spacetime diagram for light-light scattering in his 1949 lecture notes: two incoming photons annihilated to form a virtual electron-positron pair, which in turn annihilated to form two outgoing photons. (See fig. 5.6.) This diagram is striking because it has the "wrong" number of lines meeting at its vertices: as Dyson had demonstrated in his pair of articles from 1949 (and as we will see below), all diagrams for problems in QED had to have exactly three lines—two electron lines and one photon line—meeting at each vertex. For Feynman, such niceties could wait: long before he got around to treating quantities like light-light scattering in a full QED calculation for his students, Feynman doodled Feynman-like diagrams as *illustrations* of proposed physical processes.[28] Though formally

28. Feynman's unpublished lecture notes must be treated with some care. A graduate student in the class, Harold Brode, prepared the notes. Michel Baranger, who arrived at Cornell as a graduate student in 1951, remembers using the notes: "We all knew that the notes were not the same thing as Feynman himself—they were beautifully written up, but of course it would have been impossible for Harold Brode to include all the nuances of Feynman's presentation. Yet we used the notes a lot

forbidden in QED, such an intuitive physical scenario was simple enough to draw—and to draw first—in a Feynman diagram. It was precisely tricks like this that had originally convinced Dyson that Feynman merely doodled various diagrams and made up the rules for using them as he went along.[29]

Dyson's "Graphs on Paper"

After struggling through frequent personal tutorials with Feynman over the course of 1948—tutorials that were solidified during their cross-country drive that summer—Dyson began to make sense of Feynman's "machines." Moreover, he began to discern how to translate elements of these quirky, particle-based methods into the tools and calculational practices of quantum field theory, which he had practiced while studying with Kemmer in Cambridge. Other physicists hardly needed to retrace Feynman's idiosyncratic path in order to benefit from the diagrams, Dyson soon realized; in fact, the diagrammatic machinery could be lifted out of the unfamiliar, unconventional package of path integrals and positron worldlines and rederived from a thoroughly field-theoretic starting point. Dyson proceeded to do just that, after taking in Schwinger's lectures at the 1948 Ann Arbor summer school. Dyson summarized his role years later: "I allowed people like Pauli who believed in field theory to draw Feynman diagrams without abandoning their principles."[30] Already there was a difference between Feynman's and Dyson's interpretations of the diagrams: where one saw hints of real particles' actual worldlines, the other saw merely a carefully ordered schematic representation of field excitations.

The differences quickly multiplied. For one thing, the diagrams hardly claimed the same immediacy for Dyson as they did for Feynman. Unlike Feynman, Dyson grounded the diagrams point by point within the mathematics they were meant to emulate. In Dyson's first article on the diagrams—the first such article by anyone, published an entire volume before Feynman's

---

and this was another way we learned about the diagrams." Baranger interview (2001). In a similar way, Feynman cautioned Cécile Morette, at the time a postdoc at the Institute for Advanced Study, against relying on his unpublished lecture notes too strongly: "in general, I am very careless in my lectures and in checking over the notes that people make from the lectures, so I would appreciate it if they weren't quoted in detail and specific points argued." Richard Feynman to Cécile Morette, 5 June 1950, in *RPF.*

29. Dyson put it this way: "Nobody but Dick could use his theory, because he was always invoking his intuition to make up the rules of the game as he went along. Until the rules were codified and made mathematically precise, I could not call it a theory." Dyson, *Disturbing the Universe* (1979), 62.

30. Dyson, "Old and new fashions" (1965), 23.

pair of articles—Dyson relegated the diagrams to section 7, "Graphical Representation of Matrix Elements," trailing after ten lengthy pages had already derived the appropriate mathematical expressions for a field-theoretic, perturbative approach to QED. Dyson agreed that the "intuitive considerations of the sort used by Feynman," though not yet published, had been quite successful. He further explained that according to "these ideas of Feynman," "the graph corresponding to a particular matrix element is regarded, not merely as an aid to calculation, but as a picture of the physical process which gives rise to that matrix element." But this picturing should not command attention, Dyson continued: "This interpretation of a graph is obviously consistent with the methods" but would not "be discussed in further detail here."[31] In Dyson's hands, the diagrams' rigorous derivation and service as perturbative bookkeepers claimed center stage. In fact, he included only one single diagram in his first article (see fig. 5.7, *right*, below), which appeared a full six pages after he had described—in words—how the diagrams were to be drawn and used.

Dyson drew even greater contrast between his and Feynman's approaches in his 1951 lectures at Cornell. These were the lecture notes, typed up and hectographed by three graduate students, that were distributed far and wide beginning in January 1952. Whereas Feynman had introduced spacetime pictures and simple line drawings throughout his semiclassical discussion of particles in external fields, Dyson would have none of that. Instead, his students worked on problems in which they practiced manipulating covariant commutation rules, constructing Hamiltonians for the semiclassical and quantized systems, and deducing the appropriate statistics—all the rudiments of field theory, and all strictly diagram free.[32] Even when Dyson proceeded to lecture on interactions between various quantized fields—that is, on QED proper—he did so without any pictures or diagrams. No diagrams appeared in Dyson's lectures until their full derivation and the specific rules for their use first had been made explicit and clear. Tucked away more than two-thirds of the way through his lecture notes, coming only on page 127, Dyson finally introduced Feynman diagrams.

31. Dyson [A.1], 496.

32. See, e.g., Freeman Dyson, "Advanced quantum mechanics" (1951), 38, problem 5. Having introduced five different relativistic Lagrangians, covering the cases of real and complex noninteracting Klein-Gordon fields, the Maxwell field, the Dirac field, and the Dirac field interacting with the Maxwell field, problem 5 assigned students to "Work out these examples; find the field equations, the momentum conjugate to each component of the field, and the Hamiltonian function. . . . Verify that the Hamiltonian gives a correct canonical representation of the field equations as Hamiltonian equations of motion." No such assignments appeared in Feynman's lecture notes.

Once they were brought out in the open, Dyson wasted no time before explaining how the diagrams would be used in *his* classroom. He noted that Feynman "regards the graph as a picture of an actual process which is occurring physically in space-time," but this was hardly the only option available. Dyson emphasized instead for his students (and for the scores of other young physicists who would soon begin poring over his widely circulating lecture notes) that there was nothing magical, let alone physically real, about Feynman's doodles. "The graphs are just diagrams drawn on paper," Dyson explained: "Feynman by the use of imagination and intuition was able to build a correct theory and get the right answers to problems much quicker than we can. It is safer and better for us to use the Feynman space-time pictures not as the basis for our calculations but only as a help in visualizing the formulae which we derive rigorously from field-theory. In this way we have the advantages of the Feynman theory, its concreteness and its simplification of calculations, without its logical disadvantages."[33] The diagrams were to be helpful in "visualizing the *formulae*," not in visualizing actual physical processes. For Dyson, moreover, all the "logical disadvantages" of Feynman's scheme stemmed from the fact that Feynman clung stubbornly to particles as his basic entities, rather than fields. Instead, Dyson lectured his charges, "In this course we follow the pedestrian route of logical development, starting from the general principles of quantization applied to covariant field equations, and deriving from these principles first the existence of particles and later the results of the Feynman theory."[34] It might not have been flashy, but it would work—and work in a systematic way.

For establishing the diagrams' systematic use, Dyson returned to a technique first worked out by Feynman's graduate adviser, John Wheeler, and separately by Werner Heisenberg during the 1930s and 1940s: the scattering matrix, or *S*-matrix.[35] The *S*-matrix related ingoing and outgoing states, $\psi^{\text{out}} = S\psi^{\text{in}}$, where the ingoing and outgoing states were assumed to be asymptotically far

33. Ibid., 129–30.

34. Ibid., 130. The particular "logical disadvantages" of Feynman's formulation that Dyson named included the need to put in by hand the specific quantum statistics that the particles obeyed, and the fact that equations of motion for individual particles became overly complicated for systems that included several interacting particles. Field theory provided a clear contrast, Dyson lectured, since "everything follows from general principles once the form of the Lagrangian is chosen." Ibid., 129–30.

35. Feynman mentioned the *S*-matrix approach in Feynman [A.4], 771, and attributed it to Heisenberg in footnote 6 on that page. John Wheeler first introduced the *S*-matrix in 1937 as a tool for his calculations of the scattering of light nuclei. Werner Heisenberg independently seized upon the *S*-matrix in 1942–43 as a general framework upon which to build a new theory of elementary particles, which would circumvent infinities-plagued QED. See Cushing, *Theory Construction* (1990), 30–42;

from the interaction region, and hence to behave freely, subject to no interactions. Though the idea of the S-matrix was more general than the specifics of any given model—it simply correlated incoming and outgoing states—Dyson advanced one convenient way to evaluate it in terms of the interactions between quantum fields. Dyson worked in terms of the unitary time-evolution operator, $U(t, t_0)$, which Heisenberg had defined by the relation $\psi(t) = U(t, t_0)\psi(t_0)$. Given the conditions on ψ^{out} and ψ^{in}, this meant that $S = U(+\infty, -\infty)$. Unlike Feynman in his semiclassical discussion, Dyson thus treated both ψ and U as quantum-field-theoretic *operators*. The time-evolution operator obeyed the Heisenberg equation of motion:

$$i\hbar\frac{\partial}{\partial t}U(t, t_0) = H_I(t)U(t, t_0),$$

where $H_I(t)$, the interaction Hamiltonian, specified the form of the interaction between the various quantum fields.[36] This equation quantified the notion that all changes in time to the quantum state must arise from interactions; if a system suffered no interactions, $H_I = 0$, then the operator $U(t, t_0)$ would simply remain a constant. Using the relation between S and U, and working in terms of a Hamiltonian density, $H_I(t) = \int d^3\mathbf{x}\, \mathbf{H}_I(x)$, Dyson solved this equation as a power series:

$$S = \sum_{n=0}^{\infty}\left(\frac{-i}{\hbar c}\right)^n \frac{1}{n!}\int_{-\infty}^{+\infty} d^4x_1 \cdots d^4x_n P[\mathbf{H}_I(x_1)\cdots\mathbf{H}_I(x_n)].$$

The symbol P stood for the time-ordered product, which Dyson introduced to avoid double-counting the contributions to S. Any interaction among

---

Cushing, "Heisenberg's *S*-matrix program" (1986); Grythe, "Early *S*-matrix" (1982); Rechenberg, "Early *S*-matrix theory" (1989); Carson, *Particle Physics* (1995), chap. 5; and Wheeler with Ford, *Geons* (1998), 149.

36. Recall that $\hbar = h/(2\pi)$, where h is Planck's constant. Dyson borrowed from some of Schwinger's and Tomonaga's work: they had introduced the various transformations on quantum fields that would yield the "interaction picture." Schwinger and Tomonaga broke the full Hamiltonian into terms corresponding to free fields and interactions, considering the interactions to be small perturbations; they next performed contact transformations on the fields so that only the interaction Hamiltonian would produce further changes in time. In fact, Schwinger and Tomonaga had produced a relativistically invariant form of the Heisenberg equation of motion, based on the timelike hypersurfaces $\sigma(t)$. See Dyson [A.1], 489–91; and Schweber, *QED* (1994), 303–18, 323–40, 509–11.

quantum fields could now be calculated—in principle—using this infinite series for S.[37]

By the time Dyson began this work, physicists had come to assume that the interaction Hamiltonian density for QED took the form

$$\mathbf{H}_I(x) = e\overline{\psi}(x)\gamma^{\mu}\psi(x)\,A_{\mu}(x),$$

where ψ is the quantum field of the electron and A_{μ} is the photon field.[38] This form for $\mathbf{H}_I$, when combined with his infinite-series expression for S, provided Dyson a means of carefully tracking the contributions to any given perturbative order. A second-order calculation, for example, would involve only two factors of $\mathbf{H}_I$ (and hence would weigh in with coefficient e^2, one from each of the factors of $\mathbf{H}_I$); a fourth-order calculation would include four factors of $\mathbf{H}_I$, with overall coefficient e^4; and so on. A perturbative expansion in the electron's charge thus could be built up by including more and more factors of $\mathbf{H}_I$ in a given calculation. Whereas Feynman had demonstrated by example the first two steps of a perturbative expansion for one specific problem, Dyson derived the general rules for calculating any process to any given order.

Only at this point did Dyson finally introduce Feynman's diagrams. First one chose the order, n, to which the calculation was to proceed. Next one doled out a series of spacetime points, which would mark off the vertices in the emerging diagrams: "The points $x_0, x_1, \cdots x_n$ may be represented by $(n+1)$ points drawn on a piece of paper," as Dyson put it in his "Radiation Theories" article of 1949.[39] The various electron- and photon-field operators within the n factors of $\mathbf{H}_I$ could then be paired off into time-ordered products and

37. See Dyson [A.1], [A.2]; and Dyson, "Advanced quantum mechanics" (1951), 119–23; as well as, e.g., Schweber, Bethe, and de Hoffmann, *Mesons and Fields* (1955), 1:192–96. Gian Carlo Wick revisited Dyson's derivation and introduced his now-famous "Wick contraction" method in Wick [A.21], which Dyson incorporated in his 1951 lecture notes (pp. 123–26, 130–31). The power-series operator expansion is based on the ordinary algebraic identity $e^x = \Sigma x^n/n!$, where the sum runs from $n = 0$ to ∞.

38. The overbar denoted a quantum field in which the electron states had been swapped with positron states and vice versa.

39. Dyson [A.1], 495. Note that in this article, Dyson restricted attention to problems in which both the initial state and the outgoing state contained only one electron. Furthermore, this electron was assumed to interact only once with an external potential, and with itself via any number of virtual photon exchanges. The external potential in these problems always acted at the point x_0, leaving the n points $x_1, \ldots, x_n$ for vertices within the nth-order perturbative calculation of field-field scatterings. Dyson moved beyond this initial restriction for the formalism in Dyson [A.2]; and Dyson, "Advanced quantum mechanics" (1951).

their vacuum expectation values taken. These field-theoretic terms, such as $(\overline{\psi}(x), \psi(y))$ and $(A(x), A(y))$, could next be replaced (as Dyson had already demonstrated) by Feynman's free-particle kernels. Continuing to avoid talk of particles and their propagation, Dyson renamed Feynman's kernels $S_F(x - y)$ and $D_F(x - y)$ for the electron and photon, respectively.[40] As Dyson deployed them, the diagrams simply tracked these various field pairings—no more and no less. In the same imperative voice that would guide so many American hobbyists to "paint by numbers," Dyson instructed his readers how to calculate graphically: "For each associated pair of factors $(\overline{\psi}(x_i), \psi(x_k))$ with $i \neq k$, draw a line with a direction marked in it from the point x_i to the point x_k. For the single factors $\overline{\psi}(x_l)$, $\psi(x_m)$, draw directed lines leading out from x_l to the edge of the diagram, and in from the edge of the diagram to x_m. For each pair of factors $(A(x_s), A(x_t))$, draw an undirected line joining the points x_s and x_t. The complete set of points and lines will be called the 'graph' of the [matrix element] *M*; clearly there is a one-to-one correspondence between types of matrix elements and graphs.... The directed lines in a graph will be called 'electron lines,' and the undirected lines 'photon lines.'"[41] Clearly the diagrams did not claim universal familiarity at first sight; Dyson had to spell out, step by step, exactly how the diagrams should be drawn and interpreted. Certain lines would be dubbed "electron lines" and others "photon lines," such names hardly following in any obvious way from laying out random points on a piece of paper and drawing bare lines between them.

With Dyson's arrangement, the number of vertices within a given diagram was determined by, and equal to, the perturbative order to which the calculation was being pursued: a second-order calculation (to e^2) corresponded to diagrams with two vertices; a fourth-order calculation (to e^4) involved diagrams with four vertices; and so on. Furthermore, because $\mathbf{H}_I$ contained two electron-field factors for every photon-field factor, Dyson clarified that each vertex within the Feynman diagrams must have two electron lines connecting with a single photon line. Unlike Feynman, Dyson thus *derived* the form and features of the diagrams: the mathematical form of the Hamiltonian dictated the pictorial form of the diagrams. The novelty of this derivation should not be overlooked. Feynman, for one, had eschewed Hamiltonians altogether, and

40. Dyson [A.1], 494. Dyson referred to S_F and D_F as "invariant functions" or "potentials" throughout Dyson [A.1], [A.2], and his 1951 lecture notes (pp. 106–7, 110).

41. Dyson [A.1], 495. I have simplified somewhat the labels assigned to the spacetime points *x*. Dyson's instruction set for drawing the diagrams became successively more user friendly: cf. Dyson [A.1], 500–501; [A.2], 1740; and Dyson, "Advanced quantum mechanics" (1951), 127–33. On the 1950s craze for "paint by numbers," see Marling, *As Seen on TV* (1994), chap. 2.

had given no indication why vertices in diagrams should have exactly three legs. Nor was Dyson's vertex rule taken for granted in its day: one author spent several *paragraphs* in his 1953 article making explicit Dyson's relation between the mathematical form of the Hamiltonian and the number of legs that must meet in each diagram's vertices.[42] Finally, Dyson included arrows on these "directed lines" to distinguish particles (such as electrons) from antiparticles (such as positrons). By including the antiparticle arrows, Dyson could distinguish diagrams that differed by the exchange of electrons and positrons, since each of these permutations had to be included in the final calculation—a distinction that had been lost on Feynman himself during some of his earliest diagrammatic calculations.[43]

The diagrams thus sprouted for Dyson from this ordered, "connect-the-dots" pattern: unlike Feynman's pseudo-Minkowski diagrams, the "graphs" in Dyson's work had carefully circumscribed roles to play. Various topological features of the diagrams carried quantitative meaning for Dyson, who talked about "open" and "closed polygons" within the diagrams—but topology was different from Feynman's "intuition" about physical processes unfolding in space and time.[44] Spurred perhaps by his original training in mathematics, Dyson went on to quantify these topological features, producing his "degree of divergence" measure for a given diagram based on counting the numbers of closed loops, vertices, external and internal lines. Drawing on these kinds of topological arguments, Dyson demonstrated in his second article how to remove systematically the infinities that had long plagued QED—indeed, how to remove them from *any* perturbative order. Until that time, Schwinger, Tomonaga, Feynman and a handful of others had managed to tame only the infinities that arose at second order.[45]

42. Riddell [A.89], 1243–44; cf. Feynman, "Space-time approach" (1948); Feynman [A.4], esp. 770–71; and Feynman, "Mathematical formulation" (1950).

43. During Dyson's and Cécile Morette's visit with Feynman at Cornell in October 1948, Feynman had calculated the probability for the scattering of light by an external electromagnetic potential, a result that had eluded physicists for nearly a decade. Yet Feynman wrote to Dyson the following week, explaining that in fact the effect vanished (at least at lowest order in the electron's charge), because the contributions from two different diagrams exactly canceled out: in one diagram, the virtual electrons circled around the closed loop in one direction, while in the second diagram, they circled in the opposite direction. Only if one clearly distinguished between these two separate diagrams could one reproduce Furry's theorem. See Freeman Dyson to his parents, 1 Nov 1948, in *FJD;* and Schweber, *QED* (1994), 450. On Furry's theorem, see Furry, "Symmetry theorem" (1937); and Pais, *Inward Bound* (1986), 381.

44. Dyson [A.1], 495–96; see also Dyson [A.2], 1746.

45. Dyson [A.2]; Schweber, *QED* (1994); an excellent, concise summary appears in Mills, "Tutorial" (1993).

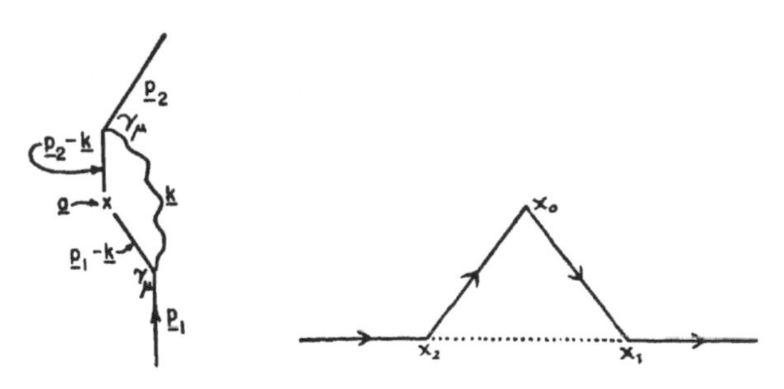

Figure 5.7. Corrections to an electron in an external potential. (Sources: *left,* Feynman [A.4], 775; *right,* Dyson [A.1], 501.)

Feynman's and Dyson's different approaches percolated through every step of their calculations. Consider, for example, the diagrams they drew to represent "the same" quantity: the second-order corrections to an electron's scattering by a classical, external field. (See fig. 5.7.) Like Feynman's diagram in figure 5.5, his figure here follows Minkowski-like spacetime propagation conventions, even though the diagram is covered with momentum labels. Dyson's "graph," on the other hand, has no clear direction of time or of the electron's propagation; its prominent feature, instead, is the "closed polygon" formed by the various field contractions.[46] Even when they were calculating the same quantity, Feynman's and Dyson's diagrams simply did not look the same. The differences between their approaches were inscribed in the diagrams they drew.

PERTURBATIVE METHODS FAIL, FEYNMAN DIAGRAMS FLOURISH

Feynman's and Dyson's differences aside, the efficiency of using Feynman diagrams for perturbative calculations within QED remained simply undeniable. As Dyson put it years later, "The calculation I did for Hans [Bethe] using the orthodox theory"—the scalar Lamb-shift calculation that Dyson undertook in autumn 1947—"took me several months of work and several hundred sheets of paper. Dick [Feynman] could get the same answer, calculating on a blackboard, in half an hour."[47] Surely the rapidity and ease with which such

46. Closer inspection of Dyson's calculation shows that he oriented his time axis horizontally, in contrast with Feynman's (and Minkowski's) vertical time axis. Note that neither Feynman nor Dyson included an explicit propagation line for the fixed, external potential, which is why one of the vertices in each of the diagrams only had two legs meeting at it. These special vertices are marked by the "x" and the label "a" in Feynman's diagram, and marked only by the spacetime point x_0 in Dyson's.

47. Dyson, *Disturbing the Universe* (1979), 54.

labyrinthine terms could be evaluated would have convinced great numbers of theorists to pick up the diagrams and march along with their own perturbative calculations. And surely this, in turn, would explain the diagrams' rapid dispersion, as charted in part 1.

And yet this simply wasn't so. Only a handful of authors published high-order perturbative calculations, trotting out the diagrams as bookkeepers for the ever-smaller correction terms within Dyson's power-series expansion. In fact, only twenty-five papers in the *Physical Review*—fewer than twenty percent of all the diagrammatic articles in that journal—made use of Feynman diagrams in this manner between 1949 and 1954. Almost all these papers, in turn, were contributed by Feynman's, Dyson's, and Bethe's students at Cornell, and by Norman Kroll's students at Columbia.[48] Moreover, these rare perturbative-bookkeeping calculations were not even the earliest diagrammatic calculations published. The remaining 80% of the diagram-filled articles put the diagrams to work for different purposes. For most diagram-wielding physicists, the problem of mesons and nuclear forces beckoned, rather than electrodynamic radiative corrections. In fact, Schwinger, Feynman, and Dyson themselves had all been concentrating on nuclear topics, rather than electrodynamics, before Lamb's experimental measurement of the electromagnetic energy-level shift caught their attention in 1947.[49]

The second major ingredient in the diagrams' dispersion thus stemmed from the fact that *most* physicists were not using Feynman diagrams for perturbative problems in QED. Rather, the diagrams were deployed overwhelmingly

48. In addition to Dyson [A.1], [A.2]; and Feynman [A.4]; see also Watson and Lepore [A.5]; Rohrlich [A.10]; Karplus and Kroll [A.11]; Wick [A.21]; Rohrlich [A.26]; Frank [A.41]; Baranger [A.45]; Brown and Feynman [A.50]; Kroll and Pollock [A.58]; Noyes [A.60]; Low [A.65]; Baranger, Dyson, and Salpeter [A.68]; Blair and Chew [A.83]; Brueckner [A.88]; Weneser, Bersohn, and Kroll [A.90]; Baranger, Bethe, and Feynman [A.93]; Chew [A.117]; Chew [A.118]; Mitra [A.131]; Visscher [A.133]; Sessler [A.134]; and Wyld [A.137]. Frank, Baranger, Brown, A. N. Mitra, and Visscher were graduate students at Cornell; Rohrlich and Sessler postdocs at Cornell. Franklin Pollock and Joseph Weneser were graduate students at Columbia working with Kroll, and Richard Bersohn was a postdoc with Kroll's group. Low wrote his paper while a postdoc at the Institute for Advanced Study; Chew worked closely with Low, once Low had arrived in Urbana. H. Pierre Noyes worked closely with both Dyson and Peierls in Birmingham, and Henry Wyld was a graduate student in Chicago working with Gell-Mann, following Gell-Mann's Institute postdoc. Most of these articles appeared between late 1951 and late 1953; a few deployed Feynman diagrams to calculate high-order corrections to meson-nucleon interactions instead of electrodynamic ones.

49. Mehra and Milton, *Climbing the Mountain* (2000), 162–71; Richard Feynman to Hans Bethe, 12 Sep 1946; and Bethe to Feynman, 16 Sep 1946, in *RPF;* Hans Bethe to Victor Weisskopf, 23 May 1947, in *HAB,* Folder 12:47; Freeman Dyson to his parents, 24 Jan 1948, 11 Feb 1948; and Dyson to Oppenheimer, 17 Oct 1948, in *FJD.*

for problems in meson-nucleon interactions—a topic of immense experimental interest, driven in postwar America by the new spate of federally funded particle accelerators.[50] Yet for problems involving mesons and nucleons, theorists quickly realized, perturbative approaches broke down. Feynman and Dyson had honed their diagrammatic techniques for the case of weakly interacting electrodynamics; the smallness of the electron's charge, which controlled the strength of its interactions with photons, prevented radiative corrections from overwhelming the more basic, lowest-order results. But mesons and nucleons interacted *strongly*—few of the specific roles for which Feynman and Dyson had fashioned the diagrams were of any use for marching through the new theoretical terrain. Nearly from the start, Feynman diagrams were picked up for new and different purposes, in a domain for which their effectiveness had not been demonstrated, or even seemed likely.

Mesons, Mesons, Mesons

Soon after the Pocono meeting of April 1948 Feynman submitted a brief report on the meeting to the fledgling journal *Physics Today.* Appearing in the journal's second issue, Feynman reported on the high-precision experiments on the splitting of electrons' energy levels in hydrogen and on deviations of the electron's magnetic moment from the value predicted by the Dirac equation. He also noted that thanks to the recent advances in QED, as achieved especially by Schwinger (giving himself second and much-reduced billing, much as the unpublished notes from the meeting would do), "all the results of the delicate experiments can now be understood." Yet altogether, his summaries of the precision atomic experiments and of the theoretical breakthroughs in QED occupied only half his report. Indeed, "The most exciting event of the year" stemmed neither from the discriminating microwave spectroscopy within Columbia's Pupin Laboratory nor from the dapper display of mathematical erudition on Schwinger's blackboard, but rather from Berkeley's monstrous 184-inch cyclotron. New particles could be created there and studied "under controlled conditions and in large numbers."[51]

50. Physicists used the term "meson" from the late 1930s through the 1950s to refer to any particle whose mass was intermediate between that of the (light) electron and the (heavy) proton. The term "nucleon" refers to both protons and neutrons; physicists chose this single term to reflect the fact that the early evidence about the strong nuclear force, which bound protons and neutrons together to form nuclei, treated protons and neutrons identically. As far as the strong force was concerned, in other words, protons and neutrons acted like the same kind of particle.

51. Feynman, "Pocono conference" (1948), 9. On the "zoo" of new particles found soon after the war, see Marshak, "Particle physics" (1983); Pickering, *Constructing Quarks* (1984), chap. 3; Brown,

"It appeared that we now know," began Feynman's report, "with varying degrees of certainty, of at least eleven so-called elementary particles." QED treated a paltry three from his list, "the familiar" electrons, positrons, and photons. Three others—protons, neutrons, and neutrinos—had been known (or at least talked about) since the early 1930s. The remainder, intermediate in mass between protons and electrons, had been dubbed simply "mesotrons," a name soon shortened to "mesons." They fell from the sky in cosmic rays, and now they seemed poised to answer experimentalists' bidding in earthly cyclotrons. Feynman was not shy about what these welterweight newcomers portended: "The conference showed that just as we were apparently closing one door, that of the physics of electrons and photons, another was being opened wide by the experimenters, that of high-energy physics. The remarkable richness of new particles and phenomena presents a challenge and a promise that the problems of physics will not be all solved for a very long time to come."[52] Like Feynman, Dyson found the prospects exciting, writing in early 1949 to a European colleague that "Certainly the present is a fine time to be doing theoretical physics: the field is wide open." A few years later, Dyson shared the common quip that it was surprising whenever an entire month passed without the announcement of a new particle discovery.[53] The new bounty of mesons even inspired Edward Teller to versify:

> There are mesons π and there are mesons μ
> The former ones serve us as nuclear glue
> There are mesons τ, or so we suspect
> And many more mesons which we can't yet detect.[54]

Scores of eager young theorists immediately began to heed what Feynman had called the new particles' "challenge."

Even before the new cyclotrons came on line, while cosmic ray experimentalists in Rome, Paris, and Bristol puzzled over the elusive tracks in their nuclear emulsions and cloud chambers, theorists had been hard at work unraveling various possible ways in which neutrons, protons, and the new mesons might

---

Dresden, and Hoddeson, *Pions to Quarks* (1989); Polkinghorne, *Rochester Roundabout* (1989); and Kragh, *Quantum Generations* (1999), chap. 21.

52. Feynman, "Pocono conference" (1948), 8, 10.

53. Freeman Dyson to Léon Rosenfeld, 7 Feb 1949, in Pauli, *WB*, 4.1:4; Dyson, "Field theory" (1953), 57.

54. Teller's handwritten poem, n.d., may be found in *HAB*, Folder 12:35. Teller was fond of composing such rhymes; see "Dr. Teller's Atomic Alphabet," reproduced in Anon., "Defense" (1957), 22.

interact. They worked in analogy to QED, using symmetry principles and phenomenological information to write down several contenders for the new interaction Hamiltonian, $\mathbf{H}_I$, which might describe the nuclear realm. Before the war was over, Wolfgang Pauli delivered a series of lectures on meson physics at MIT; just after the war, Hans Bethe gave a course on elementary nuclear theory as part of his consultant work for the General Electric research laboratories in Schenectady, New York.[55] By the mid-1940s, Pauli, Bethe, and others had pulled together various contenders for $\mathbf{H}_I$. Yet just as in QED, infinities appeared whenever theorists tried to calculate anything beyond lowest order.

There was some promise for change as soon as Tomonaga, Schwinger, Feynman, and Dyson conquered the vagaries of calculating in QED. In fact, Feynman closed his second article from 1949 with the final section "Application to Meson Theories." "The theories which have been developed to describe mesons and the interaction of nucleons," Feynman began, "can be easily expressed in the language used here."[56] Feynman was hardly alone in recognizing the potential of his diagrammatic language for meson physics. A young researcher in Glasgow drew out some of the simple translations: "The notation of Feynman is used . . . ; the fact that Feynman's notation is given for an electron-photon theory while we are thinking of a nucleon-meson theory only requires that a trivial change be made in the meaning of the symbols. (For example, $\boldsymbol{A}$, instead of being $\gamma_\mu A_\mu$ becomes $\gamma_5 \varphi$ or $\gamma_5 \varphi^*$, where φ is the usual meson field quantity.)"[57] Dyson similarly wrote to a senior colleague in February 1949 that his troupe of fellow postdocs at the Institute was "busily making calculations and hoping that the existing field theories will be enough to reduce mesons and nuclear forces to order."[58]

Immediately, young theorists began to tinker with specific features of the diagrams, tailoring Feynman's diagrams to their new mesonic applications.

55. Pauli, *Meson Theory* (1946), based on his 1944 lectures at MIT; and Bethe, *Elementary Nuclear Theory* (1947), based on his course at General Electric. Gregor Wentzel likewise updated his section on "meson theory and nuclear forces" for the 1949 English translation of his 1943 textbook: Wentzel, *Quantum Theory of Fields* (1949 [1943]), vii, 104–10. See also J. Keller, "Mesons old and new" (1949); Rosenfeld, "Meson fields" (1950); Bethe, "What holds the nucleus together?" (1953); Bethe, "Mesons and nuclear forces" (1954); Breit and Hull, "Advances in knowledge of nuclear forces" (1953); Oppenheimer, "Thirty years of mesons" (1966); Mukherji, "History of the meson theory" (1974); and Brown and Rechenberg, *Concept of Nuclear Forces* (1996), chaps. 13–14.

56. Feynman [A.4], 783.

57. Moorhouse [A.74], 959.

58. Freeman Dyson to Léon Rosenfeld, 7 Feb 1949, in Pauli, *WB*, 4.1:4. An early example of the Institute postdocs' efforts appeared as Watson and Lepore [A.5]; see also Matthews and Salam, "Renormalization of meson theories" (1951), 311.

Figure 5.8. New line types for nucleon-meson scattering. (Source: Paul Matthews to Wolfgang Pauli, 18 May 1950, as reprinted in Pauli, *WB*, 4.1:97.)

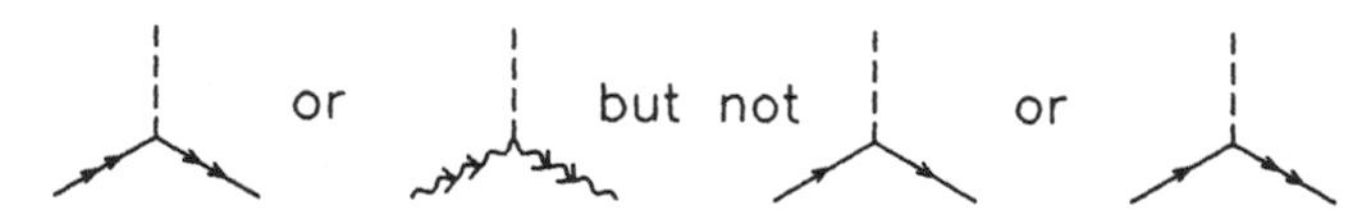

Figure 5.9. Paul Matthews's double-arrow notation. (Source: Paul Matthews to Wolfgang Pauli, 18 May 1950, as reprinted in Pauli, *WB*, 4.1:97.)

Paul Matthews, for example, wrote excitedly to Wolfgang Pauli in May 1950 with news of how his studies of the "meson-nucleon-photon mixture" were progressing. His adviser, Nicholas Kemmer, had recently handed a manuscript with some of Matthews's work to Pauli at a conference in Paris; since then, Matthews had continued fiddling with the diagrams. "I have since found a notation which is simple and reliable," he explained. "If double arrows denote movement of charge, then double arrowed lines must form continuous chains. Also nucleon lines must form continuous chains of either single or double arrows. These chains must be consistent with each other."[59] According to Dyson's rules for QED, single arrows on electron and positron lines sufficed to distinguish particles from antiparticles; electrons and positrons were the only charged particles involved in QED calculations, and the only spin-1/2 particles involved as well, so that tracking particles and antiparticles was the *same* as tracking the flow of electric charge. But in the new mesonic realm, pions could carry electric charge, even though they seemed to carry integer units of spin (and hence Dyson's antiparticle-arrow rule did not apply to them); Dyson's antiparticle arrow did apply to nucleons, on the other hand, even though these could be either charged (protons) or uncharged (neutrons). Matthews thus picked up Dyson's mantle, trying to bring order to these new diagrammatic possibilities. Naturally, he included examples in his letter to Pauli: "Thus (a) is excluded but (b) is allowed." (See fig. 5.8.) "A photon line can only appear between double arrows," he continued, concluding with the four examples of figure 5.9.[60] Like Dyson (from whom he had learned his diagrammatic

59. Paul Matthews to Wolfgang Pauli, 18 May 1950, in Pauli, *WB*, 4.1:97–98. See also Matthews to Pauli, 1 May 1950, in ibid., 95.

60. Ibid.

craft), Matthews wrote of "continuous chains" rather than Feynman's physical processes. Faced with the new meson-nucleon interactions, Matthews adapted Dyson's version of Feynman's diagrams, incrementally emending his notation to better serve his new calculational purposes.

Yet Matthews's diagrammatic efforts, like those of other eager meson theorists, quickly hit an obstacle far larger than fashioning a "simple and reliable" pictorial notation. Theorists' diagrammatic mesonic calculations faltered on a major problem: the coupling between the nucleons and pions appeared to be much larger than one. Whereas Feynman and Dyson could exploit the smallness of the electron's charge ($e^2 \sim \frac{1}{137}$) for their perturbative bookkeeping in QED, various experiments indicated strong-interaction coupling strengths, g^2, in the range between 7 and 57.[61] If theorists tried to treat pion-nucleon scattering the same way they had treated photon-electron scattering, with a long series of more and more complicated Feynman diagrams, each containing more and more vertices, then each higher-order diagram would include extra factors of the large-number g^2. Unlike the situation in QED, these complicated diagrams, with many vertices and hence many factors of g^2, would overwhelm the lowest-order contributions. In the face of such difficulties, Feynman lamented to Bethe in June 1951, "It is amazing how little predictive ability theoretical physics has" when it came to strongly coupled mesons. Citing the large value of g^2, Feynman instructed Enrico Fermi a few months later, "Don't believe any calculation in meson theory which uses a Feynman diagram!"[62]

Few young theorists heeded Feynman's warning. Instead, they continued to press Feynman diagrams into service for their studies of meson physics, even as they wrung their hands over how to approach the new interactions. The principal contender among the meson models required "so large a coupling constant to give the correct orders of magnitude," sighed Matthews and Salam in their 1951 Americal Physical Society talk, "that the perturbation expansion is invalidated and no comparison with experiment is possible."[63] Meeting after meeting, theorists echoed the same plaintive tone. Standing before a small circle of his colleagues nearly three years later, for example, Murph Goldberger outlined

61. Not only was the coupling constant far too large to support perturbative approaches, but further complications arose because different physical processes seemed to indicate different values for the coupling constant, even though only one constant, g, was supposed to fit all these sets of data. See Pais, *Inward Bound* (1986), 482–85, and Polkinghorne, *Rochester Roundabout* (1989), 40–41.

62. Richard Feynman to Hans Bethe, 7 June 1951, in *HAB*, Folder 10:28; Feynman to Enrico Fermi, 19 Dec 1951, in *RPF*.

63. Matthews and Salam, "Renormalization of meson theories" (1951), 314.

his recent calculations with Murray Gell-Mann. After displaying their diagram-filled calculations, Goldberger "readily admit[ted]," so the published summary explained, that "If the coupling constant is as large" as recent experimental data seemed to indicate, then their calculation, "being based on expansion in terms of a small coupling constant no longer holds." Goldberger remarked with characteristic sarcasm that given the large coupling constant, "I think everyone will agree that this is not a red-hot convergent series." Indeed, with the coupling constant so high, "we would never have made a perturbation theoretic expansion . . . in the first place. With such a large coupling constant we can reach no conclusion about the theory."[64] Decades later, a physicist compared this urge to apply perturbative techniques to meson-nucleon interactions, despite the swollen coupling constant, to "the sort of craniometry that was fashionable in the nineteenth century," which "made about as much sense."[65]

If there could be no reliable perturbative expansions when these new interactions between protons, neutrons, and pions were studied, then what place could there be for Feynman's stick figures? Despite Feynman's own advice not to trust Feynman diagrams in mesonic calculations, scores of young theorists ensured that there would be plenty of room for the diagrams in the new realm. In fact, over half of all the diagrammatic articles in the *Physical Review* between 1949 and 1954 applied Feynman diagrams in one way or another to meson physics, including the four earliest diagram-filled articles published after Dyson's and Feynman's own.[66] Rather than discard the diagrams in the face of the breakdown of perturbative methods, theorists clung to the diagrams' bare lines, fashioning new uses and interpretations for them. The new uses—themselves different from each other—showed only partial correspondence to Dyson's systematic rules for the diagrams' use. During the early and mid-1950s,

64. Goldberger's comments were reported by Gregor Wentzel, who presided over this session at the Rochester conference. See the transcript of the "Theoretical Session" in Noyes et al., *Proceedings* (1954), 26–40, quotations from 30, 31, and 34. See also Marvin Goldberger to Hans Bethe, 1 Feb 1954, in *HAB*, Folder 10:36. In the early and mid-1940s, some physicists began working on nonperturbative approaches to the strong nuclear force, including Tomonaga Sin-itiro and Gregor Wentzel. See Wentzel, "Recent research" (1947); and Maki, "Tomonaga" (1988).

65. J. Bernstein, "Need to know" (1983), xvii. Jeremy Bernstein completed his Ph.D. in particle theory at Harvard in 1955, having worked with Julian Schwinger and Abraham Klein; see also J. Bernstein, *Life It Brings* (1987), esp. chaps. 4–6.

66. Matthews [A.3]; Watson and Lepore [A.5]; Steinberger [A.6]; and Morette [A.8] were all dedicated to meson physics. The other diagrammatic articles, as listed in appendix A, which were about meson physics include the articles numbered 13, 15–17, 20, 22–23, 26–27, 29, 31, 36–37, 39–40, 46–47, 49, 53, 55, 57, 60, 64, 66–67, 69–70, 72–74, 76, 78, 80–81, 83–88, 92, 95–98, 100–104, 106–10, 113–15, 117–20, 122, 125, 127–34, and 136–39.

theorists leaned on the diagrams for everything from quantitative "power-counting" arguments to "intuitive" illustrations of particles' interactions.

Pictures for Power Counting

Some meson theorists tried to differentiate between the various models of the mesonic realm with the litmus test of renormalization. Dyson came to believe that renormalizability would offer the most trenchant means of sifting through the various $\mathbf{H}_I$ contenders to find the unique winning candidate, or at least to narrow the playing field. In February 1949 he opined to Léon Rosenfeld that only once "the higher-order radiative effects are taken seriously" in the meson-nucleon models would any progress be made.[67] Four years later, Pauli reported to a younger colleague on Dyson's continuing quest: "From Dyson I have heard (in different ways) that he still very much believes in the 'Renormalization-philosophy,' in the sense of a sharp distinction between renormalizable and nonrenormalizable interactions, and particularly in the postulate (which I doubt) that only the former exists in nature. From this postulate he derives the conviction that the pseudoscalar coupling for the (pseudoscalar) π-meson with nucleons (with the γ_5 and the large coupling constant) must be the right one. Bethe also seems to share this conviction."[68] Bethe did indeed share Dyson's conviction, writing a few months later in *Physics Today* that imposing the criterion of renormalizability would allow theorists to "tell the sheep from the goats" among meson models.[69] Pauli

67. Dyson to Rosenfeld, 7 Feb 1949, in Pauli, *WB*, 4.1:4. Along with the work by Matthews and Salam cited above, Fritz Rohrlich dedicated some of his early diagrammatic investigations to the question of whether—and which—meson theories could be renormalized in Dyson's manner. See Rohrlich [A.26].

68. Wolfgang Pauli to Gunnar Källén, 21 Jan 1953, in Pauli, *WB*, 4.2: 26–27. My translation. The original reads, "Von Dyson habe ich (auf verschiedenen Wegen) gehört, daß er noch sehr an die 'Renormalizations-Philosophie' glaubt, im Sinne eines scharfen Unterschiedes zwischen renormalisierbaren und nicht-renormalisierbaren Wechselwirkungen und insbesondere dem (mir zweifelhaften) Postulat, daß nur die ersten in der Natur vorkommen sollen. Aus diesem Postulat leitet er die Überzeugung ab, daß die pseudoskalare Koppelung für das (pseudoskalare) π-Meson mit den Nukloenen (die mit dem γ_5 und der großen Kopplungskonstante) die richtige sein muß. Auch Bethe scheint diese Überzeugung zu teilen." On Dyson's criterion of renormalizability, see also Schweber, *QED* (1994), 567–72. Schweber and Tian Yu Cao have explored the challenges to Dyson's earlier view of renormalizability as an arbiter among possible theories and models, given work since the 1980s on effective field theories, in Cao and Schweber, "Renormalization theory" (1993); Cao, "New philosophy of renormalization" (1993); and Schweber, "Changing conceptualization of renormalization theory" (1993).

69. Bethe, "Mesons and nuclear forces" (1954), 7.

remained unimpressed: "I remain on this rather skeptical; the arguments of these authors do not appear good to me, and I believe that Heisenberg might be closer to the truth about the mesons than is Dyson-Bethe."[70] Later that spring, he speculated further to one of the Institute's young theorists, "The *production* of mesons seems to me so much different from the production of photons that I feel the 'physical' limitations of the analogies of meson-theories with electrodynamics much more strongly than Dyson: I do not see any reason to believe that the interactions of mesons with nucleons should belong to what he defined mathematically as the 'renormalizable' type of interactions." Indeed, as early as 1951, Pauli had complained to Dyson that "the whole analogy between meson-theory and electrodynamics was and is fallacious." By the end of 1953, even Dyson's hopes were flagging. He reported to Bethe that he intended to take a break from meson theory, "to sit back for a time and look at it from a distance, in the hope of finding a radically better way of doing things."[71]

Not everyone agreed on the power of renormalizability to tell the mesonic "sheep from the goats." Still less, it turned out, did theorists agree on the more pragmatic question of how to treat higher-order corrections within their meson-nucleon studies. Sometimes theorists drew diagrams within their mesonic calculations that included closed loops. If evaluated in strict analogy to the QED case, the contributions from these diagrams would diverge: independent of the size of g^2, the closed loops would give rise to integrals that ran off to infinity. Of course, within QED, Tomonaga, Schwinger, Feynman, and Dyson had found ways to isolate these closed-loop infinities, leaving sensible results behind. Rather than pursue this complicated renormalization problem in the strongly coupled meson models, however, many meson theorists simply ignored such infinities altogether. Teller explained the procedure in verse:

> From mesons all manner of forces you get
> The infinite part you may simply forget
> The divergence is large, the divergence is small
> In meson-field quanta, there's no sense at all.[72]

Though running the risk of yielding "no sense at all," many theorists used their Feynman diagrams for a new kind of bookkeeping: they simply counted the

70. Pauli to Källén, 21 Jan 1953, in Pauli, *WB*, 4.2:26–27. My translation. The original reads, "Ich bin da doch recht skeptisch, die Argumente dieser Autoren scheinen mir da nicht gut und ich meine, daß Heisenberg hinsichtlich der Mesonen der Wahrheit vielleicht näher ist als Dyson-Bethe."

71. Wolfgang Pauli to Abraham Pais, 21 Apr 1953, in Pauli, *WB*, 4.2:117–19; emphasis in original; Wolfgang Pauli to Freeman Dyson, 18 Feb 1951, in Pauli, *WB*, 4.1:264; and Freeman Dyson to Hans Bethe, 23 Dec 1953, in *HAB*, Folder 10:18.

72. Teller's undated poem, in *HAB*, Folder 12:35.

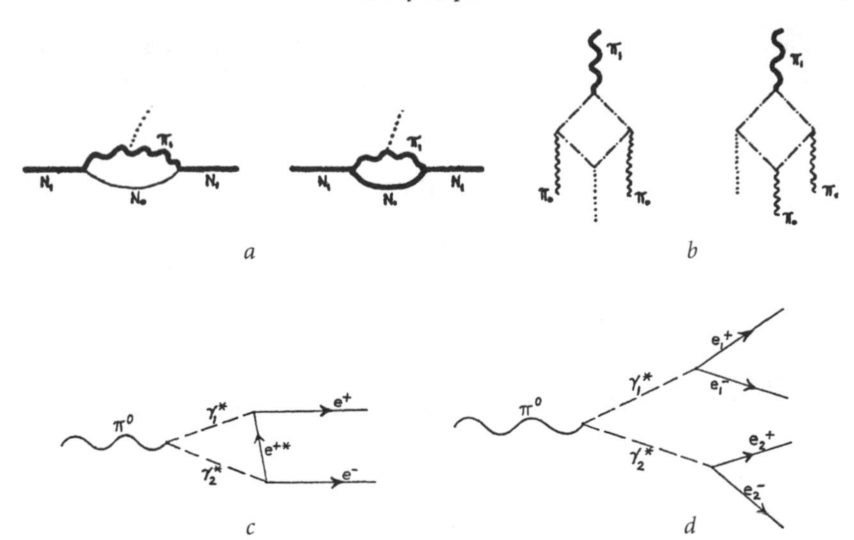

Figure 5.10. Pictures for power counting. (Sources: *a, b,* Pais [A.55], 666, 671; *c, d,* Lindenfeld, Sachs, and Steinberger [A.72], 532.)

number of different *types* of vertices in the diagrams. Their diagrams allowed them to figure out the combination of coupling constants that would appear in the full scattering amplitude *if* that amplitude were formally evaluated.

This maneuver sprang from attempts to bring order to the large variety of different particles and interactions within nuclear physics. Because protons and some of the mesons were electrically charged, they could interact electromagnetically (via photon exchange, with strength governed by the small electric charge, e) as well as via the strong nuclear force (governed by the large meson-nucleon coupling constant, g). Physicists used Feynman diagrams (such as the examples in fig. 5.10) to explore the *relative* probabilities for various competing processes. It was hoped, though never proved, that since g was so much larger than e (the source of all the calculational problems), diagrams that included more photon vertices (and hence more factors of e) would contribute—after the putative renormalization dust had settled—less strongly than would diagrams that included more meson vertices (and hence more factors of g).

Consider, for example, the diagrams reproduced in figure 5.10*c–d.* Jack Steinberger, who had first learned about Feynman diagrams from Dyson while a postdoc at the Institute for Advanced Study, teamed up with two physicists at Brookhaven National Laboratory to study the different ways pions could decay. Both of the diagrams they drew included two vertices at which photon lines and electron lines met. Without further discussion, they concluded that each diagram would therefore contribute to the pion's decay with a factor of e^2—even though the diagram in figure 5.10*c* contained a divergent closed loop, whereas the diagram in figure 5.10*d* involved no such infinities. Lumping the two diagrams together, they guessed that these electrodynamic contributions, despite the formal divergence of one of them, would weigh in several orders of magnitude less than the competing meson-mediated decay processes, and hence could be ignored.[73] Unlike Feynman or Dyson, these Brookhaven physicists relied upon particle labels, rather than spacetime coordinates or momentum states, as the most effective pictorial means of accomplishing their calculational task. Arrows, meanwhile, appeared on their electron and positron lines, though they were not applied in accordance with Dyson's careful antiparticle convention: in this case, arrows on electrons and positrons ran in the same direction, leaving the arrows' purpose unclear, while time flowed tacitly from left to right, rather than from bottom to top, as it did in Feynman's diagrams. Abraham Pais similarly ignored the closed-loop infinities in his diagrams (fig. 5.10*a–b*), making relative-weight arguments about each diagram's contribution instead.[74]

* * *

Young theorists produced a wide array of objects, each of which they labeled "Feynman diagram," during the late 1940s and 1950s. Feynman and Dyson themselves disagreed on how best to treat the new diagrams: were they to be "intuitive pictures" or merely "graphs drawn on paper"? Were pictures of particles' propagation in space and time to supersede formal derivations of the diagrams' properties, or were the diagrams little more than convenient mnemonic devices, useful for "visualizing formulae"? Even when treating the same calculation, Feynman's and Dyson's calculational approaches—and hence the diagrams they drew—showed important differences. Feynman's Minkowskian line drawings recalled the particle worldlines he talked so much about, while Dyson drew topological graphs composed of static polygons. From the very

73. Lindenfeld, Sachs, and Steinberger [A.72], 531–32.

74. Abraham Pais produced a "qualitative investigation" of why the mysterious *V* particles, recently discovered in cosmic ray experiments, were so long lived. Pais [A.55], 663. He used arguments about symmetries under parity to establish selection rules regarding what types of decays the *V* particles could undergo and then used coupling-constant arguments to approximate relative decay rates for various allowed processes—including those processes which involved closed loops as in figure 5.10*a–b*.

beginning, physicists encountered sharply contrasting visions of what the unadorned lines of Feynman diagrams were meant to represent.

Beyond the realm of weakly coupled electrodynamics, these initial seeds of dispersion quickly separated into different directions. Faced with the breakdown of perturbative methods for the strongly coupled meson interactions, theorists drove the Feynman-Dyson split wider and wider. The diagrams, so carefully honed for perturbative studies, offered little guidance in the mesonic realm. Given the bloated size of the meson-nucleon coupling constant, Dyson's elegant, rule-bound instructions for using Feynman's diagrams provided little help for charting mesons' and nucleons' interactions—indeed, Feynman himself threw up his hands and declared that his own diagrams were not to be trusted when it came to nuclear forces. The diagrams' great success in tackling the infinities of QED provided few lessons—and little incentive—to ply them in the new realm; their prior success held little promise of future rewards.

All the same, meson theorists were quick to vote with their pencils. Even as they eschewed perturbative techniques, they made extensive use of Feynman diagrams when exploring the new particles and forces. In the process, ties between the diagrams and Dyson's formal rules for their use became further and further attenuated, stretching the plastic diagrams even more than the initial differences between Feynman and Dyson had done. Young theorists such as Paul Matthews, Jack Steinberger, and Abraham Pais tinkered with the form of their diagrams, adding arrows here, deleting spacetime labels there, all in an effort to craft their diagrams to better suit their new—and still evolving—purposes. The result: between the Feynman-Dyson split and the failure of perturbative approaches for treating nuclear forces, Feynman diagrams' pictorial forms and calculational roles became more and more differentiated—a visual pastiche arising from the theorists' incessant bricolage. As we will see in the following chapter, order may yet be discerned within this hodgepodge of straight lines and squiggles, an order stemming from where and by whom these various young theorists were trained.

· 6 ·

Family Resemblances

> Consider for example the proceedings we call "games." I mean board games, card-games, ball-games, Olympic games, and so on. What is common to them all?—Don't say: "There *must* be something common, or they would not be called 'games' "—but *look and see* whether there is anything common to all.—For if you look at them you will not see something that is common to *all*, but similarities, relationships, and a whole series of them at that. . . . And the result of this examination is: we see a complicated network of similarities overlapping and criss-crossing: sometimes overall similarities, sometimes similarities of detail.
>
> LUDWIG WITTGENSTEIN, 1945[1]

The two seeds of dispersion that we examined in the previous chapter led physicists to deploy Feynman diagrams in a wide variety of ways; physicists' appropriations of the diagrams exemplified the second meaning of "dispersion," yielding a "scattered order." Yet there are patterns amid the dispersion that get lost if we focus too narrowly on individuals' diagram-laden calculations. Young theorists' uses of Feynman's diagrams, in their dissertations and in their articles, provide an information-rich road map for charting intellectual traditions and local pedagogical practices. Consider again the diagrams in figures 1.3 and 1.4. These examples share many features, but no two of them share all the same features: some use wavy lines, others squiggly or dashed lines; some lines are inclined at angles, others march straight up the page; some lines have particles labeled, while others include spacetime or momentum-state information, and so on.

The overlapping similarities illustrate an important Wittgensteinian theme: "family resemblances." Descriptive terms like "game," Wittgenstein argued, do not have a single inherent, essential meaning. Instead, the items we call "games" share some features from an overlapping set of similarities. Not all of them have the same qualities—some use balls, some involve teams, some employ points to determine winners—and yet we usually have little difficulty identifying things as diverse as golf and chess as "games."[2] In this chapter, I build on Wittgenstein's important observation to bring some order to the visual

1. Wittgenstein, *Philosophical Investigations* (1958), 31–32 (§66); emphasis in original.
2. Ibid., 31–34 (§65–72); Fogelin, *Wittgenstein* (1987 [1976]), 133–40.

cacophony of Feynman diagrams that proliferated immediately after their introduction. Just as we have difficulty defining "games," we are hard pressed to identify any inherent meaning or unique identifying marks that would qualify a given line drawing as a "Feynman diagram." The diagrams shared certain features, but no single set of shared features can be found across all diagrams. Yet the *density* of shared or overlapping features shows a remarkable local specificity: diagrams drawn by Robert Marshak's students at Rochester look more like Marshak's own diagrams than they do the diagrams from Norman Kroll's group at Columbia; diagrams drawn by Bethe's students at Cornell look more like each other than the diagrams drawn by their peers at Illinois or Iowa or Chicago. Clearly students learned from their mentors something more than just an abstract or first-principles idea of what a Feynman diagram should look like or how it might be used. As they appropriated Feynman's line drawings, students learned to recognize *relevant* similarities and differences—those differences that made a difference—depending on the specific contexts of use.

Attending carefully to theorists' training thus allows us to put flesh on Wittgenstein's useful notion: diagrams drawn by members of local pedagogical "families" shared classes of resemblances. These pedagogical family resemblances, in turn, lead to a more nuanced version of the postdoc-cascade description elaborated in part 1: we are led to study *local* schools and approaches. As the examples throughout this chapter make clear, students and mentors crafted their diagrams in pursuit of distinct types of calculation; the diagrams' pictorial forms changed in step with theorists' choices of calculational goals and pedagogical priorities. In the final section of the chapter, I consider some of the determinants behind these local choices of what to work on and how to train one's students. In the United States during the postwar years, the pressing needs of neighboring experimental groups provided a strong factor in shaping theorists' diagrammatic choices. In Europe, different institutional arrangements—including in general a further separation between theorists and experimentalists—led those few physicists who did pick up the diagrams to direct them to very different ends. The differences—in training pedigree and experimental exigency—were inscribed on the diagrams themselves.

KROLL'S PERTURBATIVE BOOKKEEPERS

Just two weeks after Dyson's first taste of fame at the January 1949 American Physical Society meeting in New York, at which news of his diagrammatic accomplishments had begun to spread, he wrote to his parents about his current plans. "During the last week I have started some very heavy calculations, which are restful as they do not require any serious thinking for the performance. They are a fairly straightforward, but important application of the theory which Feynman and I have been developing." Compared with the rigors of preparing his long second paper, these new calculations seemed relatively unexciting—a mere exercise to make sure the abstract formalism could be applied to concrete problems. "If all goes well," he continued, the new calculations "will be finished in about two weeks and 200 pages, and the result will be one little number, the fourth-order correction to the electron's magnetic moment."[3] In order to test the mettle of his new formalism, Dyson had set his sights on extending Julian Schwinger's calculation of how radiative corrections affect an electron's magnetic moment, μ. Recall that Rabi and his fellow experimentalists at Columbia University, using surplus wartime radiofrequency equipment, had measured tiny deviations from the predicted values of various electromagnetic quantities. Schwinger's original calculation from November 1947—performed, of course, without any diagrams—had shown that the virtual particles that always surround an electron would alter the value of an electron's effective magnetic moment. These deviations, moreover, could help to explain the lion's share of the Columbia group's measurements. The first round of radiative corrections, Schwinger found, would add a tiny contribution, $\delta\mu = (\alpha/2\pi)\mu_0$, to the electron's basic value of $\mu_0 = e\hbar/(2mc)$, where $\alpha = e^2/\hbar c \sim \frac{1}{137}$.[4] Armed with his new diagrammatic techniques, Dyson now proposed chasing these radiative corrections even further: how would the next-higher round of radiative corrections, proportional to e^4, affect μ?

Dyson did not labor on the new calculations alone. He recruited two of his fellow postdocs at the Institute for Advanced Study, Norman Kroll and Robert Karplus. During the past few months, Dyson had worked with Kroll and Karplus to recheck both Schwinger's and Feynman's calculations of the second-order shift to an electron's energy levels in hydrogen—the Lamb shift—to find out why Schwinger and Feynman kept getting different answers. Karplus and Kroll were particularly well suited for the work: Karplus had just arrived from his graduate studies at Harvard, where he had worked with Schwinger, and Kroll, just as he was completing his Ph.D. at Columbia, had published with Willis Lamb a lengthy paper in which they had used prewar perturbative techniques to calculate the second-order Lamb shift. Throughout the autumn,

3. Freeman Dyson to his parents, 15 Feb 1949, in *FJD*.

4. The "fine-structure constant," α, is a dimensionless number. Most often, physicists adopted a system of units in which $\hbar = c = 1$, leaving $\alpha = e^2 \sim 1/137$.

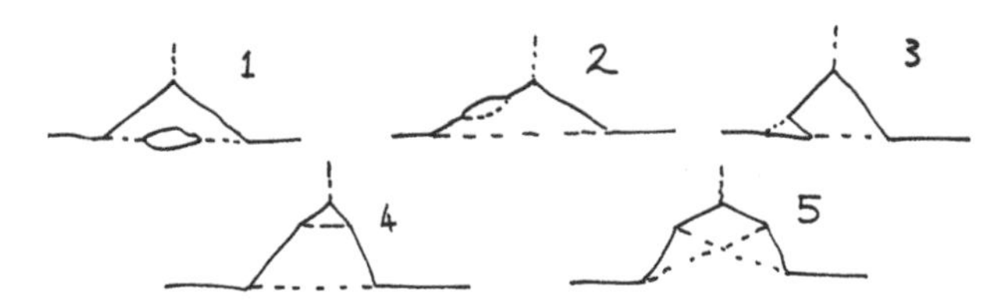

Figure 6.1. Fourth-order Feynman diagrams for the electron's magnetic moment. (Source: Freeman Dyson to Hans Bethe, 10 February 1949, in *HAB*, Folder 10:20.)

Dyson coached them both in how to redo these calculations using Feynman's diagrams, and now he enlisted them to press on with the new fourth-order calculations.[5] He reported on the group's progress to Hans Bethe in February: "Some of us here [at the Institute] have now begun a calculation of the 4th order electron magnetic moment. As you may imagine, it is quite tough. There are essentially five different contributions to be considered." (See fig. 6.1.) At this point, Dyson had already made some headway: "As [diagram] 1 is the easiest, I started with that, and it came out very nicely," he reported to Bethe. "It gives an increase of the magnetic moment by a relative amount

$$\frac{\alpha^2}{6}\left[\frac{119}{12\pi^2} - 1\right].$$

Since $119/(12\pi^2) = 1.0048$, this is not a very big effect. The way the (–1) comes and nearly cancels the whole thing out is very peculiar and beautiful."[6] As Norman Kroll remembers it, Dyson handed over all his notes for this calculation to Kroll and Karplus to get them up to speed.[7] And so they were off.

Dyson's original prediction that the entire calculation would be completed in two weeks proved overly optimistic. Whether he lost interest in the project or gallantly stepped aside to let his compatriots garner credit for the work remains unclear. Either way, goaded by Dyson's example, Karplus and Kroll worked hard on the remaining parts of the calculation, and by late July 1949 they had enough results in hand to send off a brief notice to the *Physical Review*.[8] On they toiled, tidying up their laborious calculation. Finally, in mid-October, they submitted their full analysis, announcing their work as an example of "the feasibility of Dyson's program," in which "the methods of Dyson have been

5. Freeman Dyson to his parents, 14 Nov 1948 and 4 Dec 1948, in *FJD*. See also Kroll and Lamb, "Self-energy" (1949).

6. Freeman Dyson to Hans Bethe, 10 Feb 1949, in *HAB*, Folder 10:20.

7. Kroll interview (2002).

8. Karplus and Kroll, "Fourth-order corrections" (1949). Their letter to the editor appeared in the same issue as Feynman's pair of articles.

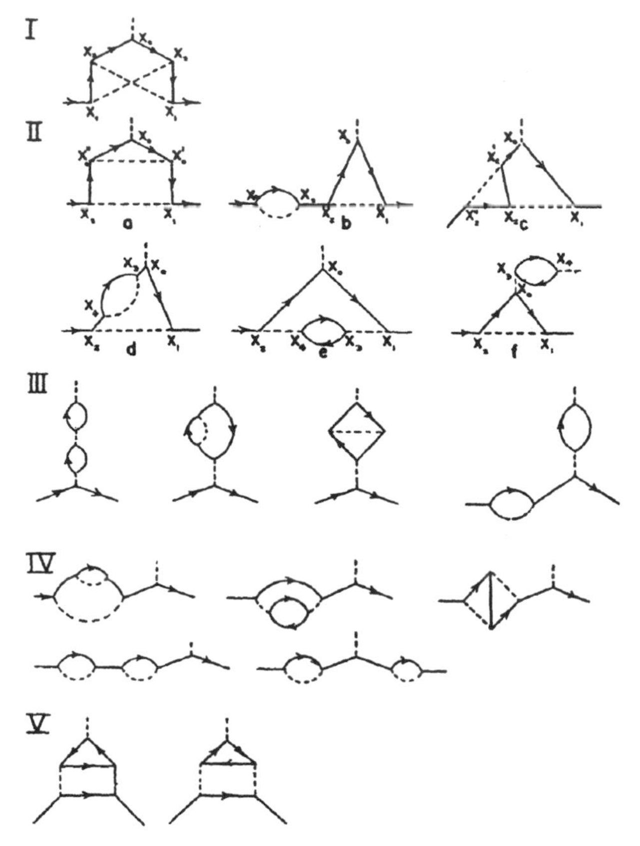

Figure 6.2. Fourth-order diagrams for the scattering of an electron in an external electromagnetic field. (Source: Karplus and Kroll [A.11], 537.)

followed quite closely."[9] They followed Dyson's methods "quite closely," indeed. Karplus and Kroll began, as Dyson had taught them, by writing out Dyson's infinite series expression for the S-matrix, complementing the mathematical expression with a full paragraph describing how S related incoming and outgoing states. Next they specified the form of the interaction Hamiltonian, $\mathbf{H}_I$, and introduced the propagation functions for electrons and photons, $S_F(x)$ and $D_F(x)$, using Dyson's notation and terminology. Only at this stage, with Dyson's field-theoretic machinery laid out in full, did the Institute postdocs introduce the five distinct classes of Feynman diagrams involved in their unprecedented calculation. (See fig. 6.2.) These classes of diagrams represented each of the topological possibilities, to order e^4, in which an electron could interact with an external electromagnetic field. The diagrams derived

9. Karplus and Kroll [A.11], 536.

from Dyson's methodical "connect-the-dots" prescription: first lay out a series of points, or vertices, and then draw all the ways in which virtual electron, positron, and photon lines could connect the vertices (subject to Dyson's basic rule that two electron or positron lines always meet one photon line at each vertex). For their fourth-order analysis, the postdocs needed to include four vertices in their diagrams—x_1, x_2, x_3, and x_4—in addition to the point, x_0, at which the external electromagnetic field interacted with the incoming electron.

Not surprisingly, given the provenance of the postdocs' calculation, their published diagrams showed more than a passing resemblance to Dyson's original diagrams, scribbled down so hastily in his letter to Bethe eight months earlier. Like Dyson, Karplus and Kroll drew their diagrams as conglomerations of polygons, with none of Feynman's Minkowskian imagery in sight. Like their mentor, Karplus and Kroll likewise applied Dyson's antiparticle arrows with care on all the electron and positron lines. As the postdocs pointed out explicitly, these arrows carried mathematical bite: precisely because the arrows in the triangles of the class V diagrams ran in opposite directions, the contributions from this pair of diagrams exactly cancelled each other. What Feynman had at first overlooked in his energetic doodling, Dyson's disciples could discern with tiny arrows and clarify with a single sentence.[10]

With all the relevant diagrams enumerated, the substantive part of the calculation could begin. Karplus and Kroll explained how each of the diagrams could be translated uniquely, using Dyson's rules, into integrals over the electron and photon propagation functions, $S_F(x)$ and $D_F(x)$. As an example, they turned to diagram II*e*—the selfsame diagram with which Dyson had begun the calculation the previous February—writing down the corresponding integral:

$$\begin{aligned} M^{\mathrm{II}e} = \frac{e\alpha^2\pi^2}{4\hbar c} \int & d^4x_0 d^4x_1 d^4x_2 d^4x_3 d^4x_4 \overline{\psi}(x_1) \\ & \times \gamma_\nu S_F(x_0 - x_1)\gamma_\mu A^e_\mu(x_0) S_F(x_2 - x_0)\gamma_\lambda D_F(x_2 - x_4) \\ & \times \mathrm{Tr}[\gamma_\lambda S_F(x_3 - x_4)\gamma_\nu S_F(x_4 - x_3)] D_F(x_3 - x_4)\psi(x_2). \end{aligned}$$

10. Karplus and Kroll [A.11], 539. Karplus and Kroll also explained that the diagrams within classes III and IV did not contribute to the electron's magnetic moment: each of these diagrams represented modifications either to the external field independent of the field's interaction with the electron (class III) or modifications of the electron independent of its interaction with the external field (class IV), and hence they could each be ignored for this specific calculation. Ibid. With this reduction, the diagrams of interest reduced to the same set that Dyson had first written about to Bethe, with only diagrams II*b* and II*f* added beyond Dyson's original reckoning.

They composed this integral piecemeal by tracing through each element of the diagram.[11] Working from the right-hand side of the diagram, they retraced the steps that the electron had followed before exiting the interaction. At the point x_1, the electron had scattered with a virtual photon; this vertex contributed a factor of $e\gamma_\mu$. To get to this vertex, the electron had traveled from x_0 to x_1—hence the factor of $S_F(x_0 - x_1)$. At x_0, the electron had interacted with the external electromagnetic field, $A^e_\mu(x_0)$; the vertex at x_0 thus contributed a factor of $e\gamma_\mu$. Before that, the electron had traveled from x_2 to x_0—so the postdocs dutifully wrote in a factor of $S_F(x_2 - x_0)$. At x_2, the electron had emitted a virtual photon—enter a factor of $e\gamma_\lambda$ for the vertex, and $D_F(x_4 - x_2)$ for the photon's propagation. En route, the photon disintegrated into a virtual electron-positron pair at the point x_4; the pair formed a closed loop, annihilating at the point x_3, where it emitted a new virtual photon. The closed-loop process contributed $e\gamma_\lambda$ for the vertex at x_4 and $e\gamma_\nu$ for the vertex at x_3, sandwiched inside the trace over the electron-positron pair's propagation functions, $S_F(x_4 - x_3)$ and $S_F(x_3 - x_4)$. Finally, the new photon, emitted at x_3, traveled from there to x_1—hence the final factor of $D_F(x_3 - x_1)$. So followed just one of the distinct mathematical contributions to be summed up to reach the final fourth-order result.

Karplus and Kroll proceeded to march through the same series of steps to evaluate the contributions from the remaining relevant diagrams—a painstaking calculation rendered feasible by using Feynman's diagrams in the strict manner that Dyson had specified.[12] In fact, writing down integrals from the diagrams turned out to be the easy step. The hard part was evaluating these integrals into numerical answers. In the bulk of their lengthy article, they drew on a host of calculational tricks and methods that Schwinger, Feynman, and Dyson had worked out for evaluating these integrals: perform a Fourier transformation to write the integrals in terms of momenta rather than space-time variables; make a particular coordinate transformation here, combine denominators there; distinguish between the finite and divergent portions of each integral; integrate over the auxilliary variables; and so on. Karplus and Kroll exhibited eighteen lines of algebra among the steps involved in evaluating the integral for diagram II*e* alone; these were only the steps they chose to

11. Karplus and Kroll [A.11], 538.

12. Rather than write out the remaining integrals, as they had done for diagram II*e*, Karplus and Kroll explained that each integral for the relevant fourth-order diagrams could be generated by systematically substituting the relevant second-order corrected terms—such as the second-order electron self-energy, photon self-energy, and vertex correction—one at a time for the lowest-order propagation and vertex functions. See Karplus and Kroll [A.11], 538–39.

make explicit, interspersed with such phrases as "On extensive rearrangement, the numerator of the integrand can be brought into the form . . . "[13] This long series of steps led them to the answer Dyson had first reported to Bethe for this diagram many months earlier; Dyson's notes in hand, Karplus and Kroll worked hard to reproduce these steps and then repeat them for the remaining diagrams. Indeed, diagram II*e* was one of the easiest diagrams to evaluate, as Dyson had originally foreseen. One of the other terms—the integral corresponding to the class I diagram—proved simply too monstrous to write out in full: after setting up the integral in Dysonian fashion, the two postdocs quoted their final result, explaining in a footnote, "The details of two independent calculations which were performed so as to provide some check of the final result are available from the authors."[14]

After fourteen densely packed pages and these two additional sheaves of unpublished calculation, Karplus and Kroll had provided the next round of tiny corrections to the basic, lowest-order result. When the dust had settled and each of the integrals had been evaluated and added together, they had demonstrated that an electron in an external electromagnetic field will behave as if it had a magnetic moment

$$\mu_{\text{eff}} = \left[1 + \frac{\alpha}{2\pi} - 2.973\frac{\alpha^2}{\pi^2}\right]\mu_0 \simeq 1.001147\mu_0$$

instead of Dirac's prewar value of μ_0 alone. More than this, they had shown how the second-order result, $\mu_{\text{eff}} = 1.001162\ \mu_0$, first calculated without diagrams by Schwinger and recalculated diagrammatically by Feynman and Dyson, was modified by the fourth-order contributions. With this extended and laborious example, for which they acknowledged "much helpful discussion with F. J. Dyson," they had demonstrated how to use the diagrams to track the ever-tiner wisps of QED's radiative corrections.[15]

13. Karplus and Kroll [A.11], 543, 546. The steps involved in evaluating diagram II*e* are reported in equations 27–32 and 52–53, many of which stretch over several lines of algebra.

14. Karplus and Kroll [A.11], 548n23.

15. Karplus and Kroll [A.11], 549. The historical significance of Karplus and Kroll's landmark calculation is hardly diminished by the fact that when a few physicists revisited the calculation almost a decade later, they found several arithmetic mistakes that had krept into Karplus and Kroll's original evaluation. See Sommerfield, "Magnetic dipole moment" (1957); Sommerfield, "Magnetic moment" (1958); Petermann, "Magnetic moment" (1957); Petermann, "Fourth order" (1957); Petermann, "Fourth order" (1958); and Terent'ev, "Application of dispersion relations" (1963). The revised value for the fourth-order terms weighed in at $-0.328\ \alpha^2/\pi^2$, rather than Karplus and Kroll's original value of $-2.973\ \alpha^2/\pi^2$, shifting the overall value to $\mu_{\text{eff}} = 1.0011596\ \mu_0$. Experiments

By the time Karplus and Kroll submitted their long article in October 1949, Kroll had begun teaching at Columbia University, while Karplus remained at the Institute for the second year of his postdoctoral study. Columbia's I. I. Rabi had been anxious to hire a young theorist; until that time, it had become common for graduate students at Columbia who were interested in theoretical physics to work with temporary visitors who happened to be in town during their sabbaticals from other places, such as Hans Bethe, Aage Bohr, and others. As early as November 1948, Rabi had offered the job to Dyson—just one month after he submitted his first article on the diagrams. Dyson had to turn down the offer with "bitterness in my heart," given his requirement to return to England for two years after his Commonwealth Fellowship. Rabi next approached Kroll with the offer. Kroll had spent only one year at the Institute and had been looking forward to a second year of postdoctoral study. But, as Kroll recently explained, Rabi "had a theory of the 'fellowship bum,' so he would not hold the position open." Kroll thus accepted the assistant professorship and arrived at Columbia for the autumn 1949 semester. Immediately he began to spread the gospel of Dyson's new techniques, offering a lecture course that fall and advising a series of Columbia's graduate students.[16]

From the start, Kroll's students worked to extend the theoretical program that Dyson, Karplus, and Kroll had begun. As Kroll put it recently, "I had my students do what I knew how to do."[17] Several of Kroll's earliest students and postdocs at Columbia worked on additional fourth-order QED calculations: whereas Karplus and Kroll had tackled the fourth-order contributions to an electron's magnetic moment, Kroll's students worked on such topics as the fourth-order contributions to an electron's self-energy and the fourth-order contributions to the Lamb shift.[18] Just as their adviser had done, Kroll's

---

performed during the 1980s, using single-electron magnetic traps instead of atomic hydrogen, measured $\mu_{\text{eff}}/\mu_0 = 1.0011596522015 \pm 4.3 \times 10^{-12}$, while Kinoshita's eighth-order perturbative calculation yielded $\mu_{\text{eff}}/\mu_0 = 1.0011596522160$. See Hughes and Kinoshita, "Anomalous *g* values" (1999); and Kinoshita and Nio, "Revised α^4 term" (2003).

16. Freeman Dyson to his parents, 4 Nov 1948 and 11 Mar 1949, in *FJD;* Kroll interview (2002). On Columbia students' need to work with visiting faculty, see also Kaiser, "Francis E. Low" (2001).

17. Kroll interview (2002).

18. Pollock, *Fourth Order Corrections* (1952); Weneser, *Fourth-Order Radiative Corrections* (1952); see also Kroll and Pollock [A.58]; and Weneser, Bersohn, and Kroll [A.90]. Richard Bersohn was a postdoc with Kroll at Columbia who repeated Weneser's calculations to provide a "quasi-independent" check on the final results. Weneser, *Fourth-Order Radiative Corrections* [1952], ii. As Pollock was completing his dissertation, Kroll learned that one of Feynman's students at Cornell, Robert Frank, was also working on fourth-order corrections to an electron's self-energy, so Kroll insisted that Pollock publish on a different topic. Kroll interview (2002); see also Frank [A.41].

students placed their work squarely in the Dysonian mold. Joseph Weneser explained in his 1952 dissertation, for example, that "The scattering calculation was carried out using the Dyson S-matrix formulation of quantum electrodynamics and the Dyson renormalization program."[19] In their dissertations, citations to Dyson's 1949 articles appeared so frequently that the students adopted short abbreviations to simplify their footnotes. (The only other paper cited as frequently in their work, for which they also adopted shorthand abbreviations, was Karplus and Kroll's article.)[20] From Kroll, they learned how to begin their calculations by drawing the relevant Feynman diagrams. Just as Kroll had learned from Dyson, the Columbia students built up their diagrams by following Dyson's systematic connect-the-dots procedure, yielding the same kinds of polygonal diagrams that Karplus and Kroll had scrutinized. (See fig. 6.3.) The students also learned to follow Kroll's example when evaluating the diagrams, translating each in turn into the appropriate integrals "immediately on application of Dyson's prescription," as one of them explained, and then working through the same series of laborious calculational steps to convert these integrals into real numbers. Students practiced rearranging the numerators of their integrals by using the commutation rules for γ_μ-matrices, combining denominators using Feynman's trick, integrating over the auxilliary variables, and the like—these are the details that fill their dissertations and articles. Over the 1950s and 1960s, Kroll trained several graduate students in the ins and outs of making these subtle higher-order perturbative calculations in QED, including Franklin Pollock (Ph.D. 1952), Joseph Weneser (Ph.D. 1952), Robert Mills (Ph.D. 1955), Eyvind Wichmann (Ph.D. 1956), Arthur Layzer (Ph.D. 1960), Daniel Zwanziger (Ph.D. 1961), and Maximillian Soto, Jr. (Ph.D. 1968).[21]

Kroll had an additional reason for pushing his students to master the intricacies of such high-order QED calculations. Columbia was home to the experimentalists who had made the first stunning observations of tiny deviations from prewar theoretical predictions for various electromagnetic quantities. Over the 1950s, they continued to refine these experiments, pushing the

---

This is why Pollock's published work focused on second-order radiative corrections to hyperfine structure, upon which he brought to bear all the same diagrammatic-calculational techniques that he had just mastered in his dissertation research.

19. Weneser, *Fourth-Order Radiative Corrections* (1952), 5.

20. Pollock, *Fourth Order Corrections* (1952), 2–3.

21. Quotation from Weneser, *Fourth-Order Radiative Corrections* (1952), 14. See also Mills, *Fourth-Order Radiative Corrections* (1955); Wichmann, *Vacuum Polarization* (1956); Layzer, *Free-Propagator Expansion* (1960); Zwanziger, α^3 *Corrections* (1961); and Soto, *Fourth-Order Radiative Corrections* (1968).

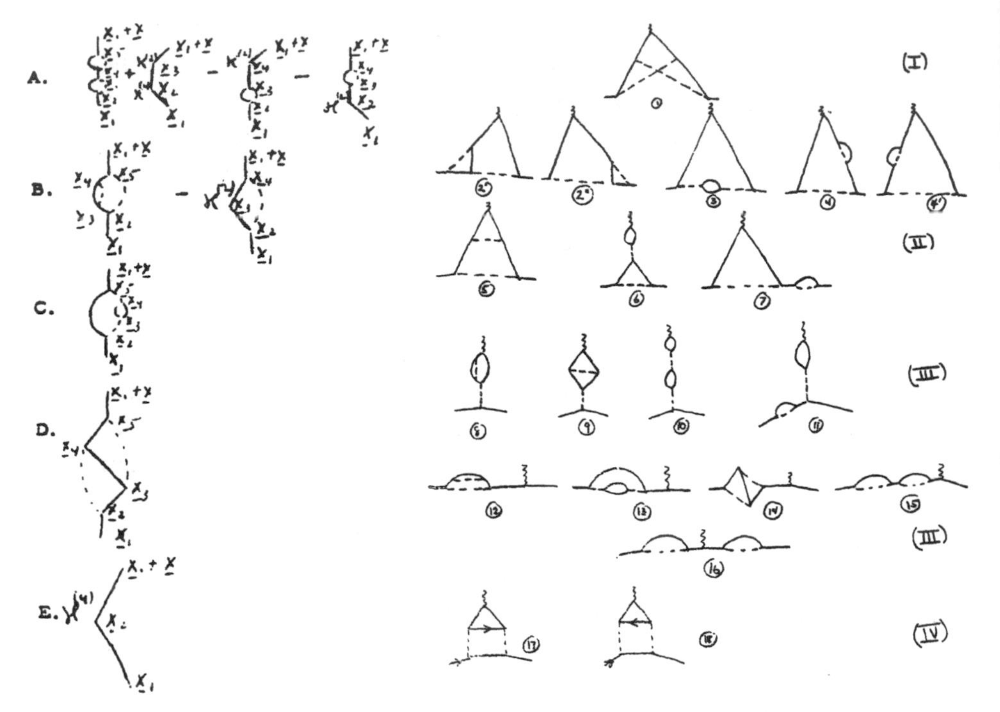

Figure 6.3. Norman Kroll's students' diagrams. *Left,* Feynman diagrams for an electron's self-energy. *Right,* Feynman diagrams for an electron in an external field. (Sources: *left,* Pollock, *Fourth Order Corrections* [1952], 8; *right,* Weneser, *Fourth-Order Radiative Corrections* [1952], 6.)

measured values of such quantities as the electron's magnetic moment and the Lamb shift to higher and higher precision. For much of this period, Columbia remained the only place where such experiments were taking place; as more and more of the country's experimental physicists turned their sights to nuclear and meson physics, pursued with the ever-growing cyclotrons, ultra-high-precision electron physics remained Columbia's in-house specialty. Columbia thus remained nearly the only place where fraction-of-percent accuracy actually mattered when it came to QED calculations, in order to compare theoretical values with freshly measured results. Kroll kept abreast of the experimentalists' work; his neighbor was Polykarp Kusch, one of the leaders of the Columbia experimental groups, and the two used to walk home from work together almost every day.[22]

With Kroll's encouragement, his students kept close watch on their fellow experimentalists' work, too: nearly all of them noted that recent experimental

22. Kroll interview (2002).

refinements motivated their calculations. The opening paragraph of Weneser's dissertation explained, for example, that "the recently completed precise measurements" of the energy levels within hydrogen and helium—recently completed at Columbia—"make possible a more severe test of the theory than can be provided by the second order calculations," and hence, "it is clear that the fourth order terms may be of an experimentally significant magnitude." Pollock similarly explained that in order to properly extract the values of fundamental constants, such as α, from recent "precision measurements"—conducted by Kusch and assistants at Columbia—one needed to quantify the contributions from radiative corrections, which he then proceeded to analyze. When reporting on his dissertation research, Eyvind Wichmann began by citing three recent experiments whose interpretation required an understanding of vacuum polarization—all three experiments having been conducted by Columbia physicists. A few years later, revised experimental values for the Lamb shift, measured by a new crop of Columbia experimentalists, pushed Kroll's students such as Arthur Layzer to scrutinize—and ultimately redo—the previous theoretical calculations. Daniel Zwanziger similarly revisited previous calculations in the light of Kusch's most recent experimental measurements.[23]

The added impetus for Kroll's students to practice making this type of diagrammatic calculation thus remained site specific. Spurred on by local experimental developments, Kroll and his students continued to press Feynman diagrams into service for higher- and higher-order perturbative calculations within QED; only in this way could they continue to test the agreement between Dyson's calculating scheme and the latest homegrown experiments. They used the progressively more complicated diagrams to make more and more accurate calculations of the known interactions between known particles—electrons and photons—at the one place where such high-precision calculations mattered most. As they practiced using Feynman diagrams for these types of calculations, Kroll's students entered the small minority of diagram users who continued to press on with higher-order QED calculations.[24] Most of their

23. Weneser, *Fourth-Order Radiative Corrections* (1952), 1; Kroll and Pollock, "Radiative corrections" (1951), 594; Wichmann and Kroll, "Vacuum polarization" (1954), 232; Layzer, "Lamb shift" (1960); Zwanziger, "α^3 corrections to hyperfine structure" (1961), esp. 1140.

24. Cornell was the other place where higher-order diagrammatic calculations within QED were pursued. Feynman himself had some interest in pushing his diagrammatic techniques to higher-order examples, and he and Bethe helped several graduate students and postdocs with these calculations, including Robert Frank, Laurie Brown, Michel Baranger, and Edwin Salpeter. Soon after Feynman left Cornell, Dyson arrived in his stead, and he also worked with some of these

colleagues, toiling away at some remove from Columbia, picked up the diagrams and fashioned uses for them outside the realm of QED.

MARSHAK'S MESON MARKERS

Whereas Kroll and his students traced their diagrammatic lineage through Dyson and the postdoc cohort at the Institute for Advanced Study, Robert Marshak's group at the University of Rochester in upstate New York learned their diagrammatic trade directly from Feynman and the growing band of diagram users at Cornell. Marshak had earned his Ph.D. under Hans Bethe's direction at Cornell in 1939, working on the nuclear interactions that powered the stars, a topic near to Bethe's heart; only two years before, Bethe had worked out one of the crucial cycles of these stellar nuclear interactions, for which he later received the Nobel Prize. After completing his degree, Marshak ventured to Rochester as an instructor, arriving on campus one week after World War II had broken out in Europe. Before long, Bethe asked Marshak to join him at MIT's Radiation Laboratory to work on radar, even before the United States had entered the war. After working on radar for a year—making frequent trips between MIT and Rochester—Marshak transferred to the Montreal facilities of the expanding Manhattan Project, before Bethe called him to Los Alamos early in 1944 to serve, like Feynman, as a deputy group leader within Bethe's theoretical physics division.[25] Marshak's personal relationships with Cornell physicists strengthened during the war years. He and his wife socialized on the New Mexican mesa with his former dissertation committee members and their wives—the Bethes, Robert Bacher and his wife—as well as with Robert Wilson (who moved to Cornell soon after the war to help lead Cornell's experimental high-energy physics program). Near the end of the war, Marshak convinced Julius Ashkin, who had worked in Feynman's subgroup at Los Alamos, to accept a teaching position in Rochester's physics department. While in Rochester, Ashkin remained in close touch with Feynman; in fact, Ashkin was the only person whom Feynman thanked in his first article on the new diagrams.[26]

After these physicists left Los Alamos for their home institutions, Marshak, Bethe, Feynman, and the others inaugurated a series of regular, informal

---

students. See Frank [A.41]; Baranger [A.45]; Brown and Feynman [A.50]; Baranger, Dyson, and Salpeter [A.68]; Salpeter [A.71]; and Baranger, Bethe, and Feynman [A.93].

25. Marshak, *Internal Constitution of Stars* (1939). On Marshak's work at the Rad Lab, see Marshak interview (1970), pt. 1, 53–54; on Montreal and Los Alamos, see ibid., pt. 2, 3–13.

26. Marshak interview (1970), pt. 2, 20–21, 36. Feynman acknowledged Ashkin's help in Feynman [A.4], 784; and in Feynman, "Mathematical formulation" (1950), 453.

meetings between the Rochester and Cornell groups. The meetings immediately facilitated discussion: even before Marshak had learned how to use Feynman's new diagrams on his own, he had been able to ask Feynman to repeat part of a calculation to determine whether certain terms could indeed be ignored as Marshak had done.[27] The meetings had a great influence on the graduate students in each group as well. Dyson wrote home excitedly about one of these joint meetings soon after he arrived at Cornell in the autumn of 1947.[28] Marshak's graduate students formed similar impressions. Morton Kaplon, for example, who completed his dissertation under Marshak in 1951, remembers that "We [at Rochester] were very fortunate in those days in being so close to Ithaca. Feynman was a frequent visitor to Rochester giving many seminars to the theoretical group. . . . He was quite dramatic playing the role of the electron moving backward and forward in time." Nor was the traffic one-way: Kaplon and several of Marshak's other graduate students made frequent trips to Ithaca to meet with Feynman. As Kaplon recalls, "Feynman was very liberal with his time. At Marshak's behest he met with the group of graduate students just to discuss physics for sessions that lasted several hours."[29] During the summers, the Rochester group also benefited from visits by recent members of Dyson's circle at the Institute for Advanced Study, such as Kenneth Case and Kenneth Watson; David Feldman took up a teaching position at Rochester in 1950 directly from his postdoc at the Institute as well. H. Pierre Noyes began teaching in Rochester's physics department in 1951, after learning about Feynman diagrams from Dyson while a postdoc in Birmingham. Yet the dominant influence remained Feynman. The diagrammatic genealogies became obvious in the students' work: whereas Kroll's students at Columbia appealed explicitly to Dyson's articles and worked within his "program," Marshak's students announced, as Albert Simon did in his 1950 dissertation, that "The basic theory and almost all the notation used in this thesis is due to Feynman."[30]

The young Rochester theorists worked in the midst of a very different experimental context than Kroll's students. Whereas the Columbia experimentalists had cornered high-precision tests of QED, Rochester's department was gearing

27. See Foldy and Marshak, "Production of π-mesons" (1949), 1493n4.

28. Freeman Dyson to his parents, 19 Nov 1947 and 27 Nov 1947, in *FJD*.

29. Morton F. Kaplon, e-mail to the author, 13 Dec 2000.

30. Simon, *Bremsstrahlung* (1950), 24. Simon thanked Case in his acknowledgements (ibid., i); Morton Kaplon thanked Watson in his dissertation: Kaplon, *Meson Production* (1951), iin15. On Case's and Watson's visits to Rochester during summer 1950, see also Case interview (2002). On Feldman's and Noyes's appointments, see *AMS*. Noyes acknowledged Dyson for "illuminating discussions" and for his lectures in Birmingham in his first diagrammatic article: Noyes [A.60], 346, 348.

up to become the first physics department outside Berkeley to house its own high-energy synchrocyclotron, which came on line with funding from the Office of Naval Research in 1949. Marshak's own interests had always tilted heavily toward nuclear physics and mesons. When he first arrived at Rochester in 1939, Victor Weisskopf, at the time the lead theorist in Rochester's department, had tried to entice Marshak to join him in chasing down the mysteries of QED's renormalization. Marshak had demurred and consequently never really learned the prewar perturbative methods for QED.[31] Nor was he caught up in the excitement at the 1947 Shelter Island conference when Columbia's Willis Lamb announced his recent measurement of a tiny shift in electrons' energy levels in hydrogen. Instead, Marshak's attention had been focused squarely on mesons. During the Shelter Island conference he famously, if tentatively, suggested what quickly became known as the "two-meson hypothesis": perhaps the meson found in cosmic ray experiments in 1937 was different from Yukawa's predicted meson, and only the latter (as yet undiscovered) particle served as the nuclear force carrier. Feynman liked Marshak's suggestion so much he suggested, with his usual sense of humor, that the new particle be called the "marshon." After the Shelter Island participants returned to their campuses from the meeting, they learned that Cecil Powell's cosmic ray group in Bristol, England, had detected a second type of meson that seemed to fit Marshak's (and Yukawa's) bill. Having only two Greek symbols on his typewriter, Powell had labeled the new particle π and the older meson, first detected a decade earlier, μ. These became known as the "pi meson" and "mu meson"—soon shortened to the "pion" and "muon."[32]

Although the name "marshon" never stuck, Marshak quickly became a recognized leader in meson physics, consulting with the experimental groups at Berkeley, Brookhaven, and Bristol, all the while emerging as the in-house consultant for Rochester's developing synchrocyclotron group. Just months after Rochester's new machine came on line and began to produce its own pions, Marshak initiated a new series of annual meetings to facilitate the exchange of information between experimentalists and theorists working in the new high-energy realm. Beginning in December 1950, the meetings convened at the University of Rochester and hence became known as the "Rochester

31. Marshak interview (1970), pt. 1, 55; and pt. 2, 47. On the Rochester synchrocyclotron, see ibid., pt. 2, 32, 37, 63.

32. Marshak and Bethe, "Two-meson hypothesis" (1947); Marshak interview (1970), pt. 2, 47–57; and Marshak, "Origin of the two-meson theory" (1985). On the 1947 detection of the pion, see Galison, *Image and Logic* (1997), 196–210; on the selection of the names π and μ, see Fitch and Rosner, "Elementary particle physics" (1995), 653.

Figure 6.4. Robert Marshak lecturing during the 1950s. (Source: AIP Emilio Segrè Visual Archives, *Physics Today* collection.)

conferences." (See fig. 6.4.) Marshak worked hard to guarantee that Rochester would be the place to study mesons.[33]

Organizing the conferences was one thing; understanding meson's interactions quite another. Feynman, Dyson, Karplus, and Kroll had all known the relevant properties of electrons and photons before they began their diagrammatic QED calculations; they had known, moreover, the basic form of these particles' interactions, as specified by the interaction Hamiltonian, $\mathbf{H}_I$. No such clarity could be found in the mesonic realm. During the mid-1940s, Pauli, Bethe and others had drawn on analogies to QED to spell out the various possible ways in which mesons and nucleons could interact. Yet with so few properties of the mesons known, the possibilities remained frustratingly open. For one thing, the pions' characteristics remained uncertain: did they carry zero or one unit of spin? Were they even or odd under parity transformations? The way in which they interacted with nucleons remained equally unclear: did

33. Marshak interview (1970), pt. 3, 32b–39; Marshak, "Rochester conferences" (1970); Marshak, "Scientific impact" (1989); and Polkinghorne, *Rochester Roundabout* (1989). The name "Rochester conference" stuck even when different campuses—both foreign and domestic—hosted the annual gatherings, beginning in 1958.

the interaction depend on the particles' spin, and if so, how? As Pauli and Bethe enumerated, the various possibilities led to eight distinct choices for the meson-nucleon $\mathbf{H}_I$, each still in the running for describing the nuclear domain.[34] Pauli and Dyson corresponded during February 1951, airing their frustrations at the variety of meson models still in play.[35] Gregory Breit sounded an even more pessimistic note two years later in a lengthy review article. The meson models came in an "almost infinite variety of modifications," introducing a "superabundance of parameters," he complained. "If one wished to be cynical," Breit chided, "one could emphasize a similarity between styles in women's hats and styles in meson theories."[36] Hardly a ringing endorsement from Yale's senior theorist. Added to these woes was the well-recognized problem that the new pions interacted strongly with protons and neutrons: whereas electrons and photons interacted weakly, as governed by the small charge of the electron, $e^2 \sim \frac{1}{137}$, pions and nucleons interacted strongly, with a coupling constant somewhere between $g^2 \sim 7$ and 57. Even if meson theorists knew which form of $\mathbf{H}_I$ described the mesonic domain, they would never be able to follow the example of perturbative QED with such a swollen coupling constant.

Marshak decided to tackle these questions head-on. Early in 1949, he developed a research program wedding Rochester's soon-to-be-completed accelerator with his graduate students' research projects. First he drew up a blueprint for what he considered to be interesting experiments for the cyclotron team to work on, many of them geared to finding answers to the nagging questions about pions' spin, parity, and other properties. In tandem, he pulled together a theoretical research group with Julius Ashkin and his growing team of graduate students. Marshak's goal was to develop a phenomenological program: how could the theorists best provide a useful framework within which to interpret

34. The main differences among the candidate $\mathbf{H}_I$'s concerned the postulated symmetry properties of the meson field and of the meson-nucleon interaction. The mesons could be either even or odd under parity transformations (which effectively replaced the field's spatial coordinates with their mirror images; physicists used the prefix "pseudo-" to denote odd-parity fields). Moreover, there were three spin-dependent ways that mesons could couple to nucleons: *scalar* couplings (independent of spin), *vector* couplings (involving either vector meson fields or derivatives of scalar meson fields), and *tensor* couplings (composed of derivatives of vector fields). Each of these couplings could be either even or odd under parity transformations. Pauli, Bethe, and others had shown that the requirement of relativistic invariance limited the total number of distinct possibilities for $\mathbf{H}_I$ to eight. See Pauli, *Meson Theory* (1946); Bethe, *Elementary Nuclear Theory* (1947); Rosenfeld, *Nuclear Forces* (1948); and Marshak, *Meson Physics* (1952), 9.

35. Wolfgang Pauli to Freeman Dyson, 18 Feb 1951, in Pauli, *WB*, 4.1:264–65.

36. Breit and Hull, "Advances in knowledge of nuclear forces" (1953), 205–6. Macalaster Hull confirmed that Breit added the biting phrase about "women's hats" to their paper in a discussion with the author on 17 May 2003 in Los Alamos, New Mexico.

and correlate the new experimental results?[37] One of the students' first steps was to learn how Feynman's new methods might help bring clarity to the nuclear realm. With Ashkin's aid, Marshak's students scrutinized Feynman's pair of articles from 1949; yet as was so often the case, careful reading of the articles alone proved insufficient. As Kaplon explains, "Extending [Feynman's methods] to meson theory was a bit tricky at the time as Feynman's articles were not totally clear. I remember making several visits to Cornell to talk with Feynman to be sure that some of the things we were doing were correct."[38] On they worked, learning more and more about Feynman's vision of how his diagrams could be used to picture physical processes.

Feynman's informal coaching sessions paid off, and soon Marshak's students were putting Feynman diagrams to work. Awash in the sea of open-ended possibilities for the meson-nucleon interaction, and unable to calculate anything beyond lowest order because of the large size of g^2, Marshak and his students had specific goals in mind when they picked up Feynman diagrams and began to calculate—goals different from those of Kroll and his Columbia students. The Rochester group wanted to fashion a useful means of differentiating between the eight different possibilities for the meson-nucleon $\mathbf{H}_I$. The way Marshak and his young team set about accomplishing this goal was to compare the lowest-order predictions of the eight contenders, scanning for *qualitative* differences between the various models' phenomenological predictions. Beginning in the spring of 1950, they fastened onto each model's predictions for the angular distribution of decay products—that is, how likely it would be to detect pions at various angles within the experimental apparatus, as they careened away from the interaction region. The predicted angular distributions, Marshak's team soon realized, depended sharply on the symmetry properties of the mesons and their interactions—symmetry properties encoded in each of the many $\mathbf{H}_I$'s. These symmetry properties led to qualitatively different predictions when one calculated the simplest, lowest-order Feynman diagrams for a given process. In one of the group's earliest diagrammatic articles, submitted in April 1950, they avowed unapologetically that working only with the lowest-order Feynman diagrams "is extremely crude, but it is thought that the qualitative features will persist in a more correct theory." By the time he finished his dissertation under Marshak's guidance in 1951, Kaplon had clearly internalized the same program, repeating virtually identical remarks in

37. Marshak interview (1970), pt. 2, 63; and pt. 4, 1–11. See also Wilson, "Marshak" (1994).

38. Kaplon, e-mail to the author, 13 Dec 2000. Both Simon and Kaplon thanked Ashkin in their dissertation acknowledgments: Simon, *Bremsstrahlung* (1950), i; and Kaplon, *Meson Production* (1951), i.

his dissertation. Never dreaming of pushing any calculation beyond g^2, they would try nonetheless to bring order to the nuclear realm.[39]

In the process, Marshak and his young collaborators refashioned Feynman diagrams, designing them for better use toward their own goal. Because the meson theorists never aimed to calculate any quantity beyond g^2, their diagrams did not need to carry any of the specific labels (such as spacetime coordinates or momentum states) that had proved so handy to those theorists who explored the terra incognita of QED's high-order correction terms. Instead, the Rochester meson theorists used the diagrams as illustrations of basic processes; then they could run down their list of the eight $\mathbf{H}_I$ contenders and calculate *each* model's prediction for the *same* physical process. Merely five pages into Marshak's 1952 textbook on these new techniques (culled primarily from his 1950 lectures), Marshak's students found a full page of these newly refurbished Feynman diagrams. (See fig. 6.5.) Calculating a given lowest-order contribution in itself was no trouble; indeed, theorists had written down analogous lowest-order terms in QED for years without the aid of Feynman diagrams. The complicated task in Marshak's meson physics was navigating the maze of competing *versions* of each lowest-order calculation. By drawing the lowest-order Feynman diagrams and labeling the *particles* involved in each specific process, they could march through each contribution in turn, calculating and recalculating each model's predictions. What mattered in these calculations—what Marshak's students spent their time practicing—was not so much the algebra of γ_μ-matrices or the means of handling large numbers of demoninators, which Kroll's students labored to master. Instead, the practiced techniques included how to draw the relevant lowest-order diagrams, take the nonrelativistic limit of the resulting expressions, and extract information about cross sections and angular distributions.[40] Marshak's students

39. Ashkin, Auerbach, and Marshak [A.13], 266–67; Kaplon, *Meson Production* (1951), 4–5. Marshak's group also pursued a complementary tactic, focusing on how the overall cross sections for various scattering processes varied with energy, since the energy dependence, like the angular distribution, depended on the form of $\mathbf{H}_I$. See Simon [A.15]; and Marshak, *Meson Physics* (1952), 16–17, and chap. 3.

40. In addition to Auerbach, Simon, and Kaplon, many of Marshak's students focused only on nonrelativistic calculations, including Warren Cheston, Lionel Goldfarb, and Norman Francis. Although not relying on Feynman diagrams, their work drew on most of the other calculational details that Auerbach, Simon, and Kaplon mastered: they, too, ran down the list of competing meson models, calculating model-dependent cross sections and angular distributions that could be compared with local experiments. See Cheston, "Interaction of a charged pi-meson" (1952); Goldfarb and Feldman, "High-energy proton-proton scattering" (1952); and Francis, "Photoproduction of π^0 mesons" (1953).

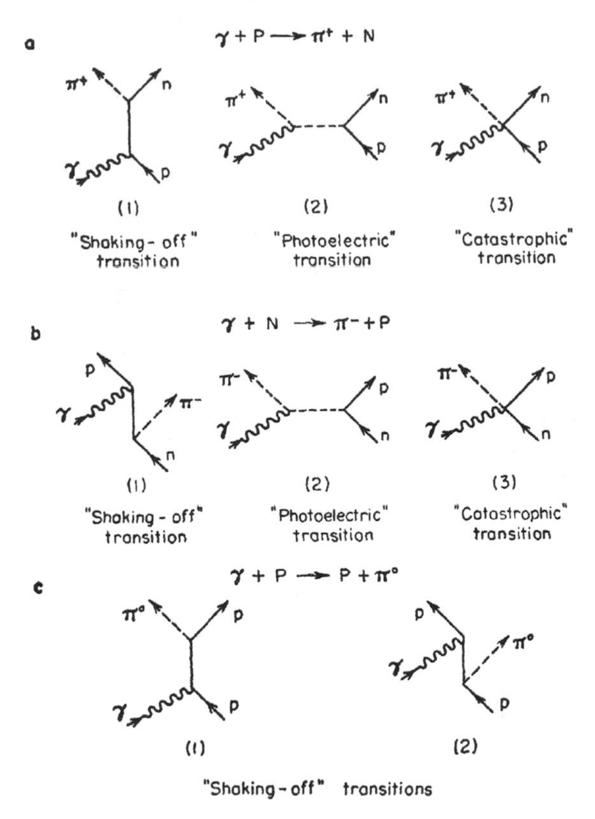

Figure 6.5. Feynman diagrams for photon-nucleon scattering. (Source: Marshak, *Meson Physics* [1952], 6.)

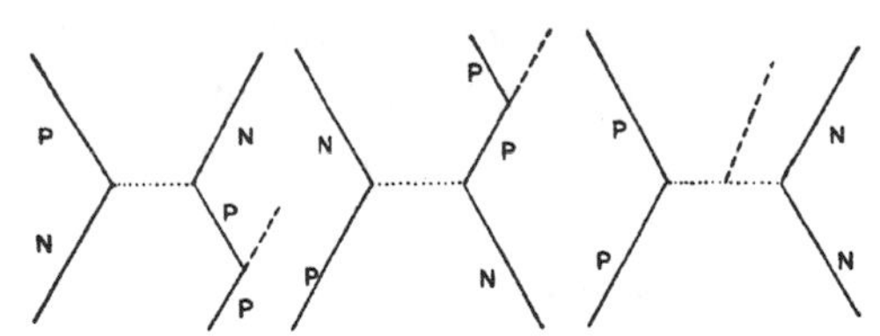

Figure 6.6. Feynman diagrams for nucleon-nucleon scattering. (Source: Simon [A.15], 574.)

threw themselves into these calculations, writing down diagrams similar to those their adviser drew. (See fig. 6.6.)

The diagrams in row *a* of figure 6.5, for example, depicted the lowest-order processes by which a positively charged pion (π^+) could be produced from the scattering of a proton (P) and a photon (γ). Running down the list of competing $\mathbf{H}_I$'s, Marshak's students could calculate the lowest-order predictions that each model would make for the angular distribution of pions produced

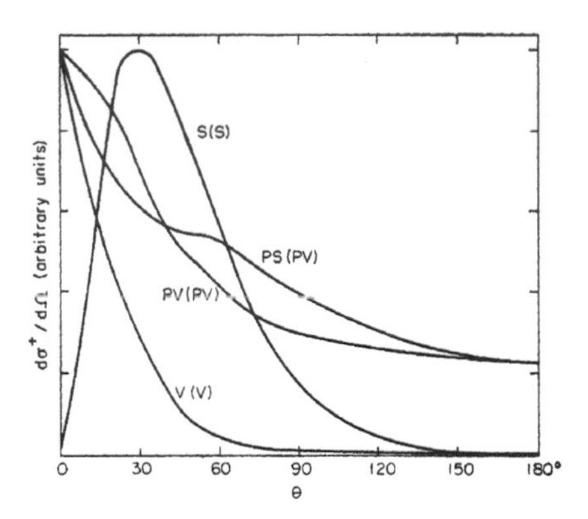

Figure 6.7. Predicted differential angular cross sections from four models of meson-nucleon interactions, for the process $\gamma + P \rightarrow \pi^{+} + N$. (Source: Marshak, *Meson Physics* [1952], 14.)

by these physical processes. (As Marshak noted, his students performed most of the calculations in his 1952 textbook; similar calculations made up their dissertations and articles.)[41] Marshak illustrated the results for four of the models in figure 6.7. These curves could then be compared with the emerging results from Rochester's own synchrocyclotron team, eventually providing—Marshak's group hoped—some means of selecting empirically between the various models.

Marshak preferred this staunchly empirical means of selection to other possible ways of sifting through the eight $\mathbf{H}_I$ contenders. As we saw in the previous chapter, Bethe and Dyson, building on Paul Matthews's and Abdus Salam's work, had begun to use renormalizability as a criterion for preferring some meson models over others.[42] Marshak and his students would have none of that. At one point, Marshak even made a bet with Bethe that the unrenormalizable pseudovector model would show the best agreement with experimental measurements; Marshak and his students pressed on to try to find further lowest-order predictions of the pseudovector model that might be compared with experiments. To the Rochester theorists, the supposed requirements of renormalization carried little weight: if Marshak's group hewed strictly to lowest-order calculations, eschewing all higher-order corrections (rendered meaningless anyway because g^2 was so large), they would never hit

41. Marshak, *Meson Physics* (1952), vi; Marshak interview (1970), pt. 4, 9. See also Ashkin, Auerbach, and Marshak [A.13]; Simon [A.15]; Kaplon [A.39], [A.53]; Simon, *Bremsstrahlung* (1950); and Kaplon, *Meson Production* (1951).

42. Freeman Dyson to Léon Rosenfeld, 7 Feb 1949, in Pauli, *WB*, 4.1:4; Bethe, "Mesons and nuclear forces" (1954), 7.

the problems of renormalizability. In the meantime, they might prove useful to their neighboring experimentalist colleagues.[43]

Marshak and his group thus used Feynman diagrams as a new kind of bookkeeper: not for a perturbative analysis of higher and higher orders in g^2, but rather as process markers for wading through a series of distinct lowest-order calculations. In doing so, Marshak's group deployed the diagrams much as Feynman did. The Rochester diagrams functioned as *pictures* of physical processes, independent of any field-theoretic derivation or Dyson's topology-minded handling of them. Indeed, the diagrams in figure 6.5 appeared fully three pages before Marshak had even introduced, let alone discussed, the eight competing forms of $\mathbf{H}_I$; much as in Feynman's articles and lecture notes, the diagrams entered up front and stood on their own, leading the various steps of the calculation.[44] Marshak even took to labeling classes of lowest-order Feynman diagrams according to the physical processes they depicted: "shaking-off transitions," "photoelectric transitions," and so on.[45] The diagrams retained some, though not all, of Feynman's Minkowskian imagery, with incoming and outgoing particles advancing along inclined paths. Upward-directed arrows, meanwhile, appeared on all external lines, including the integer-spin pion lines. In other words, Marshak's group ignored Dyson's arrow convention for distinguishing spin-1/2 particles from antiparticles. In the context of Karplus and Kroll's perturbative calculations, these antiparticle arrows had proved essential for correctly distinguishing distinct closed-loop diagrams; to Kroll and his students, as to Dyson, the tiny arrows carried specific meaning. In the Rochester meson work, by contrast, no one was calculating any closed-loop contributions, so there was hardly any need to distinguish between different kinds of closed loops; the antiparticle arrows meant nothing. In the space

43. Marshak interview (1970), pt. 4, 3; see also Kaplon, *Meson Production* (1951), 68–69, 87.

44. Like Feynman, Marshak and his students worked without the field-theoretic machinery that guided Dyson's, Karplus and Kroll's, and Kroll's students' calculations. Just as Feynman had done in his 1949 articles, Marshak's group treated the spinor fields, ψ, for the nucleons as semiclassical single-particle wave functions rather than as quantized fields decomposed into creation and annihilation operators. Marshak emphasized in the preface to his 1952 textbook that readers were expected only to have studied nonrelativistic quantum mechanics and did not require extensive knowledge of quantum field theory, let alone renormalization. See Marshak, *Meson Physics* (1952), vi.

45. Marshak explained that the "catastrophic transition" diagrams in figure 6.5, whose four-legged vertices had no correlate in ordinary QED, arose in the derivative-coupling models of meson-nucleon interactions: due to gauge invariance, any gradient term in the Hamiltonian would acquire a photon-field factor under the gauge transformation, $\partial_\mu \to (\partial_\mu - eA_\mu)$. In other words, these four-legged vertices followed directly from the form of some (but not all) of the $\mathbf{H}_I$ candidates. Marshak, *Meson Physics* (1952), 9–10. Kaplon made the same point about the unusual four-legged vertices: Kaplon [A.39], 713.

of a few short months, the antiparticle arrows had become a difference that no longer made any difference. The diagrams' pictorial forms, calculational roles, and interpreted meanings became intertwined; all three shifted as young theorists deployed the diagrams toward different ends.

Marshak and his students pushed their local use of Feynman diagrams still further, using the diagrams to study "phenomenological" models (in explicit contrast to "field-theoretic" ones). When studying nucleon-nucleon scattering, for example, Marshak introduced diagrams with different numbers of legs meeting at the vertices: having set aside each of the $\mathbf{H}_I$'s in favor of an ersatz potential, he was no longer bound by Dyson's rule relating the form of the diagrams to the structure of $\mathbf{H}_I$. Despite their muddled conceptual heritage, Marshak simply labeled these new diagrams "Feynman diagrams." Field theory or no, Marshak's students would scrutinize Feynman diagrams.[46]

The Rochester group thus carved out a new space for Feynman diagrams. Despite the failure of perturbative approaches for the strongly coupled meson interactions, suitably generalized diagrams could still prove useful for the new kinds of calculations. The Rochester theorists fashioned their diagrams by leaning heavily on Feynman's wing of the Feynman-Dyson split; indeed, they were in frequent contact with Feynman as they tinkered with his diagrams and found new uses for them. As they unglued Feynman diagrams from Dyson's stipulated field-theory rules, the diagrams they drew changed in subtle ways. Features of the diagrams that carried specific meaning for Kroll's students at Columbia—such as the antiparticle arrows, or the number of legs that could meet at a given vertex—could now be discarded without being missed. In Marshak's hands, and in the hands of his students, Feynman diagrams became a new kind of tool for their new brand of calculation.

CLIMBING BETHE'S LADDER: FEYNMAN DIAGRAMS AND THE MANY-BODY PROBLEM

Physicists soon forged a third way to use Feynman diagrams: as generators of new approximation schemes. This diagrammatic tertium quid synthesized features of both Kroll's and Marshak's group efforts: the diagrams were treated both as crucial mnemonic aids for constructing long series of more and more

46. Marshak, *Meson Physics* (1952), 53–54; see also v, 44–47. Kaplon discussed the rival approaches in *Meson Production* (1951), 62. Several other theorists adopted Feynman diagrams in similar ways when studying nonrelativistic potential scattering: Drell and Huang [A.92]; Wick, "Meson theory" (1955), esp. 345–58; and Omnès and Froissart, *Mandelstam Theory* (1963), 10–11. See also Bryan, "Marshak's nucleon-nucleon program" (1994).

complicated terms, and as physical pictures, certain classes of which would dominate a given physical process. Two groups, based at the diagrammatic strongholds of Cornell and the Institute for Advanced Study, worked out this new approach during the first half of 1951. Their efforts laid the groundwork for what would become, by the late 1950s, a flourishing industry: applying Feynman diagrams to problems involving large numbers of particles, including applications in low-energy nuclear physics and solid-state physics.[47] The topic that set these developments in motion was how to treat bound states in a relativistically invariant way. Feynman and Dyson had designed the diagrams for scattering problems: one or two particles entered the interaction region, and later a similar set of particles exited. Feynman had built his diagrammatic apparatus upon the assumption that the incoming and outgoing particles could be treated as free, noninteracting states. Dyson's formal expansion of the S–matrix was similarly predicated upon the assumption that for asymptotically early and late times, the particles behaved as free particles, subject to none of the interactions pictured in the diagrams. But what to do about particles that *never* moved far away from the interaction region, such as an electron bound in orbit around an atom's nucleus? The way each group approached the problem starkly illustrates the trickle-down effects of the Feynman-Dyson split.

At Cornell, Hans Bethe began investigating the relativistic bound-state problem with his postdoc Edwin Salpeter. (See fig. 6.8.) They approached the problem explicitly on Feynman's coattails, referring time and again to the "Feynman formalism," thanking him for his help, following his notation to the letter, and citing his two 1949 articles so often they referred to them simply as "FI" and "FII."[48] Not surprisingly, they also made great use of Feynman diagrams, borrowing many of Feynman's Minkowskian codes when drawing their stick figures. All kinds of complicated diagrams could connect two interacting

47. Both Feynman and Dyson developed interests in solid-state physics during the mid-1950s. See, e.g., Feynman, "λ transition in helium" (1953); Feynman, "Liquid helium near absolute zero" (1953); Feynman, "Superfluidity" (1957); Dyson, "Disordered linear chain" (1953); Dyson, "Spin-wave interactions" (1956); Pines, "Richard Feynman" (1989); Mehra, *Beat of a Different Drum* (1994), chaps. 17, 19; Goodstein and Goodstein, "Richard Feynman" (2000); and Dyson interview (2001). Although their research was no doubt influenced by their earlier diagrammatic work, neither made explicit use of Feynman diagrams during these early forays into solid-state material. For more on solid-state physics during this period, see Hoddeson et al., *Crystal Maze* (1992); and Hoddeson and Daitch, *True Genius* (2002).

48. Salpeter and Bethe [A.47], 1232, 1242. See also their abstract from the Feb 1951 American Physical Society meeting in New York City, at which they first announced their results: Bethe and Salpeter, "Relativistic equation" (1951).

Figure 6.8. Hans Bethe at the 1957 Rochester conference. (Source: AIP Emilio Segrè Visual Archives, Marshak collection.)

particles, Bethe and Salpeter began, emphasizing the point with some characteristic examples in figure 6.9. If they were to follow Dyson's usual prescriptions —as put into practice by Norman Kroll and his students at Columbia—Bethe and Salpeter would need to write out the integrals corresponding to each of these diagrams, grind through the evaluation of each integral, and add them all up. But for bound-state problems, the Cornell theorists reasoned, not all these diagrams should carry the same weight. Unlike in general scattering situations, they explained, "In a bound state the particles interact with each other for an infinite (or at least very long) time." Given that the probability for the two bound particles to exchange one virtual quantum was governed by the size of the coupling constant, that probability would be small for weak couplings—and hence the probability for finding two virtual quanta in "mid-flight" at the same time would be that much smaller still.[49]

The basic process of interest was therefore one in which the two bound particles traded only one virtual quantum at a time. There was an infinite

49. Salpeter and Bethe [A.47], 1234.

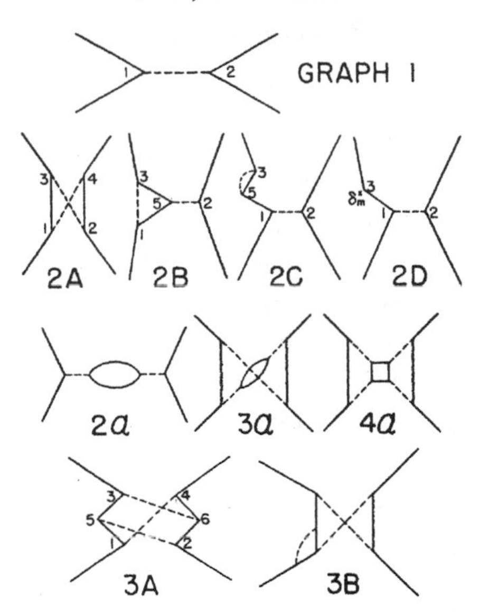

Figure 6.9. Sample Feynman diagrams for two-electron interactions. (Source: Salpeter and Bethe [A.47], 1233.)

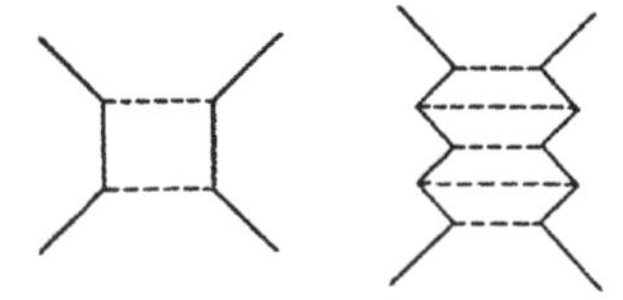

Figure 6.10. Ladder diagrams for two-electron interactions. (Source: Salpeter and Bethe [A.47], 1234.)

number of these types of diagrams—Bethe and Salpeter gave some examples in figure 6.10—and these would dominate the contributions from the diagrams within figure 6.9. Bethe and Salpeter explained that this special class of diagrams—dubbed "ladder diagrams" because of their shape—could be built up from repeated application of the basic starting diagram, "graph 1" of figure 6.9. "Although the probability for the exchange of a quantum during a small time interval is fairly small," they continued, "during the infinite time of existence of the bound state an indefinite number of quanta may be exchanged *successively.* It is just such processes that the ladder-type graphs deal with." All the infinite variety of excluded diagrams, meanwhile, involved "crossed quantum lines" or "Lamb-shift-type terms," such as the diagrams in classes 2, 3, and 4 in figure 6.9. "Such graphs refer to processes in which two or more quanta are 'in the field simultaneously,' which are indeed unimportant if the coupling constant is small."[50]

Moreover, Bethe and Salpeter could include the effects of an *infinite number* of these ladder diagrams by isolating them from the welter of general diagrams and adding up their contribution. The trick was that the contributions from the ladder diagrams could be put in the form of a geometric series.[51] If the operator $V(q)$ gave rise to the exchange of a single virtual quantum (which carried momentum q), then the next-simplest ladder, shown on the left in figure 6.10, would arise from two successive operations of $V(q)$, and so on. The effects of an infinite number of ladder diagrams could be included within the overall interaction, G. Written schematically, one had:

$$\begin{aligned} G &= V(q) + V(q)V(q) + V(q)V(q)V(q) + \cdots \\ &= V(q)[1 + V(q) + V(q)V(q) + V(q)V(q)V(q) + \cdots] \\ &= V(q)[1 + G] \end{aligned}$$

or

$$G = \frac{V(q)}{[1 - V(q)]}.$$

In this way, the effects on the kernel for the two bound particles could be evaluated by summing an infinite number of diagrams drawn from a particular subclass of all diagrams. The results of this partial resummation yielded an expression that depended only on the characteristics of the simplest, lowest-order ladder diagram. Bound-state problems, Bethe and Salpeter concluded, thus required a different kind of perturbative accounting from the kind used in the usual Dysonian prescription: a Feynman diagram of the ladder type that depicted four successive photon exchanges (and hence contributed to order e^8) should be included in a bound-state calculation long before theorists devoted even fleeting attention to the complicated e^4 and e^6 diagrams in figure 6.9. In this way, Bethe and Salpeter deployed Feynman's diagrams to great—and

50. Salpeter and Bethe [A.47], 1234; emphasis in original.

51. Salpeter and Bethe actually wrote the two-body kernel as an integral equation (eq. 9 in Salpeter and Bethe [A.47], 1234); I have adopted a simpler notation here, in terms of $V(q)$, for illustrative purposes. Dyson introduced a geometric series treatment for calculating an electron's self-energy in Dyson [A.2], 1748–51. Unlike Bethe and Salpeter's procedure, Dyson's included (in principle) *all* self-energy diagrams, rather than only a subclass. Salpeter and Bethe referred to Dyson's general procedure in [A.47], 1232; see also Schweber, Bethe, and de Hoffmann, *Mesons and Fields* (1955), 1:336–42; and Bjorken and Drell, *Relativistic Quantum Fields* (1965), 285–86.

novel—effect. Heuristic suggestions about the form of various diagrams led to a new approximation scheme, especially appropriate for treating a particular class of problems.

Salpeter and Bethe's main result—their equation for the bound-state two-body kernel—had actually been published in the *Physical Review* by two young theorists a few months before the Cornell theorists' article came out. Fresh from his MIT Ph.D., and still mistaken for an undergraduate because of his boyish looks and mannerisms, Murray Gell-Mann tackled the bound-state problem in his very first published article. He began working on the problem as soon as he arrived for his postdoc at the Institute for Advanced Study, collaborating closely with his office mate, Francis Low, another Institute postdoc. By the time Gell-Mann arrived, Dyson had left the Institute to return to England, before taking up his Cornell post, yet the group of other young postdocs at the Institute continued to pursue Dyson's special brand of diagrammatic field-theory calculations, and Gell-Mann and Low, thanking "the members of the Institute for many interesting discussions," joined in that common effort.[52] At first they were unaware of the Cornell theorists' efforts. The two groups learned of each others' work when Salpeter visited the Institute in the winter of 1951 to talk about his and Bethe's research. Low and Gell-Mann pressed Salpeter on how his equation could be derived from a field-theoretic standpoint, instead of guessed at by heuristic analogies to Feynman's semiclassical demonstrations. The Cornell postdoc, wondering about the aggressive young Gell-Mann, who kept peppering him with questions about vacuum expectation values and field-theoretic matrix elements, could supply few answers. The two groups continued to communicate following this initial exchange, though it was the Princeton duo that completed its calculations first.[53]

Comparing Gell-Mann and Low's article with Salpeter and Bethe's is very much like holding Dyson's 1949 articles up to Feynman's own papers from that year—a comparison that was hardly lost on the postdocs themselves.

52. Gell-Mann and Low [A.43], 354. Low had arrived at the Institute the previous year, 1950–51. That fall, Dyson had been back in residence, and over the year the Institute had hosted such diagram enthusiasts as Paul Matthews, Abdus Salam, Keith Brueckner, Abraham Pais, C. N. (Frank) Yang, and Maurice Lévy. During 1951–52, Low, Pais, Yang, and Lévy stayed on, while new postdocs, including John Ward, T. D. Lee, and Gell-Mann arrived. Institute for Advanced Study, *Publications* (1955).

53. Johnson, *Strange Beauty* (1999), 82–84; see also Salpeter interview (1978). Low actually collaborated with Salpeter on a brief letter to the editor, submitted just one week before Gell-Mann and Low submitted their bound-state article: Low and Salpeter, "Hyperfine structure" (1951). This was on the topic of Low's dissertation, which he completed at Columbia in 1950. See Kaiser, "Francis E. Low" (2001), 29–30.

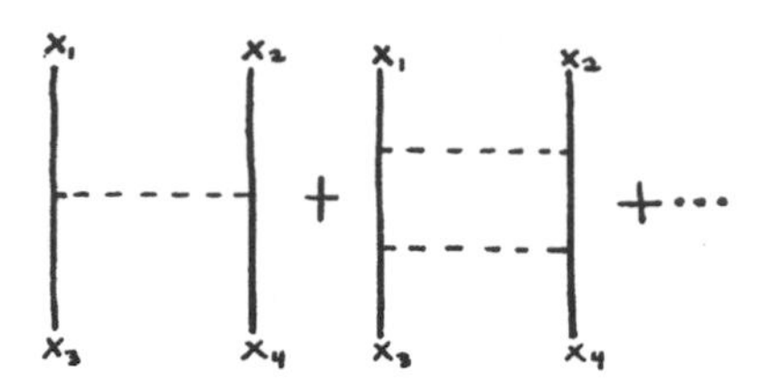

Figure 6.11. Ladder diagrams for the bound-state problem. (Source: Gell-Mann and Low [A.43], 352.)

Gell-Mann and Low emphasized that the equation for the two-body kernel that Bethe and Salpeter had "proposed" was "based on an analogy to that in Feynman's 'Theory of Positrons' and the demonstration of the equivalence to conventional field theory is incomplete. Our purpose is to provide such a demonstration."[54] The differences percolated through every step of the two groups' calculations. Like Dyson before them, Gell-Mann and Low began their discussion with a lengthy review of the governing equations of motion for quantized field operators, before rederiving Dyson's infinite-series expression for the S-matrix as a series of integrals over the time-ordered $\mathbf{H}_I$'s. Then off Dyson's shelf came the propagation functions, S_F and D_F, coming, as Gell-Mann and Low reminded their readers, from the vacuum expectation values of pairs of electron and photon quantum fields. Only at this point did they introduce Feynman diagrams (again following Dyson's pattern), describing the polygonal features of classes of diagrams, rather than reproducing Feynman's "intuitive" spacetime pictures. Restricting the infinite series of diagrams to the simplest, lowest-order contribution, Gell-Mann and Low finally demonstrated that the single-photon-exchange term, when substituted into their integral expression for the two-body propagator, produced an infinite series of ladder diagrams. (See fig. 6.11.) Even with their diagrams, the Institute postdocs advertised their allegiance to Dyson's program: the relevant features were topological. In place of Bethe and Salpeter's Feynmanesque tilted propagation lines and zigzagging particles in figure 6.10, Gell-Mann and Low's examples revealed none of this *Zitterbewegung*.

When they appeared in the autumn of 1951, the Cornell and Institute bound-state articles provided physicists with examples of ways to extend diagrammatic calculations beyond scattering problems. Soon other theorists began to work on related topics, such as how to handle problems involving many interacting particles, or the behavior of one particle in a medium of others—problems, like the bound-state problem, in which the particles continued to interact

54. Gell-Mann and Low [A.43], 350–51.

over long periods of time.[55] Kenneth Watson, a graduate of Dyson's original Institute postdoc cohort and one of the first theorists to tackle a fourth-order calculation diagrammatically, began studying many-body problems in 1952–53. He developed a series of highly formal expressions for the multiple scattering of pions within a nucleus, the upshot of which was that even inside a nucleus, one could treat the pion's interactions as a series of two-body scatterings between the pion and individual nucleons.[56] Keith Brueckner, who had first learned about Feynman diagrams from Watson, began thinking about the nuclear-matter problem within a few months. (Watson had coached Brueckner in how to use the diagrams during 1949–50, while Brueckner worked on pion-nucleon scattering for his dissertation at Berkeley. During the early 1950s, they worked together again at Indiana University, before Watson moved on to the University of Wisconsin at Madison.) Inspired by Watson's recent results, Brueckner decided to try to calculate the properties of nuclear matter by treating the interactions between pairs of nucleons within a nucleus exactly, while approximating the effects of all the other nucleons on that pair by means of an average background potential.[57]

As several senior theorists warned would happen, Brueckner immediately hit a problem with this approach. If he followed the ordinary perturbative scheme, the various terms in his expansion carried the wrong dependence on the size of the medium. (Theorists had encountered the same problem decades earlier, when they had tried to calculate properties of many-electron

55. One of the first to pick up on the new scheme was Maurice Lévy, a young French theoretical physicist in residence at the Institute for Advanced Study during 1950–52. Thanking Francis Low for help, Lévy combined the new bound-state work with another recent approximation scheme, which had been worked out independently by the Soviet theorist Igor Tamm and the American theorist Sidney Dancoff. Lévy showed that in the nonrelativistic limit of one of the popular meson models, the exchange of two mesons at a time would actually dominate single-meson exchanges. Hence, he could work within the bound-state framework—selecting one basic interaction and iterating its effects to the exclusion of all others—while working with a different starting interaction than in the Gell-Mann–Low and Bethe-Salpeter work. Lévy immediately turned to Feynman diagrams to help with his accounting. Lévy [A.57], [A.66], and [A.69]. Lévy's work spawned widespread interest, and soon dozens of papers were written making use of Feynman diagrams in the context of the Tamm-Dancoff and Bethe-Salpeter approximations, including papers [A.52], [A.59], [A.71], [A.77], [A.84], [A.85], [A.86], [A.87], [A.92], [A.93], [A.96], [A.98], [A.99], [A.105], [A.109], [A.114], [A.122], [A.125], [A.127], [A.128], [A.133], [A.135], [A.139], [B.9], [B.10], [B.11], [B.13], [B.15], [B.16], [C.45], [C.48], [C.66], [C.76], [C.85], [C.90], [C.92], [C.93], [C.96], and [C.97]. See also Schweber, Bethe, and de Hoffmann, *Mesons and Fields* (1955), 2:193–94, 199–215, 303–12.

56. Watson, "Multiple scattering" (1953); and Francis and Watson, "Elastic scattering" (1953).

57. Keith Brueckner, "Autobiographical notes" (ca. 1986), sec. 2, 15–17, in *NBL*, call number MB84. See also Brueckner and Watson, [A.96].

atoms.) As Brueckner knew, the energy of a large system, when in equilibrium, should be proportional to the number of particles in the system, N. The first two terms in his series fitted this dependence, but the third term in the perturbative series entered with N^2, and the fourth term with N^3. Not only did these terms violate the basic physical expectation (that the energy should be proportional to N); when applied to a large system with many particles, these rising factors of N would produce a new divergence, in addition to all the old problems of self-energies against which Tomonaga, Schwinger, Feynman, and Dyson had worked so hard. Unlike the theorists who had encountered this problem during the 1920s and 1930s, however, Brueckner knew about the postwar developments in renormalization, and he began trying to map his new problem onto the other framework.[58]

Brueckner pressed on with his perturbative study, showing that (at least up to fourth order) the troublesome N^2 and N^3 terms arose from what he called "unlinked" processes in which not all the particles interacted with each other; groups of interacting particles entered into the higher-order equations side by side, one group independent of the others. Moreover, if one took special care to track all the possible permutations of these unlinked processes, Brueckner showed that they exactly canceled each other—at least up to the fourth-order terms he examined. The terms that remained were all "linked"—they could not be broken up into smaller, independent groups—and these linked terms all entered with the correct coefficient, N. Brueckner thus suggested that studies of many-body problems always drop the unlinked terms.[59] In his later reminiscences, Brueckner recalled deriving this result diagrammatically, though his published presentations made no use of any diagrams. Building on the idiosyncratic notation that he and Watson had been developing for several years but that few others had adopted, Brueckner's presentation of the "linked cluster theorem," first published in autumn 1955, proved very difficult for others to follow. On top of this, Brueckner's result could at best be considered suggestive—even Brueckner conceded that his "theorem" was merely "strongly suggest[ed]" by his perturbative analysis, not really proved. As late

58. Brueckner, "Autobiographical notes," sec. 2, 15. Two other early diagram enthusiasts, Kinoshita Tōichirō and Nambu Yoichiro, likewise tried to apply the Feynman-Dyson diagrammatic techniques to many-body problems during their stay at the Institute for Advanced Study in 1953–54, treating the (filled) ground state as analogous to particle theorists' vacuum: Kinoshita and Nambu [A.111]. Though their work was not developed far enough to be immediately useful, it did lay the groundwork for Nambu's later efforts on superconductivity. Tōichirō (Tom) Kinoshita, letter to the author, 14 Jan 2004.

59. Brueckner, "Autobiographical notes," sec. 2, 15–17; Brueckner and Levinson, "Many-body problem" (1955); Brueckner, "Many-body problem," pt. 2 (1955).

as January 1957, cognoscenti still found Brueckner's work controversial: at a specialist meeting of solid-state theorists, discussion revolved around how to make sense of Brueckner's result and whether to trust it.[60]

In the meantime, Brueckner had sent out dozens of preprints of his "linked-cluster" paper, including a copy to Hans Bethe. Bethe had a sabbatical coming up and decided to spend his year in Cambridge, England, studying Brueckner's work and trying to make sense of it. He took along Brueckner's preprint and some other recent articles, delighting in the chance to return to one of his first scientific loves: low-energy nuclear physics. Bethe trained his disciplined eye carefully upon Brueckner's work, making explicit the at-times-tacit assumptions and approximations at work within Brueckner's papers and producing an unusually long excursus several months into his Cambridge stay.[61] More important, Bethe began working with a Cambridge graduate student, Jeffrey Goldstone. Bethe served as one of Goldstone's dissertation advisers; Goldstone's other adviser, Richard Eden, had been among the earliest crop of Cambridge graduate students to pick up Feynman diagrams during the early 1950s. Upon arriving in Cambridge in the autumn of 1955, Bethe suggested that Goldstone work on the nuclear-matter many-body problem. He quickly came to admire Goldstone's talents, writing a few years later that "Dr. Goldstone was one of the best graduate students I have ever had. He ranks second only to Freeman Dyson"—high praise indeed from the doyen of nuclear and particle physics.[62] It was a fitting comparison: Dyson had domesticated Feynman's diagrams, deriving self-consistent rules for their use; Goldstone likewise put Brueckner's suggestive results onto solid ground. Goldstone demonstrated that if one added a few more rules to the customary uses of Feynman diagrams, then one could prove the "linked cluster theorem" to all perturbative orders.[63]

Goldstone's first move was to use the many-body ground state as the reference, rather than the particle theorists' favored vacuum state. Because of the

60. Quotation in Brueckner, "Many-body problem," pt. 2 (1955), 41. On the difficulty of following Brueckner's notation, and the skepticism at the Jan 1957 meeting, see Schrieffer interview (1974), 17, 34.

61. On Brueckner's distribution of preprints, see Brueckner, "Autobiographical notes," sec. 2, 16–17; on Bethe's interest in the work, see Bethe interview (1972), 18–19. Filling nearly forty double-column pages, Bethe's article on Brueckner's work is surely one of the longest papers ever published in the *Physical Review:* Bethe, "Nuclear many-body problem" (1956).

62. Hans Bethe to Shaun Wylie, 21 May 1958, in *HAB*, Folder 30:1.

63. Goldstone, "Brueckner many-body theory" (1957). Professor Michel Baranger kindly supplied copies of his unpublished lecture notes from his 1980 graduate course on many-body physics at MIT; I have also benefited from Baranger interview (2001).

Pauli exclusion principle, certain types of particles (such as spin-1/2 electrons and nucleons) could not occupy the same state at the same time. In large collections of these particles, therefore, the particles stacked up in the lowest allowable energy state for the system: one particle in the lowest state, another particle in the next-lowest state, and so on, until all particles had been placed. The ground state thus consisted of this ensemble, with all allowable energy states occupied up to some highest energy, known as the Fermi energy; such an ensemble was known as the "Fermi sea." The only states to enter Goldstone's diagrams were those states that deviated from the (filled Fermi sea) background. Thus, the only items that would enter into the diagrams were excited particles (with energies above the Fermi energy) and "holes" (states that would ordinarily be filled within the ground-state Fermi sea but that happened to be empty). Like Dyson, Goldstone turned to arrows to distinguish the two possibilities: up arrows for excited particles, down arrows for holes (analogous to Dyson's arrow conventions for particles and antiparticles). Next Goldstone imposed time ordering within his diagrams. In Feynman's case, one diagram stood in for all time orderings. Feynman's canonical diagram, in figure 2.5, for example, stood in both for the case in which the electron on the right emitted a photon that was later absorbed by the electron on the left, and vice versa. Not so in Goldstone's scheme: time order mattered, and these two cases would each receive its own diagram. Finally, Goldstone announced that one must ignore the Pauli exclusion principle in intermediate states of his diagrams—the basic effects of the principle had already been taken into account by working in terms of particles and holes.

Goldstone showed how to calculate with his new diagrammatic scheme in an article published in the *Proceedings of the Royal Society* in February 1957. Though he called his diagrams "Feynman diagrams," others soon came to call them "Goldstone diagrams" or "Feynman-Goldstone diagrams." With the aid of his new diagrams, Goldstone demonstrated that all unlinked clusters did indeed cancel each other, in all perturbative orders. Moreover, Goldstone showed that Brueckner's abstruse approximation scheme for nuclear matter boiled down to a ladder approximation: what mattered most in low-density nuclear matter was the repeated interactions between the same pair of nucleons, rather than one nucleon scattering successively off of many different nucleons within the nucleus. (See fig. 6.12.) Quantities like the binding energy for nuclear matter thus could be calculated by the same kind of iterative, ladder-diagram approach that Goldstone's adviser, Hans Bethe, had pioneered five years earlier in a rather different context.

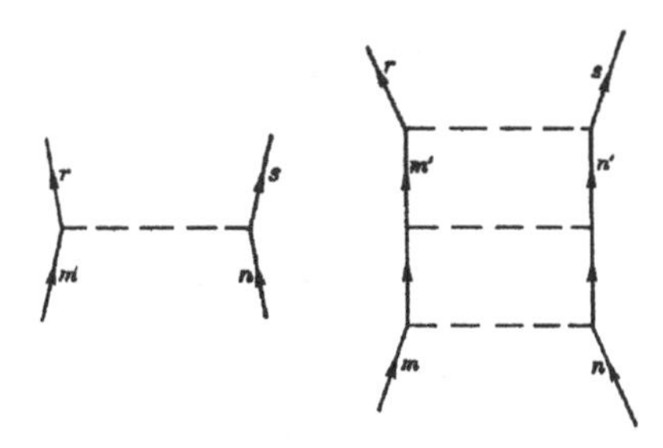

Figure 6.12. Goldstone's ladder diagrams for the Brueckner model. (Source: Goldstone, "Brueckner many-body theory" [1957], 276.)

Goldstone's work quickly brought diagrams into new branches of physics; Bethe hardly exaggerated when he explained proudly in May 1958 that Goldstone's diagrammatic efforts "led to a great simplication of the treatment of this [many-body] problem, and his method is generally used by everybody in the field."[64] Just over a month after Goldstone's paper appeared in print, two Dutch theorists, Léon van Hove and N. M. Hugenholtz, extended his work, showing in particular that with Goldstone's diagrams, Brueckner's earlier concern about correct factors of *N* was alleviated. Van Hove had first learned about Feynman diagrams as a postdoc at the Institute for Advanced Study during 1949–50; he returned for a five-year research position beginning in 1952. Yet in 1954, van Hove was invited to return to Utrecht as a professor and director of the Theoretical Physics Institute there—an attractive offer made more so by the recent souring of the Institute for Advanced Study's atmosphere amid rising McCarthyism, refracted locally through the Oppenheimer hearing during the spring of 1954. Upon van Hove's return to Utrecht, he began thinking about how to apply some of the basic ideas of renormalization to problems in solid-state physics. For example, the shielding of bare charges by clouds of virtual particles in QED might be analogous to the shielding of charged particles in a many-body medium.[65] He worked with his first graduate student in Utrecht, Hugenholtz, on applying these field-theoretic techniques to many-body systems. They were already well under way when Goldstone's paper

64. Bethe to Wylie, 21 May 1958, in *HAB*, Folder 30:1. Within a few years Goldstone's research interests had shifted from solid-state to high-energy particle theory. His most lasting contribution in this new area was the "Goldstone theorem" regarding the presence of massless particles in (most) models with spontaneously broken symmetries: Goldstone, "Field theories" (1961).

65. Van Hove, "Autobiographical sketch" (2000), xvi; Hugenholtz, "Perturbation theory" (2000); van Hove, "Energy corrections" (1955); van Hove, "Energy corrections," pt. 2 (1956).

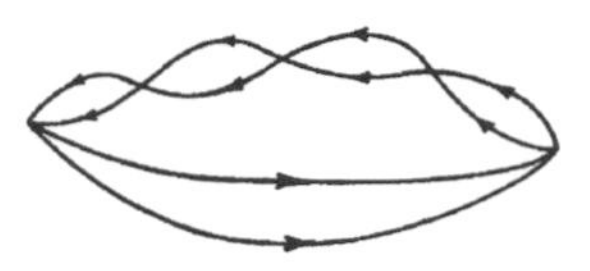

Figure 6.13. Hugenholtz's diagram of the repeated scattering of a pair of particles in the ground state. (Source: Hugenholtz, "Perturbation approach" [1957], 536.)

arrived. (They had received advanced news of Goldstone's work from Goldstone's other adviser, Richard Eden.) Though they had developed a distinct pictorial form for their diagrams—they collapsed the (nonrelativistic) interaction between pairs of particles into a point, where Goldstone, following Bethe, continued to depict these interactions as unfolding in space and time—they nonetheless were able to import Goldstone's main result about iterating one basic interaction to calculate many-body quantities.[66] (See fig. 6.13.) Brueckner himself soon championed the new approach. Having been in contact with both Bethe and van Hove, Brueckner began working with Gell-Mann to apply the new techniques to high-density gases. They showed that whereas the Bethe-Goldstone ladder approximation worked best for low-density nuclear matter, theorists should iterate a different diagram—the so-called ring diagram—when studying high-density gases, such as electrons in metals.[67]

David Pines, a young solid-state theorist, further contributed to putting many-body diagrams into circulation. Pines led an especially peripatetic life during the 1950s: earning his Ph.D. at Princeton in 1950, he spent several years as a postdoc at the University of Illinois at Urbana, working with Francis Low and John Bardeen; he then went back to Princeton, in 1954–57, as a postdoc, before spending 1957–58 at Bohr's institute in Copenhagen and the École Normale Superieure in Paris. In Copenhagen, Pines overlapped with Goldstone, who had just arrived for postdoctoral study at Bohr's institute. Pines next spent 1958–59 at the Institute for Advanced Study before returning to Urbana to join the faculty. While at the Institute in Princeton, Pines teamed

66. Hugenholtz, "Perturbation theory" (1957); Hugenholtz, "Perturbation approach" (1957); Hugenholtz and van Hove, "Theorem on single particle energy" (1958). Hugenholtz thanked Eden in "Perturbation approach" (1957), 545.

67. Gell-Mann and Brueckner, "Correlation energy" (1957). Brueckner gave a talk on his developing work at Cornell, about which Bethe reported to Goldstone in a letter of 30 Nov 1956, in *HAB*, Folder 30:1; Bethe and Goldstone thanked Brueckner for "several stimulating discussions" in Bethe and Goldstone, "Effect of a repulsive core" (1957), 566; and Hugenholtz and van Hove acknowledged several discussions with Brueckner in Hugenholtz and van Hove, "Theorem on single particle energy" (1958), 375–76.

up with Hugenholtz, by that time also a postdoc at the Institute, to write a diagram-heavy analysis of certain many-body problems, and he carried the new techniques back with him to Urbana.[68] Two years later, Pines began editing a new series of textbooks, the Frontiers in Physics series. Unlike traditional textbooks, the Frontiers in Physics books highlighted informal lecture notes, often combined with reprints of relevant articles. The aim was to get new results into graduate students' hands quickly and affordably. As one of the very first installments in the Frontiers series, Pines published his own lecture notes on the many-body problem; they became the first textbook to introduce the recent diagrammatic techniques to solid-state theorists and students. Pines also engineered publication of some of van Hove's recent diagrammatic lecture notes in the same series.[69]

Thanks to these informal lecture notes and the continued circulation of young theorists like Goldstone, Hugenholtz, and Pines, solid-state theorists put Feynman-Goldstone diagrams to work at an ever-quickening pace during the early 1960s. Drawing together lectures he gave between 1962 and 1965, Richard Mattuck wrote a textbook that he hoped would help his readers "learn enough so they no longer tremble with awe when a many-body theoretician covers the blackboard with Feynman diagrams."[70] More than the previous lecture-note authors, Mattuck codified the new approach in his textbook, *A Guide to Feynman Diagrams in the Many-Body Problem* (1967). Using Goldstone's rules for the diagrams—drawing with reference to the ground state, imposing strict time ordering, and ignoring the Pauli exclusion principle for intermediate states—Mattuck instructed physicists how to replay Bethe and Salpeter's original trick. First they would select one basic type of diagram from the morass of complicated diagrams; then they would perform a partial sum on the infinite number of these iterated diagrams, while ignoring all other diagrams by fiat. (See fig. 6.14.) Features of the diagrams drove the calculation, and not the other way around.

Pedagogical genealogies—Bethe to Goldstone, Watson to Brueckner, van Hove to Hugenholtz, Goldstone and Hugenholtz to Pines—thus put the

68. Hugenholtz and Pines, "Ground-state energy" (1959). Pines's changing institutional affiliations can be tracked by the many papers he published in the *Physical Review* during the 1950s.

69. Pines, *Many-Body Problem* (1961); van Hove, Hugenholtz, and Howland, *Many-Particle Systems* (1961). See also Pines, *Elementary Excitations* (1963); and Pines and Nozières, *Quantum Liquids* (1966). Another useful collection of reprints is Morrison, *Many-Particle Systems* (1962), which includes several of the early diagrammatic applications to many-body problems.

70. Mattuck, *Guide to Feynman Diagrams* (1967), vi.

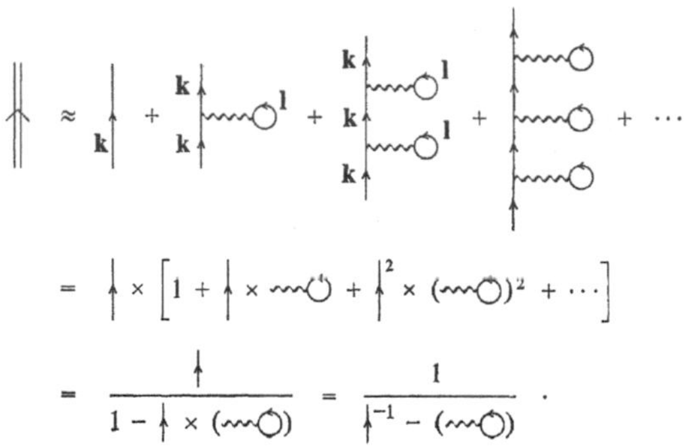

Figure 6.14. Diagrammatic partial summations. (Source: Mattuck, *Guide to Feynman Diagrams* [1967], 73.)

diagrams into circulation for studying everything from particles in a nucleus to electrons in a metal, from atoms in solids to molecules in liquids, and much more. Across this vast range of phenomena, students learned to put Feynman diagrams first.

TRAINING THEORISTS FOR HOUSE AND FIELD

Kroll's, Marshak's, and Bethe's approaches were clearly different from each other. How difficult was it for members of one group to understand what the other groups were doing? Some physicists at the time spoke as if it were *impossible* for members of such different "schools" to understand each other—thereby reaching a conclusion similar to those of sociologists such as Harry Collins, in his work on tacit knowledge and skills transfer. Consider, for example, Wolfgang Pauli's amusing formulation of how one could define various schools within theoretical physics. "*What is the definition of the 'Wigner School'?*" he underlined in a 1954 letter for emphasis: "The question can be answered in the practical American way by an *operational* definition. In order to decide, whether or not a person is a member of the Wignerschool [*sic*], you give him a paper of Schwinger. . . . If he then says, that this is very obscure, and that he can't understand it, he is a member of the Wigner school. But he, who says that it is quite understandable and clear will be excluded from the Wigner school. . . . Similar tests are certainly very popular at Harvard where the test object for membership of the *Schwinger* school is a paper of *Wigner*."[71] Pauli thus

71. This addendum, dated 30 Dec 1954, was apparently intended to be included in a letter to Léon Rosenfeld; it continued themes that Pauli and Rosenfeld had been discussing in recent correspondence. The addendum is reprinted in Pauli, *WB*, 4.2:956. Emphasis in original.

tied the *social* question of young physicists' training directly to the *epistemic* issue of understanding various calculational approaches.

Yet such ties were not always so tight in practice. Ashkin, Marshak, and Marshak's graduate student Albert Simon published a long article in 1950 in the Japanese journal *Progress of Theoretical Physics*, in which they went through the motions of a fourth-order perturbative calculation within a given meson model. With much help and personal coaching from Dyson, whom they thanked for "a lucid presentation of his method," Marshak and his group *could* perform calculations closer in kind to those of Kroll and the Columbia contingent. They worked in terms of the meson coupling g^2 (neglecting the fact that $g^2 \gg 1$) in order to demonstrate that Dyson's systematic renormalization program could fix certain types of calculations among mesons and nucleons.[72] (They restricted attention to the nonrelativistic limit, thereby simplifying the problem considerably compared to the calculations tackled by Kroll's students.) This appears to have been a one-time exercise: neither Marshak nor his students pursued mesonic calculations beyond lowest order after this early paper. Students other than Simon never sweated such details; at least, traces of such types of calculations were left in neither their dissertations nor their published articles. The differences between Marshak's and Kroll's pedagogical programs thus point to *local choices* about what to work on and what skills to drill one's students in, rather than incommensurable or incommunicable epistemic regimes.

Within the United States during this period, local experimental demands provided a major determinant for theorists' choices of what to work on and what to teach their students. Columbia students practiced making high-precision electrodynamic calculations in part because Columbia experimentalists were performing just those types of experiments. Rochester students, on the other hand, had to get ready for (and then work closely with) the department's new meson-producing synchrocyclotron. (Even Simon undertook his fourth-order mesonic calculation because of local experimental prodding. Given the meson's large coupling constant, he explained, "it is possible for the higher order corrections to be as large as the lowest order"—usually the death knell for perturbative calculations—"the hope being that if the corrections served to decrease the [theoretically calculated] cross section we would approach agreement with experiment.")[73] This is not the place to

72. Ashkin, Simon, and Marshak [C.13], quotation on 635. See also Simon, *Bremsstrahlung* (1950), 22–48.

73. Simon, *Bremsstrahlung* (1950), 22–23.

engage the much-debated topic of American physicists' "pragmatism" head-on. Building on one of Sam Schweber's key observations, however—that physics departments in the United States grouped theorists and experimentalists together, whereas most theorists and experimentalists occupied separate institutes in Europe—we can get far by noting how various groups of physicists put their Feynman diagrams to work in the United States and in Europe.[74]

The experiences of Pauli and Dyson prove especially illuminating. Both serve as latter-day de Tocquevilles: they each worked in the United States and in Europe and offered analyses of what the American experience was all about, at least within the world of theoretical physics. Unlike most young American theorists, burrowing deeper into their diagrammatic investigations of meson-nucleon scattering in the hopes of making some contact with the reams of new data streaming forth from Berkeley, Brookhaven, Rochester, Cornell, and Chicago, Pauli and his circle in Zurich scrutinized Dyson's formulation for renormalizing QED. Dyson's renormalization program relied crucially upon making a perturbative expansion of the *S*-matrix. Pauli and his European colleagues were concerned that even though any given order within the infinite series could be rendered finite, the infinite sum might still diverge. Rather than chasing the latest reports about cosmic rays and factory-produced mesons—much less the latest diagrammatic approximation schemes, such as the Bethe-Salpeter work—Pauli reported to several colleagues in December 1951 that his young Swedish protégé, Gunnar Källén, was immersed in trying "to peep behind the veil of the Dysonian power-series."[75]

Dyson remained in regular contact with Pauli during this period. Soon after returning to Birmingham from the Institute for his last semester of teaching in England, Dyson responded to Pauli's missives. "The test which I wish to be applied to my methods," he explained, "is 'Do they enable us to decide definitely the correctness or incorrectness of meson theories by a quantitative comparison with experiment?' If they can do this, I shall be very well satisfied." Dyson noted that Pauli had different ideas in mind: "You are asking, 'Do the new methods give us any new theoretical insight into the foundations of physics?' This question I am content to answer in the negative." "Of course," he concluded, "I too would like to escape altogether from series expansions, if

74. Schweber, "Empiricist temper regnant" (1986).

75. Wolfgang Pauli to Markus Fierz, 23 Dec 1951, in Pauli, *WB*, 4.1:478–81. My translation. The original reads, "hinter den Schleier der Dysonschen Potenzreihen zu blicken." Cf. Pauli to Homi Bhabha, 31 Dec 1951, in ibid., 482–83; and Källén, "Renormalization constants" (1952).

I could do it."[76] But that was not Dyson's main focus; producing quantifiable predictions from specific models was the name of the game. Pauli replied three days later that even "apart from practical questions of computation," he had a lingering "impression" that the "final form of a fundamental law of nature should *not* be formulated in such a way that a power series development is *essential*"—and to Pauli this was as true for electrodynamics as for the more computationally troublesome meson models.[77]

Dyson had the opportunity to talk more with Pauli about their rival visions of the theorist's proper task during the summer of 1951, which Dyson spent with Pauli's group in Zurich. Then (and only then), Dyson took up the problem that Pauli and Källén had elevated to central importance: Dyson began to agree with them that his formal power series would not converge, even for weakly coupled systems like QED.[78] His friends back in the States joked about his transformation behind his back. In October 1951, Fritz Rohrlich asked Bethe whether Dyson—who had just begun teaching at Cornell—was "preach[ing] his newest gospel." Bethe replied, "Dyson does not preach his newest gospel but has changed to the Anti-Christ. He is convinced now that the power series in all field theory diverges, although he finds it hard to prove this." Yet all hope was not lost for poor Dyson, as Bethe continued: "Apart from this he has got interested in the Bethe-Salpeter equation and is helping us to establish suitable boundary conditions for it."[79] Dyson's return to Ithaca quickly exorcised his demonlike possession: he dropped the series-divergence question and pressed on with the Bethe-Salpeter and similar approximations. He shepherded a large group of postdocs and graduate students in a lengthy calculation, eagerly comparing their results—calculated diagrammatically in a particular approximation—with the latest experimental news from Fermi's group in Chicago.[80] Firmly ensconced once again in Ithaca, Dyson returned his diagrammatic prowess to the problem of making contact with experimental results, nonrigorous approximations and all.

76. Freeman Dyson to Wolfgang Pauli, 15 Feb 1951, in Pauli, *WB*, 4.1:263. The "new methods" to which Dyson referred were contained in Dyson [A.33] and [A.38], preprints of which Dyson had sent to Pauli late in 1950. Their correspondence about these papers appears in Pauli, *WB*, 4.1:220, 240–42, 244–45, 250–53, 262–65.

77. Wolfgang Pauli to Freeman Dyson, 18 Feb 1951, in Pauli, *WB*, 4.1:264–65; emphasis in original.

78. See Dyson, "Divergence of perturbation theory" (1952). Dyson thanked Pauli "for valuable discussions of these problems," 632.

79. Fritz Rohrlich to Hans Bethe, 8 Oct 1951; and Bethe to Rohrlich, 10 Oct 1951, in *HAB*, Folder 12:13.

80. Dyson et al. [A.127]; see also Dyson, "Comments on Selected Papers" (1996), 17–21.

Dyson ruminated further in the pages of *Physics Today* on what he saw as the particularly American way of approaching theoretical physics, which had taken shape after World War II. He used as his platform a book review of *Mesons and Fields,* a two-volume textbook aimed at graduate students that Hans Bethe and some younger colleagues had just published. The second volume, which, as Dyson put it, "describes the experimental results [within meson physics] with theoretical analysis and commentary," appeared a few months before the first volume. "Volume 2 is an important historical phenomenon," Dyson proclaimed. "It represents in a nutshell the American philosophy in the training of physicists, the idea that there should be a wide range of common ground with which experimental and theoretical people should both be thoroughly familiar." It was "this philosophy," Dyson continued, "much more than the availability of money and equipment"—those signs of postwar American physics that were easier for the casual observer to spot—that furnished "the reason why so much good physics is done nowadays in America." The emergent American philosophy, moreover, provided ready contrast with Dyson's own undergraduate training in Cambridge, England: "When I was a student in Cambridge, an experimental physicist was considered good if he knew how to build good equipment, a theorist was considered good if he knew how to handle his mathematics. It was not generally understood that a creative experimenter is one who thinks about what he will discover with his equipment, and a creative theorist is one who thinks about what he will do with his mathematics."[81] Dyson emphasized that theorists in the United States proved their worth by remaining always useful, attuned to experimental results and conversant with the outstanding empirical questions—a lesson he remembered upon his return to Cornell in 1951.

The rift between Pauli's and Dyson's visions of the important problems in theoretical physics paled, in turn, in comparison with how most American theorists approached their craft. Marshak and his students, as we have seen, avoided all questions of renormalization altogether, whether stuck within Dyson's formal power series or not; they simply ignored all terms beyond lowest order as they scribbled their refashioned Feynman diagrams day in and day out. Jack Steinberger and his colleagues, meanwhile, gleefully compared formally divergent diagrams with straightforwardly convergent ones, weighing the number of factors of e and g from each, all in an effort to make useful estimates for their experimentalist neighbors at Brookhaven. (See fig. 5.10.) The early diagram users in England, on the other hand, showed a preference

81. Dyson, review of *Mesons and Fields,* vol. 2, by Schweber, Bethe, and de Hoffmann (1955), 27.

for formalism over data more akin to the preferences of Pauli and Källén. When Matthews, Salam, Ward, and company wielded their diagrams, it was to work out self-consistent formalisms: Could divergences be subtracted in principle from meson models; could arbitrarily high-order overlapping divergences be removed systematically? Matthews and Salam never calculated, with their elaborate formalism, mesonic quantities that might be compared with experimental data; when their Cambridge colleague, C. Angas Hurst, found a means to enumerate the number of diagrams in the Bethe-Salpeter approximation, it was as a mathematical exercise rather than to make contact with real data.[82]

During the mid-1950s, postdocs at the Institute for Advanced Study often shared a quip: there are field theorists, and there are house theorists.[83] "House theorist" was no term of praise: these were people who served "merely" as aids to experimental groups, presumably pursuing little research beyond the local experimental dictates. Within postwar American physics, however, such a division rarely existed in practice: even most field theorists worked primarily as in-house consultants to their local experimental groups. Kroll's, Marshak's, and Bethe's students learned to put their diagrams to work in just these sorts of ways. As we will see in the next chapter, the first generation of American textbooks on the diagrammatic techniques reinforced the point: the best thing a theorist could be was "useful." In the United States, young theorists learned the lesson quickly.

82. The situation in Japan formed an interesting hybrid: many young Japanese theorists worked hard to make contact with empirical data, yet after the ravages of war, few experimental groups were operational. Thus, most Japanese physicists turned to their American colleagues for news of the latest experimental results—leading at times to awkward interactions, as we have seen in chap. 4, such as those that played out in 1948 between Cornell's Kenneth Greisen and Hayakawa Satio in the newsletter *Soryūshi-ron kenkyū*.

83. J. Bernstein, *Life It Brings* (1987), 108.

PART III

Feynman Diagrams in and out of Field Theory, 1955–70

· 7 ·

Teaching the Diagrams in an Age of Textbooks

> Remember, these [scientists] are very intelligent men. Their culture is in many ways an exacting and admirable one. It doesn't contain much art . . . , though perhaps not many would go so far as one hero . . . who, when asked what books he read, replied firmly and confidently: 'Books? I prefer to use my books as tools.' It was very hard not to let the mind wander—what sort of tool would a book make? Perhaps a hammer? A primitive digging instrument?
>
> C. P. SNOW, 1959[1]

Parts 1 and 2 focused on the early years of the diagrams' dispersion. During this period a fast-growing community of young theorists began to deploy Feynman diagrams thanks largely to informal mentoring and network building throughout the United States and abroad. In part 3, we consider what happened to the diagrams during a later period, extending roughly from 1955 to 1970. This period saw the publication of several textbooks devoted to teaching the new diagrammatic techniques. It also saw an even broader widening of physicists' appropriations of the diagrams: from the mid-1950s through the end of the 1960s, theorists progressively loosened the tether of rules and derivations that Dyson had so carefully fashioned for the diagrams' use. As physicists tinkered with the diagrams, some began to find new content in the simple line drawings that neither Feynman nor Dyson had ever anticipated. In chapters 8 and 9 we will follow one particular strand of these improvisations in the work of theorists who strove, diagrams in hand, to build theoretical structures that could bring some order to the subatomic domain. Many of these diagrammatic maneuvers provided the scaffolding for what coalesced as the "*S*-matrix program." Several theorists hoped that this diagram-derived program would provide a new theory of nuclear particles' interactions that could replace the rival quantum-field-theoretic framework.

Before turning to those paper tools that would become essential to the *S*-matrix program, however, we consider in this chapter how physicists chose to

1. Snow, *Two Cultures* (1959), 14.

write about and teach Feynman's diagrams during the new age of textbooks. The editors of a recent volume dedicated to the history of scientific textbooks caricatured a commonly held view: "Boring, dogmatic, conservative . . . textbooks have a bad reputation, at least in science studies."[2] Thomas Kuhn contributed more to this dour view of scientific textbooks than almost any other analyst. On the one hand, he acknowledged that scientific education revolves more narrowly around textbooks than does training in nearly any other field. When describing "the nature of education in the natural sciences," Kuhn wrote that "the single most striking feature of this education is that, to an extent totally unknown in other creative fields, it is conducted entirely through textbooks. Typically, undergraduate *and* graduate students of chemistry, physics, astronomy, geology, or biology acquire the substance of their fields from books written especially for students."[3] While acknowledging (and, indeed, exaggerating) textbooks' central role in scientists' training, however, he scoffed equally strongly at the picture of science that most textbooks offered. Textbooks, Kuhn charged, obscured the actual routes by which scientists arrived at the knowledge that textbooks purported to present: "though texts may be the right place for philosophers to discover the logical structure of finished scientific theories," he asserted in a 1961 article, "they are more likely to mislead than to help the unwary individual who asks about productive methods." Scientific textbooks, Kuhn concluded, "are the unique repository of the finished achievements of modern physical scientists."[4]

Kuhn's charge that textbooks obscure the historical routes by which scientific knowledge develops has become something of a truism. His characterization of what scientific textbooks contain, on the other hand, requires revision. The textbooks that theorists wrote to introduce Feynman diagrams and related techniques of calculation are the *last* place one could hope to find "the finished achievements of modern physical scientists," let alone "the logical structure of finished scientific theories." Instead, textbook authors worked hard to present just those "productive methods" that they believed graduate students, and even their own colleagues, needed to practice in order to deploy them confidently in

2. Lundgren and Bensaude-Vincent, "Preface" (2000), vii; ellipsis in original. See also Brooke, "Chemical textbooks" (2000), esp. 5–6; Topham, "Textbook revolution" (2000); García-Belmar, Bertomeu-Sánchez, and Bensaude-Vincent, "Power of didactic writings" (2005); Hall, "Short course" (2005); and Park, "Quantum chemistry" (2005).

3. Kuhn, "Essential tension" (1977 [1959]), 228–29; emphasis in original.

4. Kuhn, "Function of measurement" (1977 [1961]), 180, 186. Kuhn returned to the theme of textbooks' misrepresentation of scientific practice, and of the historical routes by which scientific knowledge developed, throughout Kuhn, *Structure* (1996 [1962]).

their research.[5] Nor were the textbook authors alone in this quest. Reviewers in *Physics Today* consistently evaluated new textbooks precisely on the criterion of methodological usefulness, criticizing those books that seemed too concerned with various theories' "logical structures" or "finished achievements." If we are looking for finished theories or the logical analysis of theories' structures, one place we will *not* find them is in the diagrammatic textbooks published between 1955 and 1970.

THE POSTWAR AGE OF TEXTBOOKS

Physicists in the United States wondered how they could best spread the new techniques broadly and efficiently. During the 1950s, they received tremendous aid in this project from American publishers. Whereas notes from the 1948 Pocono meeting, Schwinger's lectures at the Ann Arbor summer school, and Dyson's 1951 Cornell lecture course circulated during the late 1940s and early 1950s by word of mouth and mimeograph, items like these began to see more formal publication over the course of the 1950s. Much like the rise of postdoctoral training, the vast expansion of physics publishing after the war represented a change in the infrastructure of how young physicists received their training and learned about new research. Also like the postdoc situation, the increases in physics publishing came at a propitious moment for the dispersion of Feynman diagrams. Timing was everything.

The changes did not take place immediately. Textbook publishers in the United States—many of them based in New York City—faced their own "Manhattan Project" as soon as World War II ended. With the aid of the G.I. Bill, record numbers of people planned to go to colleges and universities; eventually over two million veterans took advantage of the federal program by enrolling

5. Some readers might object that I am unfairly criticizing Kuhn here, given his well-known discussion of textbooks as collections of "exemplars," or shared problems. Indeed, Kuhn did emphasize that students must practice solving exemplars, but he explicitly distanced such textbook-related problem solving from the tools that scientists needed to use in their mature research. He explained, for example: "Last, but most important of all, is the characteristic technique of textbook presentation. Except in their occasional introductions, *science textbooks do not describe the sorts of problems that the professional may be asked to solve and the variety of techniques available for their solution.* Rather, these books exhibit concrete problem solutions that the profession has come to accept as paradigms, and they then ask the student, either with a pencil and paper or in the laboratory, to solve for himself problems very closely related in both method and substance to those through which the textbook or the accompanying lecture has led him." Kuhn, "Essential tension" (1959), 229; emphasis added. He elaborated on his notion of exemplars as pedagogically shared examples in Kuhn, *Structure* (1996 [1962]), esp. 43–51, 175, and 187–98.

in undergraduate and graduate courses.[6] Still facing wartime shortages in paper, machinery, binding facilities, and labor, the American publishing industry went into "emergency mode," making tremendous efforts to fill the unprecedented textbook demands. Barely through the first postwar academic year, the industry congratulated itself for having survived the "humdinger" of 1946. Both paper and labor shortages began to ease by late 1947, and by 1950 the American publishing industry was producing nearly as many new titles as the prewar highs.[7]

From the start, several publishers recognized that science publishing—including textbooks and other technical publications—could become a lucrative market in the postwar nuclear age. The G.I. Bill bloated enrollments in nearly all fields, yet the explosive rate of growth in the nation's physics departments quickly outstripped the changes in all other fields by a factor of two.[8] With graduate-level enrollments in physics skyrocketing, several presses inaugurated new textbook series on nuclear physics and related topics; others sought to increase their physics lists quickly by publishing translations of older, successful textbooks. American publishers announced twenty-five new physics textbooks in 1948 alone. Beyond the classroom, publishers produced more new science books in 1949 than any other category—more than such staples as biography, cooking, and religion.[9] Marketing tips for the new wave of scientific and technical books became a regular feature in the publishing trade journal *Publishers' Weekly*.[10] By the late 1950s, several New York presses—Interscience,

6. Geiger, *Research and Relevant Knowledge* (1993), 40–41; Olson, *G.I. Bill* (1974), 103.

7. Anon., "More books for more students" (1947), 1112. For more on the paper, machinery, and labor shortages, see also Melcher, "Editorial" (1946); Anon., "Many books postponed" (1946). Paper and labor shortages began to ease by late 1947: Anon., "Currents in the trade" (1948); Anon., "More new books published" (1950); and Anon., "Number of new books published" (1951).

8. Kaiser, "Cold war requisitions" (2002), 134–37.

9. Anon., "Currents in the trade" (1947); Anon., "Famous nuclear scientists write books" (1947); Donald McPherson (assistant manager, John Wiley) to Samuel Goudsmit, 1 Mar 1948, in *SAG*, Folder 24:256; Anon., "More books for more students" (1947), 1113–14; Anon., "Opportunities for technical book selling" (1948); Anon., "Checklist of highspots" (1948), 558, 567–68. On publishers' interest in publishing translations of foreign-language textbooks, see Donald McPherson to Samuel Goudsmit, 12 Aug 1949, and Goudsmit to McPherson, 23 Aug 1949, in *SAG*, Folder 24:256. On the publication of new science titles, see Anon., "More new books published" (1950), 220; the following year, new science titles ranked second, behind religion: Anon., "Number of new books published" (1951), 214.

10. McMahon, "Sale of technical books" (1948); Anon., "Retailing of technical and medical books" (1948); Anon., "Technical book promotion" (1949); Anon., "How technical books are bought" (1950); [Melcher], "Demand increases for technical books" (1950); Anon., "Technical and business book sales" (1951); and H. Klein, "Promoting technical books" (1951).

W. A. Benjamin, Gordon and Breach, Academic Press, Prentice-Hall—had begun to publish everything from summer school lectures, to conference proceedings, to informal lecture note and reprint series. No longer would these items circulate in haphazard, mimeographed form. From these many sources, assigned as teaching texts for classes or nabbed from libraries for self-study, students learned and practiced the fast-changing calculational tools of the trade.

Despite the postwar publishing boom, textbooks on the new diagrammatic techniques did not enter the publishing maelstron immediately. In the midst of his Cornell lecture course in the autumn of 1951, the hectographed notes from which would soon become coveted possessions, Freeman Dyson wrote home to his parents. "My work continues well," he reported. "I like my lectures more and more. And now I have arranged with a firm of publishers to write them up into a book." On Bethe's suggestion, Dyson had approached the publisher John Wiley and Sons, which had immediately agreed to publish the book, envisioning an initial print run of several thousand copies.[11] Alas, the book was not to be. A year and a half later, Fritz Rohrlich—who had been a postdoc with Dyson at the Institute for Advanced Study, and who was currently delivering his own course at Princeton—wrote to Dyson to ask about the book's progress. "Do you still have plans for working your lectures into a book? A textbook of this sort is very much needed," especially since no such textbook, treating Feynman diagrams and their uses, had appeared as late as March 1953. In the meantime, Rohrlich wondered whether Dyson could send him a copy of the 1951 notes, which continued to be "a valuable reference" while teaching; the Princeton librarians were having difficulty acquiring copies through the usual channels. Dyson wrote back, informing Rohrlich that copies of the hectographed lecture notes had run out "some time ago," and reporting that his work on converting the popular notes into a published textbook was proceeding slowly. Dyson "found it was hopeless to work seriously on the book" while teaching, so that, by the spring of 1953, "I have in fact done nothing beyond what is in the notes, and I doubt whether it will ever get written, at least not for several years." He was now "glad to leave this unrewarding task to others."[12]

11. Freeman Dyson to his parents, 2 Nov 1951, in *FJD.* Bethe had worked with Wiley for Bethe, *Elementary Nuclear Theory* (1947), and recommended the publisher to Dyson.

12. Fritz Rohrlich to Freeman Dyson, 7 Mar 1953, in *FR,* Box 23, Folder "Correspondence, 1946–54"; and Freeman Dyson to Fritz Rohrlich, 17 Mar 1953, in the same folder. Two textbooks that treated the diagrams had appeared by the end of 1953, but no translations were available for several years: Akhiezer and Berestetskii, *Kvantovaia elekrodinamika* (1953), later translated from the Russian as

As copies of Dyson's hectographed notes became more and more difficult to acquire, several other physicists began to circulate their own unpublished lecture notes, distributed, as Dyson's had been, within an informal network. Across the country, physicists inquired about the availability of each others' notes, while department secretaries sent one-dollar payments to the originating departments to cover distribution costs.[13] With publishing prospects looking better and better, however, several authors soon decided to write and publish their own, more formal textbooks. Many of these early authors, including Rohrlich, had learned the diagrammatic techniques from Dyson directly, and several used Dyson's unpublished notes as a template for their own textbooks. The textbook authors, much like Dyson himself, had little trouble finding publishers willing to take on their new projects. Often publishers approached physicists who were rumored to be thinking about writing textbooks; some of the textbook authors even had to fend off competing offers from several publishers.[14] By 1955—seven years after Feynman had first introduced his diagrams at the Pocono meeting—two American textbooks on Feynman's and Dyson's methods had appeared. Several more saw publication during the late 1950s and 1960s.

We might expect the diversity of diagrammatic appropriations, as charted in part 2, to lessen after 1955 with the publication of these diagrammatic textbooks. After all, now the rules and conventions governing Feynman diagrams' uses were bound between two covers, easy to transport from place to place, and simple to take off a shelf for review and reference. Instead, just the opposite happened: the dispersion in Feynman diagrams' pictorial forms, calculational roles, and interpreted meanings only continued to widen in the decade following the textbooks' publication. There were two main reasons why the publication of textbooks did not halt the diagrammatic diversity. First, the textbook

---

Akhiezer and Berestetskii, "Quantum Electrodynamics" (1957), and Akhiezer and Berestetskii, *Quantum Electrodynamics* (1965 [1959]); and the original edition of Umezawa, *Quantum Field Theory* (1956), which appeared in Japanese in 1953.

13. Richard Feynman to Cécile Morette, 5 June 1950, in *RPF;* S. H. Vasko to Fritz Rohrlich, 5 Dec 1952; Myrtle Farlee to Fritz Rohrlich, 19 Apr 1954 and 5 Oct 1954; Norman Kroll to Fritz Rohrlich, 5 May 1954; Richard Spitzer to Fritz Rohrlich, 28 Mar 1955; Rohrlich to Spitzer, 28 Mar 1955, all in *FR,* Box 23, Folder "Correspondence, 1946–54."

14. On publishers' practice of approaching physicists rumored to be working on textbooks, see Donald McPherson (then at Row, Peterson and Company) to Hans Bethe and Frederic de Hoffmann, 16 Dec 1953 and 18 Jan 1954, in *HAB,* Folder 9:21. Fritz Rohrlich similarly received competing offers for his diagrammatic textbook, cowritten with Josef Jauch; see the correspondence between Rohrlich and Robert Hofstadter from Nov 1953, in *FR,* Box 23, Folder "Correspondence, 1946–54."

authors focused on teaching "useful," piecemeal tools and techniques rather than overarching theories—physicists continued to adapt the diagrams' pictorial form to better suit given applications, rather than to emphasize systematic theoretical coherence. Second, the textbook authors reinscribed the original Feynman-Dyson split, continuing a tension over just what place Feynman diagrams should assume.

Even without any uniquely "proper" role for Feynman diagrams, however—or perhaps because of it—the diagrams continued to flourish. Graduate students learned about the line drawings at an ever-quickening pace, thanks in part to the spate of new textbooks. As they scribbled Feynman diagrams into their notes, practiced using them on problem sets and examinations, and drew them into their dissertations, they internalized a diagrammatic vision of the world much as their young advisers had done a few years earlier. More than ever, Feynman diagrams defined the daily practice of young theorists.

THE NEW DIAGRAMMATIC TEXTBOOKS

Useful Bags of Tricks

Franz Mandl's *Introduction to Quantum Field Theory* first appeared in 1959; editions of the textbook are still in print and in use today.[15] Two years before completing the book, Mandl mused on what ideas and techniques students and older physicists needed to master. He discussed his early plans for the textbook in a letter to Berkeley's famed experimentalist Emilio Segrè in the spring of 1957. Mandl was driven to consider writing the textbook, he explained, because he had just finished teaching two different courses on the subject over the previous two years. "The first one, lasting 45 hours, went into great detail," wrote Mandl, "and enabled me to get many points clear in my own mind." His second course, in contrast, lasted only sixteen lectures, and had been "intended for those without specialised theoretical knowledge, i.e. mainly experimentalists, though some theorists, not acquainted with field theory, also attended." Mandl hoped to repeat a version of this shorter course during Berkeley's summer school and to base a textbook on the shortened format.[16] "Now field theory is

15. Mandl, *Quantum Field Theory* (1959). Later editions include Mandl and Shaw, *Quantum Field Theory* (1984); and Mandl and Shaw, *Quantum Field Theory* (1993).

16. Franz Mandl to Emilio Segrè, 12 May 1957, in *BDP*, Folder 5:83. See also Segrè to Mandl, 27 May 1957, in the same folder. Segrè won the Nobel Prize in 1959 for his contributions to the discovery of the antiproton, based on work with Berkeley's cyclotron in 1955. See also Segrè, *Mind Always in Motion* (1993).

a difficult subject to present, either in book form or in lectures," Mandl next relayed—and, no doubt, Segrè would have agreed. "A detailed treatment (such as Schweber, Bethe and de Hoffmann, vol. 1 [1955], or Jauch and Rohrlich's 'Theory of photons and electrons' [1955]) is much too heavy going to be a useful introduction." Yet "Briefer treatments are either very terse and difficult... or hardly get to grips with the subject at all," Mandl continued. In his own short lecture course, and in his developing textbook, Mandl figured he could "get the basis [*sic*] concepts of field theory across." What were these basic concepts? They included second quantization, quantum statistics, and the like, to be sure. But Mandl's list of "concepts" mostly included *operational* techniques for making calculations: "S-matrix expansion, Feynman graphs, calculation of scattering processes," and so on.[17] These were the elements of modern theoretical physics that students and newcomers to the field had to learn and practice; these were the elements with which Mandl hoped to equip his auditors, and eventually his readers, by means of a short and useful textbook.

Much like Mandl's musings, the other textbooks on the new diagrammatic methods that were published during the 1950s and 1960s did not contain, much less enshrine, a single "theory." With very few exceptions—such as Josef Jauch and Fritz Rohrlich's *The Theory of Photons and Electrons* (1955)—almost none of the early textbooks was dedicated specifically to quantum electrodynamics. Even in this case, Rohrlich's original draft, which circulated as mimeographed lecture notes from his 1953 course at Princeton, was "primarily designed to teach the techniques and applications of the modern version of q.e.d.," as he explained to Dyson. To emphasize his focus on "techniques and applications" rather than theoretical foundations, Rohrlich titled his unpublished notes "Applied Quantum Electrodynamics."[18] The other textbook authors likewise developed a series of specific calculational techniques—termed, tellingly, a "bag of tricks" by James Bjorken and Sidney Drell in their famous pair of textbooks from the mid-1960s—to be applied to problems from across nonrelativistic quantum mechanics, QED, and various meson models.[19] Analogies to methods in single-particle quantum mechanics and nonrelativistic, semiclassical potential scattering competed for space with the conceptual apparatus of quantum field theory. Even renormalization, that triumph of diagrammatic

17. Mandl to Segrè, 12 May 1957, in *BDP*, Folder 5:83.

18. Fritz Rohrlich to Freeman Dyson, 7 Mar 1953, in *FR*, Box 23, Folder "Correspondence, 1946–54"; see also Norman Kroll to Fritz Rohrlich, 5 May 1954, in the same folder. See also Rohrlich, "Applied quantum electrodynamics" (1953).

19. Bjorken and Drell, *Relativistic Quantum Mechanics* (1964), viii. The identical preface appeared in Bjorken and Drell, *Relativistic Quantum Fields* (1965).

calculation, entered into these textbooks in hybrid forms. "Practical" discussions of how to use various cutoffs (similar to Feynman's early methods) to arrive at approximate numerical solutions to specific problems often won more space than did the "mathematical mumbo-jumbo"—the "highbrow," formal analysis of renormalizability—established by Dyson, Salam, and others.[20] We would search in vain to find any single theory developed or clarified in these textbooks. The new books were all about techniques.

As Mandl's letter illustrates, the early textbook authors worked hard to pitch their textbooks at the right level. Frederic de Hoffmann offered suggestions to the advertising manager of the press preparing *Mesons and Fields* on how best to advertise his new textbook. The subject matter of the book "is widely discussed in advanced journal publications and it is usually very difficult for a graduate student or even a fresh Ph.D. to orient himself in the field," de Hoffmann explained. "We believe that the first part of our book will breach this gap. We present the essentials of field theory and the methods developed in the last two or three years in a simple way." Sam Schweber wrote most of this first part, which introduced Feynman's and Dyson's diagrammatic methods—it eventually grew into a separate volume—and he updated his coauthor, Hans Bethe, a few months later: "I sent 1 copy of Vol. I to Art Wightman [Schweber's thesis adviser at Princeton] who has read through it and liked it. Thus it seems that we may have satisfied the very high-brows." This was good news, but not sufficient: "The question is what about the low brows and middle brows? Some graduate students here have seen it and expressed great satisfaction with it, in terms of clarity and exposition, so maybe the reviewers won't object to it too strenuously." Bethe replied that "I, who consider myself a 'middle-brow' like it very much too, so I suppose it is all right." To further emphasize that the book aimed to make the new techniques accessible, the authors recommended that the dust jacket for the first volume feature a large Feynman diagram.[21]

A series of book reviews helped to clarify and make explicit what most of the textbook authors already knew: what was most needed was catalogues of calculational techniques, "owner's manuals" for putting Feynman diagrams

20. Dyson referred to the "mathematical mumbo-jumbo of renormalization" in Dyson, review of *Mesons and Fields,* vol. 1, by Schweber, Bethe, and de Hoffmann (1956), 33, and characterized certain books as "'highbrow' expositions" of QED in Dyson, review of *Einführung,* by Thirring (1955).

21. Frederic de Hoffmann to Olaf Dahlskog (advertising manager, Row, Peterson), 25 Jan 1954, in *HAB,* Folder 9:21; Sam Schweber to Hans Bethe, 27 Sep 1954, in *HAB,* Folder 12:30; Hans Bethe to Sam Schweber, 5 Oct 1954, in *HAB,* Folder 12:30; and Donald McPherson to Hans Bethe, 1 Mar 1955, in *HAB,* Folder 10:22. Cf. Moreton, "Jackets sell technical books" (1951)—sometimes people really did judge a book by its cover.

to work, rather than any formal statements of quantum field theory's ultimate foundations or final articulation. "Clearly no book written now can expect to be the last word on the subject," explained John Polkinghorne in his review of Jauch and Rohrlich's *Theory of Photons and Electrons* (1955). "No doubt many changes will occur before a final theory is achieved, but in the meanwhile this book will serve as a most useful compendium of current knowledge." Polkinghorne emphasized that this "current knowledge" centered on how to "apply" QED to "detailed calculation[s]."[22] Dyson highlighted the same theme in his review of *Mesons and Fields*, volume 2 (1955): "The reviewer wishes in this connection to enunciate a theorem: if a book is to be published at a time T, and if it is supposed to be up-to-date to a time $T - t$, then it will inevitably be out-of-date at the time $T + t$." He added quickly, however, that his theorem hardly spelled doom for Bethe's textbook: "The question many people will ask is, why try to write a book about meson theories now, when the probability of obsolescence of ideas is so particularly acute? To this Bethe would probably reply, the purpose of such a book is to be useful, not to be immortal. And I believe he is right."[23] Textbooks had to be "useful," Bethe and Dyson agreed—and to the great majority of theorists in postwar America, at least, "usefulness" had little to do with timelessness or foundational rigor.

The reviewers had clear ideas about what made these textbooks "useful," even if not "immortal." There was a certain place, Polkinghorne judged, for books that "present[ed] the salient features" of a given subject, "exhibit[ing] its coherence and elegance" at the expense of the "incidental detail of manipulations."[24] Yet such presentations would not suffice. Books such as Umezawa Hiroomi's *Quantum Field Theory* (1956) suffered precisely because "the generality here adopted . . . has the disadvantage of making the book less suitable to be put in the hands of graduate students as their first introduction to the subject."[25] Walter Thirring's *Einführung in die Quantenelektrodynamik* (1955) similarly struck Dyson by its "adultness and breadth," features which, however,

22. John Polkinghorne, review of *Photons and Electrons*, by Jauch and Rohrlich (1956), 34.

23. Dyson, review of *Mesons and Fields*, vol. 2, by Schweber, Bethe, and de Hoffmann (1955). Bethe made a similar observation a few years earlier, in response to a proposal by Yukawa Hideki to write a textbook on meson theory. Bethe explained to Yukawa's potential publisher: "It is possible, in fact likely, that meson theory will make considerable advances over the next few years. Undoubtedly, therefore, a better book could be written three or four years from now [just when Bethe himself began to write *Mesons and Fields*!], and any book written now may become obsolete fairly rapidly. On the other hand, a summary at this time from a competent author would be very useful to all students in the field." Hans Bethe to Donald McPherson, 4 Jan 1949, in *HAB*, Folder 11:17.

24. Polkinghorne, review of *Photons and Electrons*, by Jauch and Rohrlich (1956), 34.

25. Polkinghorne, review of *Quantum Field Theory*, by Umezawa (1956).

rendered Thirring's elegant effort "unsuitable as a text-book for average students." The title of Thirring's book seemed "misleading" to Dyson: rather than "Introduction," "it had better been called 'The General Principles of Quantum Electrodynamics.'" Destined to "remain a standard reference for serious workers in quantum field theory," it did not answer the needs of students and those many workers, experimentalists and theorists alike, who required tips on techniques rather than esoteric niceties.[26] Books such as Schweber, Bethe, and de Hoffmann's first volume of *Mesons and Fields* (1955) could certainly err, at least in Dyson's opinion, on the side of "monumental dullness," casting "the theory of fields, the dragon which we fought with such high hopes in 1947–49," into "such a tame insipid beast." But at least such workaday books provided "detailed exposition of the rules for calculating observable quantities." Even if the heroism of the earlier days had begun to fade, textbooks that focused squarely on tools and techniques could help the new generation master the requisite calculational tasks.[27]

A Replay of the Feynman-Dyson Split

As the textbook authors wrestled with the general requirements of "usefulness," they struggled with the question of Feynman diagrams' proper place. The initial tension between Feynman's and Dyson's positions, with their varying emphases on "intuition" versus derivation, physical pictures versus topological indicators, made its way into practically all the textbooks on quantum field theory and particle physics that were published during the 1950s and 1960s. An advertising flyer for Schweber, Bethe, and de Hoffmann's *Mesons and Fields* (1955), for example, announced that the textbook "uses as the main basis Dyson's approach to field theory, rather than the reformulation of Schwinger and Feynman." The authors returned to this point in their preface, explaining that their presentation would follow Dyson's "rederivation" of the "Feynman formalism" from field-theoretic first principles.[28] But this "rederivation" appeared only after ninety pages had first spelled out Feynman's own pre-field-theoretic approach. Accordingly, low-order Feynman diagrams

26. Dyson, review of *Einführung*, by Thirring (1955), 22. Note that neither Umezawa nor Thirring worked in the United States at the time they produced their overly formal textbooks.

27. Dyson, review of *Mesons and Fields*, vol. 1, by Schweber, Bethe, and de Hoffmann (1956), 32–33.

28. A draft of the undated flyer may be found in *HAB*, Folder 10:22; Schweber, Bethe, and de Hoffmann, *Mesons and Fields* (1955), 1:xv; see also ix–x. As explained in the preface, the first volume was written mostly by Schweber, and the second volume mostly by Bethe and de Hoffmann (ix).

litter several of the early sections, making their debut ten chapters ahead of the diagrams' official Dysonian introduction. In keeping with their pre-field-theoretic status, these early examples omitted self-energy diagrams and other closed-loop contributions which, according to the later treatment, would have entered at the same perturbative order. More than 160 pages separated the first appearance of the diagrams from their later explanation and justification.[29]

Though diagrams appeared throughout these early sections, instructions for how to draw the diagrams, such as how to orient the time axis and how to affix arrows to distinguish electrons from positrons, appeared only in the later sections based on Dyson's derivation. Only in these later sections did the authors emphasize the one-to-one correspondence between terms in the mathematical perturbation series and diagrams with certain numbers of vertices, noting that this "correspondence is the basis of the usefulness of the concept of Feynman diagrams."[30] The diagrams' sole justification in the later sections was painted as stemming from their narrowly defined bookkeeping function, as an aid to the perturbative mathematics. Yet this strict sense of "usefulness" had not hindered the diagrams' earlier, "intuitive" introduction. Earlier in the book, several chapters in advance of their Dysonian discussion, the authors praised Feynman diagrams for "permit[ting] visualization and ease of understanding," independent of their field-theoretic correspondences.[31]

Despite his earlier objection to Schweber et al.'s approach, Franz Mandl charted a similar course in his *Introduction to Quantum Field Theory* (1959). He announced that his book concentrated "on the beautiful ideas of Dyson and Feynman: the perturbation expansion of the scattering matrix . . . and its interpretation in terms of graphs."[32] Introducing the famous diagrams in chapter 14, Mandl further explained that although Feynman's own route to the diagrams had been a "strongly intuitive" one, "We shall follow the more conventional method, due to Dyson, which does not tax the intuitive powers of the reader unduly. Instead, everything is obtained from the mathematics. In this spirit we shall look on Feynman graphs merely as a most useful aid to interpreting the mathematics."[33] This "interpretive" role cast the diagrams squarely

29. Schweber, Bethe, and de Hoffmann, *Mesons and Fields* (1955), vol. 1. Feynman diagrams are included throughout sections 8 and 9 (54–86), as well as in section 14 (151–65); the Dyson-style presentation of the field-theoretic "Feynman rules" appears in sections 18 and 19 (218–54). Examples of low-order diagrams that omit self-energy or other closed-loop contributions appear on, e.g., 80–81.

30. Ibid., 234. The same point was made on 222.

31. This earlier use was described as "intuitive" in ibid., 224; quotation on 54.

32. Mandl, *Quantum Field Theory* (1959), v.

33. Ibid., 94.

within the tradition of perturbative bookkeeping that Dyson had worked so hard to establish. Yet like Schweber, Bethe, and de Hoffmann before him, Mandl, too, made liberal use of Feynman diagrams throughout the thirteen chapters that preceded any mention of Dyson's "conventional method." Two further chapters followed before the means of drawing and calculating with Feynman diagrams in the Dysonian manner were at last explicitly laid out.[34] Again following the pattern of Schweber, Bethe, and de Hoffmann, Mandl emphasized only *after* Dyson's field-theoretic "Feynman rules" had been established that "This graphical description is a most useful way of interpreting the mathematics, particularly in complicated problems."[35] Yet Mandl used the diagrams long before he introduced the "mathematics" they were meant to "interpret."[36]

By highlighting Dyson's close-to-the-math derivation, these early textbooks revealed how to use the diagrams to streamline and simplify what would otherwise be dense and extended calculations. But by placing freestanding Feynman diagrams dozens of chapters and hundreds of pages before their official introduction, intending the diagrams to "speak for themselves," the textbooks tacitly promoted Feynman diagrams to a status beyond Dyson's carefully circumscribed roles for them. Indeed, while professing to follow Dyson's treatment of the diagrams, each of the early textbooks adopted Feynman's Minkowskian imagery—in explicit defiance of Dyson's clearly articulated objections. (See fig. 7.1.) Schweber, Bethe, and de Hoffmann made the pictorial link to Minkowski diagrams most explicit, including (as Feynman had done) time and space axes. Jauch, Rohrlich, and Mandl similarly inclined their particles' propagation lines according to the Minkowski tradition, scaling the speed of light to one. In his unpublished notes, Rohrlich had gone even further, describing particles' motion in Feynman diagrams in terms of their "past" and "future light cones"—terminology usually reserved only for Minkowski diagrams.[37] Each of these authors had gone out of his way to emphasize that they were simply following Dyson's pedagogical prescriptions, yet each intermingled Dyson's derivations with Feynman's "intuitive" diagrammatic style. The early textbooks reinscribed the Feynman-Dyson split.

34. Ibid. See, e.g., the diagrams included within chaps. 4, 11, and 13. Dyson's "Feynman rules" are derived in chap. 14.

35. Ibid., 104.

36. A later textbook similarly included a full, two-page spread of Feynman diagrams a chapter before readers were told how to use them or what was pictured in them. See Gasiorowicz, *Elementary Particle Physics* (1966), 130–31.

37. Rohrlich, "Applied quantum electrodynamics" (1953), 56.

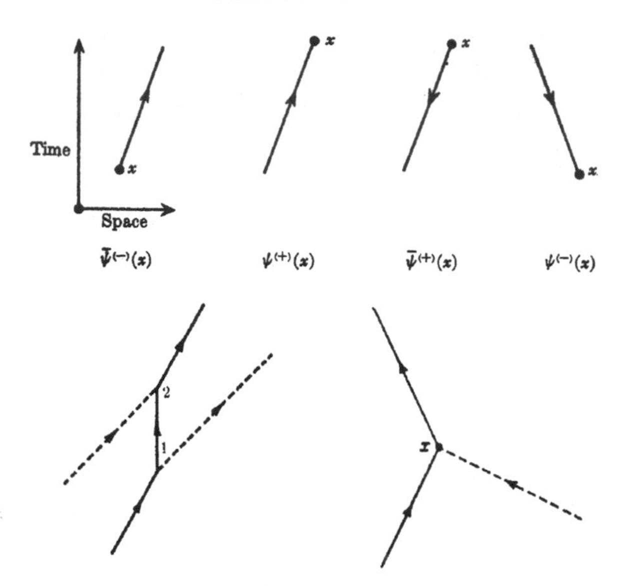

Figure 7.1. Textbook Feynman diagrams. (Sources: *top,* Schweber, Bethe, and de Hoffmann, *Mesons and Fields* [1955], 1:219; *bottom left,* Jauch and Rohrlich, *Theory of Photons and Electrons* [1955], 150; *bottom right,* Mandl, *Quantum Field Theory* [1959], 73.)

Later textbooks trafficked in the same tension. James Bjorken and Sidney Drell began their twin textbooks on relativistic quantum mechanics and quantum field theory from 1964 and 1965 with the strong statement that "one may go so far as to adopt the extreme view that the full set of all Feynman graphs *is* the theory." Though they quickly backed off this stance, they firmly stated their "conviction" that the diagrams and rules for calculating directly from them "may well outlive the elaborate mathematical structure" of canonical quantum field theory, which, they further opined, might "in time come to be viewed more as a superstructure than as a foundation."[38] Here the arrow of legitimation was turned around: Feynman diagrams, instead of being seen merely as a helpful tool based on the prior foundation of Dyson's mathematics, were placed at the forefront, likely to "outlive" the specific mathematical structures from which Dyson had derived rules for their use.

More than a prediction about future developments steered these theorists to privilege the diagrammatic over the rigorously mathematical. On the heels of their musing about field theory's ultimate foundations, they made clear the specifically pedagogical aspirations they saddled to the line drawings. Working directly in terms of Feynman diagrams, rather than "a deductive field theory

38. Bjorken and Drell, *Relativistic Quantum Mechanics* (1964), vii–viii; emphasis in original.

approach, should bring quantitative calculation, analysis, and understanding of Feynman graphs into the bag of tricks of a much larger community of physicists" than the "specialized narrow one" that practiced the recondite mathematical theory. As they continued, "we have in mind our experimental colleagues and students interested in particle physics. We believe this would be a healthy development."[39] Simple diagrams, rather than convoluted contour integrals, would serve as the ambassadors of understanding. Bjorken and Drell elevated the diagrams to a partially autonomous status, assuming that the diagrams could lead practically by themselves to an increased understanding of particle interactions.

Bjorken and Drell proceeded to take the early textbook examples to an extreme. They introduced Feynman diagrams within the first quarter of their first volume, but "delay[ed] as long as possible"—in fact to the last half of their second volume—"the enormous task of develop[ing] the formalism of quantized field theory," which would ultimately provide the "basis" for the "unambiguous rules of calculation."[40] Hundreds of pages of the first volume detailed low-order, semiclassical applications of Feynman's diagrammatic formalism, an entire volume before the quantization of free fields had even been introduced. Indeed, the Feynman diagrams they drew differed sharply between the two contexts: more Feynman-like, often with time axes and spacetime propagation conventions in the first volume, and more Dysonian, built from static polygons, in the second. (See fig. 7.2.) As Feynman had demonstrated with his articles from 1949 and 1950, it mattered little that their second volume, deriving the rules for the diagrams' use, was not even published at the time that the first diagram-laden volume hit campus bookstores' shelves. The pair of texts by Bjorken and Drell thus made tangible the split between Feynman's "intuitive," pre-field-theoretic introduction of his diagrams and Dyson's canonical field-theoretic derivation of their use.[41] Whether one worked in a semiclassical approximation and restricted attention to the lowest order within the perturbative expansion (as outlined throughout the first volume) or read in the diagrams subtle features of arbitrarily high-order correction terms (as treated

39. Ibid., viii.

40. Ibid., 78. Thus, Feynman diagrams were put to work throughout the 1964 volume beginning in chap. 6 and used continuously for the remainder of the volume (chaps. 7–10), whereas the Dyson-style "Feynman rules," derived from fully field-theoretic considerations, appeared only in §17.5, more than 180 pages into the 1965 volume.

41. The entire first volume, much like Feynman's "Theory of positrons" (1949) and the early sections of Schweber, Bethe, and de Hoffmann's *Mesons and Fields* (1955), thus restricted attention to Dirac's equation for the electron as a single-particle wave function, subject to interactions with a classical electromagnetic source. It is symbolic of the different missions of the two volumes that the

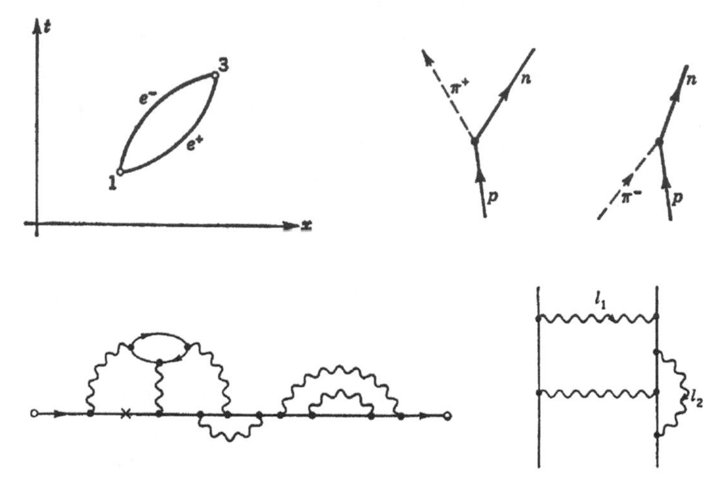

Figure 7.2. Feynmanian and Dysonian diagrams. (Sources: *top,* Bjorken and Drell, *Relativistic Quantum Mechanics* [1964], 91, 216; *bottom,* Bjorken and Drell, *Relativistic Quantum Fields* [1965], 285, 322.)

at length in the second volume), all readers of Bjorken and Drell's texts were to begin their studies diagrammatically.

Very similar foundational and pedagogical statements about the role of Feynman diagrams were voiced two years later in Richard Mattuck's influential textbook *A Guide to Feynman Diagrams in the Many-Body Problem* (1967). Note Mattuck's title: his book aimed to guide readers in using the diagrams, not in analyzing the theoretical foundations of many-body physics. Mattuck explained that rather than focus on the mathematical niceties of the underlying field theory, "I have instead concentrated exclusively on giving the reader a *feeling for the diagrams themselves* [and for] their physical significance."[42] Like Bjorken and Drell, Mattuck upheld his promise: throughout the entire book, discussion was formulated directly in terms of diagrams, with the corresponding integrals simply left unevaluated. Mattuck believed the line drawings were the key both to conceptual clarity and pedagogical effectiveness.

During the remainder of the 1960s, some particle theorists pushed harder still to invert Dyson's hierarchy and to invest Feynman diagrams with an

---

Klein-Gordon *equation* is presented in the first volume only after 180 pages of Dirac-electron examples (and is introduced solely in the service of treating scalar-meson applications in the same semiclassical way as the fermion examples), whereas the second volume *begins* with the Klein-Gordon *field,* as the simplest system for which the rules of free-field canonical quantization could be developed.

42. Mattuck, *Guide to Feynman Diagrams* (1967), viii; emphasis added.

autonomy unheard of even in Feynman's original formulation. As a final example, consider a 150-page technical report circulated in 1973 by CERN, Europe's main laboratory for particle physics. The report's title declared its purpose: "Diagrammar," a grammar of diagrams. The authors, Gerard 't Hooft and Martinus Veltman, had astounded the world's particle theorists just two years earlier by proving that quantum field theories involving far more complicated interactions than QED could still have their infinities tamed and removed systematically.[43] Fresh from their success, they announced that recent theoretical advances made it "necessary to reconsider many well-established ideas" about the relations between diagrams and perturbative calculations. The result: "Few physicists object nowadays to the idea that diagrams contain more truth than the underlying formalism."[44] It was not quite necessary, they assured their readers, to discard all earlier work. Instead, "The situation must be reversed: diagrams form the basis from which everything must be derived. They define the operational rules." With Feynman diagrams promoted to "primary objects," the entire mathematical edifice, including Dyson's beloved S-matrix, would be "defined in terms of diagrams."[45] In their detailed report, the preeminent field theorists invested Feynman diagrams with unprecedented priority; from the diagrams, all else would follow.

Despite their confident opening, 't Hooft and Veltman were hardly reporting the majority opinion of their day. Their widely circulating report, however, provides a telling endpoint for this review of the many hopes and aspirations that theorists pinned to Feynman's diagrams in the quarter-century following their introduction. For Schwinger and others, as we saw in chapter 3, the diagrams were redundant and eliminable. For Dyson, they were highly useful but, like any mnemonic device, needed to be handled with care. In Feynman's

43. These theorists provided a proof that non-Abelian gauge field theories could be unambiguously renormalized. Veltman had been 't Hooft's dissertation adviser, and the two had collaborated on several articles on the way to this proof. See in particular 't Hooft, "Renormalization" (1971); 't Hooft, "Renormalizable Lagrangians" (1971); and 't Hooft and Veltman, "Renormalization" (1972). On the swift influence of 't Hooft's proof, see Sullivan et al., "Understanding rapid theoretical change" (1980); Koester, Sullivan, and White, "Theory selection" (1982); and Pickering, *Constructing Quarks* (1984), 177–80.

44. 't Hooft and Veltman, "Diagrammar" (1973), 29–30. My thanks to Sam Schweber for bringing this report to my attention.

45. Ibid. Strictly speaking, it was "global diagrams," shorthand diagrams representing the set of all Feynman diagrams contributing to a given perturbative order, which 't Hooft and Veltman labeled the "primary objects." Veltman recently published a far more elementary textbook on quantum field theory that similarly aims to put all weight on Feynman diagrams directly, and to use them to build up the mathematical machinery of a perturbative S-matrix: Veltman, *Diagrammatica* (1994). See also Veltman, "Path to renormalizability" (1997).

hands, and in the textbooks by Schweber, Bethe, and de Hoffmann, and by Mandl, they appeared more robust and "intuitive" than their strict derivation would allow. To Bjorken and Drell, as to Mattuck, they held the key to broad pedagogical success. And to 't Hooft and Veltman, they were the only thing steady enough to cling to.

PEDAGOGY AND THE PICTURES' PLACE

Practice, Practice, Practice

The new textbooks provided young physicists a range of evaluations—in introductory statements as well as worked examples—of the diagrams' proper place and role. Of course, few students read these textbooks entirely in isolation; I know of no physicist who learned to use the new techniques well from textbooks alone. Instead, students learned about the new techniques and practiced calculating with them from the textbooks in conjunction with instructors' lecture courses—and rarely would instructors simply follow the books, word for word, as they taught their classes. A telling example of physicists' classroom appropriation of textbooks comes from Bethe's lecture notes for his Cornell course Advanced Quantum Mechanics, delivered before Dyson taught his own version of the course, and before the better-known lecture notes and textbooks had begun to circulate. Bethe relied on Walter Heitler's prediagrammatic textbook, *The Quantum Theory of Radiation* (1936; 2nd ed., 1944), supplemented by Feynman's and Dyson's 1949 *Physical Review* articles. In his notes, Bethe reminded himself which equations from Heitler's book to present, and in which order. Entire sections of Heitler's book were rearranged, alternately expanded upon or condensed as Bethe presented the relevant material for his class: "Discuss most of §6, ~~especially #1~~, leaving out proofs on p. 44, 50, 52"; "Leave out radiation reaction, p. 93, change discussion of classification, p. 97"; and so on. *Physical Review* articles in hand, Bethe next introduced the new diagrams to his class, spelling out in detail the "Compton effect à la Feynman."[46] No doubt later instructors followed a similar pattern after the diagrammatic textbooks had been published; without more sets of lecture notes like Bethe's, it is difficult to discern in detail how instructors used the textbooks to teach

46. Bethe's handwritten notes for his Advanced Quantum Mechanics course may be found in *HAB*, Folder 1:4. The earliest set of notes extant in this folder come from his 1946 course, which he revised over the years. A thirteen-page insert, entitled, "Quantum electrodynamics," which treated applications of Feynman diagrams to scattering and self-energy problems, is dated "Summer 1950." Cf. Heitler, *Quantum Theory of Radiation* (1944).

their students about the new diagrammatic techniques. What the surviving sources do demonstrate, however, is that physicists across the country increasingly trained their students to begin their calculations with Feynman's line drawings.

Often the problems assigned to graduate students involved little more than *drawing* the correct Feynman diagrams for a given process—long before the strings of associated mathematical equations would be trotted out and evaluated, students were drilled in manipulating the bare lines of the diagrams themselves. Students in Feynman's 1953 Caltech course Quantum Mechanics III (Physics 205c), for example, worked on a series of diagram-based homework problems. In one of them, they had to "Prepare diagrams and integrals needed for the radiative corrections (of order e^2) to the Klein-Nishina formula" for electron-photon scattering. The thrust of the assignment was to figure out which diagrams were relevant, and to practice translating them one-to-one into their associated integrals, as Feynman emphasized: "Do as much as possible and compare results with Phys. Rev. 85, 231 (1952)." In the reference cited, Feynman and his doctoral student, Laurie Brown, had proceeded beyond this homework step, evaluating the integrals that arose from the relevant diagrams.[47] Graduate students just beginning their diagrammatic training, however, would proceed step by step, *first* practicing how to draw the appropriate Feynman diagrams, and next practicing how to translate these diagrams into integral expressions. Another assignment in Feynman's class prepped the students to calculate electron-positron scattering to lowest order using the diagrams, giving hints about which diagrams contributed to the process. Still another asked students to "Set up the integrals for each of the two [given] diagrams"—again, without any need to evaluate the resulting mathematical expressions.[48] In Feynman's classroom, much as in his published presentations, the diagrams came first.

The diagrammatic textbooks likewise emphasized the need to practice drawing and manipulating the diagrams themselves, separate from evaluating their associated mathematical expressions. Franz Mandl, for example, included the following exercises within his *Introduction to Quantum Field Theory* (1959): "14.1. Obtain the fourth-order Feynman graphs for Compton scattering.

47. Brown and Feynman [A.50].

48. Feynman, "Quantum mechanics III" (1953), 131, 110, and 139. These notes formed the basis for Feynman, *Quantum Electrodynamics* (1961). Unfortunately, problem sets and students' notebooks from this early period are only very rarely extant today, though some of the unpublished lecture notes, such as this set from Feynman's 1953 course, included some of the assigned problems. Similarly, only a handful of the published textbooks from this period included their own printed problems.

14.2. Obtain the fourth-order Feynman graphs for electron-positron Møller scattering." Much as in Feynman's classroom, students were asked to draw the relevant diagrams, not to evaluate the resulting integrals. Nor was this considered a trivial problem: the printed solution to these exercises, and the explanation of the answers, spilled over several pages.[49] Bjorken and Drell similarly assigned readers of their *Relativistic Quantum Mechanics* (1964) the problem of drawing the appropriate Feynman diagrams to demonstrate "that a closed loop from which an odd number of photon lines emerge, vanishes." Students were thus asked to practice drawing Feynman diagrams and applying Dyson's antiparticle-arrow convention much the way Karplus and Kroll had done in their 1950 paper.[50] More commonly, lecturers and textbook authors asked students to practice writing down the associated matrix elements for given sets of Feynman diagrams. Often, as in Feynman's classroom exercises, the problems did not require students to carry the calculation through to the end, but rather to write down the diagrams' associated integrals and leave it at that.[51]

Other times, worrying that simply listing the rules for the diagrams' use would "appear very complicated at first," as Feynman put it, lecturers and textbook authors worked through example after example of how to handle the diagrams in specific calculations, collecting the final list of rules only in appendixes at the very end of their lecture notes or textbooks. Felix Villars, lecturing on the diagrams at MIT soon after his postdoctoral stay at the Institute for Advanced Study, emphasized that "The technique can only be developed with practice once a few standard tricks, which we may illustrate later, are learned."[52] Dyson likewise devoted over eight pages to laying out the rules for drawing and calculating with the diagrams in his 1951 lecture notes before spending nearly thirty pages working out a long series of examples of actual diagrammatic calculations. He lectured to students at the August 1954 Les Houches summer school in a similar fashion, teaching them the rudiments of "graphology" with a series of lengthy worked examples.[53] Rohrlich's

49. Mandl, *Quantum Field Theory* (1959), 166, 177–79.

50. Bjorken and Drell, *Relativistic Quantum Mechanics* (1964), 181.

51. See, e.g., Feynman, "Quantum mechanics III" (1953), 83, 102, and 139; Bjorken and Drell, *Relativistic Quantum Mechanics* (1964), 145, 181, 207, and 279.

52. Quotations from Feynman, "High energy phenomena" (1951), 45; and Villars, "Quantum electrodynamics" (1951), 66. Bjorken and Drell printed a six-page appendix, "Rules for Feynman Graphs," at the back of both of their textbooks: Bjorken and Drell, *Relativistic Quantum Mechanics* (1964), 285–90; Bjorken and Drell, *Relativistic Quantum Fields* (1965), 381–86.

53. Dyson, "Advanced quantum mechanics" (1951), 127–61; Dyson's handwritten notes for his August 1954 Les Houches lectures, on 37–41, 45–48. A copy of these notes is in Professor Dyson's

Applied Quantum Electrodynamics lecture notes from 1953 contained almost exclusively worked problems. He noted time and again that "The applications . . . will give sufficient examples to illustrate these statements," and that "The details of [a particular equation] will become clear in the applications which follow."[54] Students and colleagues did not pick up the diagrams and apply them following a casual perusal. Instead, young physicists developed their diagrammatic craft by practicing with sample problems and scrutinizing worked examples.

The Diagram-Density Spectrum

Some physicists feared that the diagrams' new roles and meanings, as carved out for the study of meson interactions and other nonelectrodynamic applications, could lead to confusion for students apt to try to read them as perturbative bookkeepers. Gabriel Barton, for example, counseled readers of his *Introduction to Dispersion Techniques in Field Theory* (1965) explicitly about how to treat the blob-vertex diagrams reproduced in figure 7.3. "Such diagrams must be carefully distinguished from Feynman diagrams," Barton explained. "Their interpretation is as follows: on the *left* we symbolize the *process* in question: the source j vomits out two particles."[55] Such explicit clarifications were necessary, Barton and others worried, precisely because the diagrams looked in certain ways so much like ordinary Feynman diagrams of perturbative QED. Other physicists, such as Stanley Mandelstam, avoided such potential confusion by rarely using Feynman diagrams outside a specifically perturbative context. Instead, Mandelstam relied upon a text-based notation more akin to chemical formulas (such as $\gamma + p \rightarrow \pi^+ + n$) to list the reactants involved in various interactions.[56]

Other prominent theorists, however, found the diagrams so intuitive, so ready to hand, that their use proved convenient in many different settings. By the late 1950s, theorists had established a spectrum of what we might call "diagram density." More established theorists, such as Murph Goldberger, Geoffrey Chew, and their peers, often opted not to include explicit diagrams within their research articles at all, choosing instead to talk about various

---

possession. Cf. Villars's similarly methodical exposition in Villars, "Quantum electrodynamics" (1951), 61–70.

54. Rohrlich, "Applied quantum electrodynamics" (1953), 62, 65.

55. Barton, *Dispersion Techniques* (1965), 149; emphasis in original.

56. Mandelstam, "Two-dimensional representations" (1961).

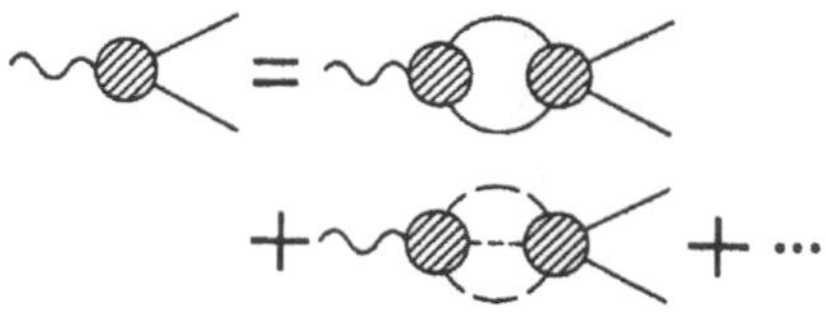

Figure 7.3. "Vomit"-blob vertex. (Source: Barton, *Dispersion Techniques* [1965], 149.)

features of the diagrams in words rather than present the actual diagrams themselves. Yet these same theorists incorporated more and more explicit diagrams as their presentations moved toward increasingly pedagogical forums: review articles included more diagrams than research articles, published lecture notes more than review articles; summer school and unpublished lecture notes included still more explicit diagrams than the published versions, and so on. The changes may be discerned from Hans Bethe's notes on the seminars given by invited theorists at Cornell: during the late 1940s and early 1950s, Feynman diagrams appeared relatively rarely in his notes (even within Cornell's bastion of diagram users); by the 1960s, hardly a single talk went by without several Feynman diagrams filling Bethe's handwritten notes.[57] As theorists stepped into the seminar hall and the classroom, the visual nature of their presentations increased dramatically.

Consider the lecture notes taken by Louis Balázs in Chew's particle theory course at Berkeley during the spring of 1961; Balázs was one of Chew's graduate students at the time. Chew had already begun developing many of the features that would eventually coalesce into his *S*-matrix program, the details of which we will examine in the next two chapters. Chew's widely used published lecture-note volume, *S-Matrix Theory of Strong Interactions* (1961), derived largely from his 1960 summer school lectures and from the course Balázs attended at Berkeley. Feynman diagrams simply cover Balázs's notes—two, three, and more diagrams appear on nearly every single page for the entire semester, sometimes stuck in the margins, other times framed by a box in the center of the page.[58] Balázs, along with Chew's other students, practiced marching through each rotation of a given scattering diagram, giving each rotated incarnation its full diagrammatic due. In figure 7.4, for example, a one-particle-exchange

57. Bethe's notes on Cornell's theory seminars may be found in *HAB*, Folders 1:4–5, 16:7–8, and 16:43–47.

58. My thanks to Professor Balázs for supplying a copy of his course notes from Physics 224B, Berkeley, spring 1961. Professor Jerry Finkelstein, another of Chew's graduate students during the mid-1960s, similarly shared copies of his study notes in preparation for his general examination, and these, too, are filled with Feynman diagrams. Cf. Chew, *S-Matrix Theory* (1961).

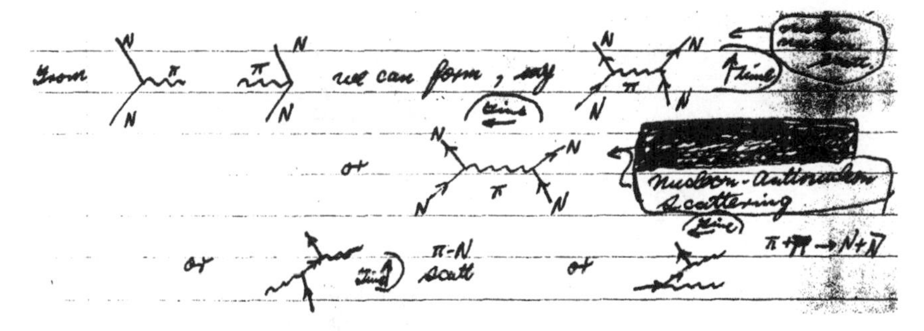

Figure 7.4. Learning to rotate the diagrams. (Source: Louis Balázs's notes for Berkeley's Physics 224B, spring 1961; pages are not numbered.)

diagram for nucleon-nucleon scattering appears at the upper right, with a time axis showing the flow of the various particles up the page (each labeled explicitly, as in Marshak's meson work); below that, with the time axis now running right to left, Balázs learned that the same diagram depicted nucleon-antinucleon annihilation. Crossed diagrams for pion-nucleon scattering could likewise be formed from the same basic diagrammatic ingredients, as Balázs scribbled near the bottom of that page. This was the pictorial bread and butter behind Chew's developing methods, yet Chew's published version of the notes from this course condensed these various diagrammatic rotations into a single, generic two-body diagram.[59]

In a similar way, students of particle theory at Princeton doodled far more diagrams in their class notes than their advisers included in their research publications. Kip Thorne filled the notes from his 1962–63 courses with Sam Treiman, Richard Blankenbecler, and Murph Goldberger with the telltale diagrams. After a lightning-fast introduction to "Feynmanology" and a short series of by-then-standard examples from perturbative QED, Thorne and his classmates let the diagrams loose on problems involving the strong nuclear interactions. One application centered on electromagnetic form factors, which described the distribution of electric charge in strongly interacting particles such as protons and pions. Beginning with a single-photon exchange approximation, Thorne learned to expand the various "blob" vertices in a series of photon and pion exchanges. (See fig. 7.5.) Again, comparison with Treiman's published lecture notes from the period reveal far more diagrams per square inch in Thorne's course notes than in Treiman's off-campus presentations.[60]

59. Chew, *S-Matrix Theory* (1961), 16.

60. My thanks to Professor Kip Thorne for providing a copy of his course notes from Sam Treiman's course Elementary Particle Physics, Princeton University, 1962–63. Cf., e.g., Treiman, "Analyticity in particle physics" (1963).

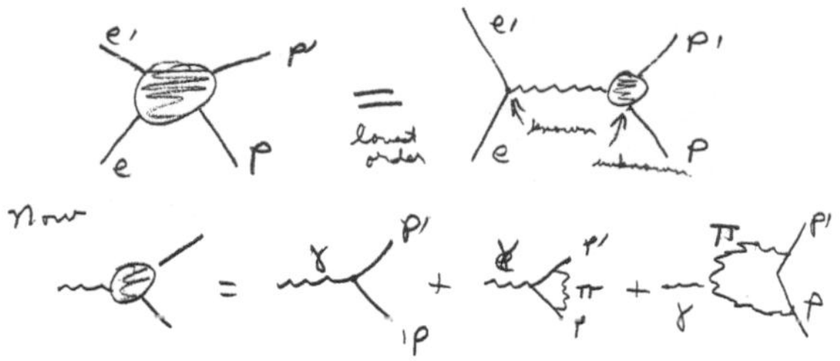

Figure 7.5. "Blob" vertices for electromagnetic form-factor calculations. (Source: Kip Thorne's 1963 Princeton course notes; pages are not numbered.)

One might expect that the relative abundance of explicit diagrams within students' unpublished, hand-copied lecture notes as compared with their advisers' published volumes would be determined by cost: printing long series of figures in published textbooks would raise the cost of the books, whereas lecturers could expend ample chalk within a lecture course to keep the diagrams ever present. Certainly the cost of reproducing so many figures may have provided one reason for the diagram-density spectrum. Yet cost alone cannot explain the phenomenon, since graduate students (usually with limited funds to spend on figure reproduction in publications) often included many more Feynman diagrams in their published research than their advisers did—and cost was certainly an issue for reproducing figures in the *Physical Review.* Henry Wyld, for example, published a long article in 1954 based on his dissertation at the University of Chicago, which he completed under the direction of Gell-Mann and Goldberger. Wyld's article appeared just a few pages after Gell-Mann and Goldberger's article on crossing symmetry, which we will examine in the next chapter. Whereas Gell-Mann and Goldberger had omitted diagrams altogether, choosing to discuss the relevant classes of diagrams in words only, Wyld printed scores of explicit Feynman diagrams within his own article.[61] He began, for example, with the full set of second- and fourth-order diagrams for meson-nucleon scattering that one would delineate according to Dyson's usual perturbative bookkeeping prescriptions; unlike Karplus and Kroll in their treatment of fourth-order diagrams for applications in QED, however, Wyld left off all details of spacetime or momentum-state information. (See fig. 7.6.) Wyld proceeded much as his advisers had done, explaining that "All conceivable Feynman diagrams for meson-nucleon scattering can be divided into two classes. In one class are all diagrams in which the incoming nucleon

61. Gell-Mann and Goldberger [A.135], 1433; Wyld [A.137], 1664.

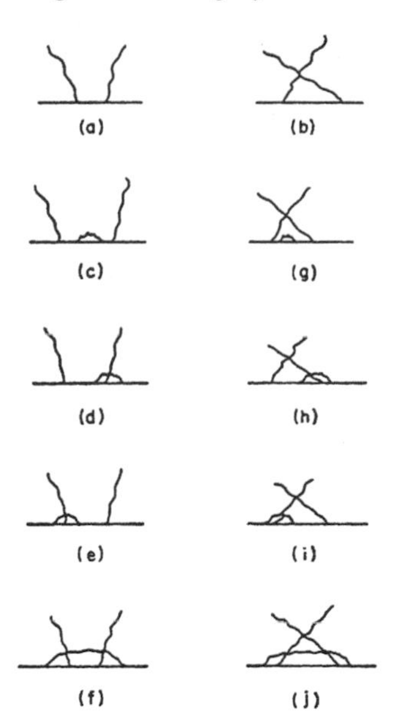

Figure 7.6. Feynman diagrams for meson-nucleon scattering. (Source: Wyld [A.137], 1664.)

line appears completely bare of virtual or real mesons at some point in the middle of the diagram." Yet whereas Gell-Mann and Goldberger had left this class of diagrams up to their readers' imagination, Wyld kept the discussion pictorially concrete: "Diagrams (*a*), (*c*), (*d*), (*e*), of [fig. 7.6] are diagrams of this class."[62] Another Chicago graduate student, R. E. Prange, printed dozens of Feynman diagrams within his 1958 article on electrodynamic dispersion relations, even though his adviser had long before stopped printing explicit diagrams within his dispersion-relations articles.[63]

In another example, William Frazer included the diagrams in figure 7.7 within his 1959 Berkeley dissertation, advised by Geoffrey Chew. Frazer explained how various terms within his adviser's famous dispersion-relations article "arose" from these diagrams, even though Chew himself had made no mention of such diagrams in the original article.[64] Most of Chew's other

62. Wyld [A.137], 1665.

63. Prange, "Dispersion relations" (1958). Prange, like Wyld, worked with Goldberger and acknowledged his help.

64. Frazer, *Electromagnetic Form Factor* (1959), 25, 52, and 58. Frazer used the diagrams in figure 7.7 to discuss three distinct terms within a scattering amplitude, derived in Chew et al., "Relativistic dispersion relation" (1957). Though Frazer's diagrams carried explicit momentum labels, their

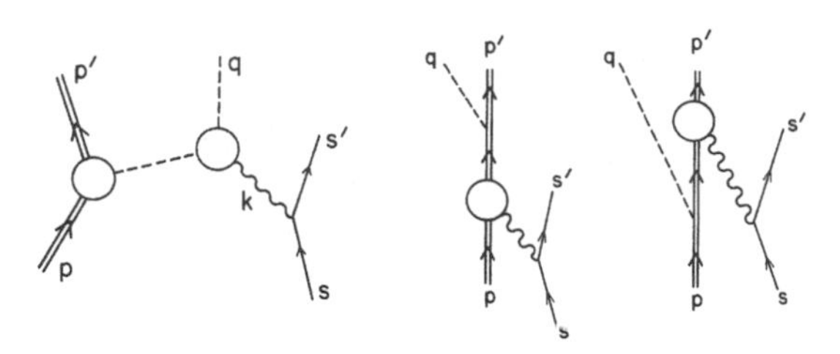

Figure 7.7. Phenomenological form factors. (Source: Frazer, *Electromagnetic Form Factor* [1959], 52, 58.)

graduate students from this period similarly made extensive use of explicit Feynman diagrams within their dissertations when treating topics for which Chew included few explicit diagrams in print.[65]

The new generation of diagram converts—those physicists who were graduate students and postdocs in the late 1950s and early 1960s, such as Balázs, Thorne, Wyld, and Frazer, who learned how to use the diagrams from members of the first diagrammatic generation, now ten or more years past their doctorates—scribbled the new diagrams just as fervently as their teachers once had done, in their course notebooks, their dissertations, and their published articles. Steven Weinberg captured young physicists' indoctrination in the diagrams particularly cogently when he explained in a 1964 article that "We will lapse into the language of Feynman diagrams when we do our 2π bookkeeping . . . , but the reader will recognize in this the effects of our childhood training, rather than any essential dependence on field theory."[66] The diagram-density spectrum thus points to a process of internalization: as theorists became older, they included fewer explicit Feynman diagrams in their research publications, even though their research was still built upon their "childhood training" in the diagrams. As soon as they stepped into a seminar

relation to the terms within the proposed scattering amplitude was distinct from their relation in an ordinary perturbative treatment. Instead, the blob-vertices corresponded to phenomenological form factors that were to be estimated by comparison with experimental data, not calculated with a series of higher-order Feynman diagrams. The diagrams' relevance for Frazer, as he explained in the accompanying text, stemmed entirely from which kind of particle was exchanged in which diagram: a meson current in the diagram on the left, nucleon currents in the two diagrams on the right.

65. See the following Berkeley dissertations: Ball, *Application of Mandelstam Representation* (1960), 5, 42; Cziffra, *Two-Pion Exchange* (1960), 13, 15, 16, 22; Young, *Two Problems of Structure* (1961), 7, 42, 51, 56; and Kim, *Production of Pion Pairs* (1961), 5, 10, 37.

66. Weinberg, "Photons and gravitons" (1964). This paper continued the work in Weinberg, "Derivation of gauge invariance" (1964); see also Weinberg, "Feynman rules" (1964).

hall or classroom, out came the diagrams by the dozen, and in they went to their students' notes, dissertations, and articles.

* * *

Graduate students needed to learn about Feynman diagrams and practice using them; manipulating the diagrams was not a skill one could acquire by passive observation alone. Several years after the diagrams had been introduced, textbooks began to be published to help students develop the new skill. Whether listening to seminars, reading review articles, or perusing the textbooks, students found themselves awash in the diagrams; in turn, they filled their notebooks with the line drawings and copied them into their dissertations. In order to help these students learn how to calculate with the diagrams, textbook authors strove to find the right level of difficulty, focusing on applications, techniques, and procedures rather than theoretical foundations or ultimate underpinnings. In their efforts to produce "useful" textbooks, the authors revisited the Feynman-Dyson split: in some places, the diagrams were presented as intuitive and simple enough to interpret that little methodological apparatus appeared necessary, while in other places, Dyson's systematic rules for drawing and using the diagrams claimed center stage. By the mid-1960s, some authors even claimed that the diagrams were more robust than the formalism they were meant to emulate.

Meanwhile, by the late 1950s theorists were often including the greatest number of diagrams when stepping into particularly pedagogical settings. In their published work, many of these same theorists—most of whom had learned how to calculate with the diagrams a decade earlier—discussed the diagrams in words only, confident that their readers would have no trouble following along. Here was a sign of just how central the diagrams had become. Physicists today practically never cite such foundational articles as Albert Einstein's 1905 paper on special relativity or Erwin Schrödinger's 1926 papers on quantum mechanics; the disappearance of explicit references to these works points to their utter ubiquity among working physicists today, rather than their unimportance. Likewise, theorists fresh out of their Ph.D. programs, as well as their still-young teachers, had learned by the late 1950s to see the world through the lens of Feynman's stick figures. Some had grown so familiar with the diagrams that they could instruct their peers to follow at least some diagrammatic discussions in their mind's eye.

· 8 ·

Doodling toward a New "Theory"

It appears to me . . . likely that the essence of the diagrammatic approach will eventually be divorced from field theory.

GEOFFREY CHEW, 1961[1]

In the spring of 1964, just a few months before the Free Speech Movement would engulf the Berkeley campus of the University of California, Jerry Finkelstein opened up a new notebook and began scribbling.[2] Finkelstein, a graduate student in Berkeley's physics department, was preparing for his qualifying oral examination, a test required for advancement to doctoral candidacy. In his tiny, quick scrawl, he summarized material culled from earlier course notes, courses that had covered several topics in theoretical particle physics—polology, single-variable and double-dispersion relations, Regge poles—many of which his adviser, Geoffrey Chew, had first pioneered or championed only a few years earlier. Sprinkled throughout Finkelstein's study notes were simple line drawings, such as those reproduced in figure 8.1. Though the diagrams looked, to the casual observer at least, simply to be Feynman diagrams, Finkelstein had learned to call the examples in his notes by different names: these were Landau graphs, polology diagrams, and Cutkosky diagrams, each of which we will examine below. As Finkelstein recalled in a recent interview, "we were using Cutkosky diagrams, and called them Cutkosky diagrams, and we thought we knew what they meant. If someone else were using Feynman diagrams, we knew they meant different things, and we didn't get confused."[3] Despite the pictorial resemblance to Feynman's famous diagrams, Finkelstein knew that

1. Chew, *S-Matrix Theory* (1961), 3.

2. Professor Finkelstein kindly provided portions of his 1964 notes. For more on the Free Speech Movement in Berkeley, see Rorabaugh, *Berkeley at War* (1989), chap. 1.

3. Finkelstein interview (1998).

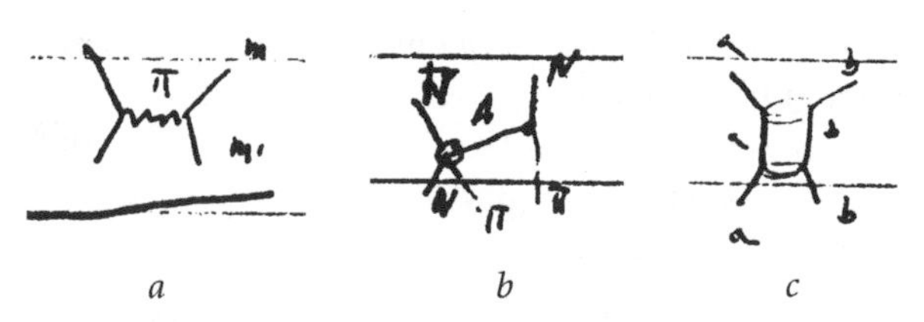

Figure 8.1. *a*, Landau graph. *b*, Polology diagram. *c*, Cutkosky diagram. (Source: Jerry Finkelstein's unpublished 1964 Berkeley study notes, 35, 39.)

the diagrams in his notes were different: they were used for different purposes and meant different things.

Finkelstein could have opened practically any volume on high energy physics during his study sessions and been inundated with figures very much like the ones he scribbled hastily in his notebook. The line drawings' ubiquity in the particle physics literature of the 1950s and 1960s might seem to confirm Bruno Latour's well-known conclusion about the pervasiveness of inscriptions and diagrams in scientific practice. As Latour argues, diagrams prove powerful for scientists because they are "immutable mobiles." The flat, two-dimensional representations can be circulated broadly, reprinted, rescaled, and recombined at very little economic cost, ensuring their easy mobility. Along their many travels, moreover, diagrams provide "optical consistency," transporting their immutable message without essential change across oceans and empires. From diagrams' mobility and immutability, Latour continues, derives their power to enroll and mobilize scientific allies. Scientists' ability to enlist diagrams to bring the faraway or the difficult-to-see into the here and now, he concludes, guarantees that much of scientific practice will revolve around their production and use.[4]

Latour is surely correct to highlight the centrality of diagrams and other types of inscriptions to daily scientific practice. Yet Finkelstein's comments about the distinctions between Feynman diagrams and Cutkosky diagrams should give us pause before equating the pictures in figure 8.1 with immutable mobiles: although certainly proving to be highly mobile, these diagrams hardly remained immutable, either in appearance, role, or meaning. As we have seen in part 2 and will continue to observe in this chapter, physicists and their students adapted the diagrams for a wide array of different purposes between Feynman's original presentation in 1948 and Finkelstein's study sessions in 1964. As they

4. Latour, *Science in Action* (1987), 227–28, 236–37; and Latour, "Drawing things together" (1990 [1986]), esp. 35–36, 44–48. On the centrality of inscriptions in general to scientific work, see also Latour and Woolgar, *Laboratory Life* (1986 [1979]).

crafted the diagrams toward different calculational ends, theoretical physicists imputed new and different meanings to the diagrams' bare lines, all the while adopting subtle shifts in the diagrams' pictorial forms as well. Rather than assuming some inherent "immutability," therefore, we will be better served by focusing squarely on the diagrams' tremendous plasticity. We must not be too hasty in leaping from the diagrams' constant presence in published texts and student notes to the conclusion that they played the same role or function in all their varied settings. Indeed, Feynman diagrams' plasticity furthered their staying power, rather than diminishing it. Like any tool, users' eagerness to tinker with the diagrams and adapt them to new purposes ensured that they would remain central to theorists' daily practice. Little about the diagrams remained "immutable."

The philosopher James Griesemer has pointed out that scientific diagrams often play roles that are "weakly generative"—helping in the process of scientific discovery but becoming eliminable thereafter—or "strongly generative"—logically equivalent to content expressed in other ways, yet holding a lasting heuristic power lost to these other systems of representation.[5] Feynman's diagrams proved to be highly generative for particle physicists working on the strong nuclear force during the 1950s and 1960s. Some theorists, based primarily in Berkeley, drew on several of the more popular uses of the diagrams that were worked out during 1955–65 to lobby for a wholesale replacement of quantum field theory itself—just over a decade after Dyson had derived the diagrams' rules and justified their use from a field-theoretic basis. These theorists, and, to varying degrees, scores of other physicists working outside Berkeley, drew on the diagrams to provide an increasingly heuristic scaffolding around which new theoretical practices—and, eventually, claims for a new theory in its own right—could be developed. Geoffrey Chew's *S*-matrix program emerged as one by-product of this decade of diagrammatic tinkering, progressively stretching Dyson's umbilical cord between Feynman's diagrams and rule-bound perturbative calculations. These improvisations began just at the moment that the first diagrammatic textbooks were hitting the shelves.

DISPERSION RELATIONS

Many particle theorists greeted the failure of perturbative techniques for the strong interactions by working with an alternative representation of particles'

5. Griesemer, "Must scientific diagrams be eliminable?" (1991), esp. 161–62; and Griesemer and Wimsatt, "Picturing Weismannism" (1989).

scatterings, in the form of dispersion relations. Though earlier precedents could be found for the dispersion-relations approach, the techniques came to dominate strong-interaction studies beginning in the mid-1950s—just when perturbative field-theoretic techniques seemed to have lost all usefulness for the strong interactions.[6] Murph Goldberger, Murray Gell-Mann, Francis Low, Geoffrey Chew, and Nambu Yoichiro—all centered in the American Midwest, and all just a few years past their doctorates—launched the new phase of dispersion-relations work.[7] A flood of review articles and summer school lectures during the late 1950s and early 1960s brought the new calculational tools to students of particle theory.[8] Dispersion relations offered theorists the hope of yielding a mathematical treatment of particles' scattering without being tied to any particular microdynamical model of particles' interactions. With the new relations, theorists hoped to implement a few general principles, such as causality and unitarity (the conservation of probability), without relying on any of the particular calculational steps of perturbative field theory. As Robert Karplus and Malvin Ruderman explained in an influential article from early 1955, the dispersion integrals depended only on "those relations among cross sections and scattering amplitudes which are independent of the underlying model"; hence they could "remove all reference to the details of the interactions."[9] Unlike Dyson, these young theorists did not begin by postulating a field-theoretic interaction Hamiltonian, $\mathbf{H}_I$, and then grind

6. Cushing, *Theory Construction* (1990), 57–63, 67–88; Goldberger, "Theory and applications of dispersion relations" (1960); J. D. Jackson, "Dispersion relation techniques" (1961); Goldberger, "Fifteen years" (1970); Cini, "Dispersion relations" (1980); and Pickering, "From field theory to phenomenology" (1989).

7. Pickering, in particular, notes these theorists' geographic overlap in "From field theory to phenomenology" (1989), 582–85. Both Goldberger (b. 1922) and Chew (b. 1924) obtained their doctorates from the University of Chicago in 1948; Gell-Mann (b. 1929) took his degree from MIT in 1951; Low (b. 1921) graduated from Columbia University in 1949; and Nambu (b. 1921) earned his D.Sc. from the University of Tokyo in 1952.

8. Early review articles include Chew, "Pion-nucleon interaction" (1959); MacGregor, Moravcsik, and Stapp, "Nucleon-nucleon scattering" (1960), esp. 324–46; and Hamilton, "Dispersion relations" (1960). Dispersion relations dominated discussion at the 1961 Solvay conference in Brussels: Stoop, *Quantum Theory of Fields* (1961). See also DeWitt and Omnès, *Relations de dispersion* (1960); Screaton, *Dispersion Relations* (1961); and Fulton, "Resonances" (1963), esp. 47–91.

9. Karplus and Ruderman, "Applications of causality to scattering" (1955), 771. Karplus, of course, had worked as a postdoc with Freeman Dyson and Norman Kroll at the Institute for Advanced Study in 1948–49, and, as we have seen in chap. 6, Karplus and Kroll published the first article to apply Dyson's perturbative use of Feynman diagrams to fourth-order corrections in QED. Ruderman, meanwhile, had been a graduate student at Caltech when Feynman arrived there in 1950 and had produced unpublished lecture notes from Feynman's early diagrammatic courses there, before accepting his post at Berkeley. See Feynman, "Quantum electrodynamics" (1950).

through perturbative approximations to an infinite series of terms. Rather, they remained agnostic about the form of the underlying dynamical laws and worked instead to correlate empirical cross sections with mathematical scattering amplitudes.

Following Karplus and Ruderman's lead, dispersion theorists treated the forward scattering amplitude, $f(\omega)$, as an analytic function of the particles' energy, ω, studying the properties of $f(\omega)$ as ω was continued throughout the complex plane.[10] Drawing on details from the mathematics of complex variables, these theorists could demonstrate—or at least suggest—that for causal processes the mathematical scattering amplitude had to obey the integral relation:

$$\mathrm{Re}[f(\omega)] = \frac{1}{\pi} P \int_{-\infty}^{+\infty} d\omega' \frac{\mathrm{Im}[f(\omega')]}{\omega' - \omega}.$$

To respect causality, in other words, the real and imaginary parts of the scattering amplitude could not be independent, but rather had to be linked by this integral relation.[11] The initial "proofs" of these so-called dispersion relations held only for massless particles; Gell-Mann, Goldberger, and Walter Thirring asserted somewhat cavalierly that it was "plausible" that such integral relations would also hold true for the more experimentally relevant massive cases.[12] For many, such "plausibility" arguments seemed sufficient—especially to people like Goldberger, who claimed to have learned what little he knew of complex-variable analysis "in the gutter."[13]

10. An "analytic" function is, loosely speaking, a function that behaves "smoothly" enough that, given its value at a particular point, it may be evaluated anywhere else within a certain domain. This domain is called the function's domain of analyticity.

11. The *P* in this expression indicates the Cauchy principal value. The particular mathematical form of causality invoked here was "microcausality," first postulated in Gell-Mann, Goldberger, and Thirring, "Causality conditions" (1954).

12. Ibid., 1613, 1616. Goldberger soon extended the derivation, still only relevant for forward-scattering, to include massive particles: Goldberger, "Causality conditions" (1955). More important, in actual applications, this equation had to be modified because the integral over Im[*f*] ordinarily diverged for large ω, making a series of "subtractions" necessary. See Goldberger, "Theory and applications of dispersion relations" (1960), 24–25; and J. D. Jackson, "Dispersion relation techniques" (1961), 8–11.

13. Goldberger, interview with Andrew Pickering, quoted in Pickering, "From field theory to phenomenology" (1989), 593. In general, the mathematics of complex analysis had not been emphasized in physics graduate students' training in the postwar United States: W. C. Kelly, "Survey of education in physics in universities of the United States," unpublished report dated December 1962, on 55–58, in *AIP-EMD*, Box 9. The early articles on dispersion relations sometimes included lengthy sections reviewing the relevant mathematical details, such as Cauchy's integral theorem and the

Theorists' next step was to expand the imaginary part of the amplitude, $\text{Im}[f(\omega)]$, in a series of intermediate states. Goldberger and his collaborators arrived at the form

$$\text{Im}[f] \propto \sum_n \langle out|n\rangle \langle n|in\rangle \delta(p_{in} - p_n - p_{out}),$$

where the sum ran over all *real* states $|n\rangle$ that could connect the incoming state $|in\rangle$ to the outgoing state $|out\rangle$, while conserving energy, momentum, and any relevant internal quantum numbers.[14] Gone was all talk of virtual particles, which had littered every single perturbative field-theoretic calculation. Instead, Goldberger and company could insert this sum into their dispersion integral for $\text{Re}[f(\omega)]$ and then make guesses—David Jackson called some of them "drastic assumptions" backed by "feeble arguments" in his summer school lectures—about which terms within the infinite sum would make the largest contributions to the integral.[15] The intermediate states $|n\rangle$ with lowest mass could usually be assumed to dominate the sum, at least in energy ranges near these masses. Enter the diagrams: given the sum over intermediate states, several dispersion theorists found it convenient to introduce Feynman-like diagrams as a new kind of bookkeeping device. (See fig. 8.2.) No longer tracking higher- and higher-order corrections within a perturbative expansion, the diagrams now were put to work to keep track of possible intermediate states $|n\rangle$. By picturing the particles included in these states, physicists could more easily determine which candidates had the lowest mass and hence warranted further scrutiny.

Like Feynman diagrams, these new doodles used different line types (squiggly lines versus straight ones) to demarcate particle types. Whereas Kroll's students at Columbia had used the diagrams to track radiative corrections at the all-important vertices, the new diagrams' form shifted to better serve their

---

residue theorem, before proceeding to particle physics applications, as in Karplus and Ruderman, "Applications of causality to scattering" (1955), 772–73.

14. This equation follows from the unitarity of the S-matrix, $S^{\dagger}S = 1$: one may relate the scattering amplitude, f, to the S-matrix as $S = 1 + iCf$, where C contains kinematic and phase-space factors; requiring $S^{\dagger}S = 1$ leads to $\text{Im}[f] \propto f^{\dagger}f$, or, in explicit matrix notation, the sum over intermediate states written above.

15. J. D. Jackson, "Dispersion relation techniques" (1961), 50. Jackson was reviewing Goldberger's and Sam Treiman's work on pion decay, which he described as "a classic example of both the good and the bad aspects of dispersion relations in action." Cf. Goldberger and Treiman, "Decay of the pi meson" (1958).

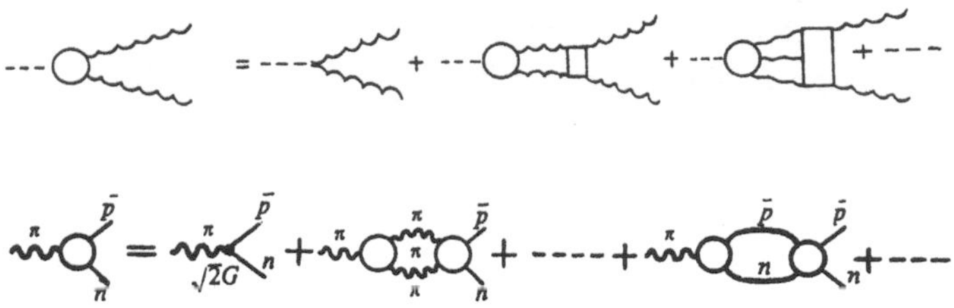

Figure 8.2. Illustrative Feynman-like diagrams for dispersion relations, showing the lowest-mass, real intermediate states. (Sources: *top,* Goldberger, "Theory and applications of dispersion relations" [1960], 97; *bottom,* J. D. Jackson, "Dispersion relation techniques" [1961], 48.)

new purpose, focusing attention on the particles exchanged *between* vertices. These exchanged particles were the states $|n\rangle$ to be singled out from the infinite sum. Unlike information about the exchanged particles, details of what occurred at the vertices would have to be governed by some underlying model of the microinteractions, such as a field-theoretic interaction Hamiltonian, $\mathbf{H}_I$—but dispersion-relations theorists proceeded without reference to any such interaction terms. Vertices became simply "blobs," as Jackson and others called the circles that appeared in the diagrams where particle lines came together. Without a specific Hamiltonian determining the microdynamics, furthermore, Dyson's rule fixing the number of legs that could meet at the vertices likewise could fall by the wayside. These pictures, with different numbers of particle lines entering and exiting blob-vertices, literally had no meaning within perturbative field-theoretic calculations. Instead, the new diagrams became tailor made for illustrating the relevant exchange terms within dispersion relations.

Physicists' approach to dispersion relations revealed interesting geographic contours. Armed with their dispersion-relations integrals and blob-vertex diagrams, theorists in the United States made themselves useful by treating the reams of experimental data pouring from the new accelerators; dispersion relations became a phenomenological exercise. In fact, nearly all the proponents of dispersion relations working in the United States trumpeted this close contact with experiments as an especially strong feature of their program. Several articles containing theoretical treatments of specific dispersion relations were published in conjunction with accompanying experimental articles.[16] Theorists in

16. Goldberger, Miyazawa, and Oehme, "Application of dispersion relations" (1955); with Anderson, Davidon, and Kruse, "Pion-proton scattering" (1955), 339–43; and Davidon and Goldberger, "Comparison of spin-flip dispersion relations" (1956). The famous Chew-Goldberger-Low-Nambu papers from 1957 likewise received instant attention from experimentalists: Chew et al., "Application

Europe often treated the topic quite differently. Cambridge-trained Richard Eden, John Polkinghorne, and John Taylor, the Göttingen theorists Harry Lehmann, Kurt Symanzik, and Wolfhart Zimmermann, and the Soviet mathematical physicists Nikolai Bogoliubov and Dmitrii Shirkov saw in dispersion relations an opportunity to build a new axiomatic version of quantum field theory. They would start with only a handful of basic physical principles (causality, unitarity, analyticity, agreement with the symmetries of special relativity, and so on), and aim to derive all other results from scratch, while avoiding the many problems with perturbative expansions and infinite renormalization constants that had already irked Pauli, Källén, and others. This group labored at length to find rigorous derivations of the dispersion-relations integrals: in working through the derivations, perhaps they would learn more about the structure of their favored axioms.[17]

Two sets of mimeographed lecture notes throw the differences into greatest relief. Richard Eden's visiting lectures at the University of Maryland in spring 1961 included an extensive list of references to the Cambridge group's work on the analytic structure of singularities in dispersion relations and to the German and Soviet groups' efforts at formal derivations, along with a separate list of references under the heading "Applications." Unlike the earlier references, nearly all the citations in the "Applications" section were to papers by American physicists—papers that applied dispersion relations to real data several years before any of the Europeans' proofs or derivations had been written.[18] Francis Low returned to this division at a Canadian summer school two years later, remarking that "the axiomists have been studying functions of many complex variables and have lost contact with physics in the sense of the comparison of theory and experiment." Rather than "trying to prove dispersion relations from the axioms of field theory," Low charted a different, "heuristic" course: "one may assume they are true and see what experimental consequences they

of dispersion relations" (1957), and Chew et al., "Relativistic dispersion relation" (1957); McDonald, Peterson, and Corson, "Photoproduction of neutral pions" (1957); Uretsky et al., "Photoproduction of positive pions" (1958); and Frazer, *Electromagnetic Form Factor* (1959), 23–24. See also Goldberger's and Gell-Mann's comments at the 1961 Solvay conference: Goldberger, "Single variable dispersion relations" (1961), 180; Gell-Mann's remark appears within a discussion reprinted in Stoop, *Quantum Theory of Fields* (1961), 177.

17. Reviews and extensive references may be found in Taylor, "Special topics" (1960); Hagedorn, "Introduction" (1961); and Eden, "Lectures" (1961). See also Eden et al., *Analytic S-Matrix* (1966); Wightman, "General theory of quantized fields" (1989); and Cushing, *Theory Construction* (1990), chap. 3.

18. Eden, "Lectures" (1961), R.1–R.4.

lead us to." For this work, as far as Low was concerned, nothing was more handy than Feynman diagrams: he had always found the Feynman-Dyson diagrammatic approach "the most useful," even in this newer terrain. "In general the series of Feynman diagrams has been an enormously helpful guide," and that was good enough for Low. All the banter about axiomatic foundations was largely beside the point, Low made clear, lecturing to his students that "In general one should not take sides too strongly but instead work with that system of equations which is most convenient to the problem at hand." More often than not, in Low's experience, Feynman diagrams had proved "most convenient."[19]

The successes scored by young dispersion-relations theorists—correlating much of the overflowing empirical data within a mathematical framework that promised to incorporate both causality and unitarity—provided a ready precedent for most of the work within particle theory that followed throughout the remainder of the 1950s and 1960s. Drawing refashioned diagrams while eschewing perturbative techniques, implementing general principles while ignoring questions of ultimate foundations, and focusing steadily upon the pressing demands of domestic data came more and more to define what it meant to do particle theory in the United States. Severed from much of Dyson's original rules and context for their use, Feynman diagrams could nonetheless be put to work by theorists in their ongoing quest to tame the strongly interacting nuclear realm.

CROSSING TO A NEW REPRESENTATION

The diagrams used for dispersion relations were distinct from those of Dyson's or Kroll's perturbative calculations. Yet only in a restricted sense could we say that physicists read new content in the diagrams that they had not read there before: by and large, dispersion theorists used the diagrams to keep track of which types of particles had been exchanged between which other particles—one of several functions for which physicists had relied upon the diagrams from the beginning. Physicists quickly put the diagrams to work in more generative ways. Some theorists began to see information in the diagrams that had not been seen there before. Physicists drew new conclusions, or at least extracted suggestive proposals, from arguments based on what they characterized as the structure of the diagrams themselves. Moreover, relations between these new

19. Low, "Dispersion relations" (1963), L2–L4. See also Cini, "Dispersion relations" (1980); Pickering, "From field theory to phenomenology" (1989); and Schweber, "Some reflections" (1989), 684–87.

diagram-derived proposals and the supposedly general dictates of quantum field theory were often left unexamined or unresolved.

Crossing Symmetry

Only a few months after Gell-Mann, Goldberger, and Thirring inaugurated the new age of dispersion relations in 1954, Gell-Mann and Goldberger introduced another use for Feynman diagrams. The two theorists, then at the University of Chicago, calculated the scattering of low-frequency light off of spin-1/2 particles in three different ways, showing that (to first order in the photons' energy) classical electrodynamics, quantum mechanics, and quantum field theory all produced the same results.[20] Though their third calculation manifestly fell within the framework of quantum field theory, they emphasized that, much as in the new dispersion-relations work, "we need not specify the nature of the fields and interactions in detail." Instead, they would argue directly, based on the structure of Feynman diagrams: "We shall employ the language of Feynman diagrams, but it should be emphasized that no use is made of perturbation theory."[21] Gell-Mann and Goldberger thus explicitly cut the diagrams loose from Dyson's perturbative tether, yet they put the diagrams to work in their calculation more directly than the early meson theorists and dispersion theorists had done. Where the meson theorists and dispersion theorists employed Feynman diagrams as simple illustrations of relevant processes, Gell-Mann and Goldberger drew new conclusions based on the form of the diagrams themselves.

Their article, submitted in September 1954 and published three months later, provides a telling indication that Feynman diagrams had definitely entered theorists' toolkits by that time: Gell-Mann and Goldberger described the diagrams in words but felt no compunction to include the accompanying figures in pictorial form. They began their field-theoretic calculation by enjoining their readers, "We may picture the entire sequence of Feynman diagrams in the following way: The nucleon line proceeds from the beginning to the end of the diagram, emitting and absorbing, in between, various virtual quanta that belong to its self-field."[22] All possible diagrams for the scattering of light

20. Gell-Mann and Goldberger [A.135].

21. Gell-Mann and Goldberger [A.135], 1435.

22. Gell-Mann and Goldberger [A.135]. Goldberger outlined crossing symmetry, described as here in terms of pairs of crossed Feynman diagrams, in his contribution to the Jan 1954 Rochester conference. See Wentzel's report on the "Theoretical Session" in Noyes et al., *Proceedings* (1954), 26–40. Chew et al., "Relativistic dispersion relation" (1957), 1346, similarly discussed crossing symmetry by invoking classes of Feynman diagrams in words, without printing the accompanying pictures.

off of this nucleon could be divided into two classes. Class *A* would include those diagrams, with any arbitrary number of virtual particles and vertices, that featured a stretch of "bare" nucleon line between the absorption of the incoming photon and the emission of the outgoing one; that is, the diagrams in class *A* would include all kinds of fancy radiative corrections at the vertices, but not along the internal nucleon line. Class *B* included all other diagrams, in which the internal nucleon line was interrupted by virtual quanta in between the two external-photon vertices. (Their graduate student, Henry Wyld, introduced these distinct classes with his long list of fourth-order Feynman diagrams in fig. 7.6.) Their distinction between the two classes of diagrams was precisely the one that Gell-Mann had learned while a postdoc at the Institute for Advanced Study; this division had been central to his work with Francis Low on bound states, ladder diagrams, and the Bethe-Salpeter equation.

Continuing to avoid perturbative methods, the Chicago theorists derived a symmetry of the full scattering amplitude based on these classes of Feynman diagrams: the scattering amplitude (whatever its final form) had to remain the same if one swapped, in each diagram, the incoming photon with the outgoing one. This symmetry, soon dubbed "crossing symmetry," arose because the amplitude could be written as the sum of two general terms—one term corresponding to the infinite number of Feynman diagrams in which absorption of the incoming photon occurred before emission of the outgoing one, and the other term corresponding to the infinite number of diagrams in which emission preceded absorption. Gell-Mann and Goldberger termed these latter examples the "crossed diagrams" of the first set. Feynman had included examples of both kinds of diagrams when working perturbatively. In his second paper from 1949, for example, he included the diagrams in figure 8.3 for the lowest-order contribution to Compton scattering.[23] Yet neither Feynman nor anyone else at the time had drawn any general conclusions about how generic such pairs of crossed diagrams would be, or what mathematical requirements this diagrammatic symmetry would entail. Whereas the presence of both diagrams occasioned no particular fanfare in Feynman's presentation, Gell-Mann and Goldberger seized upon this diagrammatic pairing, generalized it to all perturbative orders for any arbitrary interaction, and derived consequences for the total scattering amplitude that followed from the structure of these diagrams. The symmetry of the amplitude, they explained, "depends only on

23. Physicists had been puzzling over "exchange" terms, such as this pair of crossed terms, since the 1920s. See Carson, "Exchange forces," pt. 1 (1996); Carson, "Exchange forces," pt. 2 (1996).

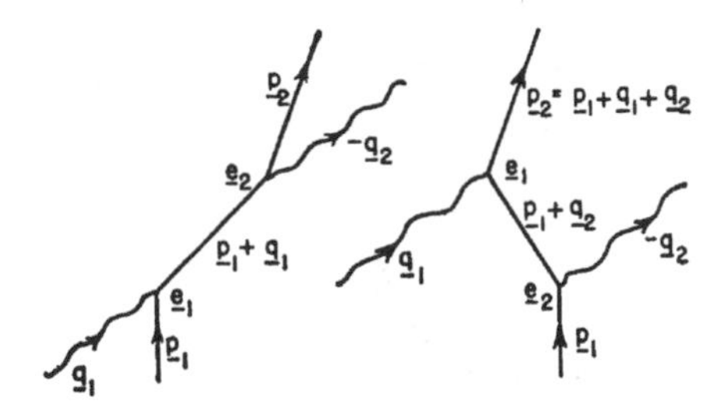

Figure 8.3. Crossed Feynman diagrams for Compton scattering. (Source: Feynman [A.4], 775.)

the fact that for every 'crossed' diagram there is a corresponding 'uncrossed' one and vice versa."[24] The diagrams became constitutive of the symmetry.

The Mandelstam Representation

Goldberger repeatedly characterized crossing symmetry simply as a property of scattering amplitudes in quantum field theory, with few or no additional implications for dispersion relations.[25] In the hands of other theorists, however, when combined with insights from dispersion relations, crossing symmetry began to play a much more central role. In his 1960 summer school lectures, Michael Moravcsik spelled out some of the new steps that had been taken during 1958–59. Discussing a more stylized pair of crossed diagrams (fig. 8.4) than Feynman's early example, Moravcsik explained that "the second diagram is just the first one rotated by 90° and some of the physical particles interchanged." Treating these diagrams now as simple rotations of each other, rather than as fundamentally separate diagrams, Moravcsik encouraged his auditors to speculate along with him: the two orientations of the diagram might share the same analytic properties as one extended variables such as the momentum transfer and energy throughout the complex plane. "Such relationships between the two kinds of diagrams," Moravcsik continued, provided "the incentive for the double dispersion relation approach."[26] In this way, Moravcsik grounded discussion of Stanley Mandelstam's new approach to dispersion relations in the structure of crossed Feynman diagrams.

In the usual approach to dispersion relations, as described above, theorists explored the analytic properties of scattering amplitudes as functions of either

24. Gell-Mann and Goldberger [A.135], 1435–36.

25. See, e.g., Goldberger, "Theory and applications of dispersion relations" (1960), 110, 120.

26. Moravcsik, "Practical utilisation" (1961), 119.

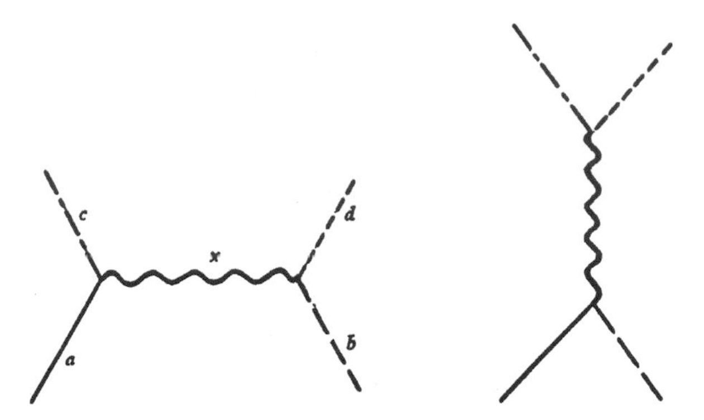

Figure 8.4. A general pair of crossed diagrams. (Source: Moravcsik, "Practical utilisation" [1961], 119.)

the energy, while keeping the momentum transfer fixed, or the momentum transfer while keeping the energy fixed. In other words, the original work on dispersion relations during 1954–58 focused on single-variable relations. While a postdoc at Columbia University, Stanley Mandelstam developed a new representation for scattering amplitudes that allowed for the *simultaneous* analytic continuation of the energy and momentum-transfer variables throughout the complex plane. This new representation quickly became known by the twin monikers "Mandelstam representation" and "double-dispersion relations."

Mandelstam first presented his results at the American Physical Society meeting in Washington, D.C., during May 1958.[27] In the audience was Geoffrey Chew, who had been struggling in frustration to extend dispersion relations into a double-dispersion form. As one of Chew's former students later reminisced: "Thus, when Stanley gave a cryptic ten-minute paper at the Washington meeting of the American Physical Society in 1958, Geoff was uniquely appreciative of its significance. The folk tale has it that Geoff whisked Stanley out of the meeting room and directly onto a plane to Berkeley!"[28] Mandelstam's absorption into Chew's growing group at Berkeley occurred only slightly less quickly than in this kidnapping story, and Mandelstam and Chew immediately

27. The abstract from his presentation is included in the schedule for the meeting, printed in the *Bulletin of the American Physical Society* 3 (1958): 216. Nambu had produced similar relations in 1955: Nambu, "Structure of Green's functions" (1955).

28. Frazer, "Analytic and unitary S-matrix" (1985), 2–3. In a recent interview, Professor Mandelstam chuckled upon hearing this characterization of his 1958 encounter with Chew: "That story is not quite accurate. I had applied to Berkeley [for a postdoc] before the APS meeting. It's probably true that I got the job because of meeting Chew at the APS meeting, though." Mandelstam interview (1998).

began a close and fruitful collaboration. In fact, while Mandelstam's original papers remained notoriously abstract and difficult to understand for most theorists—years later Goldberger exclaimed, "I have never understood a word that Stanley says," even though Mandelstam was "almost always right"—Chew championed the new work and quickly spread the results far and wide.[29] Chew reported on Mandelstam's new representation in July 1958 at a conference at CERN, included it prominently in his 1959 review article, and dedicated all his 1960 summer school lectures to the new techniques. As theorist John Polkinghorne put it years later, "There was a saying at the time that 'there is no God but Mandelstam and Chew is his prophet.'" Nearly every one of Chew's graduate students between 1959 and 1963 completed dissertations based on the Mandelstam representation, while his postdocs and collaborators wrote additional informal lecture notes on the topic.[30]

Much as Moravcsik characterized it in his 1960 summer school lectures, Mandelstam elevated crossing symmetry to a central role in dispersion relations. If the crossed Feynman diagrams were really just rotations of a single diagram, then perhaps one single analytic function would describe each of these processes. Mandelstam suggested, for example, that the scattering amplitude for pion-nucleon scattering ($\pi + N \rightarrow \pi + N$) also described the crossed process in which a nucleon-antinucleon pair created two pions ($N + \overline{N} \rightarrow \pi + \pi$). The second reaction resulted by rotating the initial Feynman diagram. As Mandelstam concluded from these diagrammatic reshufflings, the two reactions "are now treated on an equivalent footing."[31]

The square of the energy for each of these reactions, Mandelstam continued, would equal the square of the momentum transfer in the crossed reaction, and vice versa. (This is what Moravcsik had termed the "incentive" for generalizing the single-variable dispersion relations.) Mandelstam thus wrote down what he called the "simplest possible" generalization of the single-variable integral

29. Goldberger, "Fifteen years" (1970), 690. Goldberger attributes his own attention to Mandelstam's work to a continuing correspondence with Chew.

30. Quotation from Polkinghorne, "Salesman of ideas" (1985), 23. See Chew, "Nucleon and its interactions" (1958); Chew, "Pion-nucleon interaction" (1959); Chew, "Double dispersion relations" (1960); Chew, "Double dispersion relations" (1961); Chew, *S-Matrix Theory* (1961), chap. 3; Omnès and Froissart, *Mandelstam Theory* (1963); Frautschi, *Regge Poles* (1963), chaps. 4–9; and Moravcsik, *Two-Nucleon Interaction* (1963), esp. chap. 5.

31. Mandelstam, "Determination" (1958), 1346. The antinucleon arose much as in Feynman, "Theory of positrons" (1949), by replacing a nucleon moving forward in time with an antinucleon traveling backward in time. Mandelstam noted that the same Feynman diagram contained a third reaction, in which the outgoing pion in the diagram's original orientation scattered with the incoming nucleon.

relations.[32] In addition to the dispersion theorists' single integrals over energy, Mandelstam added the additional terms:

$$\frac{1}{\pi^2}\sum_{i\neq j}\int ds_i'\int ds_j'\frac{\rho(s_i', s_j')}{(s_i'-s_i)(s_j'-s_j)},$$

where s_i represented the square of the energy in one "channel," or orientation of the Feynman diagram, and the square of the momentum transfer in one of the "crossed" channels.[33] The spectral functions $\rho(s_i, s_j)$ would be nonzero in only one of three nonoverlapping regions, and the amplitude could be evaluated in any of these three regions, or physical channels, by means of analytic continuation from one physical region to another. In this way, the single scattering amplitude could be studied as the invariants s_i were continued throughout the complex plane.

At the time Mandelstam developed his double-dispersion-relations representation, no one could prove that this representation followed from quantum field theory. (To date, no one has ever produced such a proof.) Only with much effort had single-variable dispersion relations been suitably "derived." In fact, post hoc derivations were given for the single-variable relations by theorists such as Nikolai Bogoliubov only in 1958, years after they had been exploited for a variety of phenomenological studies. These difficult, abstruse proofs remained valid, moreover, only for limited ranges of the momentum transfer, and thus did not apply to Mandelstam's work, as he readily conceded in each of his papers on the subject.[34] Instead, Mandelstam demonstrated that the scattering amplitudes calculated from fourth-order Feynman diagrams according to Dyson's rules for ordinary perturbative calculations could be cast into Mandelstam's general form. While hardly a proof that the double-dispersion relations held for arbitrary scattering processes, these early demonstrations at least showed that the two approaches might prove compatible. In the same way, Mandelstam went on to establish which Feynman diagrams corresponded

32. Mandelstam, "Analytic properties" (1959), 1743.

33. The energy for each channel was the energy in the center-of-momentum reference frame. The Mandelstam kinematic variables were usually written as (s, t, u), though Chew consistently wrote them in the more symmetric form, (s_1, s_2, s_3). See Chew, *S-Matrix Theory* (1961), 10–22.

34. Mandelstam, "Determination" (1958), 1344–45; Mandelstam, "Analytic properties" (1959), 1741–42. Speaking at the 1961 Solvay conference in Brussels, Mandelstam explained of his new double-dispersion relations, "The view will be taken that they are probably consequences of quantum field theory, though our mathematical tools are not yet sufficiently powerful to carry out the proof." Mandelstam, "Two-dimensional representations" (1961), 209.

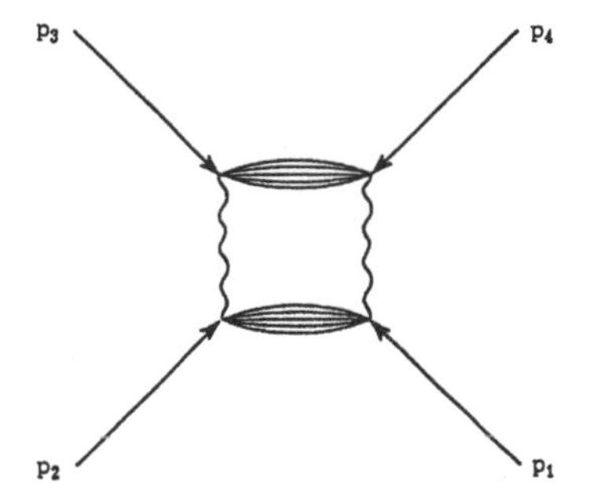

Figure 8.5. A simple Cutkosky graph. (Source: Chew, *S-Matrix Theory* [1961], 26.)

to which terms in the double-dispersion relations, and to use the diagrams to calculate the boundaries of the regions in the variables s_i for which the various spectral functions ρ_{ij} remained nonzero.

Within a year, this last task had been further routinized by Richard Cutkosky, who introduced generalized Feynman diagrams and gave a simple recipe for how to use them to solve for the boundaries of the spectral functions.[35] (See fig. 8.5.) Cutkosky revealed that the spectral functions $\rho(s_i, s_j)$ could be evaluated most simply by replacing all internal propagators by mass-shell delta functions—$(p^2 - m^2)^{-1} \rightarrow \delta(p^2 - m^2)$—that is, by substituting a *new* translation rule in place of Dyson's standard rule in figure 3.2. The lowest-order Feynman diagrams to contribute to the double-variable spectral functions were fourth-order "box" diagrams, since these were the simplest diagrams that included products of at least two propagators. With the mounting popularity of Mandelstam's representation, these "Cutkosky graphs" appeared with greater and greater frequency, both in print and in students' notebooks, such as Jerry Finkelstein's study notes (fig. 8.1*c*).

In both crossing symmetry and the Mandelstam representation, Feynman diagrams thus proved to be at least weakly generative, prompting and enabling new developments. Just as many theorists in the United States had proudly proclaimed the close connection between single-variable dispersion relations and experimental data, so too did they greet Mandelstam's diagram-inspired innovation. When Chew introduced Mandelstam's work at the July 1958 CERN conference, he reassured his listeners that "This work will perhaps seem baffling

35. Cutkosky, "Singularities" (1960). Though published in the first volume of a less central journal than the *Physical Review,* Cutkosky's method received immediate attention, in part because Chew discussed the work and included the original article with other reprints in his widely influential collection, *S-Matrix Theory* (1961).

to the experimenters, but I assure you that we shall come quickly to an important practical application."[36] The "practical application" opened up by Mandelstam's multichannel analysis of Feynman diagrams was that physicists could now *calculate* data for pion-pion scattering, even though all attempts to conduct pion-pion scattering experiments had failed; pions were too unstable to be used as targets in conventional scattering experiments. By flipping Feynman diagrams around, however, Mandelstam had elevated pion-pion scattering to the same status, governed by the same analytic function, as the more familiar pion-nucleon case. On the new view, the data for the pion-pion interaction were already included in—or could be extracted from—the results of garden-variety pion-nucleon scattering experiments. Within a single year, Mandelstam's multichannel interpretation of Feynman diagrams passed from being a theorist's curiosity to what the Berkeley experimentalist Luis Alvarez called "the bread and butter of the experimentalists."[37] Inspired by Mandelstam's example, and now in constant dialogue with him, Chew pressed this diagram-laden work still further.

FROM BOOKKEEPERS TO POLE FINDERS: POLOLOGY AND THE LANDAU RULES

The dispersion theorists dissociated Feynman diagrams from their specific mathematical meanings within perturbative calculations. In the cases of crossing symmetry and the Mandelstam representation, theorists took heuristic guidance from the diagrams, using them to undergird new features of general scattering amplitudes; these new uses drew force from specific mathematical associations with the diagrams, though these associations were not always the ones that Dyson and others had codified so carefully in figure 3.2. As the examples in this section demonstrate, other theorists—and at times even these same theorists—returned to Dyson's original translation rules for the diagrams but analyzed these mathematicodiagrammatic expressions to very different ends.

"The Semiphenomenological Black Art of Polology"

Geoffrey Chew introduced a new use for Feynman diagrams in the summer of 1958, just a few months after meeting Stanley Mandelstam and hearing

36. Chew, "Nucleon and its interactions" (1958), 95–96.

37. Luis Alvarez, reporting at the 1959 Kiev conference on high-energy physics, as quoted in Polkinghorne, *Rochester Roundabout* (1989), 79.

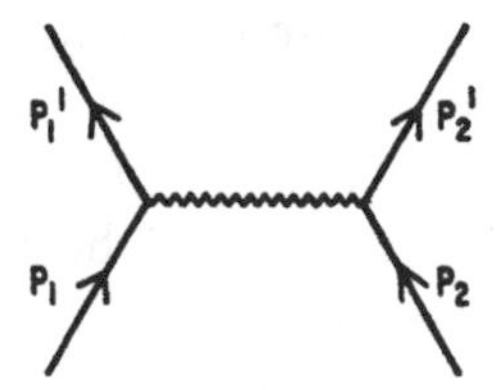

Figure 8.6. Single-pion exchange. (Source: Chew, "Proposal" [1958], 1381.)

about his double-dispersion work. In early July, Chew took advantage of his role as rapporteur for a theoretical strong-interaction session at a conference at CERN to announce what he called a new "conjecture." Based on the successes of dispersion relations and the excitement Chew sensed in Mandelstam's recent work, Chew found it "reasonable" to assume that singularities or poles in scattering amplitudes—those places at which the amplitudes diverged rapidly toward infinity—would occur only at the values of exchanged particles' masses. At the time of the meeting, he was uncertain whether such a "particle-pole conjecture," as it soon came to be known, could be reconciled with quantum field theory. But field theory or no, Chew appealed to his listeners that the conjecture "leads to definite and non-trivial experimental predictions" and therefore deserved further attention.[38]

In an article submitted a few weeks later, Chew continued his discussion. Here he explained that "A motivation for our conjecture" concerning the location of poles in scattering amplitudes "can be given in terms of Feynman diagrams."[39] Returning to the usual Dyson rules for evaluating Feynman diagrams within perturbative calculations, Chew explained that the lowest-order diagram for nucleon-nucleon scattering, akin to Feynman's figure 5.2 for the case of electron scattering in QED, would feature a single pion exchanged between the two incoming nucleons. (See fig. 8.6.) For years, the practice of keeping only the amplitude associated with the lowest-order Feynman diagram had been known as the "Born approximation," named for Max Born, one of the founders of quantum mechanics. Following Dyson's original rules for the diagrams' use, as codified in figure 3.2, Chew wrote the Born term for figure 8.6:

$$g^2 \frac{N}{\Delta^2 + \mu^2},$$

38. Chew, "Nucleon and its interactions" (1958), 94.
39. Chew, "Proposal" (1958), 1381.

where g, as usual, was the coupling constant between nucleons and pions. Each factor within the numerator of Chew's expression, which I have abbreviated as simply N, followed from the spin and symmetry properties of pions and nucleons.[40] The critical part for Chew's conjecture, however, lay in the denominator. The change in momentum of one of the incoming nucleons before and after emitting a pion, called the momentum transfer, obeyed $\Delta^2 = (p_1 - p_1')^2$; by the conservation of energy and momentum, this was equal to the change in momentum of the other nucleon upon absorbing the pion, $\Delta^2 = (p_2 - p_2')^2$. The remaining term in the denominator was the square of the exchanged pion's mass, μ^2.

What would happen, Chew asked his readers, if one extrapolated this expression to the point where $\Delta^2 = -\mu^2$? Here, at the physical mass of the pion, the mathematical term corresponding to the single-pion exchange diagram would become infinite. "A little thought about other diagrams," Chew continued, "shows that none of them becomes infinite for $\Delta^2 \rightarrow -\mu^2$."[41] The original attempts to study pion-nucleon interactions perturbatively had failed because lowest-order contributions like this one quickly became overwhelmed by higher-order diagrams, given $g^2 \gg 1$; that is, the Born approximation was notoriously poor for interactions involving large coupling constants. By making his new extrapolation to the pole in this expression, however, Chew demonstrated that the Born term would in fact dominate all other contributions in the vicinity of $\Delta^2 \sim -\mu^2$, regardless of the value of g^2. Chew explained further in a review article the following year that in using this approach, based upon the usual translation rules between Feynman diagrams and equations, "no statement is being made about the validity of perturbation theory or even about the legitimacy of the concept of an interaction proportional to the product of local fields. We are simply giving a recipe that is convenient because the rules of perturbation calculation are familiar."[42] Perturbative expansions might be dead for the strong interactions, but, as Chew demonstrated, that was no reason to discard Feynman diagrams.

In this way, Chew built his particle-pole conjecture directly upon lowest-order Feynman diagrams. He relied upon their standard mathematical translation (he used, in other words, precisely the translation rules summarized by Jauch and Rohrlich in fig. 3.2) but analyzed this diagram-equation pair

40. Ibid.; see also Chew, "Pion-nucleon interaction" (1959), 43. Following Mandelstam's work from 1958–59, the momentum transfer, written here as Δ^2, was more commonly denoted by either $-t$ or $-s_2$.

41. Chew, "Proposal" (1958), 1381.

42. Chew, "Pion-nucleon interaction" (1959), 42.

in a new way and toward a new end.[43] In perturbative field-theoretic calculations, the pole singled out by Chew was of no consequence, because the momentum transfer, Δ^2, could *never* equal $-\mu^2$ in any physical situation.[44] In ordinary perturbative analyses, therefore, there was nothing particularly interesting about the point $\Delta^2 = -\mu^2$. To Chew, on the other hand, several years' work with dispersion relations had already brought home the lesson that analytic functions were meant to be extended and extrapolated all over the complex plane. Reading Feynman diagrams from this vantage point, if figure 8.6 dominated the contributions from all other Feynman diagrams near the unphysical point $\Delta^2 = -\mu^2$, this information alone would prove valuable in the service of bending and twisting the analytic amplitude throughout the complex plane.

Within months, Chew expanded upon this work with his former colleague from Urbana, Francis Low. (Low had first taught Chew the ins and outs of Feynman diagrams and renormalization just a few years earlier.) Again, a lowest-order Feynman diagram undergirded their demonstration that the all-important pole in the scattering amplitude occurred when the momentum transfer, Δ^2, was set equal to the negative mass squared of the exchanged particle. This diagram allowed Chew and Low to extract information on such processes as pion-pion scattering—at that time completely unobservable in the laboratory—from data of a more accessible, familiar sort, such as pion-nucleon scattering. In other words, Chew and Low again began with a lowest-order Feynman diagram, this time for the process $N + \pi \to N + \pi + \pi$, an interaction that could be studied experimentally. (See fig. 8.7.) They went on to demonstrate that this interaction contained information about the experimentally unfeasible pion-pion scattering, $\pi + \pi \to \pi + \pi$, if one took the general output F in figure 8.7 to be two outgoing pions; in that case, the vertex on the right would show two pions (one of them the pion exchanged between the incoming pion and the nucleon) scattering into two pions. Making the same extrapolation that Chew had introduced earlier, $\Delta^2 \to -\mu^2$ and again analytically continuing the momentum transfer to an unphysical region allowed Chew and Low to extract information about the unobserved pion-pion

43. In fact, Chew cited Jauch and Rohrlich's textbook in "Proposal" (1958), 1381, and his graduate students made use of this textbook as a standard reference, as indicated in the bibliographies of their dissertations.

44. As Chew noted in his article, $\Delta^2 = -\mu^2$ corresponded to a scattering angle (in the center-of-momentum frame) of $\cos\theta = 1 + \mu^2/2k^2$, where k was the momentum of the incoming particles. Because scattering angles for any physical situation were limited to $|\cos\theta| \leq 1$, the pole at $\Delta^2 = -\mu^2$ lay outside the so-called physical region.

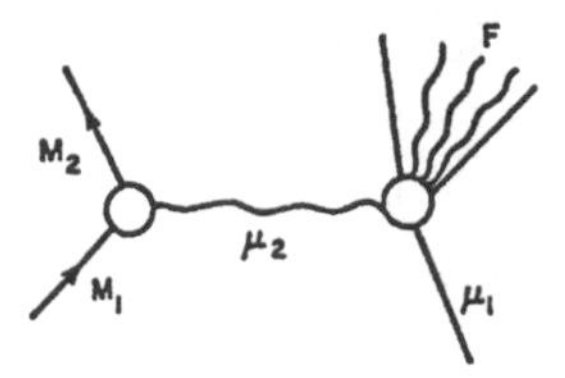

Figure 8.7. Lowest-order diagram from which information about pion-pion scattering could be extracted. (Source: Chew and Low, "Unstable particles" [1959], 1648.)

scattering.[45] The particle-pole conjecture and these methods from 1958–59 for singling out the most singular contributions to scattering amplitudes were soon labeled "polology."

This method of squeezing experimentally useful information out of lowest-order Feynman diagrams while circumventing the failure of perturbative methods attracted immediate attention. By the 1960 Rochester conference, a group of Berkeley experimentalists had studied more than seven hundred events of the type $N + \pi \rightarrow N + \pi + \pi$, and their preliminary analysis revealed a plausible value for the pion-pion cross section.[46] The same summer, Michael Moravcsik taught members of the 1960 Edinburgh summer school how to put the Chew-Low methods to work within his discussion entitled "Practical Utilisation of the Nearest Singularity in Dispersion Relations."[47] Sidney Drell's 1961 review article on high-energy interactions featured the Chew-Low extrapolation procedure.[48] Richard Dalitz adopted the Chew and Chew-Low approaches in his own 1962 text on the strong interactions.[49] In fact, long after theorists had stopped emphasizing these methods, the 1959 Chew-Low "semiphenomenological black art of polology" was still guiding the experimentalists who gathered in Tallahassee, Florida, for the 1973 "International Conference on π-π Scattering and Associated Topics."[50]

45. Chew and Low, "Unstable particles" (1959), 1640–48.

46. This work was summarized by Julius Ashkin in his rapporteur presentation: Ashkin, "Experimental information" (1960).

47. Moravcsik, "Practical utilisation" (1961). Moravcsik dedicated all his fourth lecture exclusively to the Chew-Low work (130–36). A thorough derivation of the pion-pion cross section from the full interaction cross section for N-π scattering was also given in Källén, *Elementary Particle Physics* (1964), 167–77. Cf. Jacob, "Strong-interaction processes" (1964), 85–99.

48. Drell, "Peripheral contributions" (1961).

49. Dalitz, *Strange Particles* (1962), 103–5, 156, 164–69. As Dalitz explained in his acknowledgments, this text originated in his summer school lectures at the Tata Institute, Bombay (178).

50. The phrase "semiphenomenological black art of polology" comes from Moravcsik, "Thirty years" (1985), 26. On the 1973 pion-pion scattering conference, see Williams and Hagopian, *π-π Scattering* (1973). Denyse Chew likewise confirmed the importance of the Chew-Low work to

In this new guise, Feynman diagrams served as pole finders: Chew and his postdocs and students put lowest-order, single-particle exchange diagrams to work, isolating the poles that would dominate the full set of complicated, higher-order diagrams.[51] The function of the diagrams in this new work differed from Dyson's usage, and yet the diagrams—suitably reinterpreted—remained what Moravscik called a "very valuable tool." The polology work thus demonstrates that even when physicists obeyed Dyson's rules for translating between Feynman diagrams and mathematical expressions (fig. 3.2), there was no guarantee that they would exploit this relationship in the way Dyson had originally intended.[52]

Lev Landau and the "Pompous Burial"

The Soviet physicist Lev Landau read a different lesson in the successes of Chew-Low polology. What impressed Landau was not that the new use of Feynman diagrams allowed theorists to reopen contact with experimentalists while sidestepping the vagaries of perturbative expansions. Landau concluded that no simple renaissance for the Born approximation would suffice, and that quantum field theory should be abandoned entirely for studies of the strong interaction. Pulling no punches, he announced in the summer of 1959 that quantum field theory was "dead." Landau expounded as rapporteur at an international conference on particle physics in Kiev that summer that "psi-operators [describing quantum fields] contain fundamentally unobservable quantities and should vanish from the theory." Furthermore, since the field-theoretic interaction Hamiltonians "can be constructed only with the help

---

experimentalists during the early and mid-1970s: Denyse Chew, e-mail to the author, 7 Sep 2003. The Chew-Low method was also revived in 1968–72 in the form of "multiperipheral models." In these models, the collisions of two particles into jets of many outgoing particles were analyzed in terms of a series of single-particle exchange poles. See, e.g., Chew, *Multiperipheral Dynamics* (1971); and Cushing, *Theory Construction* (1990), 161–64.

51. Chew, *S-Matrix Theory* (1961), esp. chaps. 3 and 4; and Moravcsik, "Practical utilisation" (1961). Chew's graduate student Peter Cziffra worked closely with Moravcsik on applications of Chew's polology scheme. See Cziffra et al., "Modified analysis" (1959); and Cziffra and Moravcsik, "Determination" (1959).

52. Moravcsik, "Practical utilisation" (1961), 118. Cf. Roland Barthes's famous declaration of the "death of the author" and the simultaneous "birth of the reader": Barthes, "Death of the Author" (1977). By the mid-1950s, Dyson himself was helping to take charge in pushing nonperturbative studies within the analytic dispersion-relations framework. See esp. his widely influential work in Castillejo, Dalitz, and Dyson, "Low's scattering equation" (1956), which introduced what quickly came to be known as "CDD" poles, named for the authors' initials.

of psi-operators," the entire field-theory approach to the strong interaction "is dead and should be buried, though pompously, considering its historical merits."[53] Previous theorists had grown frustrated with field-theoretic approaches when treating the strong interactions because perturbative methods had proved intractable. Landau complained instead that the problem lay with the theory's foundations, not the niceties of its calculational apparatus. Landau attacked the very core of Dyson's beloved approach: neither quantum fields nor the mathematical description of their dynamics impressed Landau for the study of strong-interaction particle physics. Yet rather than discard Feynman diagrams, which Dyson had worked so hard to fashion for the study of quantum fields, Landau lifted the diagrams out of their field-theoretic setting and sought to build a new theory directly from them.

Landau discussed his emerging strategy with Werner Heisenberg over the course of 1958. In late November, Landau summarized what he took to be the main thrust of Heisenberg's recent work:

1. The ψ-operators are dead. From now on they cease to exist.
2. Feynman's diagram technique remains, but must be essentially reformulated, since the Green's functions of free particles have lost meaning in a theory without a Hamiltonian equation.

"That is how your ideas look from my point of view," Landau concluded. "This is more a program than a theory. The program is magnificent, but it still must be carried out. I believe that this will be the main task of theoretical physics. The task unfortunately seems to be a gigantic one."[54] By the time the Kiev meeting convened a few months later, Landau had made this task his own.

Landau soon published the seeds of his alternative program. His introduction made clear his new plan for the line drawings: "As has become clear recently, a *direct study of graphs* is the most effective method of investigating the location and nature of the singularities of vertex parts." Such a direct study of Feynman diagrams, Landau reiterated, brought the theorist "beyond the framework of the current theory. An assumption is thereby automatically made that there exists a non-vanishing theory in which ψ-operators and Hamiltonians are not employed, yet graph techniques are retained." Indeed, Landau

53. Landau, "Analytical properties" (1960), 97–98. For more on Landau's renouncement of quantum field theory throughout the mid- and late 1950s, see Cushing, *Theory Construction* (1990), 129–33, 167–72; Cao, *Conceptual Developments* (1997), 215–16, 220; Brown and Rechenberg, "Landau's work" (1990); and Gross, "Chasing the Landau Ghost" (1990), 97–111.

54. Lev Landau to Werner Heisenberg, 21 Nov 1958, as quoted in Brown and Rechenberg, "Landau's work" (1990), 73–74. On Heisenberg's program in particle theory during this period, see Carson, *Particle Physics* (1995), chap. 5.

conceded, "the employment of the graph technique is . . . solely consistent, since the problem becomes meaningless if the graph technique is rejected."[55] Eschewing all the while Hamiltonians and quantum fields, Landau began his discussion in medias res. He fixed upon the usual perturbative amplitude associated with an arbitrary Feynman diagram, containing several internal lines and vertices. Borrowing Feynman's famous trick for combining denominators, Landau wrote the amplitude that would accompany a Feynman diagram with n internal lines:

$$f(n) = \int_0^1 d\alpha_1 \cdots d\alpha_n \int d^4k_1 \cdots d^4k_m \frac{\delta(\alpha_1 + \cdots + \alpha_n - 1)}{[F(\alpha_i, k_j, p_k) + i\epsilon]^n},$$

where k is the momentum of an internal line, and p the momentum of an external line.[56] The function F was defined in the usual way as

$$F(\alpha_i, k_j, p_k) = \sum_{i=1}^{n} \alpha_i (q_i^2 - m_i^2),$$

where the q's were linear combinations of the k's and p's—just those combinations required by the conservation of energy and momentum at each vertex. The form of these equations was very nearly the same as Khalatnikov, Abrikosov, Galanin, and Ioffe had labored to teach the stubborn Landau five years earlier—except now Landau simply ignored all details of the numerator of his integral. Although such numerator details had been crucial to his young countrymen's calculations during the mid-1950s, Landau could now ignore them with impunity. As Chew's polology work had demonstrated, such details did not affect the positions of the amplitude's poles and hence played no role in Landau's discussion.[57]

Like Chew before him, Landau proceeded to analyze his equations—which had been generated using Dyson's association between mathematical amplitudes and Feynman diagrams—in a novel way. In only a few lines of algebra, Landau demonstrated that the only poles for this generic amplitude occurred

55. Landau, "Analytic properties" (1959), 181; emphasis added.

56. Ibid. Cf., e.g., Schweber, Bethe, and de Hoffmann, *Mesons and Fields* (1955), 1:268.

57. Dropping the numerator terms quickly became common. Frederik Zachariasen, for example, lectured in his 1962 summer school session, "Forgetting the Dirac spinors and the other inessential details, the scattering amplitude is . . . " Zachariasen, "Theory and application of Regge poles" (1963), 41. Some theorists did point out, however, that the numerator terms were crucial for evaluating the residue of these poles as well as determining the strength of the singularities in the massless limit: Kinoshita, "Mass singularities" (1962); Lee and Nauenberg, "Degenerate systems" (1964).

either when $\alpha_i = 0$ or when $q_i^2 = m_i^2$. The case $\alpha_i = 0$ meant that the ith internal line within the arbitrary Feynman diagram had been contracted away, erased from the original diagram altogether. The second case, $q_i^2 = m_i^2$, meant that poles occurred when the virtual particles, depicted by a diagram's internal lines, were assigned the momentum they would carry as real, physical particles. In Landau's hands, virtual particles—the mainstay of all field-theoretic interactions—vanished from the scene; the only relevant content arose from the propagation of real particles, with $q_i^2 = m_i^2$. The conditions $\alpha_i = 0$ or $q_i^2 = m_i^2$ were soon dubbed the "Landau rules."[58]

Much as Chew had conjectured for his audience at CERN the previous summer, Landau thus demonstrated that the only poles of the scattering amplitude occurred for physical particles.[59] Meanwhile the two Landau rules, $\alpha_i = 0$ and $q_i^2 = m_i^2$, each spawned a new form of Feynman-like diagram. The $\alpha_i = 0$ case featured ordinary Feynman diagrams with one of the internal lines removed; Landau introduced several examples of these new diagrams, dubbed "reduced diagrams," in his 1959 article. John Taylor included five different reduced diagrams, each derived from the same Feynman diagram, in his lectures at the University of Rochester the next year. "These reduced diagrams are quite different in appearance from the original Feynman diagrams," Taylor explained, "because an arbitrary number of lines can meet at the one corner [vertex]."[60] Reduced diagrams often featured vertices connecting four lines in one place, whereas the other vertices still pictured three lines connecting in one place—even though both kinds of vertices were meant to connect the same kinds of particles. (See fig. 8.8.) Like the diagrams brought in to illustrate dispersion relations, these reduced graphs therefore carried no meaning,

58. From the fact that the q's were linear combinations of the k's and the p's, Landau derived a third rule, which followed from the condition that $\partial F/\partial k_j = 0$: $\Sigma_j \alpha_i q_i = 0$, where the sum ran over all j closed internal loops. This enabled Landau to construct a new geometric means of determining the physical and unphysical regions for any of the poles, in direct analogy with graphic solutions for such things as Kirchhoff's rules for the flow of current in a closed circuit. See Landau, "Analytic properties" (1959), 184–89. These new geometric diagrams were dubbed "dual diagrams." For more on the dual diagrams, see Polkinghorne, "Analytic properties" (1961), 66–67, 80–82; and Eden et al., *Analytic S-Matrix* (1966), 26–33, 50–80. Most of Landau's results were worked out independently by James Bjorken in his 1959 dissertation at Stanford; see Bjorken and Drell, *Relativistic Quantum Fields* (1965), chap. 18.

59. A few years later, a team of British theorists demonstrated that in fact a new class of singularities, not treated by Landau, existed for these same complex amplitudes. These new singularities arose from the intricacies of analytic continuation and the distortions of integration contours at infinity in the complex plane. See Fairlie et al., "Singularities" (1962); cf. Cushing, *Theory Construction* (1990), 131.

60. Taylor, "Special topics" (1960), 66.

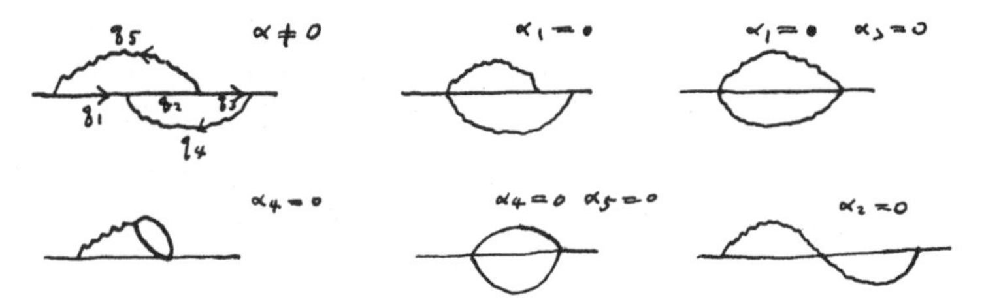

Figure 8.8. Feynman diagram (*top left*) and its associated reduced diagrams. (Source: Taylor, "Special topics" [1960], 66.)

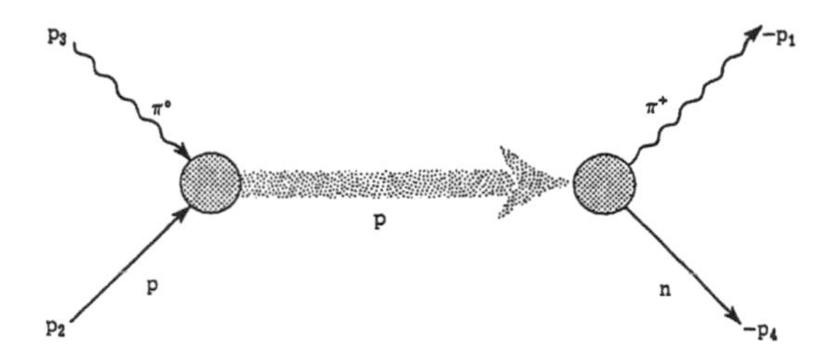

Figure 8.9. Landau graph. (Source: Chew, *S-Matrix Theory* [1961], 16.)

and could be put to no use, in ordinary perturbative calculations, for which a specific interaction Hamiltonian would restrict the form of diagrams' vertices. Growing out of Dyson's rules, suitably bent to a new purpose, these diagrams remained incompatible with their field-theoretic cousins.

The other set of diagrams to come from Landau's work, corresponding to the rule $q_i^2 = m_i^2$, received the name "Landau graphs." Unlike the reduced diagrams, these diagrams were drawn identically to lowest-order, one-particle-exchange Feynman diagrams. The Landau graphs stood in for an infinite number of Feynman diagrams, all of which shared the same intermediate exchange. All perturbative corrections at vertices were simply ignored and subsumed under a single Landau graph.[61] (See fig. 8.9.) Much like the Feynman diagrams used in Chew-Low polology, these Landau graphs highlighted only the exchange term, picturing which intermediate state would lead to a pole in the associated amplitude. Vertices in these newly fashioned pole finders, just as in the early dispersion-relations work, became simply "blobs."

Unlike the early meson and dispersion-relations work, or crossing symmetry and the Mandelstam representation, Chew's and Landau's work from 1958–59 began with the usual mathematical expressions that Dyson had worked so

61. See, e.g., Landau, "Analytic properties" (1959); Chew, *S-Matrix Theory* (1961), chap. 3; Polkinghorne, "Analytic properties" (1961); Moravcsik, "Thirty years" (1985); and Chew, "Particles as *S*-matrix poles" (1989).

hard to associate, one to one, with Feynman diagrams. But Chew and Landau quickly put the diagrams to work toward very different ends than those for which Dyson had originally labored. Certain mathematical details, such as the Dirac γ_μ matrices cluttering the numerators within Kroll's students' equations, were no longer relevant to Landau's cause and could be dropped without being missed. In the new work, proceeding without any reference to Hamiltonians, virtual particles, or quantum fields, there could be a proliferation of diagrammatic techniques that shared many features of Dyson's Feynman diagrams, but not all of them—such as Landau's reduced graphs with more legs meeting at vertices than the usual field-theoretic rules would allow. In Chew's hands, and in Landau's, these Feynman and Feynman-like diagrams now served a new purpose: they were pole finders, rather than radiative-correction markers. Starting from the same mathematical relations, Chew and Landau put the diagrams to a new use.

CHEW THE PROGRAM BUILDER: NUCLEAR DEMOCRACY AND THE BOOTSTRAP

As he doodled his way toward a new approach to the strong interactions, Landau left no doubt about the status of quantum field theory: it was dead and useless. The gauntlet having thus been thrown down with such fanfare, Landau quickly gained an ally for the campaign against quantum field theory in Geoffrey Chew.[62] Announcing at a conference at La Jolla in June 1961 that quantum field theory was "sterile with respect to the strong interactions," and therefore "destined not to die but just to fade away," Chew quickly rose as the charismatic leader of a Berkeley group dedicated to overturning field theory and replacing it with an autonomous S-matrix theory. Though, like Landau, still unclear about the particular shape his new approach would ultimately take, Chew built much of his developing program upon a scaffolding of Feynman diagrams.[63]

At one level, Chew's and his students' diagrammatic improvisations were just further examples of the piecewise adaptation we have examined

62. Many read these developments at the time in terms of a "gauntlet" being thrown. See, for example, Murph Goldberger's remarks at the twelfth Solvay conference in 1961: Goldberger, "Single variable dispersion relations" (1961), 179.

63. A preprint of Chew's talk at the 1961 La Jolla conference is quoted in Cushing, *Theory Construction* (1990), 143. See also Gell-Mann, "Particle theory" (1987). This portion of Chew's unpublished talk was incorporated verbatim in Chew, *S-Matrix Theory* (1961), 1–2. For more on Chew's S-matrix program, see especially Cushing, *Theory Construction* (1990); Cao, "Reggeization program" (1991); and Gordon, *Strong Interactions* (1998).

throughout this chapter and part 2. Consider, for example, the Berkeley theorists' treatment of Regge poles. The young Princeton theorist Tullio Regge had investigated the analytic behavior of scattering amplitudes within nonrelativistic potential scattering as one made the angular momentum (J) into a complex variable; in other words, Regge studied how solutions to the Schrödinger equation behaved in the presence of a fixed, external potential, as one continued J throughout the complex plane.[64] Chew and several postdocs tried to import analyticity in this third variable into their treatment of high-energy scattering. Their inspiration for pirating Regge's move came in part from Mandelstam's treatment of both energy and momentum transfer as complex variables. By considering now the angular momentum to be a complex variable as well, Chew and his postdocs found tantalizing hints that scattering amplitudes' poles at specific values of J corresponded uniquely to particles of specific masses, M, and that these J-M pairings reproduced large classes of the newly discovered strongly interacting particles—at least if one treated some of the recent mass measurements somewhat generously. Perhaps, they began to wonder, the particles of the microworld could correspond *only* to the so-called Regge poles.[65]

As they had already become for polology studies, Feynman diagrams quickly became part of the daily currency of these Regge-pole investigations. When calculating the details of the analytic continuation in angular momentum, physicists once again returned to the Born approximation and its lowest-order Feynman diagrams. Frederik Zachariasen, for example, who worked with Chew's group as a postdoc in the early 1960s, walked students through some of these calculations. He used as his example the ρ meson, a heavy, unstable particle that decayed relatively rapidly into pairs of pions; the ρ had been discovered by Berkeley experimentalists working at the famous cyclotron in 1960, though its existence actually had been predicted by a pair of Chew's graduate students a few years earlier.[66] The ρ meson became a true "exemplar," or

64. Regge, "Analytic properties" (1958); Regge, "Analytic behavior" (1958); Regge, "Complex orbital momenta" (1959); and Regge, "Bound states" (1960). Cf. Chew, Gell-Mann, and Rosenfeld, "Strongly interacting particles" (1964); Frautschi, *Regge Poles* (1963); Omnès and Froissart, *Mandelstam Theory* (1963); and Cushing, *Theory Construction* (1990), chap. 6.

65. Chew made it clear that his appropriation of Regge's work did not square entirely with Regge's own ideas on the subject, which, after all, had nothing to do with high-energy particle scattering. Chew introduced the new Reggeism during his 1962 summer school lectures by explaining, "Motivation? We don't know what was in Regge's mind when he considered continuation in angular momentum, but its value in the S-matrix framework is related to subtractions, so let's talk about subtractions." Chew, "Strong-interaction" (1963), 9.

66. The discovery of the ρ meson as a resonance in pion scattering was announced in Erwin et al., "Evidence for a π-π resonance" (1961), having been predicted earlier in Frazer and Fulco,

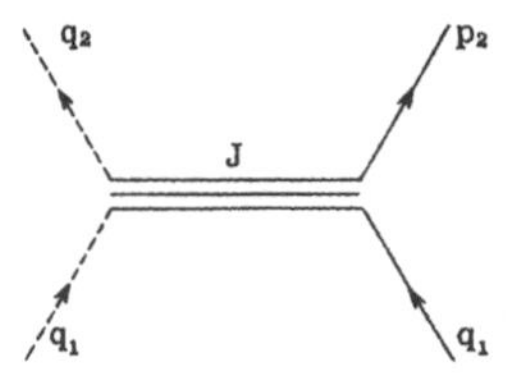

Figure 8.10. "Spinology" diagram. (Source: Zachariasen, "Theory and application of Regge poles" [1963], chap. 10, 38.)

model system, for Chew's group in Berkeley, appearing at the center of many of Chew's programmatic investigations. Zachariasen explained to his summer school auditors that the ρ could be exchanged between pairs of pions, and that this exchange would give rise to a force, much the way that electrons and positrons could exchange photons, giving rise to the electromagnetic force. As had happened so many times before, Zachariasen introduced a slightly altered Feynman diagram to accompany these new Regge-pole calculations. (See fig. 8.10.) The new "spinology" diagrams used multiple lines to indicate the angular momentum of the exchanged particle, since this was now the quantity of most direct importance to the associated mathematical manipulations.[67] Zachariasen's diagrammatic improvisation continued in the by-then-long tradition, dating back to Paul Matthews's multiple-arrowed scribblings in his letters to Wolfgang Pauli from 1950 (see figs. 5.8 and 5.9), of tinkering with the pictorial form of the diagrams to bring out those elements most relevant for the accompanying calculations.

The Reggeism and associated spinology diagrams proved to be only one portion of Chew's developing program. Of much larger consequence was the pronouncement, which Chew delivered at every opportunity beginning in the early 1960s, that theorists' usual descriptions of the subatomic realm rested on a false dichotomy. The single most novel conjecture of Chew's developing S-matrix program, and its most radical break with the usual approach, was that all nuclear particles should be treated "democratically"—Chew's word. Part of the motivation for this self-proclaimed democratic move derived from the

---

"Effect of a pion-pion resonance" (1959); and Frazer and Fulco, "Partial-wave dispersion relations" (1960). Frazer completed his dissertation under Chew's direction in 1959, and Chew helped to advise José Fulco's dissertation, completed in 1962 in Buenos Aires. Years later, Frazer recalled that "Geoff advised us every step of the way" with this work, "but generously decided not to put his name on the paper." Frazer, "Analytic and unitary S-matrix" (1985), 4.

67. Zachariasen, "Theory and application of Regge poles" (1963); Zachariasen, "Lectures on bootstraps" (1965).

proliferation of new particles that had been discovered throughout the 1950s in the big accelerators. Chew and his postdoc Steven Frautschi argued in 1961, for example, that the superabundance of new particles signaled the need for a change in theory: "The notion, inherent in Lagrangian field theory, that certain particles are fundamental while others are complex," Chew and Frautschi explained, "is becoming less and less palatable... as the number of candidates for elementary status continues to increase."[68] The traditional field-theory approach, against which Chew now spoke out, posited a core set of "fundamental" or "elementary" particles that acted like building blocks out of which more complex, composite particles could be made. Chew and his collaborators argued against this division into "elementary" and "composite" camps—no particle was inherently more "fundamental" or special than any others. Deuterons, for example, usually treated as bound states of more "elementary" protons and neutrons, were to be analyzed instead within Chew's program exactly the same way as one treated protons and neutrons themselves; the "democracy" extended, in principle, all the way up to uranium nuclei. Chew bestowed on this notion the colorful term "nuclear democracy."[69]

Chew's democracy sprang from more than just the embarrassment of riches delivered by particle experimentalists over the previous decade. His move to deny the very categories of "elementary" and "composite" particles stemmed in large part from his active reinterpretation of Feynman diagrams. For Chew, his and Landau's work on polology—restricting all interesting physics to physical (rather than virtual) particles—seemed to portend a major break from field-theoretic approaches, both conceptually and ontologically. Whereas theorists had lectured to their students for a decade that the lines within Feynman diagrams could represent only elementary particles—that is, the particles whose associated quantum fields appeared in the governing interaction-Hamiltonian—Chew countered that this distinction was not borne out by the diagrams themselves. As Chew explained in his 1961 published lecture notes, Landau's modified rules for treating Feynman diagrams "contain not the slightest hint of a criterion for distinguishing elementary particles [from composite ones].... If one can calculate the *S* matrix without distinguishing elementary particles, why introduce such a notion?"[70] He returned

68. Chew and Frautschi, "Principle of equivalence" (1961), 394.

69. Murray Gell-Mann later suggested the term "nuclear egalitarianism" in place of "nuclear democracy," arguing that "democracy" implied actual voting and participation, whereas "egalitarianism" implied only the notion of equal treatment. Gell-Mann, "Particle theory" (1987).

70. Chew, *S-Matrix Theory* (1961), 5. See also Landau, "Analytic properties" (1959), 181, 191; and Landau, "Analytical properties" (1960), 99.

to this point during his 1962 summer school lectures in Cargèse, Corsica. "One can't say that this [Landau's diagrammatic pole-searching method] is 'derived' from field theory," Chew began. "In the Feynman diagram approach of Landau one must include in the diagrams all intermediate lines corresponding to 'composite' as well as 'elementary' particles (resonances as well as stable particles), and no justification can be given for considering composite particles in a perturbation approach."[71] Yet if the diagrams' lines looked the same for any kind of particle, Chew began to ask, and if Landau's rules for handling these lines made no distinctions between composite or elementary particles, then why should the accompanying mathematical expressions continue to be based upon such a dichotomy?

The other key diagrammatic ingredient behind Chew's "nuclear democracy" came from crossing symmetry and Mandelstam's multichannel analysis of scattering amplitudes. Particles that in one orientation of a Feynman diagram looked like the fundamental building-block constituents of more complicated particles would appear, upon various rotations of the diagrams, as either the exchanged particle responsible for a force between other particles or the end-state composite of other constituents. Frederik Zachariasen and Charles Zemach, two young theorists working with Chew, built directly upon this work in 1962, including a series of Feynman diagrams within their study of the dynamics of the ρ meson.[72] (See fig. 8.11.) A "democratic" reading of the diagrams undergirded their work. Figure 8.11 *a*, for example, depicted a ρ meson being exchanged between two incoming pions and thereby giving rise to a force between them. Figure 8.11 *b*, on the other hand, showed two pions coming together to form a new bound-state composite particle, the ρ meson, which, being unstable, later decayed into a new pair of pions. If two pions could create a ρ meson, however, then interactions like that shown in figure 8.11 *c*, in which a ρ meson acted just like an "elementary" particle, scattering with an incoming pion, had to be considered as well; in figure 8.11 *c*, meanwhile, it was a pion that appeared as the exchanged force-carrying particle. The theorists' usual labels of "elementary," "force carrier," and "composite" swapped places with each rotation of a given Feynman diagram, Chew and his young colleagues charged. The only consistent theoretical framework, they therefore maintained, was a "democratic" one that made no distinctions between "elementary" and "composite" particles.

71. Chew, "Strong-interaction" (1963), 8.

72. Zachariasen and Zemach, "Pion resonances" (1962).

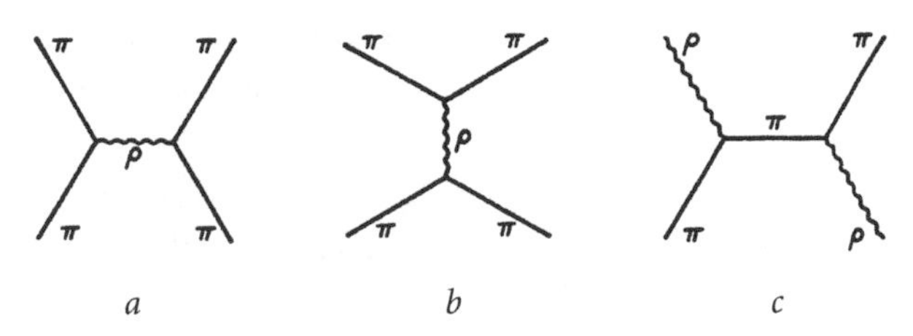

Figure 8.11. Bootstrap diagrams. (Source: Zachariasen and Zemach, "Pion resonances" [1962], 850–51, 857.)

From this reinterpretation of Feynman diagrams, a short step led Chew to the essential point of his *S*-matrix program: not only were composite particles to be treated equivalently to elementary ones, but all (strongly interacting) particles could in fact be seen as composites of each other.[73] Diagrammatic initiatives akin to those in figure 8.11 were central to the new scheme. Elevating crossing symmetry—itself derived somewhat heuristically during the mid-1950s from pairs of rotated Feynman diagrams—to foundational status, Chew explained his notion of a particle "bootstrap": "One channel [of a Feynman diagram] provides forces for the other two—which in turn generate the first."[74] Consider again the case of the ρ meson, one of the earliest successes for Chew's bootstrap model. In its force-carrying mode, as in figure 8.11 *a*, the exchange of the ρ would create an attractive force between the two incoming pions. Drawn together by this attractive force, the two pions could combine to produce a resonance—a new bound state or composite particle, as in figure 8.11 *b*. Chew and his young collaborators looked for self-consistent solutions such that the force-carrying process produced a resonance whose properties were precisely those of the force-carrying particle. If such a solution existed, then the ρ

73. Chew's dissertation adviser, Enrico Fermi, had explored the idea that pions could be bound states of nucleon-antinucleon pairs—a possibility allowed because the (relativistic) binding energy of the strongly coupled nucleons could offset most of their rest mass. Fermi and Yang, "Are mesons elementary particles?" (1949). A similar balancing of (negative) binding energy with large rest masses became a staple of Chew's bootstrap approach.

74. Chew, *S-Matrix Theory* (1961), 32; see also Chew, "Nuclear democracy" (1964). Chew's former postdoc and partner in the early bootstrap work, Steven Frautschi, explained simply in lectures from 1961–62 that "Bootstrap calculations lean heavily on 'crossing.'" Frautschi, *Regge Poles* (1963), 176. The bootstrap notion was illustrated with the aid of crossed Feynman diagrams in Chew, "Nuclear democracy" (1964), 134, 136, and also in a text by Chew's former student: Frazer, *Elementary Particles* (1966), 134. For more on Chew's "bootstrap," see Chew, "Bootstrap" (1968); Chew, "Hadron bootstrap" (1970); Chew, "Impasse" (1974); Freundlich, "Theory evaluation" (1980); Chew and Capra, "Bootstrap physics" (1985); Cushing, "Is there just one possible world?" (1985); Cushing, *Theory Construction* (1990), chap. 6; and Cao, "Reggeization program" (1991).

could, all by itself, bring the pions together so that they could produce a ρ. The ρ meson, in this series of diagrams, would have pulled itself up by its own bootstraps.[75]

The detailed treatment of the ρ meson example drew on many of the calculational techniques that Chew and others had established throughout the mid- and late 1950s. First they related the real and imaginary parts of the amplitude corresponding to figure 8.11 *b*, just as in the earlier dispersion-relations work:

$$\mathrm{Re}[f_{\mathrm{res}}(\omega)] = \frac{1}{\pi} P \int d\omega' \frac{\mathrm{Im}[f_{\mathrm{res}}(\omega')]}{\omega' - \omega},$$

where f_{res} was the amplitude for creating a ρ meson in a resonant bound state (corresponding to figure 8.11 *b*). The amplitude f_{res} would in general be a function of the energy, ω, as well as the mass of the produced resonance, m_ρ, and the coupling between the resonant state and its ingredients, $g_{\pi\pi\rho}$. Next they derived from the requirement of unitarity the expression

$$\mathrm{Im}[f_{\mathrm{res}}] \propto \sum_n f_{an}^* f_{nb},$$

(much as Goldberger had derived before), where the infinite sum included all real intermediate states that could connect the incoming state (the incoming pair of pions) with the outgoing state (the pair of pions stemming from the decay of the short-lived resonance, ρ).

Whittling down the infinite sum to what they assumed would be, at least within a first approximation, the single most dominant term, Chew and his young collaborators next made the same move as Chew had been making for years with the born-again Born approximation: they made a polology assumption, choosing the term corresponding to the single-particle exchange of figure 8.11 *a*. This one-particle exchange would give rise to a particular force, yielding the amplitude f_{force}, a known function of ω, m_ρ, and $g_{\pi\pi\rho}$. Substituting this single term for the infinite sum yielded

$$\mathrm{Im}[f_{\mathrm{res}}] \propto |f_{\mathrm{force}}|^2,$$

75. See Chew and Frautschi, "Unified approach" (1960); Chew and Frautschi, "Principle of equivalence" (1961); Chew and Mandelstam, "Low-energy pion-pion interaction" (1960); Chew and Mandelstam, "Low-energy pion-pion interaction," pt. 2 (1961); and Chew, Frautschi, and Mandelstam, "Regge poles" (1962). See also the references cited in the previous note, as well as Zachariasen, "Lectures on bootstraps" (1966), and references therein.

from which, combined with their dispersion-relations integral, the real part of f_{res} could also be obtained. In this way, by constructing their calculation along dispersion-theoretic and polology lines, the Berkeley theorists could approximate f_{res} (associated with fig. 8.11 *b*) based entirely on the form of the input force, f_{force} (coming from fig. 8.11 *a*).

But if a bound-state resonance really were produced as in figure 8.11 *b*, then the amplitude for this process, Chew and his young charges knew, would have to have a specific mathematical form, known as the Breit-Wigner resonance formula, F_{res}, a known function of ω, m_ρ, and $g_{\pi\pi\rho}$. Setting the real and imaginary parts of f_{res} (as calculated with the series of steps outlined above) equal to the real and imaginary parts of F_{res} (taken "off the shelf" from long-standing studies of resonance scattering) gave them two equations for the two unknowns, m_ρ and $g_{\pi\pi\rho}$. Zachariasen emphasized that these physical parameters were uniquely determined: "There are no parameters to be adjusted in obtaining these results."[76] Considering the severity of the many assumptions behind this "simple-minded calculation," the numerical values that Zachariasen first found ($m_\rho \sim 350$ MeV and $g^2_{\pi\pi\rho} \sim 35$) were not so very different from the recent experimental values ($m_\rho = 750$ MeV and $g^2_{\pi\pi\rho} = 13$). Working with Charles Zemach, Zachariasen broadened the polology assumption a bit, including three other particle-exchange contributions within the sum in addition to the single-ρ exchange for the input force, and arrived at a self-consistent solution much closer to the experimental values ($m_\rho = 740$ MeV, $g^2_{\pi\pi\rho} = 11$).[77] Though Chew had been pursuing bootstrap calculations such as this one for the better part of two years, Goldberger confided many years later that it was only after seeing this "elementary calculation of Zachariasen" that he could feel "comfortable" with the whole bootstrap idea.[78]

Chew and Frautschi wondered if *every* particle arose in this way: their "bootstrap" conjecture held that every strongly interacting particle was a composite particle, composed of just those other particles that were brought together by exchanging the first particle. Chew elaborated on this point within his 1964 lecture notes: "The bootstrap concept is tightly bound up with the notion of a democracy governed by dynamics. Each nuclear particle is conjectured to be a bound state of those S-matrix channels with which it communicates, arising from forces associated with the exchange of particles that communicate with 'crossed' channels. . . . Each of these latter particles in turn owes *its*

76. Zachariasen, "Self-consistent calculation" (1961).
77. Ibid.; Zachariasen, erratum (1961); and Zachariasen and Zemach, "Pion resonances" (1962).
78. Goldberger, "Fifteen years" (1970), 690.

existence to a set of forces to which the original particle makes a contribution."[79] The bootstrap thus offered Chew the ultimate nuclear democracy: elementary particles deserved no special treatment separate from composite ones; in fact, there might not even exist any "aristocratic," elementary particles, standing above the composite fray. As Chew explained in lectures from 1964, "every nuclear particle should receive equal treatment under the law"—a "democratic" program, as we will see in chapter 9, that became remarkably successful during the 1960s.[80] The successes came from reading new content in Feynman diagrams, drawing conclusions about what the diagrams "really" portrayed that were increasingly at odds with a decade's worth of diagrammatic calculations.

DIAGRAMMATIC BOOTSTRAPPING AND THE EMERGENCE OF NEW THEORIES

Dyson's work did not sway other theorists from casting their own, often divergent readings of Feynman diagrams. As we have seen throughout this chapter and the previous ones, the diagrams did not carry their own meaning in some unchanging, unproblematic way; theorists worked hard to affix meaning to the diagrams with their problem sets, lecture notes, review articles, and textbooks. Theorists toyed with the pictorial form of the diagrams themselves, as certain features began or failed to make a difference: Marshak could drop Dyson's antiparticle-arrow convention, just as dispersion theorists could allow any number of lines to meet at their "blob" vertices. Chew and his crew could derive from the diagrams' bare lines a series of new ontological commitments and calculational strategies. For all these reasons, Bruno Latour's talk of "immutable mobiles" seems off the mark in this case: it is precisely the diagrams' mutability that most requires attention and explanation.

Not everyone greeted the spate of new diagrams with glee. Douglas Hofstadter, who worked as a graduate student with Chew's former postdoc Michael Moravcsik in the early 1970s, recalled recently his aesthetic revulsion from the diagrammatic menagerie: "I remember looking at all these diagrams. People would draw all sorts of diagrams. For me, a diagram has to have some clarity to it. It has to really mean something. Feynman diagrams meant something. I knew that I could understand what a Feynman diagram was. But there were all these other diagrams that people would draw with blobs, and all sorts of

79. Chew, "Nuclear democracy" (1964), 106; emphasis in original.
80. Ibid.

complicated lines. Sometimes there would be double lines connecting blobs, and triple lines, and sometimes dotted lines, and sometimes wiggly lines. And it was never clear what those lines meant, or how they corresponded to calculations or anything. It was all very kind of hocus pocus. It was sort of like, these are semi-Feynman diagrams, but they're not quite. And everything was blurry. There was never a sense of precision. . . . Everywhere I turned, I found ugliness, and arbitrariness, and vagueness." He decided to leave graduate school—indeed, to leave physics altogether. He turned his attention more squarely to aesthetics in mathematics and art, soon winning the Pulitzer Prize for his best-selling study, *Gödel, Escher, Bach* (1979).[81] Obviously Hofstadter's reactions were extreme. To many other young theorists at the time, such as Chew's graduate student Jerry Finkelstein (whose diagrammatic jottings appear in fig. 8.1), the various types of diagrams could be distinguished with practice and implemented in their advisers' expanding programs.[82]

Feynman diagrams and Feynman-like diagrams remained ever present in the particle physics literature of the 1950s and 1960s—perhaps surprisingly so, considering all the challenges to their original perturbative usefulness. In the early dispersion-relations work, theorists invested the diagrams with a new mnemonic role, similar in kind to—if different in details from—the role for which Feynman and Dyson had first designed them. Soon, however, theorists found inspiration for new ideas in the diagrams. They used Feynman's simple line drawings, suitably adapted, as heuristic guides into the unknown territory of the strong interactions, putting the diagrams to work, reinterpreting their meaning, and gleaning new insights into the behavior of subatomic matter. Goldberger and Gell-Mann detected a new feature of scattering amplitudes—crossing symmetry—in the diagrams' lines; Mandelstam, Chew, and Low pushed these diagrammatic rotations further to "find" (and work hard to extract) data on difficult-to-study phenomena from diagrams depicting well-known interactions. Some theorists, such as Landau and Chew, eventually proclaimed that a new theory would arise from this hodgepodge of diagrammatic improvisation. Though a coherent articulation of this hoped-for theory always seemed just beyond reach, they pressed on with their

81. Interview with Douglas Hofstadter by Stephen Gordon, May 1997, as quoted in Gordon, *Strong Interactions* (1998), 57; Hofstadter, *Gödel, Escher, Bach* (1979). Hofstadter's father, Robert Hofstadter, won the Nobel Prize in Physics in 1961 for his Stanford experiments on elastic electron-proton scattering.

82. Carleton DeTar, who completed his dissertation under Chew's direction in 1970, similarly recalled recently how little trouble Chew's students seemed to have had in keeping the various diagrams straight: DeTar, e-mail to the author, 21 Sep 2003.

diagrammatic machinations, convinced that a few more diagram-based innovations would topple quantum field theory altogether. In its place, they hoped, would be a new theory, jury-rigged from a base of reinterpreted Feynman diagrams. Note the irony here: where Dyson had *derived* Feynman diagrams' form and use from quantum field theory, Chew refashioned the diagrams as a tool with which to *eliminate* quantum field theory altogether.

With Landau, Chew put his reliance on Feynman diagrams up front, severed as they were from their field-theory umbilical cord. He emphasized in his 1961 lecture notes that many of the new results of his developing program were "couched in the language of Feynman diagrams," even though, contrary to first appearances, they did not "rest heavily on field theory." "It appears to me," he further prophesied, "likely that the essence of the diagrammatic approach will eventually be divorced from field theory" altogether.[83] The divorce began early in the game, even though his program still lacked its own clear foundations. The following year, still lacking a "logical structure" for his emerging theory, Chew again trumpeted the primary, generative roles of Feynman diagrams in his new program. "What is the basis for saying there are no other singularities? We can look at Feynman diagrams. A complete recipe for determining the singularities of any Feynman graph has been given by Landau, and a prescription for calculating the discontinuities by Cutkosky." He similarly explained in his 1962 summer school lectures that his "particle-pole" conjecture, which he had begun to articulate with his 1958 polology work, "grew out of field theory, particularly from Feynman graphs," but could now be formulated without any recourse to quantum field theory.[84] Chew derived inspiration from the form of the diagrams and sought to build a new theory around the diagrams' bare lines—rather than the other way around.

From the beginning, Feynman diagrams thus played a "bootstrapping" role for Chew's breakaway program, much like the particles eventually "seen" in them.[85] If the diagrams simply tracked particles' propagation and scattering,

83. Chew, *S-Matrix Theory* (1961), 3. Cf. ibid., vii–viii, 15–16, 23–25. Feynman diagrams, and their *S*-matrix theory generalizations in "Landau" and "Cutkosky diagrams," appear frequently throughout Chew's own textbooks and those of his former students and collaborators. See, e.g., Chew, *S-Matrix Theory* (1961), 16–17, 24–26, 38, 42–43, 75, 82–83; Omnès and Froissart, *Mandelstam Theory* (1963), 76, 82–83, 85–86, 88, 114, 120–21; Frautschi, *Regge Poles* (1963), 70–72, 93, 174, 192; and Frazer, *Elementary Particles* (1966), 51, 103, 105, 112–15, 124, 128, 131, 134, 154, 161.

84. Chew, "Strong-interaction" (1963), 6–7.

85. Chew himself used the term "bootstrapping" in a similar way in Chew, "Strong-interaction" (1963), 8–12. Chew's former graduate student, Louis Balázs, made this bootstrapping role of the diagrams in Chew's program explicit when he lectured on bootstrap techniques at the 1965 summer school in Dalhousie, India, just a few years after completing his Berkeley doctorate. Professor Balázs

and if rotations of the diagrams showed particles swapping roles from "elementary" to "force carrier" to "composite," then particles no longer had to bear any relation to the descriptions that theorists had previously developed for them. Gone were all ties to "psi-operators," Hamiltonians, and virtual particles. Gone too was all talk of perturbative expansions, and of the diagrams' role in keeping track of higher-order correction terms. Gone, in other words, were the rules, roles, and motivations for Feynman diagrams that Dyson had labored so hard to secure a decade earlier.[86] Feynman diagrams, carefully exhumed from their field-theoretic associations, became the girders, beams, and buttresses around which Chew and his growing band aimed to build their new theory.

Year after year, Chew's active group produced dozens of articles, conference talks, and dissertations, even though they had no single "theory" to replace the already-abandoned quantum field theory. Their lack of a replacement theory hardly ground the group to a halt; no one sat around waiting for a coherent or complete theory before diving into diagrammatic calculations. Rather, Feynman diagrams' pliability kept Chew's group busy for years, tinkering and improvising with the tools at hand. What Chew eventually proclaimed as his long-sought new "theory" in the late 1960s emerged from a series of heuristic, piecemeal diagrammatic improvisations.[87] The tools came first. From them, Chew hoped all else would follow.

---

kindly shared copies of his unpublished lecture notes. He delivered a similar set of summer school lectures at the Tata Institute in Bombay: Balázs, "Lectures on bootstraps" (1966).

86. See Chew, "Particles as S-matrix poles" (1989), 601.

87. See, e.g., Chew, *Analytic S Matrix* (1966).

· 9 ·

"Democratic" Diagrams in Berkeley and Princeton

[Geoffrey Chew was] the [Thomas] Jefferson of nuclear democracy.

RUDY HWA, MARCH 1998[1]

In October 1961, several of the world's leading theoretical physicists gathered in Brussels for the twelfth Solvay conference. An elite, by-invitation-only conference series begun fifty years earlier, the Solvay conferences had in the beginning hosted some of the seminal discussions and debates over the budding quantum theory and quantum mechanics.[2] The physicists in attendance at the Brussels meeting included several members of the old guard: Niels Bohr, Werner Heisenberg, Paul Dirac, Léon Rosenfeld, Eugene Wigner, and Yukawa Hideki, among others. The leading minds behind the wartime marriage of physical theory and nuclear weapons, such as Robert Oppenheimer and Hans Bethe, were also there, along with the postwar heroes of quantum electrodynamics—Richard Feynman, Julian Schwinger, Tomonaga Sin-itiro, and Freeman Dyson. Yet in between dining on fine food and sipping even finer wines, the physicists kept returning to the work of the young American theorist Geoffrey Chew.[3] Time and again, they referred to a strongly worded, memorable talk Chew had delivered just a few months before, in La Jolla, California.[4]

1. Rudy Hwa, interview with Stephen Gordon, Mar 1998, as quoted in Stephen Gordon, *Strong Interactions* (1998), 21.

2. Niels Bohr reviewed the history of the Solvay conferences during this 1961 meeting: Bohr, "Solvay meetings" (1961). See also Mehra, *Solvay Conferences* (1975).

3. Lists of invited speakers and auditors are included in Stoop, *Quantum Theory of Fields* (1961), 7–10. Abraham Pais describes the exceptionally fine meals and wine as a highlight of the Brussels meeting in Pais, *Tale of Two Continents* (1997), 379–80.

4. Feynman remarked on "points [which] have been emphasized by Chew in a remarkable speech at the conference in La Jolla, California this year," in Feynman, "Present status" (1961), 88. Similar references at the Solvay conference to Chew's La Jolla talk appear in Gell-Mann, "Symmetry

In La Jolla, Chew had cast off his final reservations and announced that quantum field theory should be abandoned for studies of the strong interaction. Quantum fields, virtual and elementary particles, and Lagrangians—the entire ontological basis and calculational machinery upon which quantum field theory rested—struck Chew as useless for understanding the strong forces that kept nuclei bound together. In words his colleagues still recalled vividly that October, Chew had cast the entire field-theory assembly aside as "sterile," destined simply to "fade away."[5] As we have seen in chapter 8, Chew aimed to erect a new program based directly on new ways of reading Feynman diagrams. Buoyed by some early calculational successes, Chew and the *S*-matrix program fast became leading lights in theoretical strong-interaction particle physics. Ten months after his "call to arms" in La Jolla, Chew was elected a member of the National Academy of Sciences, an honor he attained before his thirty-eighth birthday; that same year he was awarded the Howard Hughes Prize by the American Physical Society. There followed a string of coveted invited papers at National Academy and American Physical Society meetings. "Such lectures invariably drew capacity crowds," Chew's department chair gloated to a dean in 1964, "since Chew is generally recognized as the outstanding exponent of a particular approach to the theory of elementary particles known as the S-matrix."[6]

That same year, Murray Gell-Mann and George Zweig independently put forward the idea of "quarks" (Zweig called them "aces") to try to make sense of the flood of new particles that had been discovered since the end of the war. From today's vantage point, quarks seem the very antithesis of Chew's *S*-matrix approach: elementary entities from which more complicated particles, such as protons and neutrons, are built. Yet when Gell-Mann introduced the idea in 1964, he took pains to emphasize quarks' conceptual compatibility with Chew's bootstrap program. Nearly a decade later, in fact, Gell-Mann still insisted that quarks were "fictitious entites," and as such "there is no need for any

---

properties" (1961), 132, 142; Goldberger, "Single variable dispersion relations" (1961), 179–80, 192–95; and Mandelstam, "Two-dimensional representations" (1961), 214–15, 222–24.

5. A preprint of Chew's talk at the 1961 La Jolla conference is quoted in Cushing, *Theory Construction* (1990), 143. See also Gell-Mann, "Particle theory" (1987); and Chew, *S-Matrix Theory* (1961), 1–2. In a recent interview, Chew likened his strongly worded talk at the 1961 La Jolla meeting to a "coming out of the closet" speech: Chew interview with Stephen Gordon, Dec 1997, quoted in Gordon, *Strong Interactions* (1998), 32.

6. Burt Moyer to Dean W. B. Fretter, 30 Dec 1964, quoted in Birge, *History* (1970), vol. 5, chap. 19, 50. On Chew's election to the NAS and his other invited lectures, see ibid., chap. 19, 49–52; Anon., "Physics Professor Wins Prize" (1963); and Schmidt, "'Basic' particle" (1963).

conflict with the bootstrap." Zweig learned even more directly the influence that Chew's program held. Not long after introducing his idea of "aces," he was turned down for a faculty position; a senior theorist in the department had decried that such deviations from nuclear democracy and the bootstrap could come only from a "charlatan."[7] By the mid-1960s, Chew and his fast-growing group at Berkeley had changed the way most theoretical physicists approached the strong interactions. Their work helped to define the field into the early 1970s.

Drawing on his own ways of manipulating Feynman diagrams, Chew argued emphatically against the theorists' usual division of the microworld into "elementary" and "composite" camps. Though the idea of "nuclear democracy" received repeated tellings throughout Chew's 1961 lectures, it took several more years before Chew could, by his own lights, escape "the conservative influence of Lagrangian field theory." His updated lectures, published in 1964, were distinguished from the older ones, Chew explained, by their "unequivocal adoption of nuclear democracy as a guiding principle." He began by contrasting at some length "the aristocratic structure of atomic physics as governed by quantum electrodynamics" with the "revolutionary character of nuclear particle democracy." Chew left his students and readers with little doubt: "My standpoint here . . . is that every nuclear particle should receive equal treatment under the law."[8]

All this bluster about "conservative" field theory versus "revolutionary" nuclear democracy might inspire a knee-jerk zeitgeist interpretation: as he sat in Berkeley, with the 1964 Free Speech Movement taking flight all around him, a particular vision of "democracy" floated freely from Telegraph Avenue into the Radiation Laboratory. It is no coincidence, one might conclude, that "democracy" was in the air among increasingly radical Berkeley activists, and therefore also among the seemingly rarefied project of Chew and his students.[9] Like most knee-jerk reactions, however, this one is both too hasty and misplaced. James Cushing, for one, has argued persuasively against this simple reading—whatever Chew was up to, his physics sprang from more than this

7. Gell-Mann, "Schematic model" (1964); Gell-Mann, "Symmetry group" (1964); Zweig, "SU(3) model" (1964). Gell-Mann's comments at the 1972 conference are quoted in Gross, "Asymptotic freedom" (1997), 205–6; Cushing notes Zweig's trouble finding a job in Cushing, *Theory Construction* (1990), 280. See also Johnson, *Strange Beauty* (1999), 209–14, 225–27, 234.

8. Chew, "Nuclear democracy" (1964), 104–6.

9. On the Free Speech Movement in Berkeley, see especially Rorabaugh, *Berkeley at War* (1989), chap. 1; and Gitlin, *Sixties* (1993 [1987]), 162–66.

vague "spirit of the times." Indeed, as we have seen in chapter 8, many roots of his diagram-derived "democracy" extended back to work from the mid-1950s on dispersion relations, crossing symmetry, the Mandelstam representation, and polology. Chew's work on nuclear democracy cannot be read, in other words, simply as another instantiation of the strong Forman thesis, with Chew "capitulating" or "accommodating" his physics to a particular "hostile external environment."[10]

We are nonetheless faced with a puzzle: why did Geoffrey Chew see "democratic" lessons in Feynman's unadorned diagrams? Feynman diagrams had been used in a host of different ways since Feynman and Dyson had introduced them; the diagrams themselves did not dictate how physicists would use and interpret them. Given this plasticity, what lay behind Chew's specific reading of them, let alone the fervor with which he turned the diagrams against their field-theoretic birthplace? Why, moreover, did physicists working outside Chew's immediate group in Berkeley attribute still different meanings to Chew's "democratic" diagrams? If we take a wider view of Chew's activities, intriguing questions and associations arise. Stemming from his earliest objections to unfair treatment of scientists and other academics during the late 1940s, through his developing political activism throughout the 1950s, and his efforts during the late 1950s and early 1960s to change the way physics graduate students were trained, certain meanings of "democracy" recurred for Chew: no one should be singled out as special, either for privileges or for penalties; all should be entitled to participate equally. In short, as he described his beloved nuclear particles in 1964, all should receive "equal treatment under the law." Weaving in and out of a series of different contexts, these particular elements of "democracy" found explicit expression again and again from Chew. Laboring to ensure a democracy among particles was only one way in which Chew's work took shape after the war.

Chew noted in his 1961 lectures that more than anything else, "general philosophical convictions" (the details of which he left unstated) helped to guide him in his democratic reading of Feynman diagrams, and in his conclusion that no particles were truly elementary.[11] In these lecture notes as elsewhere, Chew struggled to find a vocabulary that would support the conceptual breaks he aimed to make with quantum field theory. Finding the right

10. Cushing, *Theory Construction* (1990), 217; cf. Gordon, *Strong Interactions* (1998), 35–37; and chap. 8, above. Paul Forman's famous argument regarding the acceptance of an acausal quantum mechanics in Weimar Germany may be found in Forman, "Weimar culture" (1971).

11. Chew, *S-Matrix Theory* (1961), 4.

terminology was no mean feat: "The language just didn't exist in physics," he later recalled. He remembers being "annoyed" by other physicists' sloppy terminology—it made him "gag"—"because the language wasn't there, there were no words, there was no way to say it."[12] The language he did produce to express his new physical concepts—his recurring incantations of "nuclear democracy," "equal treatment," and "equal participation"—therefore had to come from somewhere other than the staple repertoire of his fellow physicists. Guided in part by Chew's conscious choices of vocabulary, I attempt to clarify what he called his "general philosophical convictions" in this chapter, and study their evolution over time. To unpack Chew's influential diagrammatic practices, we must begin not with his 1961 talk in La Jolla, but with the battles fought in Berkeley over domestic anticommunism beginning in 1948.

GEOFFREY CHEW: A SCIENTIST'S POLITICS OF DEMOCRACY IN 1950S AMERICA

Geoffrey Chew, born in 1924, came of age in a generation of American theoretical physicists just past that of Richard Feynman and Julian Schwinger. Growing up in Washington, D. C., where his father worked in the U.S. Department of Agriculture, Chew graduated from high school at the age of sixteen. He completed his college education with a straight-A record from George Washington University in 1944. One of his undergraduate advisers, George Gamow, helped to arrange for Chew to head straight for Los Alamos, where he joined Edward Teller's special theoretical division, working on early ideas for a hydrogen "superbomb." Entering graduate school in February 1946 at the University of Chicago, Chew completed his doctorate in fewer than two and a half years, under the tutelage of Enrico Fermi. Ph.D. in hand, he and his fellow graduate student Murph Goldberger headed off to postdocs under Robert Serber at Berkeley's Radiation Laboratory. Before long, Berkeley's physics department took notice of Chew, who was already developing a reputation for both brilliance and clarity. The physics department appointed Chew to an assistant professorship, to begin in the autumn of 1949.[13] Chew's transit

12. Chew, interview with Stephen Gordon, Dec 1997, as quoted in Gordon, *Strong Interactions* (1998), 31–32. Chew explained in one of his popular articles on his physics program, "The finding of appropriate language is the essence of the game." Chew, "Impasse" (1974), 94.

13. These biographical details are taken from Raymond Birge to Dean A. R. Davis, 27 Feb 1949, in *RTB*, and A. C. Helmholz to Dean Lincoln Constance, 25 Mar 1957, in *RTB*, Box 40, Folder "Letters written by Birge, January–May 1957." See also Birge, *History* (1970), vol. 5, chap. 19, 43–51; and DeTar, Finkelstein, and Tan, *Passion for Physics* (1985).

Figure 9.1. Geoffrey Chew, ca. 1960. Note the plethora of Feynman-like diagrams on the blackboard. (Source: E. O. Lawrence Berkeley National Laboratory, courtesy AIP Emilio Segrè Visual Archives.)

from the East Coast to the West Coast included some of the best stops along the way for young physicists at the time, and by the age of twenty-five he had arrived, this physicist's manifest destiny complete. (See fig. 9.1.)

This simple picture of professional progress grew complicated, however, nearly as soon as Chew was hired. Chew became increasingly engaged with political issues in the late 1940s, continuing with greater intensity throughout the 1950s. His activities took him on an extended orbit, beginning in Berkeley and, seven years later, bringing him back there again. His reasons for leaving Berkeley in 1950 can be understood only by considering the political situation in which physicists found themselves soon after World War II, and how Berkeley physicists in particular encountered the early years of the cold war. The fast-moving descent into McCarthyism affected daily life in Berkeley's Radiation Laboratory and Department of Physics, shaping hallway discussions, straining old friendships, and altering many young physicists' career paths.

Politics and Physics at Berkeley, 1949–54

Few American physics departments experienced the pains of transition to the postwar political scene more abruptly, or more publicly, than the Berkeley campus of the University of California. The House Un-American Activities Committee (HUAC) turned its sights on atomic espionage directly on the heels of its sensational probe of alleged Communists in the film industry.[14] Its first stop: Berkeley's Radiation Laboratory, built up to international prominence during the 1930s by Ernest O. Lawrence. During the war, Lawrence's famed laboratory on the hill had been staffed with teams endeavoring to separate the scarce, fissionable uranium-235 isotope from its more common cousin, uranium-238. Some of this wartime staff, HUAC began to insinuate nearly as soon as Chew arrived at the Rad Lab as a postdoc in 1948, had been "red." Though no hard evidence of espionage at the Rad Lab was ever uncovered, five former Rad Lab employees were convicted of contempt of Congress, and one for perjury, based on their testimony. Each lost his job immediately upon being cited by Congress.[15]

The politics of domestic anti-Communism invaded physics departments far beyond Berkeley's in the ensuing months and years. Physicists across the country debated the new proposal, in 1949, to require full background security checks for all recipients of Atomic Energy Commission graduate student fellowships. Thirty-four Berkeley physics graduate students wrote to the *San Francisco Chronicle* to protest what they saw as a proposed exclusion from training of "those among us who hold unpopular viewpoints," since "education in a democracy must be available to everyone."[16] The next year, after the final, contentious establishment of the National Science Foundation, prominent

14. The literature on HUAC and McCarthyism is, of course, vast. On McCarthyism and American higher education in particular, see esp. Schrecker, *No Ivory Tower* (1986); Diamond, *Compromised Campus* (1992); Hershberg, *James B. Conant* (1993), chaps. 19, 21–23, 31; Chomsky et al., *Cold War and the University* (1997); Hornby, *Harvard Astronomy* (1997), esp. chap. 2; Wang, *American Science* (1999); Badash, "Science and McCarthyism" (2000); Schweber, *In the Shadow of the Bomb* (2000); Oreskes and Rainger, "Science and security" (2000); B. Bernstein, "Elusive Robert Serber" (2001); McCumber, *Time in the Ditch* (2001); Jerome, *Einstein File* (2002); and Herken, *Brotherhood of the Bomb* (2002). On the politicization of American scientists before World War II, see Kuznick, *Beyond the Laboratory* (1987).

15. House Un-American Activities Committee, *Hearings* (1949). See also the San Francisco–area newspaper clippings in *BDP,* Folder 4:12; Beck, *Contempt of Congress* (1959), 65–70; Schrecker, *No Ivory Tower* (1986), 126–48; and Herken, *Brotherhood of the Bomb* (2002), chaps. 5–7.

16. Letter to the editor, *San Francisco Chronicle,* 27 May 1949. See also Raymond Birge to Lyman Spitzer, Jr., 26 May 1949, in *RTB;* [Rabinowitch], "'Cleansing' of AEC fellowships" (1949); Anon., "Fellowship program" (1949); Anon., "Loyalty tests cause cut" (1950); Anon., "Curtailment of AEC

scientists such as Berkeley's iconic physics department chair, Raymond Birge, objected to congressional discussion of a proposal that would have forbidden NSF grants from going to members, past or present, of any organization listed on the attorney general's "subversive" list.[17]

With the outbreak of fighting in Korea that June, department chairs such as Birge found more and more of their time devoted to draft deferments for their students and young faculty, and to problems involving personnel security clearances.[18] Berkeley physicists, like their colleagues across the country, found themselves routinely denied passports for foreign travel, and foreign physicists experienced insulting delays or rejections of their applications for visitors' visas.[19] The Rad Lab, since 1947 under the auspices of the Atomic Energy Commission, could no longer have foreign scientists conduct any work there, classified or unclassified, whether paid or not.[20]

Not all the reactions to these fast-moving events were glum. Five years after the HUAC investigation of purported espionage at the Rad Lab, Berkeley student reporters concluded their five-part series on the Radiation Laboratory in December 1953 with the lighthearted story "Espionage at the Rad Lab—Naw!" After describing dozens of "scattered instruments painted a bright red with the white letters 'USSR' printed on them," the reporters explained that the letters stood for "United States surplus reserve," an old joke up at the laboratory. The gag, and visitors' stunned reactions to it, could still evoke "guffaws" from

fellowship program" (1950); Anon., "Loyalty tests for science students?" (1950); and Wang, *American Science* (1999), chap. 7.

17. On the debates over the founding of the NSF, see Kevles, "National Science Foundation" (1977); Kevles, *Physicists* (1995 [1978]), chaps. 11–12; Reingold, "Vannevar Bush's New Deal" (1987); and Wang, "Liberals" (1995). On the issue of the attorney general's list of "subversive" organizations and NSF grants, see Raymond Birge to Robert G. Sproul, 14 Mar 1950, in *RTB*.

18. On draft deferments, see Raymond Birge to R. C. Gibbs, 10 August 1950, in *RTB;* Birge to Local Board no. 62, Santa Clara County, 8 June 1953, in *RTB;* and Kaiser, "Cold war requisitions" (2002). The concerns with draft deferments for physics students persisted well after fighting had ceased in Korea; see the correspondence from 1958 in the *AIP-EMD,* Box 4, Folder "Scientific Manpower Commission, Washington, D.C." On security clearance troubles, see Birge to K. K. Darrow, 11 Jan 1955, in *RTB;* and Wang, *American Science* (1999), passim; see also Schrecker, *Age of McCarthyism* (1994), 37–40, 150–64; and Yarmolinsky, *Case Studies* (1955).

19. Raymond Birge to K. K. Darrow, 26 May 1955; Birge to Congressman Francis Walter, 15 July 1955; and Birge to Senator Harley Kilgore, 17 Nov 1955, in *RTB.* The Federation of American Scientists focused on passport and visa problems, and many of their efforts were reported in *BAS.* A special issue of *BAS,* vol. 8 (Oct 1952), was dedicated especially to these problems.

20. Raymond Birge to Walter Thirring, 8 Jan 1952, in *BDP,* Folder 5:117. On the establishment of the AEC national laboratory system, see Hewlett and Duncan, *Atomic Shield* (1969), chap. 8; Seidel, "Home for big science" (1986); and Westwick, *National Labs* (2003).

"the six foot, amiable, loose-jointed physicist" who gave these student reporters their tour.[21] Yet the levity could go only so far. Berkeley's Raymond Birge remained more somber, hesitating before accepting his nomination as vice president of the American Physical Society. Writing in May 1953 to the society's secretary, Karl Darrow, Birge recalled a former "time when the American Physical Society was concerned only with physics. At the present time, however, I am afraid it is concerned almost as much with politics as it is with physics and I must say I do not like politics." Almost exactly one year later, just as Birge was preparing to assume the presidency of the society, he and his Berkeley department were stunned over the decision by the Atomic Energy Commission to deny J. Robert Oppenheimer, Berkeley's former star theorist and the world-famous "father of the atomic bomb," his security clearance.[22]

These many developments shaped physicists' experiences of the late 1940s and 1950s across the country, ranging from graduate students' protests to the hand-wringing of the president of the American Physical Society. At Berkeley, however, all these events took shape in the shadow of a local situation that dominated hallway discussions and faculty meetings for the better part of a decade: the "loyalty oath" controversy at the University of California, sparked in the spring of 1949 when the university's board of regents imposed a new anti-Communist oath on all university employees.[23] Under the new oath, everyone from janitorial staff, to graduate student teaching assistants, to tenured faculty had to swear that they were not members of the Communist Party; signing the oath had to be witnessed by a notary public. Letters to faculty soon revealed that their "reappointment," and the payment of their salary, was now conditional upon their signing the oath—even for faculty who believed tenure had removed

21. Littlewood and Garretson, "Espionage in the Rad Lab—Naw!" (1953). It is interesting to note that this was the only segment of the five-part series which Birge clipped and saved with his other newspaper clippings, which may be found in *BDP,* Folder 4:12.

22. Raymond Birge to Karl K. Darrow, 22 May 1953, in *RTB;* Atomic Energy Commission, *Oppenheimer* (1954); P. Stern, *Oppenheimer Case* (1969); Major, *Oppenheimer Hearing* (1971); B. Bernstein, "'In the matter of J. Robert Oppenheimer'" (1982); and Herken, *Brotherhood of the Bomb* (2002). Of course, not all members of Berkeley's department were shocked by the news; some had even lobbied behind the scenes to ensure the outcome. On Berkeley involvement in and reactions to the hearing, see Anon., "Oppenheimer conflict" (1954); Raymond Birge to Edwin A. Uehling, 28 Mar 1955, in *RTB;* Birge, *History* (1970), vol. 5, chap. 17; N. Davis, *Lawrence and Oppenheimer* (1968), chaps. 8–10; Alvarez, *Alvarez* (1987), 179–81; and Helmholz with Hale and Lage, *Faculty Governance* (1993), 152–57, 276–79.

23. Gardner, *California Oath Controversy* (1967); and Schrecker, *No Ivory Tower* (1986), 116–25. See also Stewart, *Year of the Oath* (1950); Birge, *History* (1970), vol. 5, chap. 19; and Helmholz with Hale and Lage, *Faculty Governance* (1993), 96–97, 152–57.

all questions of reappointments.[24] A vocal minority of the faculty, including several professors who had fled European dictatorships before and during World War II, began to decry the oath as an infringement upon academic freedom. Hundreds of faculty members at first refused to sign the oath in protest, further strengthening the resolve of certain key regents that the faculty could not be trusted to govern themselves. Crowds of fifty to two hundred professors met each week at Berkeley's faculty club throughout the academic year 1949–50 to discuss tactics and strategies.[25] Yet when North Korea invaded South Korea in late June 1950, and the United States entered the conflict, most of the remaining holdouts simply signed the oath. Dismissing a faculty committee's recommendations, the regents then fired the remaining thirty-one "nonsigners" on 25 August 1950. Even though none of those fired had ever been accused by the faculty or the regents of Communist Party membership or sympathy, their fates drew national attention to the question (and, as many proclaimed, the danger) of Communists in the classroom.[26]

The ensuing seven-year battle between the fired nonsigners and the regents, weaving in and out of the California State Supreme Court, left no department on the California campuses untouched. Yet few departments felt the full brunt of the controversy more, or in more ways, than Berkeley's Department of Physics. It is difficult to overestimate the effects of the oath controversy on daily life within the department. The department secretary had to rush off lists of graduate students who had signed the oath after an initial "oversight," to prevent their termination as course graders or teaching assistants. The examining committees for several students' dissertation defenses had to be rearranged at the last minute once the regents ruled that faculty nonsigners could no longer serve on such committees. Birge circulated a memorandum to the department's faculty in early April 1950, cautioning them against putting their personal opinions regarding the oath in writing, or even of discussing them "at a meeting of a fairly large group."[27] Writing to the university's president,

24. See the forms and notices in *BDP,* Folder 3:41; Gardner, *California Oath Controversy* (1967), 52–54.

25. Gardner, *California Oath Controversy* (1967), 87; see also Segrè, *Mind Always in Motion* (1993), 235–36; and Serber with Crease, *Peace and War* (1998), 171.

26. Gardner, *California Oath Controversy* (1967), chaps. 5–6; Schrecker, *No Ivory Tower* (1986), 120–22.

27. O. Lundberg, university controller, memo to "chairmen of departments, administrative officers, and others concerned," 27 Nov 1950; RLY [Rebekah Young], physics department secretary, to O. Lundberg, 30 Nov 1950; M. A. Stewart, associate dean of the Graduate Division, memo to physics department graduate advisers, 14 Dec 1950, all in *BDP,* Folder 3:41; Rebekah Young to Robert

Robert Sproul, in the midst of the controversy, Birge found himself wondering whether the "wave of hysteria now sweeping the country," as evidenced locally by the loyalty oath disaster, might even put "the entire democratic structure of this country . . . in some danger."[28] Whether interpreted ultimately as an anti-Communist witch hunt, a principled fight over academic freedom, or a local power play between faculty and administration, the results of the oath controversy on mundane daily life could be felt palpably by students and faculty alike. Young physicists at the Rad Lab felt the fallout from the oath controversy equally viscerally: postdocs who had not signed the oath received notes on their desks on 30 June 1950, informing them that they had to turn in their badges and keys, clear off their desks, and leave by the end of that day.[29]

More than these many types of participation in the controversy, Berkeley's physics department fell victim to the loyalty oath by losing six faculty members within one year. Two professors, Harold Lewis and Gian Carlo Wick, allowed themselves to be fired in August 1950 when the regents finally dismissed all remaining nonsigners. By June 1951, four other professors—Robert Serber, Wolfgang "Pief" Panofsky, Howard Wilcox, and Geoffrey Chew—had resigned in protest. Included among the losses were *all* the department's theoretical physicists. The very first to resign over the issue from the physics department, and perhaps the first to resign from the entire university, was Geoffrey Chew.[30]

Geoffrey Chew and the Politics of Democracy

Though no one on the faculty could know it at the time, Chew's official letter of appointment to his assistant professorship had arrived exactly one week after the regents had secretly passed the new loyalty oath.[31] Chew, who still

---

Serber, 18 July 1951; and Serber to Young, 25 July, 1951, in *BDP*, Folder 3:4; Birge, "Memorandum to members of the physics department staff," 6 Apr 1950, in *RTB*.

28. Raymond Birge to Robert Sproul, 14 Mar 1950, in *RTB*.

29. Serber with Crease, *Peace and War* (1998), 171–72; Segrè, *Mind Always in Motion* (1993), 234–37; Steinberger, "Particular view" (1989), 311; Steinberger, "Early particles" (1997), xxxix–xl.

30. Birge suggests that Chew was "apparently" the first professor to resign from all the University of California over the oath controversy in Birge, *History* (1970), vol. 5, chap. 19, 45. See also Pauli to Erwin Panofsky, 23 Oct 1950, in Pauli, *WB*, 4.1:179.

31. Birge notes that Dean A. R. Davis sent the official letter of appointment to Chew on 1 Apr 1949 (Birge, *History* [1970], vol. 5, chap. 19, 45); the regents enacted the new loyalty oath in their meeting on 25 Mar 1949. While still a postdoc at the Rad Lab, Chew had delivered several talks on his research to the physics department, both formal and informal, so that Birge could introduce Chew as already "well known" at the September 1949 departmental meeting. See Birge's handwritten notes, "First dept. meeting, Wed., Sept. 28, 1949," in *BDP*, Folder 2:4.

maintained Q clearance to work on classified nuclear weapons projects, refused to sign what he called, in a letter to Robert Oppenheimer, "the objectionable part of the new contract," which seemed to Chew to threaten "the right of privacy in political belief." He became further frustrated with what he saw as weak attempts by the rest of the faculty to fight the oath.[32] At the end of his very first year of teaching, Chew took a further step, becoming the first person to resign from the physics department over the loyalty oath issue. As he explained to Birge in July 1950, one month before the regents finally dismissed the remaining nonsigners, Chew had decided "to get away from an intimidating and precarious situation."[33]

The university fired a physics department teaching assistant for asserting his right, under the Fifth Amendment of the U.S. Constitution, to refuse to answer questions that might incriminate him when called to testify before HUAC. The act of dismissing this graduate student, Chew told Birge, had shown beyond doubt that the regents seemed bent on removing from the faculty the "right" to "maintain its own qualifications." The regents' actions with the oath, furthermore, aimed at nothing less than to "root out the last resistance" among the faculty. The few signs of "faculty solidarity" with the nonsigners had all but vanished when fighting broke out in Korea in June 1950. As Chew pressed Birge one month later, "in a war-time situation, what security can a non-conformist have?" With the outbreak of fighting, the few remaining nonsigners on campus "have now become lepers who must keep out of sight." On top of this, Chew reported that the Radiation Laboratory, which was "the chief stimulus" of Chew's scientific work, had made it clear that it "does not welcome non-signers. Even if I were allowed to maintain my affiliation [with the laboratory], the unsympathetic atmosphere would not be pleasant. This would be a more subtle form of intimidation."[34] Though Chew found it a difficult decision to resign, he left Berkeley in July 1950 and accepted a position at the University of Illinois in Urbana. He was promoted from assistant to associate professor there within one year and became a full professor at Illinois in 1955 at the age of thirty-one.[35]

32. Geoffrey Chew to J. Robert Oppenheimer, 11 May 1950, quoted in Birge, *History* (1970), vol. 5, chap. 19, 45; Birge notes Chew's frustration with Academic Senate resolutions regarding the oath in ibid.

33. Geoffrey Chew to Raymond Birge, 24 July 1950, in *RTB*, Box 5, Folder "Chew, Geoffrey Foucar, 1924–."

34. Ibid.

35. Birge, *History* (1970), vol. 5, chap. 19, 47. The University of Illinois also had a standing loyalty oath requirement at the time Chew accepted his job there, but one which did not mention the Communist Party or any other group by name. Birge and University of California president Sproul

A few months after he left Berkeley, and after the regents dismissed the thirty-one remaining nonsigners, Chew reported on the struggle in the *Bulletin of the Atomic Scientists.* Like most of the faculty at the University of California, Chew objected that the faculty had been singled out and subjected to a more specific loyalty oath than that required for any other state employees. The fact that the regents then went beyond this, to threaten and eventually dismiss tenured faculty, constituted a further violation of "the cornerstone of academic freedom." He refrained from detailing any of his own experiences, reviewing instead the positions taken by the university president, the Academic Senate, and various factions within the board of regents. The controversy had been fanned in part, Chew explained, by what he called "fundamentalists," people who struck principled stands on questions like academic freedom even though they had lived without complaint since 1942 with an official university policy excluding Communists from teaching there. The "moral of this very sad story," Chew concluded, was that more explicit procedures needed to be defined for tenure. The rights and roles of the faculty, senate, and regents needed similar attention and explication to guarantee that faculty would not be singled out for special treatment again. The procedures of due process might then guard against a repeat of "the present sad and humiliating situation."[36]

Over the course of the 1950s, while teaching at Illinois, Chew became more and more active in what has been called "the atomic scientists' movement." Soon after arriving on campus, Chew founded Urbana's local branch of the Federation of American Scientists (FAS), a national organization dedicated to moderate and liberal causes. As Jessica Wang has detailed, by the late 1940s the FAS had become a largely bureaucratic organization, collecting information about some of the more severe abuses of McCarthyism and lobbying certain legislators for reform. In part because of pressure from HUAC and the FBI,

---

found it ironic that Chew would agree to go to Illinois, but Chew explained that in Illinois this was "the same oath required of all state employees [and] no one feels it to be a restriction on his political activity. . . . The intent of the trustees, therefore, does not seem inimicable to academic freedom." In other words, as far as Chew was concerned, the Illinois oath did not single out faculty for special treatment or unfair scrutiny. Chew to Birge, 24 July 1950.

36. Chew, "Academic freedom" (1950), 336. The objection that faculty were singled out for more scrutiny than other people was common among Berkeley faculty. Chew noted this in passing on 334 but gave it a more extended discussion in his letter to Birge of 24 July 1950. See also Birge, *History* (1970), chap. 19; Gardner, *California Oath Controversy* (1967), chap. 3; and Schrecker, *No Ivory Tower* (1986), 122–23.

the FAS had begun to refrain from its earlier pattern of public demonstrations by the time Chew joined the group, choosing instead the route of "quiet diplomacy."[37]

Chew participated directly in this FAS diplomacy, both on Illinois's campus and, soon, as a visible leader within the national organization. After founding the local FAS branch, Chew immediately began inducting friends and colleagues, such as his fellow physicist Francis Low. As Low recalls, Chew strode up to him soon after he arrived in Urbana and asked simply, "Okay, are you ready to join FAS?" "I was happy to do it," Low continues. "I thought it was a good organization. Geoff's position was very good, and I was happy to take part in it. It was a serious time." Under Chew's direction, the group organized monthly meetings on campus and hosted speakers to talk about things like the Fifth Amendment. On Chew's initiative, they also became a clearinghouse for campuswide complaints about unfair treatment, such as problems obtaining passports.[38] Soon Chew's activities extended beyond Illinois's campus. In November 1955 he testified before a U.S. Senate subcommittee as chair of the FAS Passport Committee. The FAS objected to the State Department's unwritten policy of denying passports to scientists for political reasons, and of further denying the passport applicants any means of due process or appeal. Usually the passports were denied with no reasons given, and appeals were greeted with delays reaching into months and even years. One of the most famous cases at the time concerned Linus Pauling, who received a passport only in 1954—after more than two years of attempts—when he applied to go to Sweden to receive his Nobel Prize in Chemistry.[39] In his testimony before the Senate Subcommittee on Constitutional Rights, Chew cited several lesser-known cases to lobby for more fair treatment. Passports, he urged, must be "recognized as a right of the U.S. citizen, not merely a privilege."[40] Both the loyalty oath and the passport situation convinced Chew that only "well-defined

37. See esp. Wang, *American Science* (1999). On the founding and early years of the FAS, see also Smith, *Peril and a Hope* (1965).

38. Low interview (2001); Kaiser, "Francis E. Low" (2001), 71–72.

39. See Anon., "Summary of testimony of Linus Pauling" (1956), 28.

40. Chew, "Passport problems" (1956), 28. This article includes Chew's testimony from 15 Nov 1955. See also Anon., "FAS Congressional activity" (1956). The specific items Chew lobbied for were conspicuously absent in all kinds of hearings from this period, having been denied to witnesses in HUAC hearings, local security-clearance boards, and often even university committees. See Wang, *American Science* (1999); Schrecker, *No Ivory Tower* (1986); and Schrecker, *Many Are the Crimes* (1998). See also [Rabinowitch], "How to lose friends" (1952); Weisskopf, "Visas" (1954); Anon., "American visa policy" (1955); and Toll, "Scientists urge lifting travel restrictions" (1958).

channels," operating under due process, could protect the equality and rights of academics. Just as he had explained in his report about the California loyalty oath, Chew labored to make it clear for the committee of senators during his 1955 testimony that academics, and scientists in particular, should not be singled out for "special privileges," nor should they be subject to special scrutiny or bias.[41] Clear and unambiguous procedures needed to be established so that disagreements would all be settled fairly, providing equal treatment to all those affected. With these safeguards, Chew believed, scientists could participate in a democratic America as citizens, all equal under the law.

PEDAGOGICAL REFORMS: "SECRET SEMINARS" AND "WILD MERRYMAKING"

Chew's pedagogical efforts in the years following his congressional testimony showed a specific resonance with his more overtly political activities. Much as he lobbied with the Federation of American Scientists, Chew worked hard to make certain that graduate students could work in such a way that none was singled out unfairly, and that all were encouraged to participate equally. His activities with his own graduate students shaped his approach to enlisting collaborators for his *S*-matrix program. Years later, when his program lay largely abandoned by most particle physicists, Chew continued to assess the turnaround in the language of democratic participation.

Chew's "Little Red Schoolhouse" in Berkeley

While neglecting several large issues surrounding tenure, academic freedom, and the legality of state-imposed loyalty oaths, the California Supreme Court ruled in favor of the dismissed nonsigners in October 1952, ordering that the regents reappoint them.[42] This court decision, however, left lingering the question of back pay, so that for several people the oath controversy lumbered on. This last issue was settled by the court, again in the nonsigners' favor, only in the spring of 1956. Yet as early as 1951, certain senior professors in Berkeley's

41. Chew, "Passport problems" (1956), 28.

42. Their reappointment remained conditional, however, upon their signing a new statewide loyalty oath, the so-called Levering oath, which was even more explicitly anti-Communist than the original university oath had been. The key difference was that the Levering oath was imposed upon all state employees, so that university faculty were no longer singled out for special treatment. See Gardner, *California Oath Controversy* (1967), 250, 253–54; and Schrecker, *No Ivory Tower* (1986), 123–25.

physics department began to consider how best to lure Chew back to Berkeley. Soon after the first court decision had been handed down, Birge tried to entice Chew to return. Reluctantly, Chew decided instead to stay at Illinois, which had made him a very generous counteroffer upon hearing of Berkeley's actions. Still excited by the prospect of returning to Berkeley's stimulating campus, however, Chew spent the spring semester of 1957 in Berkeley as a visiting professor. Eager to keep Chew in Berkeley, the new department chair, Carl Helmholz, did some impressive financial gymnastics to convince the administration that it could afford to hire Chew as a full professor. Helmholz's schemes worked, and Chew accepted an appointment as a full professor, beginning in the 1957–58 academic year.[43]

Immediately Chew began advising a large and growing group of graduate students within the department. The group was especially large considering that Chew and all his students were theoretical physicists, for whom working in large groups was still relatively unusual. Often ten or more students would be under Chew's wing at a time, and Chew would be engaged in collaborative work with four or five of them; postdocs and research associates kept the group even larger.[44] A steady stream of Chew's students completed Berkeley dissertations beginning in 1959, often with four or five students finishing each year.[45] In choosing to train his students in this manner, Chew followed a pattern similar to that set by Robert Oppenheimer at Berkeley in the 1930s, in which the students worked collectively, discussing their research projects regularly with the entire group. The large, close-knit group format, on the other hand, contrasted starkly with Julian Schwinger's approach, for example, who famously advised ten or more Harvard graduate students at a time during the

43. See the handwritten notes between Francis Jenkins, Robert Brode, and Raymond Birge, undated, ca. Apr 1951, in *BDP*, Folder 5:25; on Chew's 1953 offer to return to Berkeley, see Geoffrey Chew to Raymond Birge, 21 Apr 1953, in *RTB*, Box 5, Folder "Chew, Geoffrey Foucar, 1924–"; and on his 1957–58 appointment, see A. C. Helmholz to Dean Lincoln Constance, 25 Mar 1957, *RTB*, Box 40, Folder "Letters written by Birge, January–May 1957." Helmholz's financial twists and turns become clear in both Helmholz to Constance, 25 Mar 1957; and A. C. Helmholz to Chancellor Clark Kerr, 5 Mar 1957, in *BDP*, Folder 1:26.

44. Many of Chew's former colleagues and students recalled that Chew's group at this time was unusually large, and that he still made time to work carefully with each of them. Birge, *History* (1970), vol. 5, chap. 19, 51; Frazer, "Analytic and unitary *S*-matrix" (1985), 7; Perry, "My years" (1985), 15; Frautschi, "My experiences" (1985), 44; DeTar, "What are the quark and gluon poles?" (1985), 77; Gross, "Uniqueness of physical theories" (1985), 128; Jones, "Deducing T, C, and P invariance" (1985), 189; Frazer interview (1998); Finkelstein interview (1998); Wichmann interview (1998); Stapp interview (1998).

45. See the list of Chew's former graduate students, together with their years of graduation, in Frazer, "Analytic and unitary *S*-matrix" (1985), 7–8.

1950s and 1960s but met only rarely with any of them individually, and never with the whole group.[46]

Whereas Oppenheimer could often intimidate students and colleagues alike with his notoriously sharp tongue, Chew's students uniformly recall a much more encouraging adviser, one who, in the words of a former student, "treat[ed] us as full partners in a common effort."[47] In a further gesture of equality, Chew regularly joined the group for informal lunches in the Rad Lab cafeteria.[48] Chew took Oppenheimer's pedagogical model a step further when he instituted what came to be known as the "secret seminar." His entire group of students met weekly with him to hear presentations from one of the students; often the meetings took place at Chew's house. It was "secret" because faculty (other than Chew) were actively discouraged from attending: the goal was to make certain that no graduate students were intimidated from participating equally with their peers. From deep within Lawrence's sprawling Radiation Laboratory, the original site of American "big science," Chew carved out what one of his former students described as a "little red schoolhouse."[49] (See fig. 9.2.)

This "little red schoolhouse" approach also shaped how Chew and some of his colleagues organized a special conference on the strong interactions, held

46. On Oppenheimer's pedagogical approach, see Serber, "Early years" (1967); Serber with Crease, *Peace and War* (1998), chap. 2; Smith and Weiner, *Robert Oppenheimer* (1980), chap. 3; Kevles, *Physicists* (1995 [1978]), 216–19. In 1958, Schwinger was technically advising sixteen Harvard graduate students. See "1958–59 Department Lists," in *HDP, Correspondence, 1958–60*, Box A–P, Folder "1958–59 Department Lists." Bryce DeWitt, who completed his dissertation under Schwinger in 1949, discussed Schwinger's style with me during several discussions (at Boston University, Mar 1996, and at the University of Notre Dame, July 1999); William Frazer raised the contrast between Chew and Schwinger during Frazer interview (1998).

47. Gross, "Uniqueness of physical theories" (1985), 128.

48. Both Carleton DeTar and Steven Frautschi recalled the lunches during their interviews with Stephen Gordon: Gordon, *Strong Interactions* (1998), 27–28.

49. Some of Chew's former students remember another reason for why the "secret seminars" were labeled "secret": often student speakers would be arranged at the last minute, leaving no opportunity to advertise the topic of the seminar far in advance. Details on Chew's "secret seminars" come from Frautschi, "My experiences" (1985), 44; Capella et al., "Pomeron story" (1985), 86–87; Frazer interview (1998); Finkelstein interview (1998); Wichmann interview (1998); Mandelstam interview (1998); Carleton DeTar, e-mail to the author, 21 Sep 2003; and Chung-I Tan, e-mail to the author, 23 Sep 2003. The term "little red schoolhouse" comes from an interview between Carleton DeTar and Stephen Gordon, May 1997, quoted in Gordon, *Strong Interactions* (1998), 29. Several more of Chew's former students with whom Gordon spoke also recalled Chew's "secret seminar." In the 1930s, Berkeley's physics department held weekly informal seminars, attended by faculty and graduate students alike, though this single department-wide meeting disappeared after World War II: Helmholz interview (1998); Shugart interview (1998).

Figure 9.2. Geoffrey Chew delivering an informal lecture at Berkeley, 1961. (Source: Lawrence Berkeley National Laboratory.)

in Berkeley in December 1960. As handwritten notes from an early planning meeting reveal, Chew, Carl Helmholz, Donald Glaser, and the other members of the committee wanted their conference to bring "new people up to date" on the status of strong-coupling particle physics. As underlined in these notes, the conference was to be "non-exclusive." Minutes from this planning meeting likewise noted that graduate students' research "should be strongly represented" at the conference.[50]

Meeting these goals would not be easy: physics conferences on special topics were rarely aimed at bringing nonspecialists up to speed, let alone highlighting the contributions of graduate students. The Berkeley committee gained some help from the well-known MIT theorist Victor Weisskopf, who had worked in the Federation of American Scientists with Chew throughout the 1950s. As a

50. See the handwritten notes, dated Mar 1960, entitled "Special Meeting APS," in *BDP*, Folder 1:39, and the typed minutes from a planning meeting held on 4 Mar 1960, in the same folder. The handwritten notes are likely to be by either Carl Helmholz, chair of the department at this time and head of the conference-planning committee, or by Howard Shugart, who was a secretary to the conference-planning committee; the notes appear to match Helmholz's handwriting.

member of the planning committee for the special meeting in Berkeley, Weisskopf lobbied hard for additional funding from the National Science Foundation, so that the Berkeley conference could include these younger participants. "You can well understand and I am sure you agree," Weisskopf urged, "that such conferences with open attendance are very important for the stimulation of young people or other people who are new in the field." Such openness was especially needed in particle physics, as Weisskopf continued, "The field of high-energy physics is, as you know, very strongly in the hands of a clique and it is hard for an outsider to enter. The Rochester conferences were the only conferences that dealt with that subject, and they limited it to invited people only. The Berkeley conference is supposed to break this custom."[51] "Open" meetings, intended equally for newcomers and students and for members of a "clique," stood out as unusual in 1960. It took work on the part of Chew, Helmholz, Weisskopf, and the others to break the mold and keep their meeting "non-exclusive," just as Chew labored to do with his "secret seminar." The meeting attracted about three hundred physicists.[52]

The theoretical portion of this special conference focused almost exclusively on recent developments by Chew, Stanley Mandelstam, and Richard Cutkosky on a new framework for approaching particle physics. Coming just six months before Chew's more outspoken break with quantum field theory at the La Jolla meeting, the material discussed at the Berkeley meeting helped form the core of Chew's emerging *S*-matrix program. As Chew came to articulate more and more explicitly, the *S*-matrix approach relied on several general principles but eschewed much of the field-theoretic formalism.[53]

51. Victor Weisskopf to J. Howard McMillen, 14 Mar 1960, in *RTB*, Box 29, Folder "Weisskopf, Victor Frederick, 1908–". As it turned out, the NSF refused any financial aid, because the Berkeley meeting was under the auspices of the American Physical Society; additional funding for the meeting was provided by the Atomic Energy Commission and the United States Air Force. See the typed report "Conference on Strong Interactions," undated, in *BDP*, Folder 1:39. Like Chew, Weisskopf became quite active with the FAS during the 1950s; while Chew chaired the Passport Committee, Weisskopf headed the Visa Committee. See Weisskopf, "Report on the visa situation" (1952); and Weisskopf, "Visas" (1954).

52. The attendance figure is included in a postconference report, unsigned and undated, in *BDP*, Folder 1:39.

53. Schedules and reports on the presentations at the 1960 Berkeley meeting may be found in *BDP*, Folder 1:39. The general principles upon which Chew and collaborators hoped to build their non-field-theoretic *S*-matrix theory included analyticity, unitarity, Lorentz invariance, and crossing symmetry, not all of which are independent of each other. See esp. Chew, *S-Matrix Theory* (1961); and Cushing, *Theory Construction* (1990), chaps. 5–7.

The *S*-Matrix and a Double Democracy: Diagrams and Practitioners

The independence of the *S*-matrix program from many of the esoteric niceties of field theory received a doubly "democratic" spin from Chew, as he championed the new approach and quickly spread its gospel far and wide. First, Chew began to argue for his concept of "nuclear democracy"—that all nuclear particles, subject to the strong nuclear force, should be treated equally, without being divided into "elementary" and "composite" factions. As we have seen in chapter 8, Chew argued for this democratic treatment and his related notion of the bootstrap based largely on his idiosyncratic treatment of Feynman diagrams. Chew and his postdocs saw in the rotated diagrams of figure 8.11, for example, the ρ meson move from force carrier, to bound-state composite, to seemingly elementary particle. Arguing from the structure of these Feynman diagrams, read against the specific backdrop of dispersion relations, crossing symmetry, and polology, Chew taught his many students to treat all nuclear particles the same way.

Chew's "democratic" sentiment did not halt with the new interpretation of Feynman diagrams; Chew fashioned his *S*-matrix work as "democratic" in a second sense. In addition to "democratic" diagrams, Chew championed a "nuclear democracy" among practitioners as well. Consider his remarks near the close of an ebullient invited lecture at the 1962 New York meeting of the American Physical Society: "I am convinced that a wild period of merrymaking lies before us. All the physicists who never learned field theory can get in the game, and experimenters are just as likely to come up with important ideas as are theorists. They may even have an advantage over us."[54] Chew returned to this theme in a lecture at Cambridge University in 1963, reporting that "the less experienced physicists hav[e] an advantage in working with a new framework. (The inverse correlation of productivity with experience in a situation like this is remarkable.)"[55] Meanwhile, Chew delivered special lectures and seminars on the new material especially for experimentalists at Berkeley.[56]

54. Chew, "S-matrix theory" (1962), 400.

55. Chew, "Dubious role" (1963), 538. This article reproduces Chew's 1963 Rouse Ball Lecture at Cambridge.

56. Owen Chamberlain in particular recalled these lectures by Chew; see Chamberlain, "Interactions" (1985), 13. Chew's unusual abilities and interest in instructing experimentalists were by this time long standing. As a visiting professor at Berkeley in 1957, Chew gave a seminar on the pion-nucleon interaction which drew an unusual number of experimentalists from the Radiation Laboratory and from Livermore in addition to the graduate students enrolled in the course. Carl Helmholz reported that the experimentalists "are getting considerable benefit from it even though

The 1963 *S*-matrix textbook by Chew's colleagues Roland Omnès and Marcel Froissart, *Mandelstam Theory and Regge Poles*, similarly carried the subtitle *An Introduction for Experimentalists*.[57]

As he traveled around the country and beyond, feverishly working his campaign of "wild merrymaking," few could miss his obvious charisma and enthusiasm. As John Polkinghorne later recalled, "We used to call Geoff Chew 'the handsomest man in high energy physics.' I know of at least one senior secretary in a British physics department who kept a photograph of him near her desk. That frank and open face, with just a hint of his one-eighth Burmese ancestry, and his tall commanding figure, made him one of the few theorists in the pin-up class."[58] Rumors spread far and wide that Chew had given up a potential career in professional baseball to work in particle physics. His personal charm and enthusiasm made Chew into an effective salesman. Polkinghorne attests that "Geoff was definitely a man from whom one would be happy to buy a used car." His talks "were always eagerly awaited," given "their inspirational and encouraging tone."[59]

His enthusiasm quickly suffused Berkeley's department, encouraging his graduate students and postdocs to participate in his *S*-matrix campaign. Louis Balázs, who completed his Ph.D. under Chew's direction in the mid-1960s, reminisced recently that "it was an exciting experience being one of Chew's graduate students at UC-Berkeley in the early 1960s. New ideas were being discussed and developed continually and vigorously, particularly by the postdocs, and it seemed we were on the threshold of a new era in Physics." Those who were pushing the boundaries of the "new era" seemed to be drawn to Berkeley, if they weren't there already: "There was a constant stream of distinguished visitors," Balázs continued, "who seemed to be eager to learn about the new developments in Berkeley." And just as Chew had announced so exuberantly at the 1962 meeting, Balázs too recalled that "even graduate students found that they could make new independent contributions at that time." Another former graduate student, William Frazer, concurs, emphasizing that Chew worked hard to make sure that no students felt intimidated. "He really made

---

the subject is quite abstract and mathematical." See Helmholz to Constance, 25 Mar 1957. William Frazer also discussed Chew's informal seminar for experimentalists during Frazer interview (1998).

57. Omnès and Froissart, *Mandelstam Theory* (1963). Froissart worked with Chew's group in Berkeley during the early 1960s.

58. Polkinghorne, "Salesman of ideas" (1985), 23. For a similar analysis of the role of charisma in modern physics, see Thorpe and Shapin, "Who was J. Robert Oppenheimer?" (2000).

59. On Chew's baseball ambitions, see Chamberlain, "Interactions" (1985), 12–13; Perry, "My years" (1985), 14–16; and Gordon, *Strong Interactions* (1998), 15.

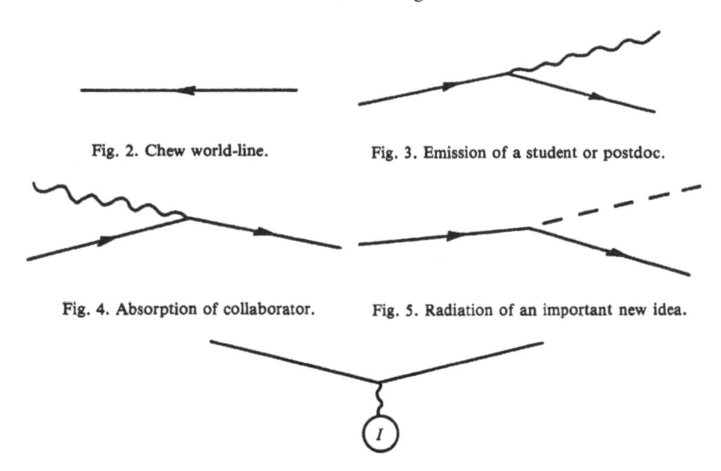

Figure 9.3. William Frazer's revised "Feynman rules." Compare with Dyson's "Feynman rules" in figure 3.2. (Source: Frazer, "Analytic and unitary S-matrix" [1985], 1–3.)

a wonderful atmosphere for us to work in."[60] Frazer turned to Feynman diagrams to capture the dynamics of Chew's active group. First he redefined what each line meant, substituting his own "Feynman rules" for the set Dyson had originally codified. (See fig. 9.3.) Next he laid out Chew's long series of interactions with students and collaborators. (See fig. 9.4.) Theorists of all stripes clung to Feynman diagrams as their own special totems.

Chew reflected in several places on the best way to train all these potential S-matrix contributors. As early as 1961 he noted that students who had not learned quantum field theory, and the usual ways of using and interpreting Feynman diagrams, seemed to fare best when approaching the new S-matrix material. He assured readers of his 1961 lecture note volume that "It is . . . unnecessary to be conversant with the subleties of field theory, and a certain innocence in this respect is perhaps even desirable. Experts in field theory seem to find current trends in S-matrix research more baffling than do nonexperts."[61] The same sentiment appeared five years later, when Chew remarked in the preface of his 1966 textbook, "No background in quantum field theory is required. Indeed, as pointed out in the preface to my 1961 lecture notes, lengthy experience with Lagrangian field theory appears to constitute a disadvantage when attempting to learn S-matrix theory."[62]

60. Louis Balázs, letter to the author, 6 Aug 1998; Frazer interview (1998); see also Perry, "My years" (1985), 15–16.

61. Chew, *S-Matrix Theory* (1961), vii–viii.

62. Chew, *Analytic S Matrix* (1966), v.

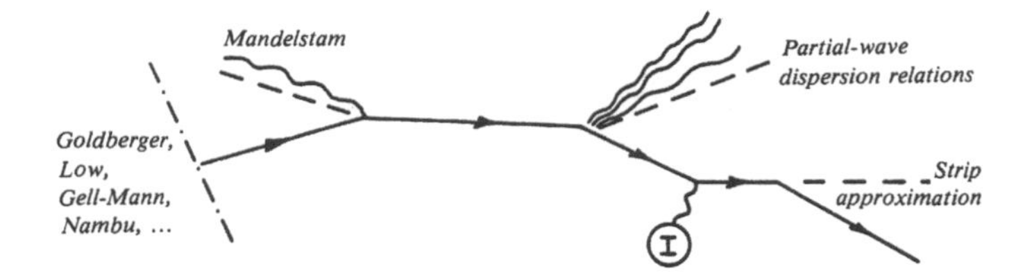

Fig. 7. The 1958–61 period: development of partial-wave dispersion relations.

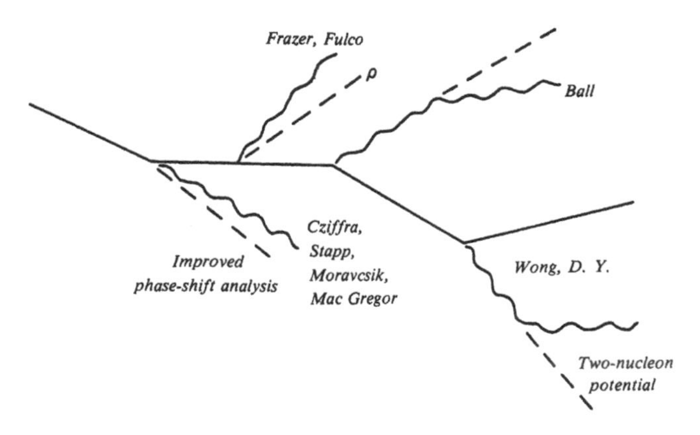

Fig. 8. The 1959–60 shower of students, ideas.

Figure 9.4. Frazer's version of Geoffrey Chew's career. (Source: Frazer, "Analytic and unitary *S*-matrix" [1985], 3–4.)

Chew's own students largely followed this prescription. William Frazer, Chew's first student to complete his dissertation after Chew's return to Berkeley, worked through Eyvind Wichmann's "very rigorous field theory course" as a graduate student, though he found his work with Chew to be much more interesting. (Wichmann had recently begun teaching at Berkeley, after receiving his Ph.D. under Norman Kroll's direction at Columbia.) "For the student," Frazer recalled recently, "life was a bit schizophrenic. Either you read on your own the very dry mathematical structure of axiomatic field theory, or you tried to follow the more exciting material."[63] Later students, such as Jerry Finkelstein, recalled that "there were quite a few of us in Chew's group who really did not spend a lot of time on field theory. I'd taken a course in it during my first or second year of graduate school, but that was all."[64] Ramamurti Shankar, another of Chew's students from this time, recently put a humorous turn on this pedagogical approach: "I had a choice: either struggle and learn field

63. Frazer interview (1998).

64. Finkelstein interview (1998).

theory and run the risk of blurting out some four letter word like φ^4 in Geoff's presence or simply eliminate all risk by avoiding the subject altogether. Being a great believer in the principle of least action, I chose the latter route. Like Major Major's father in *Catch 22*, I woke up at the crack of noon and spent eight and even twelve hours a day not learning field theory and soon I had not learnt more field theory than anyone else in Geoff's group and was quickly moving to the top."[65] David Gross, who completed his dissertation under Chew's direction in 1966, paraphrased an often-heard refrain from colleagues who didn't realize that Gross, since the 1970s a preeminent field theorist, had come from Chew's group: "Funny—you don't look Chewish."[66] One would search in vain to find much use of specifically field-theoretic techniques in the sixty dissertations by Chew's graduate students.[67]

So much for Chew's immediate circle in Berkeley; with several well-known texts on particle physics already in print, how best could their physics of "nuclear democracy" spread beyond Chew's large, but geographically limited, Berkeley group? Just as politicians debated a purported "missile gap" with the Soviets, Chew and his collaborators in the early 1960s faced a "textbook gap": given that so many physicists conceivably could participate in developing the "democratic" *S*-matrix program, the challenge became reaching these students, experimentalists, and other theorists and delivering the *S*-matrix message. Toward this goal, Chew and many of his *S*-matrix students and collaborators delivered many sets of summer school lectures, sometimes, like Chew in 1960, delivering the same set of lectures at two different schools in the same summer.[68]

Chew and his postdocs also began to publish their lecture notes in inexpensive editions, nearly as soon as the lectures had been delivered. These books reveal much about the quest to attract students to the *S*-matrix team. Most of the important *S*-matrix textbooks were published as part of the Frontiers in Physics series, which began to publish collections of lecture notes and reprints in 1961. Chew's 1961 *S-Matrix Theory of Strong Interactions*, based on his 1960 summer school lectures, was one of the first books to be included in the new series. By 1964, *S*-matrix tracts constituted nearly one-third of *all* the books

65. Shankar, "Effective field theory" (1999), 47.

66. Gross, "Uniqueness of physical theories" (1985), 128. Gross shared the 2004 Nobel Prize in Physics for some of his field-theoretic contributions.

67. Dissertations by Chew's students from 1959 to 1983 may be found in the Berkeley Physics Department Library.

68. Chew, "Double dispersion relations" (1960); Chew, "Double dispersion relations" (1961). Cf. Heilbron, "Earliest missionaries" (1985).

in the Frontiers series, even though the series was meant to treat all aspects of physics and not only particle theory.[69]

These books were rushed into print. *Regge Poles and S-Matrix Theory* (1963), by Chew's postdoc Steven Frautschi, stemmed from lectures Frautschi had delivered once at Cornell University in 1961–62 and that he augmented for delivery at a June 1962 summer school. The lectures by Maurice Jacob and Chew in their 1964 *Strong Interaction Physics* also had been delivered only once, during the academic year 1962–63. As the series editor David Pines explained, the series was intended to feature just this kind of "rough and informal" lecture notes, rather than cater to polished, final monographs. The very production of the books likewise reflected their mission, as Pines explained: "Photo-offset printing is used throughout, and the books are paperbound, in order to speed publication and reduce costs. It is hoped that the books will thereby be within the financial reach of graduate students in this country and abroad."[70] The progress of *S*-matrix theorists toward establishing an axiomatic foundation for their new work can be read immediately from the material form of their books. Chew's early reports on the developing program in 1961 and 1964 were printed within books that had not been carefully typeset, were printed on inexpensive paper, and were hurried into print. Chew's first textbook to treat the newly proposed axiomatic version of his *S*-matrix program, *The Analytic S-Matrix* (1966), on the other hand, was not published as part of the Frontiers in Physics series. Its professional typesetting and glossy pages provide clear contrast with the earlier volumes.[71]

From summer school lectures to Frontiers in Physics volumes, the task of creating and maintaining a community of *S*-matrix theorists involved far more than publishing research articles in the *Physical Review*. Chew was well aware that the dissociation of Feynman diagrams from their older meanings could

69. Lists of the titles published in the Frontiers in Physics series were included in the front of each of the books within the series. The *S*-matrix books within the series published between 1961 and 1964 included Chew, *S-Matrix Theory* (1961); Omnès and Froissart, *Mandelstam Theory* (1963); Frautschi, *Regge Poles* (1963); Squires, *Complex Angular Momenta* (1963); and Jacob and Chew, *Strong-Interaction Physics* (1964). During this same period, only two books were included in the series that focused on quantum field theory for particle physics.

70. See "Editor's foreword," by David Pines, dated Aug 1961, which appears within each of the volumes in the series. Frautschi's book, for example, was reproduced from hand-typed originals. Professor Chew mentioned his friendship with David Pines, also a physicist at the University of Illinois at Urbana, as one of the reasons he decided to publish his *S*-matrix lecture notes in this series. Geoffrey Chew, e-mail to the author, 11 May 1998.

71. This reading of the material production of Chew's textbooks is inspired by Cressy, "Books as totems" (1986); Johns, *Nature of the Book* (1998); and Secord, *Victorian Sensation* (2000).

be difficult to grasp for students who had already mastered the other means of calculating diagrammatically. His *S*-matrix program promised opportunities for those students, experimentalists, and other theorists who had not become overly attached to field-theoretic approaches. Inexpensive pedagogical resources, such as lecture notes and reprint collections, could serve to broaden this base of "democratic" *S*-matrix practitioners. There is some evidence that this aggressive textbook campaign worked: one physicist explained years later that, although he had not been a direct student of Chew's, he had learned particular calculational details of the *S*-matrix program, together with some sense of its special "philosophy," from reading Chew's 1961 *S-Matrix Theory of Strong Interactions.*[72]

The Language of Democracy for a Program in Decline

Chew did more than just assemble a democratic team of contributors; he drew an increasing contrast during the 1960s and into the early 1970s between the kind of work his *S*-matrix program fostered and that of the field theorists. The field theorists sought to explain the phenomena of particle physics based on a small set of basic or unit interactions taking place between a core set of "fundamental" or "elementary" particles. Chew mocked this approach of the "fundamentalists," arguing that such an "aristocratic" arrangement of fundamental particles could not provide an adequate framework for describing the strong interactions.[73] Much as in his 1950 article describing the loyalty oath controversy, Chew reserved the term "fundamentalist" for colleagues espousing a position at odds with his own "democratic" ideals.

During the late 1960s, Chew's own program appeared to many physicists to have lost its original focus, becoming mired in details and complexity. Treating the ρ meson within a democratic bootstrap framework was one thing, most physicists began to agree by the end of the 1960s, but moving beyond such a simple system to more complicated calculations had become much more frustrating. Quantum field theory, meanwhile, again attracted many particle physicists' attention, now augmented with a new emphasis on gauge symmetries and the quark hypothesis.[74] Chew lamented that this turn of events

72. Mueller, "Renormalons" (1985), 137.

73. Chew used the term "fundamentalists" throughout Chew, "Hadron bootstrap" (1970); and Chew, "Impasse" (1974). See also Chew, "'Bootstrap'" (1968).

74. Cushing, *Theory Construction* (1990), chaps. 6–7; Gordon, *Strong Interactions* (1998), chaps. 4–5; Pickering, *Constructing Quarks* (1984), chaps. 4–8; Pais, *Inward Bound* (1986), chap. 21; Cao, *Conceptual Developments* (1997), chaps. 8–11; Hoddeson et al., *Rise of the Standard Model* (1997).

sprang more from physicists' "psychology" than from stubborn experimental data. The trouble, as Chew wrote about it in the early 1970s, was that the "fundamentalists" dreamed "of the press conference that will announce to the world a dramatic resolution of their quest." Yet unlike the night thoughts of these fundamentalists, the *S*-matrix program was "the cumulative result of many steps stretching out over decades."[75]

Progress for the *S*-matrix camp, Chew explained, was necessarily a gradual game of constructing more and more partial models, incorporating the effects of more and more particle exchanges and interactions, all under the rubric of several general principles (such as crossing symmetry, causality, and unitarity). Chew's program was therefore built upon the assumption, he wrote in 1974, "that there will gradually develop a more and more dense coverage of the nuclear world by interlocking models, no single model having preeminent status. Such a pattern might be characterized as a 'democracy of models.'"[76] Chew reminded physicists in 1970 that "even though no press conference was called," the stepwise construction of these interlocking models within the *S*-matrix program had already scored several "breakthroughs," a fact Chew attributed to the "brilliant collective achievement of the high-energy physics community." Reinforcing this point with a history lesson, Chew recalled the "precedent of classical nuclear physics": "This model enjoyed an aristocratic status for almost thirty years, but eventually it was democratized."[77]

This notion of constructing interlocking models through a series of "collective achievements" fitted well with Chew's "secret seminar" approach to training graduate students. His students' dissertations reveal this close-knit, mutually buttressing approach: Bipin Desai, completing his dissertation in April 1961, built directly upon work by James Ball in his own dissertation, filed almost exactly one year earlier; Yongduk Kim's dissertation from June 1961 in turn drew explicitly on the dissertations by Desai and Ball.[78] This pattern continued throughout the 1960s. The abstract from Shu-yuan Chu's May 1966 dissertation, for example, explained that "The method [employed in the dissertation] is an extension of [C. Edward] Jones' proof in the single-channel case, making

75. Chew, "Hadron bootstrap" (1970), 24; and Chew, "Impasse" (1974), 124.

76. Chew, "Impasse" (1974), 124.

77. Chew, "Hadron bootstrap" (1970), 25; Chew, "Impasse" (1974), 124.

78. Ball, *Application of Mandelstam Representation* (1960); Desai, *Low-Energy Pion-Photon Interaction* (1961); and Kim, *Production of Pion Pairs* (1961). All these dissertations drew heavily upon Frazer and Fulco, "Effect of a pion-pion resonance" (1959); and Frazer and Fulco, "Partial-wave dispersion relations" (1960).

use of an explicit expression of the determinant D constructed by [David] Gross"—Jones completed his dissertation in 1964 with Chew, and Gross submitted his dissertation a few months after Chu.[79] Citations to work, both published and unpublished, by other graduate-student members of Chew's group, and acknowledgments of extended discussions with fellow students, fill nearly every one of Chew's students' dissertations. Chew's pedagogical ideal of equal participation melded seamlessly with his program for the piecewise construction of "interlocking models" by a "collective" of researchers.

By the 1970s, with his program all but abandoned, Chew bewailed the failure of his collective vision to attract those physicists who insisted instead on searching for a single "glamorous-sounding fundamental entity": "Few of the stars of the world of physics are content with the thought that their labors will constitute only a fraction of a vast mosaic that must be constructed before the complete picture becomes recognizable and understandable."[80] These "stars" of physics, now nearly all within the "fundamentalist" camp, routinely embraced only models based on Lagrangian field theories, Chew noted. They accorded only these models "special status," failing to "consider on an equal footing" models not derived from such a field-theoretic basis. The task of the collective-minded *S*-matrix theorist, on the other hand, remained "to view any number of different partially successful models without favoritism."[81]

The late 1960s and early 1970s were a difficult time to be a young physicist in the United States, over and above Chew's frustrations with the fate of "nuclear democracy." With the onset of détente and dramatic cuts in defense spending, U.S. physicists rapidly slid into the worst job shortage the profession had ever witnessed. Enrollments in the placement service registries of the American Institute of Physics tell the grim tale: in 1968, nearly 1,000 applicants fought for 253 jobs; the next year, almost 1,300 competed for 234 jobs. Only 63 jobs were on offer at a 1970 American Physical Society meeting, at which 1,010 young physicists were looking for work; 1,053 students competed for 53 jobs at the 1971 meeting.[82] Anecdotal evidence suggests that students who were steeped too heavily in *S*-matrix methods, to the exclusion of field-theoretic techniques, felt the crunch especially hard. Although many of Chew's students from this later period have gone on to successful careers in theoretical physics, several *S*-matrix students left the field after earning their Ph.D.'s to become medical

79. Chu, *Study of Multi-channel Dynamics* (1966).

80. Chew, "Impasse" (1974), 124.

81. Chew, "Hadron bootstrap" (1970), 27.

82. Kaiser, "Cold war requisitions" (2002), 33–35.

doctors or lawyers; others were denied tenure at places like MIT. Although it is difficult to disentangle the root causes of these theorists' difficulties from the overwhelming across-the-board cutbacks in physics, several physicists to this day continue to associate *S*-matrix training with job-placement difficulties during the late 1960s and early 1970s.[83]

In frustration as in triumph, Chew spoke of his program in distinctly "democratic" tones. On the one side, in the early 1960s, it seemed open to all: everyone could participate equally, and none were singled out for special privileges. When the program fell into neglect, the very language of Chew's complaints likewise came laden with the tropes of democratic participation. Field theorists spoke of an "aristocracy" of particles, and likewise granted "special status" only to certain kinds of models. *S*-matrix theorists, on the other hand, strove for an equality of particles, models, and practitioners, all judged "without favoritism" as members of a collective. "Patience" should be the order of the day, Chew wrote, not the yearning for press conferences and the special privileges (such as increased government funding) that such singular attention could foster.[84] As he lobbied for fair treatment of academics under a controlling state in the cold war, Chew likewise tried to produce a community of peers within high-energy physics, not various "cliques" singled out for special treatment or splintered between idea-producing theorists and fact-checking experimentalists. Each contributor to the "vast mosaic" of the *S*-matrix program should be an equal partner under the law.

THE VIEW FROM PRINCETON

Traces of Chew's outspoken stance on the failure of quantum field theory and of the need to treat all nuclear particles democratically may be found throughout his students' dissertations. Peter Cziffra, writing a year before Chew's famous La Jolla talk, echoed his adviser's attitude when he opened his dissertation by citing how "stymied" ordinary perturbative field-theoretic methods remained when treating the strong interactions. Well into the campaign for democracy, Akbar Ahmadzadeh reminded readers of his dissertation that insisting with the field theorists on a strict division between "elementary" and "composite" particles often leads to "absurd conclusions" when studying the strong interactions; instead, all particles should be treated as bound states, "on

83. Several physicists drew these connections during their interviews with Stephen Gordon: Gordon, *Strong Interactions* (1998), 50, 53–54.

84. Chew, "Impasse" (1974), 125.

an equal footing."[85] Henry Stapp, a research associate at the Rad Lab when Chew returned there in the late 1950s, pursued an axiomatic foundation for the *S*-matrix program in the early 1960s. "Early on," Stapp has recalled recently, "I was not really in close touch with Chew; I picked up the *S*-matrix ideas by osmosis, since Chew's ideas were permeating the area."[86]

Despite the group's successes in spreading the word via summer school lectures and Frontiers in Physics volumes, however, their ideas did not "permeate" in the same way to all physics departments. Princeton's department, in particular, provides a telling contrast with Chew's Berkeley. By the early 1960s, Princeton boasted a large and active group of theorists working on many aspects of particle physics. In fact, the Princeton group, as much as Chew's group in Berkeley, championed and extended many of the nonperturbative, diagram-based tools—such as dispersion relations, crossing symmetry, and polology—which we examined throughout chapter 8. Though a large group of theorists there pursued topics that now fell under the *S*-matrix rubric, their work did not share Chew's zeal for a "nuclear democracy." Comparing the Princeton group's work with that of Chew's group highlights which elements of Chew's diagrammatic "democracy" remained particular to his Berkeley group.

One of the leaders of Princeton's group shared many early stops with Chew along a common trajectory. Murph Goldberger had been a graduate student with Chew in Chicago immediately after the war. The two became fast friends, sharing office space, arranging social outings together, completing their dissertations at the same time, and moving together to Berkeley's Rad Lab as postdocs in 1948. Following his postdoc, Goldberger took a job back at the University of Chicago, while Chew taught for one year at Berkeley and then resigned over the loyalty oath. Once Chew landed in Urbana, he and Goldberger were again in proximity; they struck up an active collaboration during the mid-1950s, together with Francis Low and Murray Gell-Mann, also both in the Midwest at the time. Just when Chew left Urbana to return to Berkeley, Goldberger left Chicago to take a position at Princeton, starting in February

85. Cziffra, *Two-Pion Exchange* (1960), 4; Ahmadzadeh, *Numerical Study* (1963), 2–3.

86. Stapp interview (1998). See Stapp, "Derivation of the CPT theorem" (1962); Stapp, "Axiomatic *S*-matrix theory" (1962); Stapp, "Analytic S-matrix theory" (1965); and Stapp, "Space and time" (1965). Stanley Mandelstam, a close collaborator of Chew's and an architect of many *S*-matrix techniques, resisted following Chew and his group into a renouncement of field theory. As early as the Dec 1960 Berkeley conference, a conference report noted, Chew's presentation based on Mandelstam's work "did not evoke Mandelstam's full assent." "Conference on Strong Interactions," 10, in *BDP*, Folder 1:39; Mandelstam interview (1998). See also Cushing, *Theory Construction* (1990), 131–32, 145.

1957. Though now separated by the breadth of the continent, Goldberger and Chew continued to correspond.[87]

At Princeton Goldberger joined Sam Treiman, who had earned his Ph.D. from Chicago four years after Goldberger and Chew; a year and a half later, Richard Blankenbecler joined the group as a postdoc and later as a faculty member. Goldberger, Treiman, Blankenbecler, and their many graduate students spent much of their time during the late 1950s and 1960s on topics that Chew would have labeled part of his *S*-matrix program—just the sort of "interlocking models" that Chew hoped would bring clarity to the strong interaction. Goldberger and Treiman investigated the decay of unstable particles without resorting to field-theoretic Lagrangians; students completed dissertations on the analytic structure of scattering amplitudes and on how to incorporate unitarity, two of the key general principles from which Chew aimed to construct his autonomous *S*-matrix theory.[88] On the surface, these all sound as if they could have been completed by Chew's students in Berkeley. Yet when Goldberger reported in 1961 on the group's many accomplishments, he categorized all their work, with his characteristic sense of humor, as "the engineering applications of quantum field theory." "This work," Goldberger continued, "is complementary to the purer aspects of quantum field theory." Just when Chew was announcing his clear and decisive break with quantum field theory in La Jolla, the Princeton group celebrated the close fit between their research and Chew's nemesis.[89]

The Princeton group's research was no closer to "engineering" than any of Chew's work; it seemed more "applied" only when compared with the work streaming out of Princeton's *other* group of theoretical particle physicists, headed by Arthur Wightman. Wightman championed an axiomatic approach to quantum field theory, akin to that of the Cambridge, Göttingen, and Moscow theorists who sought to derive dispersion relations rather than apply them to experimental data.[90] No doubt with reference to Chew's standing-room-only invited lectures before the American Physical Society, the Princeton group

87. Annual report, 1956–57, pp. 1–2, in *PDP-AR;* Goldberger, "Fifteen years" (1970); Goldberger, "Francis E. Low" (1983); Goldberger, "Passion for physics" (1985). See also Pickering, "From field theory to phenomenology" (1989).

88. Goldberger and Treiman, "Decay of the pi meson" (1958); annual report, 1958–59, 2, 7; annual report, 1960–61, 12–13; annual report, 1962–63, 4; annual report, 1963–64, 66; annual report, 1964–65, 79–81; annual report, 1965–66, 4–5; and annual report, 1967–68, 23–24, all in *PDP-AR*. See also Treiman, "Connection" (1989); and Treiman, "Life in particle physics" (1996).

89. M. L. Goldberger, "An outline of some accomplishments in Theoretical Physics," in annual report, 1960–61, 12–13, in *PDP-AR*.

90. See Wightman, "General theory of quantized fields" (1989).

reported in 1966 that "There are presently two approaches to relativistic quantum theory. These are axiomatic field theory and dispersion or S-matrix theory. Notwithstanding some passionate claims, there is not yet any evidence that the two are really different. . . . Both theoretical approaches have been and will continue to be pursued actively at Princeton."[91] Together with Wightman, Goldberger and Treiman erected a "big-tent" approach to quantum field theory: "engineering applications" based on *S*-matrix calculational techniques would coexist peacefully with the more "pure" investigations into axiomatic structures and foundations. At Berkeley, meanwhile, there were no longer any senior field theorists to respond to Chew's challenge; the loyalty oath had made certain of that, when figures such as Gian Carlo Wick were fired as nonsigners. Years later, Chew mused on how he might have remained more "intimidated," and less likely to dismiss quantum field theory outright, if Wick had still been in Berkeley.[92] The same might be said of his work with Francis Low during the mid-1950s in Urbana. Although Low credits Chew with having provided most of the original ideas during their fruitful collaboration, it is likely that if Chew and Low had remained together in Urbana, Chew's later flamboyant pronouncements about the death of quantum field theory would have been tempered by Low's impressive and authoritative grasp of field-theoretic methods.[93]

In keeping with the double-barreled approach to field theory at Princeton, and in clear contrast with the approach of Chew's group in Berkeley, Goldberger's and Treiman's graduate students actively studied quantum field theory as an essential part of their training. Stephen Adler, who completed his Ph.D. under Treiman's direction in 1964, remembers auditing Wightman's course, since Wightman's work "was seen as undergirding" the calculational techniques of dispersion relations and the *S*-matrix program. "We were doing an evasive physics," Adler continued: since no one knew how best to treat the strong interactions, "we used whatever methods we could. . . . Dispersion relations were a tool, but we also learned field theory methods because they were useful for treating symmetries."[94] Other former Princeton students

91. Annual report, 1965–66, 4–5, in *PDP-AR*.

92. Geoffrey Chew, interview with Stephen Gordon, Dec 1997, quoted in Gordon, *Strong Interactions* (1998), 33; see also 32–35.

93. Low attributed the main originality of their collaboration to Chew in Low interview (2001); see also Kaiser, "Francis E. Low" (2001), 71–72.

94. Adler interview (1999). This emphasis upon semiphenomenological tools rather than overarching theory construction became a hallmark of Treiman's group and helped shape Adler's later work on current algebras; see Adler and Dashen, *Current Algebras* (1968); and Treiman, Jackiw, and Gross, *Lectures on Current Algebra* (1972); cf. Pickering, *Constructing Quarks* (1984), 108–14; and Cao, *Conceptual Developments* (1997), 229–46.

from this time similarly recall an emphasis upon learning quantum field theory.[95]

This peaceful coexistence of *S*-matrix-style calculations with axiomatic field theory led to a different appraisal of Chew's *S*-matrix program from the one "permeating" the Berkeley area. Adler recalls that when Chew came to Princeton to give a talk, "he sounded very messianic." Adler was hardly alone in comparing Chew's active campaigning to religious indoctrination. In fact, many of Chew's colleagues and former collaborators, now working at a distance from Chew, began to characterize Chew's vigorous pronouncements as religious proselytizing. Goldberger wrote to Murray Gell-Mann in January 1962 that Chew had become "the Billy Graham of physics." He added playfully: "After his talk I nearly declared for Christ. Since I had already changed from Jewish to Regge-ish, it was the only thing I could think of." Years later, John Polkinghorne recalled that Chew proclaimed his ideas "with a fervour" like that "of the impassioned evangelist. There seemed to be a moral edge to the endeavor. It was not so much that it was expedient to be on the mass-shell of the *S* matrix as that it would have been sinful to be anywhere else." Hardly ringing of a democratic political campaign, Chew's efforts sometimes struck his former colleagues as runaway zealousness.[96] Whereas Chew characterized his own reports as "indecently optimistic"—"I would be masking my feelings if I were to employ a conventionally cautious attitude in this talk," he exclaimed at his invited talk before an American Physical Society meeting in 1962—others criticized the rhetorical zeal of the movement. "[S]eldom has there been so much emotion attendant upon the formulation of a new and useful, but abstract, mathematical technique as is embodied in the *S*-matrix approach to the strong interactions," sniffed one reviewer in *Physics Today*.[97]

95. John Bronzan, e-mail to the author, 15 May 1997, and E. E. Bergmann, e-mail to the author, 16 May 1997. The same is drawn out in the course notes taken by Kip Thorne when Thorne was a graduate student at Princeton in the early 1960s. See in particular Thorne's notes from Properties of Elementary Particles, a course given by Val Fitch (spring 1963); Elementary Particle Physics, taught by Sam Treiman (spring 1963); Intermediate Quantum Mechanics and Applications, taught by Goldberger (fall 1963); and Elementary Particle Theory, taught by Blankenbecler (spring 1964). All notes in the possession of Professor Kip Thorne; my thanks to Professor Thorne for sharing copies of these notes.

96. Adler interview (1999); Marvin Goldberger, letter to Murray Gell-Mann, 27 Jan 1962, as quoted in Johnson, *Strange Beauty* (1999), 211; Polkinghorne, "Salesman of ideas" (1985), 24–25. Francis Low similarly remarked that Chew's later efforts seemed "religious" in character: Low interview (2001).

97. Chew, "*S*-Matrix Theory" (1962), 394–95; A. Stern, "Third revolution" (1964), as quoted in Cushing, *Theory Construction* (1990), 175–76.

As Adler recalled, Chew announced in his Princeton lecture "a great hope, but at the same time, Treiman was always a bit skeptical of any grand theory."[98] Indeed, Treiman was skeptical. He delivered his own lecture on *S*-matrix material ten months after Chew predicted a "wild period of merrymaking." In contrast with Chew's clear enthusiasm, Treiman focused instead on how even the most promising-looking "partial results and insights" remained "all tangled up with approximations which have inevitably to be introduced and which vary in style and severity from one application to another, one author to another."[99] Treiman was no critic of approximations, even "severe" ones, per se. His 1958 work with Goldberger on pion decay relied, in another reviewer's words, on "drastic assumptions" and "feeble arguments." Even in Treiman's own estimation, his had been "hair-raising approximations" that he and Goldberger remained "quite unable—apart from hand waving—to justify."[100] It wasn't the recourse to approximations that irked Treiman about Chew's program; it was the vehemence with which Chew pitted his work against field theory.

Treiman further dismissed as "amusing" the bootstrap hypothesis, trumpeted by Chew as the logical conclusion of nuclear democracy.[101] As we have seen in chapter 8, the goal of the bootstrap work was to find a single self-consistent solution that would show that strongly interacting particles might each produce the forces that led to their own production; each strongly interacting particle might thus be able to "pull itself up by its own bootstraps." What had captured the imagination of the Berkeley group left Treiman simply unimpressed. Throughout his 1962 lectures, for example, Treiman emphasized the "conjectural" basis of the bootstrap work, and "indulg[ed]" in what he characterized as "pessimistic remarks" regarding the bootstrap program.[102] When it came to relations between the *S*-matrix program and quantum field theory, Treiman was not Chew, and Princeton was not Berkeley.

These differences led to some subtle reinterpretations of the meaning of *S*-matrix work, and not only of its relative importance. Princeton's L. F. Cook,

98. Adler interview (1999).

99. Treiman, "Analyticity in particle physics" (1963), 149.

100. J. D. Jackson, "Dispersion relation techniques" (1961), 50; Treiman, "Life in particle physics" (1996), 16. As Treiman himself later remarked of this calculation, "no one but Goldberger and I would have had the effrontery to do what Goldberger and I did." Treiman, "Connection" (1989), 388.

101. Treiman, "Analyticity in particle physics" (1963), 163.

102. Ibid., 143. Blankenbecler and Goldberger similarly characterized Chew's bootstrap work as a collection of "interesting speculations" lacking a "physical basis." Making explicit reference to Chew's 1961 La Jolla talk, they dismissed the entire discussion as having merely a "religious nature." Blankenbecler and Goldberger, "Behavior of scattering amplitudes" (1962), 784.

another faculty member in the Goldberger-Treiman-Blankenbecler group, reported on his research on Chew's beloved bootstrap mechanism. The diagram-derived bootstrap, deflated as at best "amusing" during Treiman's 1962 lecture, remained to Chew, Frautschi, and most other members of the Berkeley group the sought-for culmination of nuclear democracy, a kind of holy grail for the equal treatment of all nuclear particles. To them, it meant that there were no "elementary" particles; each was a bound-state composite of others. Yet to Cook, surrounded as he was by Princeton's field theorists, elementarity remained central. In his 1965 gloss, "The bootstrap embodies a philosophy which supposedly enables one to calculate" various parameters for "*elementary* particles." Cook's work went on to emphasize the lack of agreement between Chew's favorite topic and existing experimental data.[103] Even when turning to topics most central to Chew's "democratic" campaign, Princeton's theorists clung to the language of "elementary" particles, rather than following Chew's group in their vocal break with the "fundamentalists." No matter how completely Chew's democratic vision might have permeated the Berkeley area, that vision did not command a single, unchanging interpretation from physicists at some remove from Berkeley.

CONDITIONS OF DIAGRAMMATIC POSSIBILITIES

Geoffrey Chew produced an unusual reading of Feynman diagrams during the late 1950s and early 1960s. He built his influential *S*-matrix program around a scaffolding of reinterpreted Feynman diagrams and worked hard to spread the new diagrammatic techniques far and wide. Theorists working elsewhere, such as Princeton, did not follow Chew in reaching the same conclusions about what the diagrams' simple lines portended. In one sense, this is nothing new: we have encountered several examples of a strong localism in physicists' treatment of Feynman diagrams, deriving from whether young physicists received their training at Cornell, Columbia, Rochester, or other institutions. Chew's prominent *S*-matrix program likewise drew on intellectual resources specific to his time and place—conceptual ingredients that extended beyond the narrow province of theoretical physics, and that assumed special salience for Chew's group in Berkeley.

103. L. F. Cook, in annual report, 1964–65, 79–80, in *PDP-AR;* emphasis added. Each of the courses on particle theory that Kip Thorne attended as a graduate student at Princeton during this period was labeled Elementary Particle Physics or Elementary Particle Theory, further reinforcing this nonbootstrap view of the field. See also Eden et al., *Analytic S-Matrix* (1966); and Matthews, "Some particles are more elementary than others" (1962).

Historians have studied several examples in which scientists framed details of their work in explicitly political language. One hundred years before Geoffrey Chew began advocating a "nuclear democracy" among subatomic particles, the Swiss-French chemist Charles-Frédéric Gerhardt suggested a "chemical democracy" in which all atoms within molecules would be treated as equals, against which the German organic chemist Hermann Kolbe countered with a hierarchical, "autocratic" model of molecular structure.[104] More recently, several Soviet theoretical physicists, including Yakov Frenkel, Igor Tamm, and Lev Landau, drew explicitly on their life experiences under Stalin's rule (often including prison terms) when describing solid-state physics, such as the "freedom" of electrons in a metal and particles' other "collectivist" behavior.[105] During the postwar period, the Japanese particle theorist Sakata Shōichi, thinking along explicitly Marxist lines, favored a strict hierarchy among subatomic particles, finding in such an arrangement appropriate base-superstructure relations.[106] Social metaphors abound within the physical sciences.

Perhaps it bears emphasizing that we need not agree that Chew's many activities were inherently "democratic." He maintained Q clearance for several years after his wartime Los Alamos work; such clearance already implied unequal access to research and resources.[107] His "secret seminar" discouraged other faculty members from participating equally. The 1960 Berkeley conference served as a stepping stone for his own *S*-matrix work, featuring primarily the research of his collaborators. Several friends and former collaborators saw quasi-religious zealotry, not a democratically minded enrollment campaign, in Chew's efforts to interest others in his *S*-matrix program. It is not clear what his thoughts or actions were during the 1964 Free Speech Movement.[108] And so on. Nor is it clear whether Chew was working with a fully articulated or

104. Rocke, *Quiet Revolution* (1993), 208, 325. Consider also the attempts by German physiologists and physical scientists to keep their dreams of political unity alive after the crushing defeat of 1848 by pursuing methodological and epistemological unity in the sciences: Anderton, *Limits of Science* (1993); and Lenoir, "Social interests" (1988).

105. Kojevnikov, "Freedom, collectivism, and quasiparticles" (1999); cf. Hall, *Purely Practical Revolutionaries* (1999). See also Kojevnikov, "David Bohm" (2002).

106. Hirokawa and Ogawa, "Shōichi Sakata" (1989); Maki, "Elementary particle theory" (1989). My thanks to Yamazaki Masakatsu for discussions of Sakata's work, as well as for bringing these articles to my attention.

107. On this point, see esp. Oreskes and Rainger, "Science and security" (2000). See also Geoffrey Chew to Raymond Birge, 21 Apr 1953, in *RTB*, Box 5, Folder "Chew, Geoffrey Foucar, 1924–"; and Goldberger, "Francis E. Low" (1983), xii.

108. In fact, by the late 1960s Chew sometimes struck his graduate students as politically conservative, at least as compared to the increasing radicalism of late-sixties Berkeley. Chung-I Tan, e-mail to the author, 23 Sep 2003.

constant political ideology. His father had worked in the Department of Agriculture while Chew was growing up in Washington, D.C., which was hardly a politically neutral bureaucracy during the New Deal; perhaps Chew had some ingoing interest, based on his early years, in political issues. Yet whether or not we agree on how "democratic" Chew's efforts ultimately were, or whether they stemmed from a clear and consistent ideology, one point remains crucial: Chew and many of his students and colleagues saw his program for strong-interaction particle physics as specifically "democratic" at the time, and special for that reason.

How, then, are we to interpret the self-proclaimed "democratic" work of Geoffrey Chew? For one thing, he used the same language over and over again ("fundamentalists," "special status," "without favoritism," "equal partners"); models, particles, and collaborators were all "democratized." This was a vocabulary Chew had honed over a decade filled with angst and activism, in front of regents and senators, years before he began to apply it to Feynman diagrams and ρ mesons. Rather than asking "how much did politics affect Chew's physics," or "what complicated admixture of politics and culture and society interacted in which complicated ways to produce Chew's ideas in physics," we can thus build on Chew's curious continuity of language to turn the question around: why did Chew's appropriation of Feynman diagrams emerge in the form that it did, at the time and place that it did? What were the conditions, in other words, that made "nuclear democracy" an intellectual possibility—indeed not just a possibility but the dominant set of techniques for strong-interaction particle physics throughout the 1960s? Why, moreover, was the work subject to so many competing interpretations by physicists further and further removed from Chew's immediate group in Berkeley?

Phrased this way, we are no longer driven to squabble over competing ledger sheets, trying to count up "how much political factors determined Chew's physics." Instead, when we ask, "why then? why there?" certain plausible connections relate Chew's choices for what to work on and what to lobby for. Feynman diagrams, those staple tools of theorists' instrumentarium, appealed to Chew with a usefulness and immediacy far beyond their narrow field-theoretic definitions—much as they did to scores of other theorists throughout the 1950s and 1960s, as we have seen throughout the previous chapters. Deciding that the diagrams' "true" meaning was not dictated by field theory alone, Chew had some choice in how he would use and interpret them—just as scores of other theorists chose to read and interpret the diagrams in still different ways, toward different calculational ends. As Chew devoted more and more time to political and pedagogical alternatives to what he saw as infringements against equal treatment, perhaps a similarly democratic reading of Feynman diagrams, and of the particles they purported to describe, carried a certain salience for Chew. Such linkages, at least, would certainly help to explain why Chew produced his fervently "democratic" reading of Feynman diagrams amid the many competing interpretations other theorists developed for the diagrams. In this sense, "nuclear democracy" seems thoroughly enmeshed within Chew's time and place—it bears the marks of McCarthy-era Berkeley.[109]

These specific resonances, no doubt aided but not uniquely determined by the particular environment of late-1940s and 1960s Berkeley, further help to explain why many other physicists who worked on *S*-matrix material did different things with it. To Chew and his many postdocs and graduate students, "nuclear democracy" and the bootstrap meant that quantum fields and virtual particles simply did not exist, and that there was no such thing as an "elementary" particle. Yet many young theorists, such as Sam Treiman's graduate students at Princeton, completed dissertations that treated the *S*-matrix program neither as a self-consciously "democratic" pursuit nor as a raised-fist competitor to "aristocratic" quantum field theory—indeed, to them the bootstrap promised at best merely one more technique for analyzing the properties of the truly "elementary" particles. Pieces of Chew's new calculational machinery were picked up and taught at various places, often bundled with still different theoretical tools and deployed toward different calculational ends. Just as Chew appropriated Feynman's diagrams toward new ends, so too did many theorists outside Berkeley convert Chew's diagrammatic program into their own.

109. Regarding possible ties between Berkeley's political culture and his physics program, Chew responded recently that he "had never thought about it," though such connections are "a possibility worth considering." Chew, interview with Stephen Gordon, Dec 1997, quoted in Gordon, *Strong Interactions* (1998), 37. Links between his pedagogical efforts and his theoretical approach to particle physics strike a similar chord in Chew, who recalls that "I might have had some idea of that," though "it's hard to recapture the way one was thinking in an earlier period," which, after all, occurred nearly four decades ago. Chew interview (1998). More recently, Professor Chew wrote that an earlier draft of this chapter was "perceptive and accurate." Geoffrey Chew, letter to the author, 19 Aug 1999.

· 10 ·

Paper Tools and the Theorists' Way of Life

You sit there and say: why isn't everybody doing S-matrix;
another guy says: why isn't anybody doing field theory?
The real problem is: *why is nobody solving anything?*

RICHARD FEYNMAN, 1961[1]

One basic premise undergirds the argument of this book: if we are to make sense of changes and developments in modern theoretical physics, we must attend to much more than the construction and selection of "theories." My aim has been to focus attention on how most theorists have spent most of their time—and at once we have been drawn into a world of *calculations,* rather than worldviews, paradigms, or theories. Theorists pursue calculations with a host of theoretical tools and calculational techniques; their day-to-day work shows distinct similarities to the domain of equipment and instruments that historians and sociologists have already become accustomed to analyzing for the laboratory and field sciences. Much as in those other areas, theoretical physicists must work hard to learn how to wield their tools; rarely if ever do the tools of theory seem natural or obvious on their own. Theorists spend much of their time practicing how to use these paper tools, building up a cache of experience closer in kind to craft skill or artisanal knowledge than to explicit textual information. Moreover, once physicists learn to deploy paper tools and make calculations with them, the range of applications to which they apply the tools can widen dramatically. As with any tool, improvisation and bricolage can lead to applications that had never been envisioned by the tool's inventors. Physicists deployed Feynman diagrams across domains and specialties that appear distinct when viewed through the lens of paradigms or reigning theories: from problems in electrodynamics to nuclear forces and many-body systems.

1. Feynman during discussion at the 1961 Solvay conference in Brussels, as reprinted in Stoop, *Quantum Theory of Fields* (1961), 178; emphasis in original.

To make sense of the recent history of theoretical physics, we must therefore attend carefully to physicists' training and to their appropriation of the (shifting) tools of their trade.

I began this book by emphasizing two distinct meanings of "dispersion" that frame these twin aspects of tool use in modern physics. The first meaning, "to put into circulation," captures the work required for new generations to develop facility with the diagrammatic tools. Feynman diagrams began to travel because of several key changes in physicists' pedagogical infrastructure during the years after World War II. Most important in the United States was the rise of postdoctoral training for theorists, with its built-in emphasis upon circulation (putting into practice Oppenheimer's dictum, "The best way to send information is to wrap it up in a person"). While the cascade of postdocs streaming out of the Institute for Advanced Study brought the new techniques to rapidly expanding cadres throughout the United States, Dyson's work with students in Cambridge and Birmingham had a similar effect in Great Britain. New circulation mechanisms within the expanding Japanese university system, put in place under orders of the American occupation forces, spread the diagrammatic techniques throughout Japan. In the Soviet Union, meanwhile, the cold war precluded any means of learning about the new developments directly from the tool's creators. Only under cover of the top-secret H-bomb project could a few young theorists, struggling for months on end, manage some limited facility with the diagrams based on close scrutiny of Feynman's and Dyson's articles. Feynman diagrams then lay relatively dormant until extended face-to-face contact was reestablished between physicists in the United States and the Soviet Union.

New types of texts began to circulate as well, reinforcing the trade in postdocs. Preprints and mimeographed lecture notes became more and more common. The booming postwar graduate enrollments (at least within the United States) convinced publishers that physics could be lucrative, and soon a spate of new textbooks hit the market. Within a few years, hybrid forms of pedagogical texts emerged, such as the Frontiers in Physics series, which combined informal lecture notes with reprints of key articles, all packaged inexpensively and produced very rapidly. These textual interventions reinforced, but usually could not replace, the circulation of people. Within the physicists' pedagogical field, Feynman diagrams spread most often by causal, local interactions—*this* person in the diagrammatic network mentoring *that* student in the new techniques.

The second meaning of dispersion, "to put in scattered order," emphasizes the diagrams' plasticity as physicists refashioned and redeployed them for new applications. Local schools emerged: the reasons for calculating diagrammatically, and the types of diagrams enlisted, showed remarkable diversity of form, purpose, and use. Two key ingredients drove this second kind of dispersion: the ingoing split between Feynman and Dyson about how the diagrams should be drawn and used, and the shift in most high-energy physicists' interests away from weakly coupled electrodynamic interactions to the strongly coupled nuclear force. Feynman looked on these developments bemusedly, cautioning Enrico Fermi not to trust any meson calculation that used Feynman diagrams; yet most young theorists paid little heed to such cautions, scribbling their diagrams more and more often for problems even the diagrams' inventor considered beyond the tool's proper domain. The dispersion of diagrammatic forms that resulted can be understood by putting flesh onto Wittgenstein's notion of "family resemblances": mentors and students crafted new uses for the diagrams together, in the service of locally selected research topics. Students' diagrams at Rochester showed greater similarity with each other (and with those of their adviser, Robert Marshak) than with the diagrams scribbled so assiduously by their peers at Cornell, Columbia, Berkeley, or Cambridge.

In the United States, the topics for which physicists deployed the diagrams were usually influenced by local experimental demands. The old ruse among Institute postdocs that "there are house theorists and there are field theorists" ultimately rang hollow: even most field theorists were in close contact with neighboring experimental groups and tailored their diagrammatic calculations accordingly. Physicists' appropriations of Feynman diagrams reveal a widening gulf during the postwar years between physicists in the United States and Europe over what "theoretical physics" should be and what "theorists" should be trained to do. Would young theorists use Feynman diagrams as adjuncts in a quest for "ultimate foundations," as Wolfgang Pauli urged in Zurich? Or would they scribble down diagrams helter-skelter as part of a patchwork of approximations, heuristics, and rules of thumb, trying to squeeze out numbers that could be compared with the latest overflowing experimental data, as Kroll's, Marshak's, and Bethe's students learned to do? Only a rare few, including Dyson, attempted to do both—playing Pauli's "foundations" game while summering in Zurich (a postlapsarian detour that left Dyson "the Anti-Christ," as far as his friends Bethe and Rohrlich were concerned) and returning to diagrammatic phenomenology when safely ensconced back on American shores. The diagrams supported all these appropriations and more; physicists in each setting had to choose what they valued, how they would calculate, and what techniques their students should master.

Figure 10.1. "Feynman diagrams in the Amazon jungle." (Source: Mattuck, *Guide to Feynman Diagrams* [1976], 120.)

The diagrams' dispersion leads to two final questions, one historical and the other historiographical. First, given physicists' tremendous variety of uses and interpretations for the diagrams, why did they remain so central to physicists' practices for so long? Second, what does an emphasis on paper tools and their dispersion portend for the way we interpret scientific theories and the recent history of theoretical physics? I take up these questions in turn.

WHY DID THE DIAGRAMS STICK? INCULCATION AND REIFICATION

Why did physicists continue to calculate with Feynman diagrams even as the diagrams' original purposes and motivations—the perturbative evaluation of weakly coupled interactions—seemed to theorists such as Geoffrey Chew to be as "sterile" as an "old soldier"? Within perturbative studies of QED, the diagrams' persistence requires little explanation: as Karplus and Kroll's landmark fourth-order calculation made clear, Feynman diagrams enabled theorists to tackle calculations that few had even dreamed possible before. A cartoon from *Physics Today* captured this use of the diagrams. (See fig. 10.1.) As Richard Mattuck explained when he reprinted the cartoon in his textbook, it was "possible in principle" to attempt "perturbation theory without diagrams, just as it is possible to go through the jungles of the Amazon without a map. However, the probability of survival is much greater if we use them."[2]

2. Mattuck, *Guide to Feynman Diagrams* (1976), 120.

Yet most theorists did *not* use the diagrams for perturbative studies of QED. For these other topics, physicists had few rules to guide them and fewer reasons for optimism that the diagrams—fashioned from the start to be perturbative bookkeepers—could serve any useful function at all. Indeed, one Nobel laureate recently wondered why his fellow physicists have clung so tenaciously to Feynman diagrams. Looking back over fifty years of diagrammatic calculation, Princeton's condensed-matter theorist Philip Anderson asked if his profession had been "Brainwashed by Feynman?" Anderson recalled decades of attempts to apply Feynman's diagrammatic tool almost blindly to problems in nuclear, particle, and condensed-matter physics—several attempts yielding dismal records of disappointment—instead of using "a much more varied toolkit of concepts and techniques."[3] Thus ease of use cannot be the whole story behind the diagrams' persistence, since "convenience" cannot explain the *leading* role that many theorists gave to the diagrams when redefining their meanings and making their break from Dyson's original diagrammatic packaging.

Some clues behind the diagrams' persistence can be found by looking at broadly shared features of physicists' training after the war, features that become obscured if we focus too narrowly on any single training center. Microhistorical scrutiny proved crucial when we considered this book's first two questions: how did the diagrams spread, and for what were they used? Yet physicists belonged to a larger community at the same time, over and above their specific pedigrees from Cornell, Cambridge University, or Tokyo Bunrika Daigaku; elements of this broader background played decisive roles in each of these far-flung places, their many site-specific variations notwithstanding. To understand why Feynman diagrams remained so central to so many physicists for so long, we must examine those pedagogical elements that physicists had in common.

Feynman diagrams' peculiar persistence across so many disparate departments invites comparison with several issues raised by art historians about why various artistic styles persist over long periods of time and across vast geographic regions. One classic focus has been the construction of "realism" in depictions of the natural world. Ernst Gombrich and Nelson Goodman both concluded that "realism" in art, as exemplified by the Renaissance turn to linear perspective, remained merely one kind of *convention* for rendering images and could lay no claim to being an especially "accurate" or "natural" kind of representation of the world. Gombrich famously attacked the notion

3. Anderson, "Brainwashed by Feynman?" (2000), 11–12.

of an "innocent eye" that could passively depict the natural world on a canvas. Instead he insisted that an artist will "tend to see what he paints rather than to paint what he sees"; artists begin with "a guess conditioned by habit and tradition."[4] Taking his cue from Gombrich, Goodman proceeded to even stronger conventionalist conclusions: "Realistic representation, in brief, depends not upon imitation or illusion or information but upon inculcation.... [R]ealism is a matter of habit."[5] Joel Snyder has chided Goodman, arguing that "familiarity does not explain the habit, the ease with which we 'pick it up,' or its strength. We have all kinds of habits, some of which are easy to break and others of which are like second nature." W. J. T. Mitchell has similarly emphasized that chalking everything up to "habit" says nothing about what makes certain habits special, lasting, or "realistic."[6] While likely agreeing with Goodman that "realism" in art derives from conventions and not from some privileged, unmediated means of capturing the world "as it is," these critics urge us to continue where Goodman left off, seeking out why only certain habits get accorded "realist" status while so many others do not. If realism arises always and only from habit, can we specify how and why some habits get singled out as realistic?

Within science studies, several people have likewise debated whether scientific diagrams and illustrations simply picture the world as it is or remain "social constructions." After Bruno Latour and Steve Woolgar highlighted the central importance of "inscriptions" to scientific work, others produced case studies to demonstrate diagrams' constructed status: computer-enhanced images of quasars cannot, any more than Renaissance botanical woodcuts, claim to speak in an unmediated way about the outside world.[7] This is surely correct, but it opens up as many questions as it seems to answer. If diagrams, much like any other means of communication, never provide direct mirrors of an underlying world, what do they provide? How and why do scientists put them

4. Gombrich, *Art and Illusion* (1969 [1960]), 86, 89, 298.

5. Goodman, *Languages of Art* (1976 [1968]), 32, 37–38. Cf. Goodman, *Ways of Worldmaking* (1978).

6. J. Snyder, "Picturing vision" (1980), 223; Mitchell, *Picture Theory* (1994), 352–53. See also Gombrich, "Image and code" (1981), 16–17; Mitchell, *Iconology* (1986), chap. 2. Cf. Freedberg, *Power of Images* (1989), esp. chap. 15.

7. Latour and Woolgar, *Laboratory Life* (1986 [1979]). See also Rudwick, "Visual language" (1976); Lynch, "Material form of images" (1985); Fyfe and Law, *Picturing Power* (1988); Lynch and Woolgar, *Representation* (1990); Taylor and Blum, *Pictorial Representation in Biology* (1991); Baigrie, *Picturing Knowledge* (1996); Jones and Galison, *Picturing Science* (1998); and U. Klein, *Tools* (2001). Many similar questions are raised in Miller, *Imagery in Scientific Thought* (1984); and Miller, *Insights of Genius* (1996).

to work? If always mere "constructions," why are some constructions seen as more useful—indeed, as more heuristic, generative, and longer lasting—than others? Do judgments about specific diagrams' usefulness vary over time or across scientific communities?[8]

These discussions from art history and science studies provide important insights into Feynman diagrams' tenacity during the 1950s and 1960s, even as their original meanings, theoretical embeddings, and associations with other calculational practices came and went. Part of the explanation for physicists' persistent use of the diagrams derives from pedagogical inculcation of the sort highlighted by Gombrich and Goodman. Much of the relevant pedagogical material had already become widely dispersed, standardized, and even "second nature" for physicists in many parts of the world. Yet as Gombrich, Snyder, and Mitchell contend in their critiques of Goodman, appeals to "habit" or "inculcation" alone do not explain the physicists' continued reliance upon these particular paper tools, rather than other tools that were also available at the time. Some physicists interpreted Feynman diagrams as capturing reality more directly or completely than other visual tools did, regardless of how standard, familiar, or easy to use the other tools were. Attributions of "realism" for these physicists thus did not stick to all diagrammatic habits equally. Feynman diagrams held, for this community at this time, certain realist associations not shared by other habit-forming diagrammatic tools.

Conventions and Pedagogical Assimilation

Gombrich suggested that "All representations are grounded on schemata which the artist learns to use."[9] The same held true for physicists learning to use Feynman diagrams. Talk of "conventions" abounds. Schweber, Bethe, and de Hoffmann instructed readers of *Mesons and Fields* (1955), for example, that the "direction of increasing time is supposed to be upward" in Feynman diagrams, which dictated how the "arrow convention" on the electron and positron lines would be established; the action of an external potential at a given point was

8. Cf. Pang, "Visual representation" (1997); Elkins, "Art history" (1995); and Elkins, *Domain of Images* (1999). Some roles of scientific illustrations and diagrams as heuristic, and often generative, metaphors have been explored in Rudwick, "Visual language" (1976); Griesemer and Wimsatt, "Picturing Weismannism" (1989); Gooding, "Theory and observation" (1990); Cambrosio, Jacobi, and Keating, "Ehrlich's 'beautiful pictures'" (1993); U. Klein, "Techniques of modeling" (1999); U. Klein, *Tools* (2001); and U. Klein, *Experiments, Models, Paper Tools* (2003). On scientific diagrams' traffic among heterogeneous communities, see Mitchell, *Last Dinosaur Book* (1998); and Dumit, *Picturing Personhood* (2004).

9. Gombrich, *Art and Illusion* (1969 [1960]), 156.

"sometimes exhibited explicitly by representing the external potential by a wavy line with a cross at the end"; the contraction of meson fields would be "represent[ed] . . . by a dashed line joining x_1 and x_2," and so on. Franz Mandl similarly introduced the diagrams in his 1959 textbook amid repeated talk of the "conventions"—his word—according to which they should be drawn. Kenneth Ford labeled one of these conventions an "artistic trick" in his book from 1963. A later text explained, "We shall agree to represent a free operator $A_\mu(x)$ by an (undirected) dotted line leading from the vertex x beyond the boundaries of the diagram," with similar conventions "agreed upon" for the other operators.[10] Every aspect of the diagrams—which points to connect, what line types to use, how to affix the arrows and positions of crosses—had to be built up for the students. Like seventeenth-century artists learning to draw landscapes and human faces from copybooks filled with specific visual schemata, students of these physics textbooks followed step-by-step instructions for how to draw Feynman diagrams.[11]

The diagrams may have been merely "conventional," but their conventions were not chosen at random. The visual features followed almost trivially from an earlier visual-representational tool that had been standardized long before the 1950s, and that was taught to students from their first courses in college physics: Minkowski diagrams for special relativity.[12] Physicists far beyond Feynman adopted Minkowskian conventions for their Feynman diagrams, orienting their time axes vertically and scaling the speed of light to one. (See fig. 10.2.) Just as in Feynman's momentum-space diagrams, the tilted propagation lines in these examples played no role whatsoever in the accompanying calculations; unlike in Minkowski diagrams, the angle of propagation no longer carried quantifiable information. The tilted lines were evolutionary remnants, like the human appendix—they help to identify genealogical ancestors but no longer had any discernible function.

Nor did these genealogies always remain tacit. Josef Jauch and Fritz Rohrlich, for example, explained in a footnote within their *Theory of Photons and Electrons* (1955): "In this and the following figures the [Feynman] diagrams are stylized. Although an interpretation in terms of world lines is sometimes

10. Schweber, Bethe, and de Hoffmann, *Mesons and Fields* (1955), 1:219–20, 225; Mandl, *Quantum Field Theory* (1959), 73–74, 96–97; Ford, *World of Elementary Particles* (1963), 199–200; Akhiezer and Berestetskii, *Quantum Electrodynamics* (1965 [1959]), 308. See also Umezawa, *Quantum Field Theory* (1956), 227; and Schweber, *Relativistic Quantum Field Theory* (1961), 448.

11. Cf. Gombrich's discussion in *Art and Illusion* (1969 [1960]), chap. 5. See also Edgerton, "Florentine 'disegno'" (1984).

12. Goldberg, *Understanding Relativity* (1984), pt. 3.

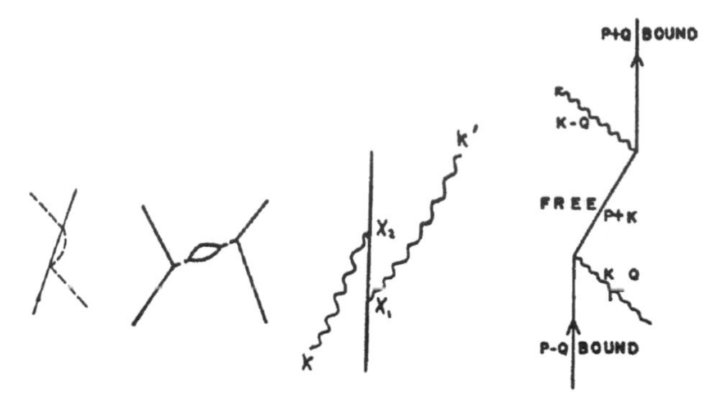

Figure 10.2. Repetition of Minkowski conventions in Feynman diagrams. (Sources, *left to right:* Rohrlich [A.26], 671; Anderson [A.112], 706; Low [A.65], 54; and Levinger [A.61], 661.)

given, this fact is irrelevant here and will be ignored. Consequently, external photon lines are not always drawn at 45° to the time axis (which is upward), and external electron lines are drawn parallel to the time axis even when these electrons are not meant to be at rest. Similarly, internal photon lines are often drawn perpendicular to the time axis, even though instantaneous interaction is not necessarily implied."[13] With the prior conventions for depicting particles' propagation so firmly established, Jauch and Rohrlich feared that the "stylized" look of their Feynman diagrams could be deceiving. A similar footnote from Ernest Henley and Walter Thirring's *Elementary Quantum Field Theory* (1962) cautioned readers: "These diagrams should not be taken too literally, since the concept of a classical path does not apply to virtual particles."[14] These cautions and caveats aside, however, several other physicists actively encouraged the associations between Minkowski and Feynman diagrams. Richard Mattuck talked openly about the close connections between the two types of diagram in his well-known 1967 textbook. Physicists also exploited these similarities when lecturing to undergraduate classes and audiences of nonphysicists. Popularizations by Kenneth Ford, Stanley Livingston, and Feynman proceeded

13. Jauch and Rohrlich, *Theory of Photons and Electrons* (1955), 149.

14. Henley and Thirring, *Elementary Quantum Field Theory* (1962), 146. This is the only source I have found that explicitly raised Bohr's original objection to Feynman, that classical paths could not be attributed to quantum particles. Thirring was at the time the director of the Institute for Theoretical Physics at the University of Vienna; his Bohr-style caution may be a remnant of the Continental physicists' deep-going absorption of the Copenhagen interpretation of quantum mechanics, which, during the 1920s, had famously forsaken attempts to visualize quantum phenomena in space and time. See Miller, *Imagery in Scientific Thought* (1984); and Chevalley, "Physics as an art" (1996). My thanks to Cathy Carson for raising this interesting point, and to Edward MacKinnon for bringing Chevalley's article to my attention.

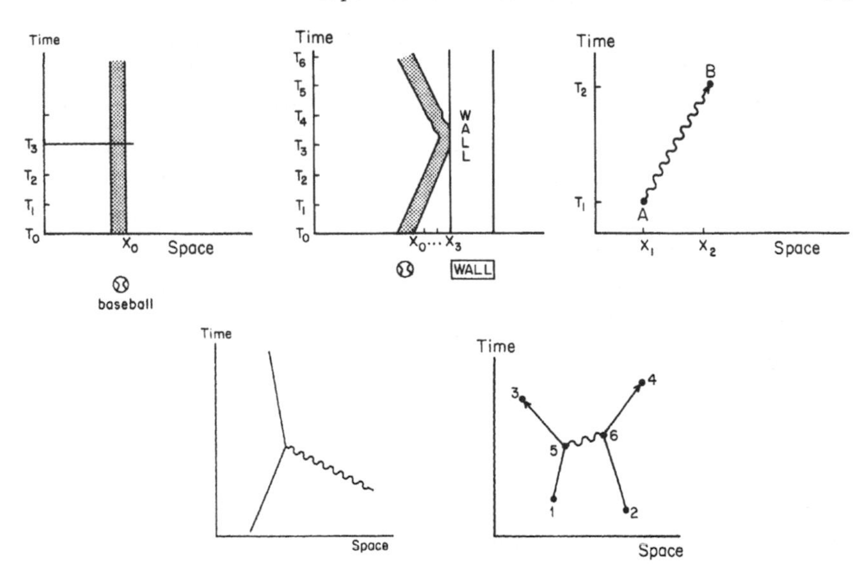

Figure 10.3. Minkowski and Feynman diagrams. (Source: Feynman, *QED* [1985], 86–88, 92, 94.)

seamlessly from Minkowski diagrams to Feynman diagrams.[15] (See fig. 10.3.) These popular books hardly found much place in graduate students' curricula. Yet they illustrate that the assimilation of Feynman to Minkowski diagrams was not a difficult or wholly foreign notion for theoretical physicists at the time.

Feynman diagrams were consistently drawn and taught as being of a piece with the reigning pictorial standards for studying particle trajectories through space and time. The mnemonic device simply was not innocent of physicists' prior inculcation in the visual practice of depicting particles' paths, regardless of the distinct meanings attributed to the stick figures in different contexts. If we were to follow Nelson Goodman, perhaps our story would be complete: the resilience displayed by Feynman's diagrams in the early 1960s, when physicists such as Geoffrey Chew aimed to build a new theory from Feynman diagrams while replacing quantum field theory altogether, can be explained by the diagrams' easy assimilation into a long visual tradition for treating particles' propagation. This long tradition carried tacit, visual "baggage" far more general than the specific role Feynman diagrams had been designed to play in QED. Habit and inculcation, in other words, may have been enough to

15. Mattuck, *Guide to Feynman Diagrams* (1967), chaps. 2–3; Ford, *World of Elementary Particles* (1963), 191–201; Livingston, *Particle Physics* (1968), 204–5; and Feynman, *QED* (1985), chap. 3.

secure Feynman diagrams a central place in theoretical physicists' heads and hands, even as their connections with other elements of the theorists' toolkit came and went. With Goodman's critics, however, we may be skeptical that "habit" alone is sufficient to explain the diagrams' tenacity. In just this period of upheaval among theoretical approaches to the strong nuclear force, a second diagrammatic tool was developed. Although these other diagrams offered equally strong ties to pedagogical inculcation and habit, they suffered quite a different fate. The comparison between Feynman diagrams and these "dual diagrams" thus can highlight those features specific to Feynman diagrams that contributed to their staying power.

Not Just Any Habit: The Case of Dual Diagrams

Soon after Geoffrey Chew and others called for a wholesale replacement of quantum field theory, a team from England developed a pictorial device to complement Feynman diagrams in the study of particles' interactions. John Polkinghorne, in particular, then saw to it that the "dual diagrams" became well known. Polkinghorne focused his summer school lectures from 1960 on the new diagrams, several articles were dedicated to the diagrams and their use, and they featured prominently in the textbook from 1966 by Polkinghorne and his Cambridge colleagues, *The Analytic S-Matrix.*[16]

Dual diagrams offered what Polkinghorne repeatedly called a "picturesque geometrical construction" for finding and characterizing certain physical properties of generic interactions. In the *S*-matrix work, as championed by Chew and his colleagues, the overriding concern was charting the regions in which the scattering amplitude behaved as an analytic function. Feynman diagrams, as interpreted according to Lev Landau's new rules for them, provided one means of trying to find and characterize these analytic properties (as we have seen in chap. 8), but, as Polkinghorne and his collaborators soon discovered, the diagrams could not be relied upon to catch all the subtle features alone.[17] Instead, a dual diagram could be introduced for any given Feynman

16. Landau had introduced dual diagrams in "Analytic properties" (1959), 184, 186, 188; see also Taylor, "Analytic properties" (1960); Okun' and Rudik, "Finding singularities of Feynman graphs" (1960); Polkinghorne, "Analytic properties" (1961); Polkinghorne, "Analyticity and unitarity" (1962); Fairlie et al., "Singularities" (1962); and Eden et al., *Analytic S-Matrix* (1966), 60–62, 73–76, 104–8.

17. Chew, *S-Matrix Theory* (1961); Eden et al., *Analytic S-Matrix* (1966), chap. 2; and Cushing, *Theory Construction* (1990), chap. 5. The insufficiency of Feynman diagrams to determine these analytic properties (especially in the face of "anomalous thresholds") was discussed in Polkinghorne, "Analyticity and unitarity" (1962); and Fairlie et al., "Singularities" (1962).

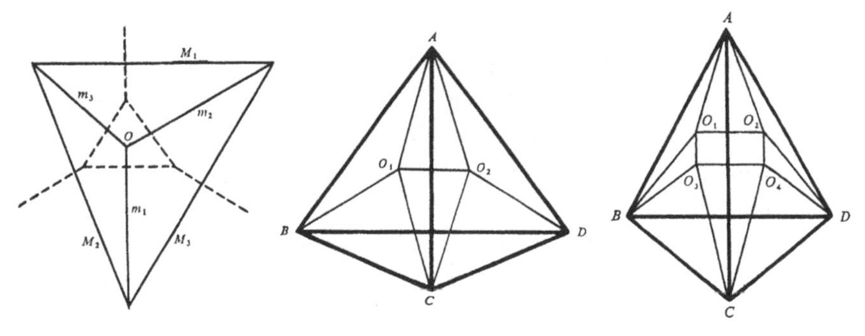

Figure 10.4. Dual diagrams. The left figure shows the dual diagram (*solid lines*) associated with the superimposed Feynman diagram (*dashed lines*). The middle and right figures show examples of more complicated dual diagrams. (Source: Polkinghorne, "Analytic properties" [1961], 67, 80.)

diagram: for every vertex in the Feynman diagram there would appear a closed loop in the dual, and vice versa. (See fig. 10.4.) With the aid of the dual diagrams, Polkinghorne demonstrated, one could analyze in a geometric fashion which particles carried how much momentum in a given interaction and thereby determine those specific processes that would dominate in particles' scattering. The dual diagrams could thus be used to isolate mathematical anomalies that had been missed when physicists had relied solely on Feynman diagrams. Dual diagrams, as Polkinghorne announced time and again, offered *S*-matrix theorists the ability to survey and understand more completely the mathematical structure associated with any given process. Where Feynman diagrams alone could lead one into trouble, dual diagrams offered a way out.[18]

As Landau demonstrated, and as noted explicitly in the pedagogical writings of Polkinghorne, the rules for building the dual diagrams were exactly analogous to Gustav Kirchhoff's rules for electric circuits.[19] The structural equivalence is significant: students learned how to study the current flowing through elementary circuits, using simple circuit diagrams and Kirchhoff's rules, even before they learned about Minkowski diagrams and special relativity—and long before they began to tackle particle physics. Polkinghorne emphasized

18. Cf. James Bjorken's development in his 1959 dissertation of "reduced graphs" (first introduced by Landau) to study these functions' analytic properties: Bjorken and Drell, *Relativistic Quantum Fields* (1965), chap. 18.

19. Polkinghorne, "Analytic properties" (1961), 83; Eden et al., *Analytic S-Matrix* (1966), chap. 2. Bjorken extended the analogy to elementary circuit theory quite a bit further when relating the closed loops of Feynman diagrams and "reduced graphs" to Kirchhoff's treatment of closed circuits: Bjorken and Drell, *Relativistic Quantum Fields* (1965), 220–26.

out loud and at length the long-standing pedagogical tradition from which dual diagrams derived: in studying the flow of momentum around a given vertex, students only had to apply the same rules and skills that they had learned in their earliest studies in physics. Whereas the connection between Feynman diagrams and Minkowski diagrams was often a hushed one within textbooks, Polkinghorne and others worked hard to make the visual and calculational associations between dual diagrams and circuit diagrams clear.

Yet despite the attention dedicated to dual diagrams, these alternative diagrams never entered the physicists' toolkit the way Feynman diagrams did. Geoffrey Chew, for example, never used dual diagrams in his research articles, lecture notes, or textbooks; none of his many students at Berkeley drew or even mentioned dual diagrams in their dissertations. The diagrams are historically significant, therefore, for at least two reasons. First, they did not *look* like Feynman diagrams; it was not predetermined that particle theorists should continue to rely so exclusively on Feynman diagrams. A contrasting diagrammatic technique was put forward, shown to be more effective than Feynman diagrams for certain crucial tasks, and yet still not exploited. Second, the untimely death of the dual diagrams illustrates just the point that Gombrich, Snyder, and Mitchell sought to append to Goodman's discussion: invoking habit, training, and inculcation alone is incapable of discriminating between competing kinds of visual depictions, or explaining why only certain visual tools get accorded special status. Dual diagrams should have been just as easy to teach to young theorists as Feynman diagrams: where one drew on Kirchhoff's laws, the other drew on Minkowski diagrams. Yet the dual diagrams, while easy to "acquire," did not interest most theorists. Thus we must ask what was *particular* about Feynman diagrams for the community of nuclear and particle theorists at midcentury, which might have further contributed to their staying power.

From Mnemonic to Mimetic: Reifying Feynman Diagrams

Historians have produced many examples of representational schemes that were introduced as mnemonic devices but that gained, often for the next generation of practitioners, an added sense of realism. Whether describing stratigraphic columns in eighteenth-century geology, Faraday's force lines in early nineteenth-century electromagnetism, chemical formulas in 1840s organic chemistry, indicator diagrams from 1860s steam engines, pictures of antibodies in turn-of-the-century immunology, or models of the earth's crust in 1920s isostasy, representations that had been developed as convenient ways to talk about the world came to be treated as pictures of how the world really was.[20] Some physicists, too, saw in Feynman diagrams more than just a convenient mnemonic device. Stephen Gasiorowicz proclaimed in his 1966 textbook, for example, that Feynman diagrams "not only serve as a simple device that keeps track of all the terms to be calculated, but also have associated with them a definite physical picture of the process. Such pictorial representations are sometimes useful in stimulating guesses of possible important contributions in a certain energy range."[21]

The next year, Richard Mattuck pressed the point even more strongly in his textbook. He included a special section to address the question of how to interpret Feynman diagrams. Under the boldface subheading "The 'quasi-physical' nature of Feynman diagrams," Mattuck explained that "Because of the unphysical properties of Feynman diagrams [that is, their depiction of 'virtual' processes that do not conserve energy], many writers do not give them any physical interpretation at all, but simply regard them as a mnemonic device for writing down any term in the perturbation expansion." This kind of nominalism was not sufficient for Mattuck: "However, the diagrams are so vividly 'physical-looking,' that it seems a bit extreme to completely reject any sort of physical interpretation whatsoever. As Kaempffer points out, one has to go back in the history of physics to Faraday's 'lines of force' if one wants to find a mnemonic device which matches Feynman's graphs in intuitive appeal. Therefore, we will here adopt a compromise attitude, i.e., we will 'talk about' the diagrams as if they were physical, but remember that in reality they are only 'apparently physical' or '*quasi-physical.*'"[22] Mattuck drew on F. A. Kaempffer's discussion rather freely. Whereas Mattuck summarized Kaempffer's description of Feynman diagrams' "intuitive appeal," Kaempffer himself had labeled their attraction a "propagandistic persuasiveness." In fact, Kaempffer had outlined the historical analogy to Faraday's work to serve as a "timely warning against all too literal acceptance of mental images based mainly on a fabric of *conventions*, however consistent that fabric may appear." Kaempffer warned that just as Maxwell was led to a literal acceptance of an ether by Faraday's pictures of

20. Rudwick, "Visual language" (1976); Gooding, "Novel observations" (1986); Gooding, "Mapping experiment" (1990); Gooding, "Imaginary science" (1994); Nersessian, "Reasoning from imagery" (1988); U. Klein, "Techniques of modeling" (1999); U. Klein, *Tools* (2001); U. Klein, *Experiments, Models, Paper Tools* (2003); Nye, *Chemical Philosophy* (1993), chap. 4; Brain and Wise, "Muscles and engines" (1994); Cambrosio, Jacobi, and Keating, "Ehrlich's 'beautiful pictures'" (1993); and Oreskes, *Rejection of Continental Drift* (1999), chap. 6.

21. Gasiorowicz, *Elementary Particle Physics* (1966), 134.

22. Mattuck, *Guide to Feynman Diagrams* (1967), 72; emphasis in original.

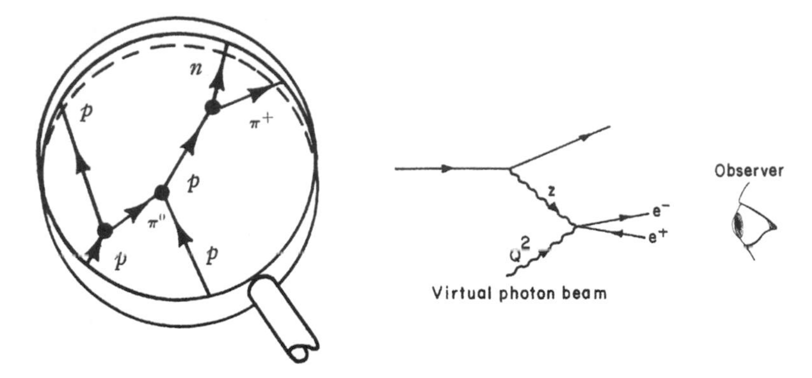

Figure 10.5. Observers watching Feynman diagrams. (Sources: *left*, Ford, *World of Elementary Particles* [1963], 221; *right*, Quigg, *Gauge Theories* [1983], 231.)

force lines, "similar temptations are lurking behind Feynman's graphs."[23] Kaempffer found such "temptations" potentially misleading; Mattuck delighted in the conventionalized diagrams, which looked so real they simply could not be of merely mnemonic value. Other physicists made this claim of realism visually. (See fig. 10.5.) Unlike the participants in these heated discussions, no one ever pushed for a physical interpretation of dual diagrams.

From where did the strong assertions and impassioned denials of Feynman diagrams' *reality* come? One locus to consider is the inundation of nuclear and particle physicists during exactly this period with nuclear-emulsion and bubble-chamber photographs. A "zoo" of unanticipated particles poured out of the postwar accelerators. To physicists in the 1950s and 1960s, photographs of these new particles from nuclear emulsions and bubble chambers were simply inescapable: they were published with articles in the *Physical Review*, on the covers of *Physics Today*, cataloged in huge atlases, reprinted on the dust jackets and frontispieces of textbooks and popularizations, and pictured on slides for public lectures. Even theorists with little connection to the experimental groups producing the pictures were awash in these photographs and their frequent reprintings.[24] Moreover, there developed in this period a highly

23. Kaempffer, *Concepts in Quantum Mechanics* (1965), 209; emphasis in original.

24. As a small sample, see Bernardini et al., "Nuclear interaction" (1950); Pickup and Voyvodic, "Cosmic-ray collisions" (1951); the cover of *PT*, vol. 8 (Aug 1955); and the dust jackets of Yang, *Elementary Particles* (1962); and Frazer, *Elementary Particles* (1966). The tradition of photographing particle tracks extends back to the turn of the twentieth century with C. T. R. Wilson's cloud chamber, but it was only after World War II that such photographs were produced so rapidly and in such volume, by the hundreds of millions, in European and American bubble chambers: Galison, *Image and Logic* (1997). For more on the atlases of cloud-chamber and bubble-chamber photographs, see also Galison, "Judgment against objectivity" (1998).

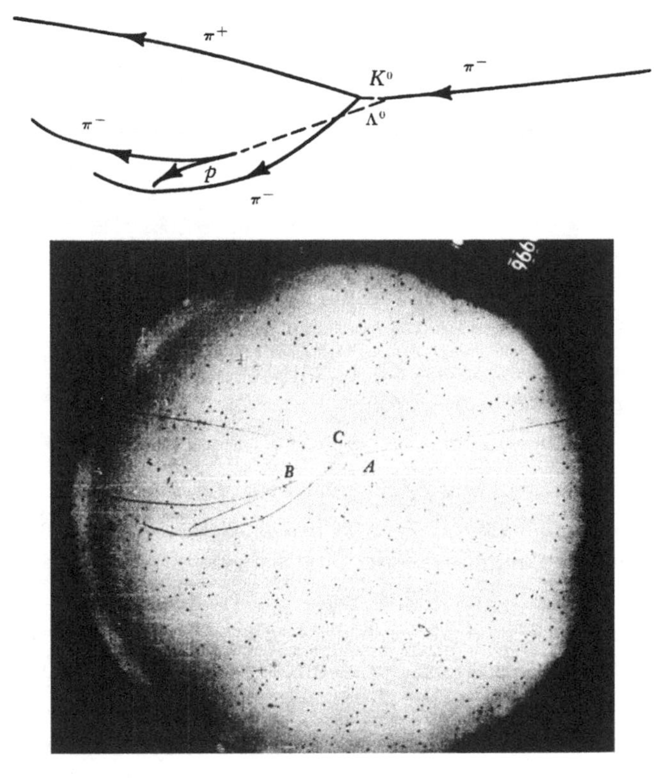

Figure 10.6. Bubble-chamber photograph together with schematic reconstruction. (Source: Ford, *World of Elementary Particles* [1963], 176.)

schematized tradition of reconstructing particles' paths, as photographed with the new detectors. These stick-figure reconstructions appeared with great frequency during the early 1960s in textbooks and lecture notes on high-energy physics. (See fig. 10.6.) While the nuclear-emulsion photographs were often published with only minimal lines and arrows superimposed, bubble-chamber photographs were usually published alongside stick-figure reconstructions of the specific events of interest.[25] Other times, physicists simply sketched these reconstructions freehand in their lecture notes and textbooks.[26] (See fig. 10.7.)

25. See, e.g., the undergraduate textbooks by Robert Leighton, *Principles of Modern Physics* (1959), 642–48, 683–87; and Novozhilov, *Elementary Particles* (1961), 40, 59, 61, 97, 102, 111, 133–34, 137, 154, 172. See also the popular books by Yang, *Elementary Particles* (1962), 22, 24, 30, 33, 45, 47; Ford, *World of Elementary Particles* (1963), 20–21, 24–25, 176, 182–83; and Gouiran, *Particles and Accelerators* (1967), 71, 82–83, 85–87, 104–5.

26. See, e.g., Novozhilov, *Elementary Particles* (1961), 139, 148, 156, and 176; Levi Setti, *Elementary Particles* (1963), 102, 109–10; Hamilton, *Elementary Particle Physics* (1964–65); and Livingston, *Particle Physics* (1968), 159, 164, 191.

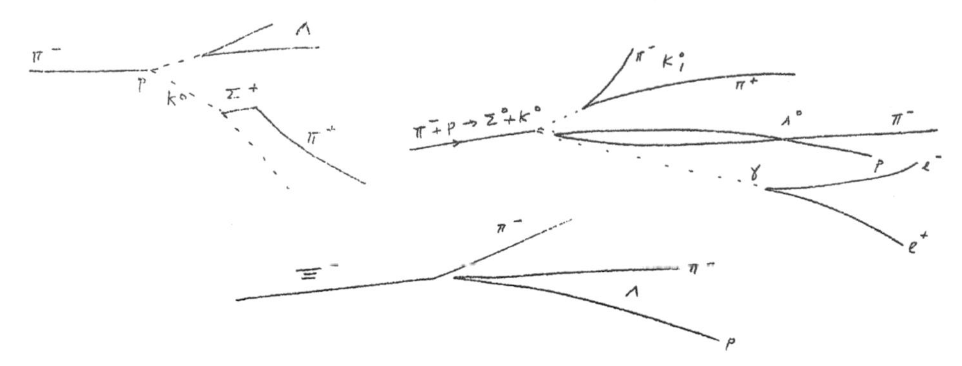

Figure 10.7. Freehand reconstruction of bubble-chamber photographs. (Source: Levi Setti, *Elementary Particles* [1963], 52, 60.)

At least one textbook introduced Feynman diagrams directly in between these photographs and their line-drawing reconstructions, tacitly placing Feynman diagrams within a continuous series of visual evidence about real particles' behavior.[27]

Just as few physicists made explicit comparisons between Feynman diagrams and Minkowski diagrams during this period, fewer still made any claim that what was pictured in Feynman diagrams and in these schematic bubble-chamber reconstructions was "the same thing." Yet the shared visual schemata suggest some possible connections. The bubble-chamber reconstructions were built from two key ingredients: vertices and propagation lines. The Feynman and Feynman-like diagrams that spread throughout the 1950s for meson calculations and studies of the strong interaction were *not* the high-order loop corrections (for which the diagrams had first been invented), but rather lowest-order, and most frequently single-particle, exchange diagrams. And what were the visual ingredients of these particular classes of Feynman diagrams? Nothing but vertices and propagation lines. (See fig. 10.8.)

The Feynman diagrams most often employed for tasks beyond QED were strikingly similar to the most heavily relied-upon evidence of actual particles and their interactions. The association of "realism" with Feynman diagrams in the 1950s and 1960s, based on their similarity to "real" photographs of "real"

27. Robert Leighton, *Principles of Modern Physics* (1959), 642–48, 683–87. Michael Lynch has highlighted the practice of publishing photographs of biological cells together with sparse line-drawing reconstructions as a means of bolstering the ontological authority of the latter while bringing epistemological clarity to the former: Lynch, "Mechanical reproduction" (1991). See also Miller, *Insights to Genius* (1996), 406–9.

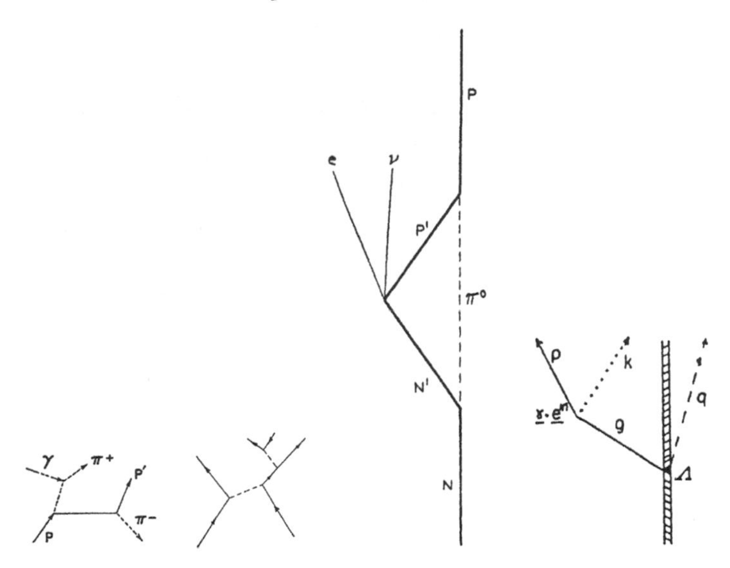

Figure 10.8. Feynman diagrams resembling freehand bubble-chamber photograph reconstructions. Compare with figure 10.7. (Sources, *left to right:* Lawson [A.100], 1273; Feldman [A.130], 1697; Finkelstein and Moszkowski [A.129], 1696; and Cutkosky [A.123], 1224.)

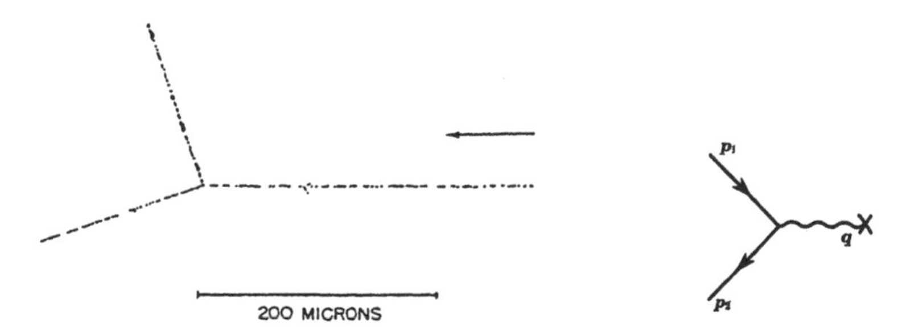

Figure 10.9. *Left,* nuclear emulsion track. *Right:* Feynman diagram. (Sources: *left,* Bernardini et al., "Nuclear interaction" [1950], 924; *right,* Schweber, Bethe, and de Hoffmann, *Mesons and Fields* [1955], 1:224.)

particles, helped Feynman diagrams stand out for many physicists. Unlike dual diagrams, Feynman diagrams could be read as more immediately related to real particles and processes, and hence less bound up with any particular abstract formalism. (See fig. 10.9.) No one had to proclaim that Feynman diagrams were "the same" as nuclear emulsions, bubble-chamber photographs, or their stylized reconstructions for visual affiliations to be made.

Feynman Diagrams Today

Feynman diagrams are even more ubiquitous today than during the 1950s or 1960s. Walking through the halls of nearly any physics department in the world, one will encounter the diagrams on physicists' blackboards, on their overhead transparencies for conference talks, and in their preprints, articles, lecture notes, and textbooks. To many physicists, the central place of Feynman diagrams today appears to be a natural or straightforward consequence of fast-moving events during the early 1970s. Following the discovery of "asymptotic freedom," according to this interpretation, Feynman diagrams once again seemed to be an unproblematic tool to be deployed much the way Dyson had originally elaborated twenty-five years earlier. If it was surprising that Feynman diagrams had played so many roles during the confusing and uncertain 1960s—so this line continues—there should be little surprise that they became steady fixtures during the 1970s and 1980s. Certain elements of this familiar narrative can indeed go a long way toward explaining the diagrams' ubiquity today. Yet matters are not quite so simple.

During the early and mid-1950s, several groups found that radiative corrections within QED would affect the measured value of a particle's electric charge. Pairs of virtual electrons and positrons, always popping into and out of existence, would effectively shield part of the original particle's charge. (See fig. 2.2.) The amount of shielding would depend on the distance scale: if the original particle were probed at very short distances, there would be fewer virtual pairs to shield the particle's charge, and the measured charge would be greater than if probed at longer distances.[28] Most physicists assumed that this behavior would carry over for other forces, including the strong nuclear force. The surprising result, first found by David Gross and Frank Wilczek at Princeton and independently by H. David Politzer at Harvard in 1973 (for which they shared the 2004 Nobel Prize), was that for a special class of models the original charge would be "antiscreened." This behavior became known as "asymptotic freedom": the strength of particles' interactions would *decrease* at small distances, approaching zero in the limit of infinitesimally short distances (or, equivalently, asymptotically large momenta), while growing at large distances. In the short-distance/large-momentum regime of these models, the

28. Stueckelberg and Petermann, "Normalisation des constantes" (1953); Gell-Mann and Low [A.124]; the brief papers by L. D. Landau, A. A. Abrikosov, and I. M. Khalatnikov, published in *Doklady Akademii Nauk SSSR* in 1954 and reprinted in Landau, *Collected Papers* (1965), 607–25; and Bogoliubov and Shirkov, *Theory of Quantized Fields* (1959 [1957]).

strong force became weakly coupled—and hence, at long last, the strong force finally became amenable to perturbative analysis.[29]

The discovery of asymptotic freedom might make it seem obvious that Feynman diagrams would now litter physicists' scratch pads and blackboards even more densely than before. After all, Dyson had domesticated Feynman's diagrams in the first place as aids for making perturbative calculations. Yet asymptotic freedom did not herald a straightforward return to Dyson's original techniques. For one thing, writing down the so-called Feynman rules proved formidable in the new models. To this day, translating Feynman diagrams into mathematical expressions remains highly gauge dependent, and selecting a suitable gauge can often prove difficult. Expressions that may ultimately be renormalizable often appear unrenormalizable in various gauges; "ghost" states proliferate and must be handled with care so that the final answers in a given calculation refer to physically meaningful quantities rather than gauge-dependent artifacts. Physicists spent *years* trying to work out the appropriate generalizations of Dyson's diagrammatic rules for the new kinds of gauge-theoretic models.[30] (See fig. 10.10.) Asymptotic freedom on its own did not make the actual use of Feynman diagrams any more automatic or obvious than had earlier appropriations of the diagrams.

Ernst Gombrich puzzled over the persistence of representational styles in his *Art and Illusion.* "Even after the development of naturalistic art," he wrote, "the vocabulary of representation shows a tenacity, a resistance to change, as if only a picture seen could account for a picture painted. The stability of styles in art is sufficiently striking to demand some such hypothesis of self-reinforcement."[31] So too with Feynman diagrams. Physicists at midcentury read Feynman diagrams against a confluence of visual associations. Minkowski diagrams and their pictorial conventions had long become standard features

29. The key difference from the QED case is that the force-carrying particles, or "gluons," carry nonzero charge for the force that they mediate. Thus, virtual gluon pairs must be included alongside virtual quark pairs when the effects of radiative corrections on the interaction's strength are calculated; and because gluons are bosonic while quarks are fermionic, the two contributions enter with opposite signs. If the number of types of quarks is small, then the negative-sign gluon contributions dominate. Gross and Wilczek, "Ultraviolet behavior" (1973); Politzer, "Reliable perturbative results" (1973); Cao, "New philosophy of renormalization" (1993); Schweber, "Changing conceptualization of renormalization theory" (1993); Shirkov, "Historical remarks" (1993); and Gross, "Asymptotic freedom" (1997).

30. Veltman, "Path to renormalizability" (1997); 't Hooft, "Renormalization" (1997).

31. Gombrich, *Art and Illusion* (1969 [1960]), 315; see also 3–30. Lüthy, "Atomist iconography" (2002), presents a fascinating study of the long-term persistence of pictorial conventions in atomic descriptions of matter.

Vertices:

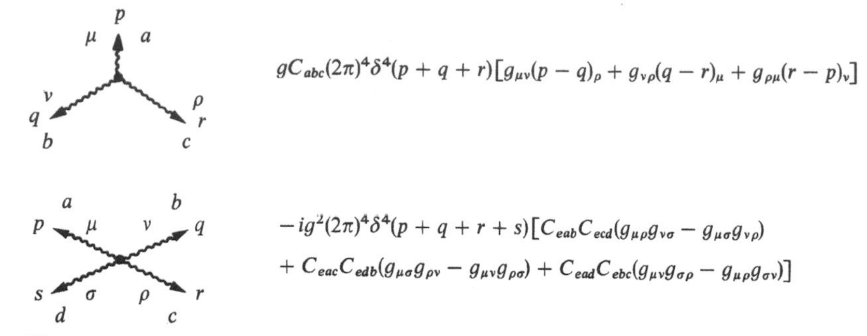

$$gC_{abc}(2\pi)^4\delta^4(p+q+r)[g_{\mu\nu}(p-q)_\rho + g_{\nu\rho}(q-r)_\mu + g_{\rho\mu}(r-p)_\nu]$$

$$-ig^2(2\pi)^4\delta^4(p+q+r+s)[C_{eab}C_{ecd}(g_{\mu\rho}g_{\nu\sigma} - g_{\mu\sigma}g_{\nu\rho}) + C_{eac}C_{edb}(g_{\mu\sigma}g_{\rho\nu} - g_{\mu\nu}g_{\rho\sigma}) + C_{ead}C_{ebc}(g_{\mu\nu}g_{\sigma\rho} - g_{\mu\rho}g_{\sigma\nu})]$$

Ghost-vector vertex:

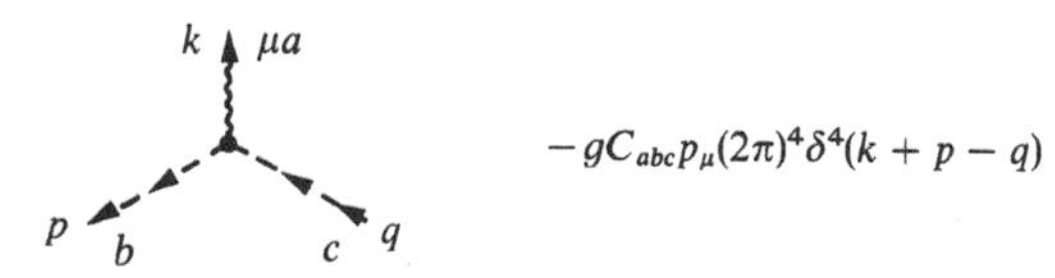

$$-gC_{abc}p_\mu(2\pi)^4\delta^4(k+p-q)$$

Fermion-vector vertex:

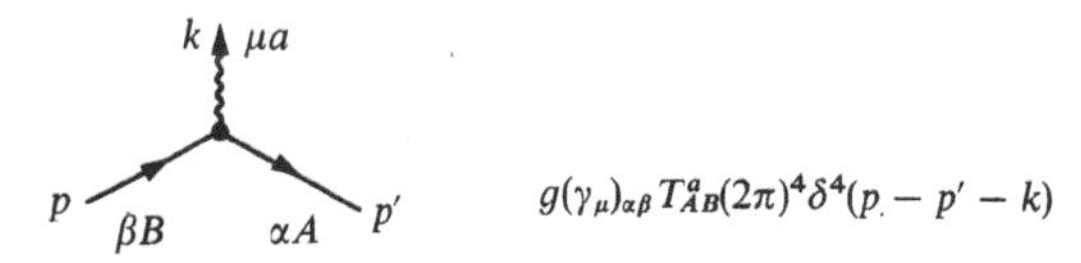

$$g(\gamma_\mu)_{\alpha\beta}T^a_{AB}(2\pi)^4\delta^4(p-p'-k)$$

Scalar-vector vertices:

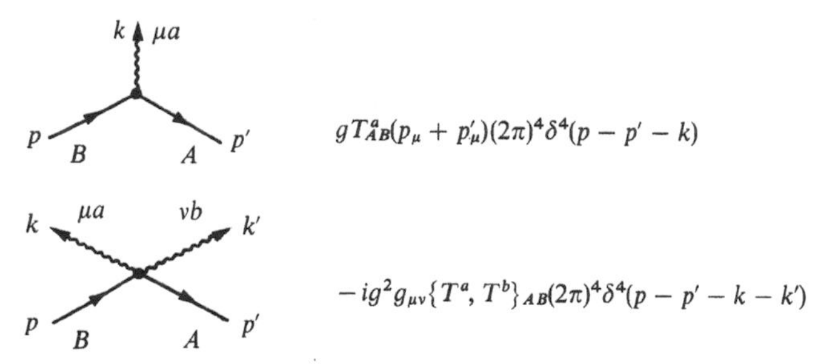

$$gT^a_{AB}(p_\mu + p'_\mu)(2\pi)^4\delta^4(p-p'-k)$$

$$-ig^2g_{\mu\nu}\{T^a, T^b\}_{AB}(2\pi)^4\delta^4(p-p'-k-k')$$

with T^a antihermitian.

Figure 10.10. Feynman rules for gauge field theories. Compare with figure 3.2. (Source: Itzykson and Zuber, *Quantum Field Theory* [1980], 700.)

of young physicists' training: students at Ithaca, Berkeley, Tokyo, and Moscow practiced drawing Minkowski diagrams in similar ways by the 1940s. Yet other visual devices put forward at this time, such as dual diagrams, were equally easy to acquire, sharing an explicit tie to Kirchhoff's rules for elementary circuit theory. Despite this habit-forming potential, dual diagrams quickly faded from view, while Feynman diagrams continued to define physicists' practice.

Unlike the dual diagrams, Feynman diagrams could evoke in an unspoken way the scattering and propagation of real particles, with "realist" associations for those physicists already awash in a steady stream of nuclear-emulsion and bubble-chamber photographs. In more recent decades, physicists have turned to Feynman diagrams with more of Dyson's original motivation in mind. Though relying less than earlier theorists had on Minkowskian codes or similarities to bubble-chamber photographs, physicists nonetheless had to work hard to fashion Feynman's diagrams into reliable tools.

IN SEARCH OF THE VANISHING SCIENTIFIC THEORY

Twenty years ago, the sociologist Andrew Pickering offered a trenchant observation: no matter how hard historians, philosophers, and sociologists might look, they will never find a quark in the midst of the materials they study. They will find notes and calculations, diagrams and plots of data, memoranda, letters, articles, and textbooks—including materials that physicists considered significant for whether quarks exist—but not the phenomena themselves. His main point: rather than "putting the phenomena first," we should focus on the materials upon which physicists drew in forming their interpretations and beliefs.[32] A similar caution should guide our studies of recent theoretical physics. Try as we might, we will never come across a "theory" in the flotsam and jetsam of our sources—and thus we should be wary of letting the categories of "theory construction and selection" direct our historical analysis. Instead, when we inspect the materials with which theoretical physicists have worked, night and day, we see tinkering and appropriation of paper tools—tools fashioned, calculations made, approximations clarified, results compared with data, interpretations advanced, analogies extended to other types of calculations or phenomena, and so on. "Theories" do not appear, nor is it clear where they might even be found.

First a few clarifications. I am interested here in the empirical question of whether or where historians of postwar physics might find scientific theories—especially in the case of American physicists, who seem to have spent so little effort trying to articulate specific theories in the first place. The broader philosophical question of whether scientific theories *ever* exist (at least in any forms resembling common philosophical depictions) remains beyond my present scope. Furthermore, as we have seen throughout this study, physicists often

32. Pickering, "Against putting the phenomena first" (1984). Pickering's example in this article was weak neutral currents, though a similar methodological impulse guides his study of the history of quark models: Pickering, *Constructing Quarks* (1984).

used the term "theory" in loose ways, such as when referring to collections of techniques, as in "perturbation theory." With this usage, nearly everything that theorists did would qualify as a "theory." "Perturbation theory," "quantum field theory," and similar terms denote *frameworks* for making calculations or modeling phenomena. Both are collections of (ever-changing) methods rather than stable sets of axioms—a point that many physicists have acknowledged in the light of recent developments in "effective field theories."[33] I am not after this loose usage of the word "theory," but rather the ways that historians and philosophers have tended to use the term. When discussing the dispersion of paper tools such as Feynman diagrams, "theories" hardly enter at all.

This framework runs at odds with two of the major movements within the history and philosophy of science, which dominated studies for much of the twentieth century. To the logical positivists, such as Rudolf Carnap and Carl Hempel, scientific theories existed as stable objects to which scientists could point and between which they could select. At theories' core was a set of axioms from which statements about phenomena could be deduced and compared—with the aid of various "correspondence rules" or "bridge principles"—with the data of the senses. Elaborate formalisms emerged that were supposed to govern the selection among competing theories: candidate theories could be enumerated seriatim, T_1, T_2, T_3, and the big guns of logical calculi brought to bear.[34] To the antipositivist critics, including Norwood Russell Hanson, Thomas Kuhn, and Paul Feyerabend, the primary entity that determined what scientists could measure or analyze was the ingoing worldview or paradigm. Paradigms, and the theories they supposedly incorporated, could not be selected between as freely as in the logical positivists' accounts—to Kuhn and his followers, tremendous conceptual ruptures greeted the cessation of one paradigm and the Gestalt-like imposition of another. Yet at the center of scientists' work, in the antipositivists' account, was a set of communally shared theories—hence Hanson's and Kuhn's emphasis upon the theory-ladenness of observations and the supposed incommensurability of concepts belonging to rival theories or paradigms.[35] To both the logical positivists and their

33. Weinberg, "Changing attitudes" (1997), 41–42; Cao, "New philosophy of renormalization" (1993); and Schweber, "Changing conceptualization of renormalization theory" (1993).

34. Frederick Suppe subjected this "received view" to its most extensive autopsy: Suppe, "Search for philosophic understanding" (1977). See also Boyd, "Confirmation" (1991); and Sklar, *Scientific Theory* (2000).

35. Hanson, *Patterns of Discovery* (1958); Hanson, *Concept of the Positron* (1963); Kuhn, *Structure* (1996 [1962]); and Feyerabend, *Against Method* (1975). Cf. Rouse, *Knowledge and Power* (1987), chap. 2.

antipositivist critics, the construction and selection of theories seemed central to the scientific enterprise.

More recently, philosophers and historians have challenged both of these notions of scientific theories. Inspired in part by Nancy Cartwright's and Ian Hacking's work, several scholars have highlighted the importance of modeling and simulation in modern science—activities that fitted rather poorly with either the logical positivists' or the antipositivists' claims about the structure of scientific theories.[36] In place of the logical positivists' deductive-nomological view of theories (often called the "syntactic view"), which cast scientific theories as axiomatic systems made up of governing laws and deduced consequences, many of these authors describe scientific theories as collections of models (the "semantic view"), focusing on the ways in which scientists fashion models and use them to mediate between concepts and the world. The question remains whether any of these views of scientific theories helps to make sense of physicists' appropriations of Feynman diagrams. For example, do either the "theories as axiomatic systems" or the "theories as collections of models" frameworks adequately describe Geoffrey Chew's diagrammatic *S*-matrix program?

James Cushing and Tian Yu Cao have analyzed the history of physicists' efforts to make sense of the strong nuclear force during the 1950s and 1960s in terms of the construction and selection of rival theories. To Cushing, Chew's *S*-matrix program—often labeled the "*S*-matrix theory" in Cushing's analysis—was in "competition" with quantum field theory; offshoots from each branch became "radically different competitor[s]" between which physicists sought to select. Cao similarly examines physicists' "attempts at reconciling quantum field theory with *S*-matrix theory."[37] Yet were there really two rival theories between which physicists strove to choose? Did physicists at the time really face two objects—T_1 and T_2—and devise means of either "reconciling" them or pitting them against each other so that only one vanquishing theory remained? As we have seen in chapter 8, Chew's enthusiastic claims on behalf of a new

36. See esp. Cartwright, *How the Laws of Physics Lie* (1983); Hacking, *Representing and Intervening* (1983); Giere, *Explaining Science* (1988), esp. chap. 3; Nersessian, "Theoretician's laboratory" (1993); Oreskes, Shrader-Frechette, and Belitz, "Verification, validation, and confirmation" (1994); Cartwright, Shomar, and Suárez, "Tool box of science" (1995); Cartwright, "Models" (1997); Cartwright, *Dappled World* (1999); Giere, *Science without Laws* (1999); Nersessian, "Model-based reasoning" (1999); Sismondo, *Modeling and Simulation* (1999); Morgan and Morrison, *Models as Mediators* (1999); E. Keller, "Models of and models for" (2000); and Oreskes, "Quantitative models" (2003).

37. Cushing, *Theory Construction* (1990), xv, 169–73, 255–56, 262, 289; Cao, "Reggeization program" (1991).

"theory" emerged post hoc, after a long stream of diagrammatic improvisation. New ways of manipulating Feynman diagrams came first, and only late in the game did Chew bundle many of these tool-based techniques together to call them a new "theory." Moreover, it remained a "theory" without clear axioms or foundations, as Chew repeatedly made clear throughout the early and mid-1960s (pace the "theories-as-axioms" view). Perhaps, Cushing and Cao might have countered, Chew's S-matrix program constituted a "theory" of the other kind, namely a collection of models. Yet here, too, "theory" talk obscures more than it clarifies. Even the semantic view of theories assumes that theories are in some sense "out there"—delimitable objects at which scientists can point and between which they can choose. As we have seen in chapter 9, however, theorists outside Chew's orbit rarely recognized Chew's program as a single "theory." Treiman's and Goldberger's graduate students at Princeton, for example, treated Chew's work as yet another collection of handy calculating techniques through which they could sort—more yard sale than theory. The tools that worked could stay (and be further reinterpreted), with little attention to the tethers that might have bound them together in Chew's packaging. While certainly an improvement over the deductive-nomological view, even the semantic view of theories offers only limited help for making sense of physicists' uses of Feynman diagrams.

Where else might we hope to find theories—of either the logico-axiomatic or the collections-of-models kind—during the decades after World War II? Certainly not in American textbooks: as we have seen in chapter 7, textbooks during the 1950s and 1960s routinely threw together techniques of mixed conceptual heritage, encouraging students to apply an approximation based on nonrelativistic potential scattering here, a lowest-order Feynman diagram there. As James Bjorken and Sidney Drell announced in the preface to their twin textbooks from 1964 and 1965, techniques such as Feynman diagrams would likely "outlive the elaborate mathematical structure" of quantum field theory, which would come to be "viewed more as superstructure than foundation." Hence their emphasis on equipping physicists with durable "bags of tricks."[38] Coherent theories were certainly not bound into the pages of physicists' textbooks.

Nor do we find theories articulated in research articles or lengthy reviews from the period. Many physicists offered a specific reading strategy for wading through their publications. Chew, Goldberger, Low, and Nambu introduced

38. Bjorken and Drell, *Relativistic Quantum Mechanics* (1964), viii; and Bjorken and Drell, *Relativistic Quantum Fields* (1965), viii.

their famous 1957 dispersion-relations work by instructing their readers, "The algebra in [section 3 of the article] is quite complicated. We advise the interested reader (as opposed to the dedicated one) to read up to and including Eq. (3.18), by which time *the method of calculation should be clear.*" Two years later, Chew gave a similar reader's guide to the literature in a review article: "For the reader who wishes to see all the essential steps in a complete and yet economical derivation of the pion-nucleon dispersion relation, the following use of the published literature is recommended": readers were to read half of one article, skip to a different paper, and then, "If any strength remains, read the Dyson paper." The interest in this work, Chew emphasized, was in ways to use dispersion relations "as a *tool* for strong coupling physics." Such tools did not derive from full-blown theories: "At present it remains true that the methods of implementation of dispersion relations are elementary and quite unrelated to the sophisticated mathematical techniques required for their derivation." Or, as Murph Goldberger put it somewhat more strongly a few years later, "The derivation of non–forward scattering dispersion relations, first given in an unbelievable tour de force by Bogoliubov [in 1958, several years after such calculations had become common], are ugly, involved, unrewarding, and uninstructive."[39] At least for these American theorists, tools trumped derivations (let alone "theories") every time; nor did the former follow in any natural way from the latter.

We would similarly search in vain to find theories lurking around physicists' crowded conference halls or summer school lecture circuits. Goldberger echoed Chew's sentiments about the independence of tools from theories at the 1961 Solvay conference in Brussels. Put on the defensive from the previous day's discussion, Goldberger forthrightly acknowledged that in either the Lagrangian or axiomatic formulations of quantum field theory, one could produce "the whole theory . . . based on only the finest postulates." On the other hand, Goldberger jousted, the axiomaticians "are very hard pressed to compute the Klein-Nishina formula" for electron-photon scattering, and the "Lagrangian people" remained similarly handicapped for any quantitative discussion of the strong interactions. Goldberger remained resolute: "Although the axioms, if you like, of the dispersion approach have not been stated in what we might imagine to be their final form, no one is really ever in much doubt as to how to proceed." Chew agreed. In his lectures at the Cargèse summer school the next year he explained that "Although the logical structure of the

39. Chew et al., "Application of dispersion relations" (1957), 1337; emphasis added; Chew, "Pion-nucleon interaction" (1959), 33; emphasis added; and Goldberger, "Single variable dispersion relations" (1961), 192.

theory is not yet clear, in practice one already knows what to do in any specific situation." An exasperated Richard Feynman responded to what he sensed was too much "theory talk" at the 1961 Solvay conference, chastising his fellow theorists: "You sit there and say: why isn't everybody doing S-matrix; another guy says: why isn't anybody doing field theory? The real problem is: *why is nobody solving anything?*"[40]

Even for Chew, tools trumped theories and functioned independently of them. He began his 1964 lectures by reminding listeners that in his 1961 *S-Matrix Theory of Strong Interactions,* the "hope was expressed that an axiomatic basis for *S*-matrix theory would soon develop. Since that time," Chew now reported, "there has been progress in the direction of axioms but a solid foundation is lacking." Yet that was hardly reason to delay publication of his new lecture notes: "Strong interaction physics nevertheless has continued to evolve, even without axioms, and my contribution to this book may be regarded as a sequel to the previous lectures written in the same spirit. Students seeking a well-defined theory will not find it here; they will find a certain point of view."[41] The "point of view" that followed centered squarely upon new diagrammatic procedures, culminating in nuclear democracy and bootstrap calculations. Chew's former postdoc Steven Frautschi captured this approach best in his 1963 lecture notes: "There is a story about the student who wanted to know, 'Can one prove the Mandelstam representation from field theory?' He went to Weisskopf who responded, 'Field theory, what is field theory?' Then he sought out Wigner who said, 'Mandelstam, who is Mandelstam?' Finally our persistent student found his way to Chew, repeated his question, and heard, 'Proof, what is proof?'"[42]

A few years later, another theorist spelled out even more explicitly the roles of paper tools and theoretical techniques for practitioners in the field. Richard Eden worked as a theorist at the Cavendish Laboratory in Cambridge, England; he had been working on topics in strong-interaction physics since the early 1950s and had teamed up with several British colleagues to pursue work similar to that of Chew's Berkeley group. One year after publishing a long textbook with his colleagues, *The Analytic S-Matrix* (1966), Eden wrote another textbook, *High Energy Collisions of Elementary Particles* (1967). He began by surveying the theoretical terrain; this was a terrain marked by specific tools and techniques, rather than by particular theories. "Theoretical methods for

40. Goldberger, "Single variable dispersion relations" (1961), 179–80; Chew, "Strong-interaction" (1963), 7; and Feynman as quoted in Stoop, *Quantum Theory of Fields* (1961), 178; emphasis in original.

41. Chew, "Nuclear democracy" (1964), 104.

42. Frautschi, *Regge Poles* (1963), 63.

studying high energy collisions can be divided into the following categories," he explained:

> (1) Standard techniques such as those of relativistic kinematics or of helicity state analysis.
> (2) Methods that make use of rigorous consequences of quantum field theory. . . .
> (3) Methods that have been developed from assumptions not yet fully proved from fundamental axioms but which involve techniques whose importance will surely remain even if there is some change in their motivation. These include dispersion theory and the Mandelstam representation, and Regge theory using complex angular momentum.
> (4) Theoretical models which can be expected to apply to a limited range of experimental results and to evolve as one begins to understand more clearly the essential approximations involved. These include the peripheral model and simple approximations to Regge theory.
> (5) The use of symmetries. These include crossing symmetry, the use of spin and isospin groups and of SU3.
> (6) Special approximations, for example those based on perturbation theory, perhaps with the addition of final state interactions.[43]

Little in the way of specific theories jumps out from the list. He went on to emphasize the importance of many of these methods as exemplars: "Although [the Mandelstam representation] has not been proved from fundamental axioms," for example, "it provides a useful illustration of analytic properties and crossing symmetry relations, many of which would remain correct even if some aspect of the Mandelstam representation (such as the subtraction assumption) turns out to be incorrect." Students should learn to use the Mandelstam representation because that set of techniques provided important practice in exploiting analyticity and crossing symmetry—not because the Mandelstam representation was itself a "theory," much less because it had any clear relation with any given "theory." The same was true, for Eden, of Regge poles: "Although some of the special assumptions may not survive critical analysis, it is certain that the methods of complex angular momentum will remain amongst the most useful techniques for parameterising and studying high energy collisions."[44] Certain techniques were likely to remain "useful," come what may—and hence, students ought to busy themselves practicing how to put those techniques to work.

Cushing's and Cao's accounts of a clash between the *S*-matrix program and quantum field theory, and of physicists' need to select between rival "theories,"

43. Eden, *High Energy Collisions* (1967), 2; cf. Eden et al., *Analytic S-Matrix* (1966).

44. Eden, *High Energy Collisions* (1967), 3–4.

were colored by later developments. A conventional story often told—with more than a whiff of Whiggish triumphalism—is that a new kind of quantum field theory rose up and defeated Chew's *S*-matrix program during the early 1970s, culminating by the end of the decade in the "Standard Model" of elementary particle physics.[45] The Standard Model is indeed formulated in a quantum-field-theoretic framework and might best be described as a collection of models sharing some common features and covering a broad range of phenomena. Yet the rise of the Standard Model testifies to no simple return to a given theory. Rather, since the 1980s the term "quantum field theory" has been used to pick out and label a set of skills distinct from those common when Dyson first began to yoke Feynman diagrams to a field-theory framework. A new generation of graduate courses and textbooks on quantum field theory now treats the topic as a playground for teaching new calculational techniques; Feynman diagrams take their place within a reconfigured theoretical toolbox. Students in today's courses practice how to construct gauge-covariant derivatives, how to manipulate global and local gauge transformations, how to derive field equations from path integrals, and so on. The day-to-day work of being a theorist today, and the specific techniques drawn upon in making calculations, bears limited resemblance to the grab bags of techniques pulled together in the earlier field-theoretic textbooks.[46] The label "quantum field theory" might be the same, but how that label is given operational meaning for students and practitioners has changed dramatically. Today's "quantum field theory" has its own set of paper tools.

At the level of calculational tools, moreover, the Standard Model hardly vanquished its supposed competitor, Chew's *S*-matrix program. The problems with which physicists have struggled for decades—for example, how to calculate the structure and behavior of nuclei based on the strong forces binding protons and neutrons together—remain no better answered by the Standard Model than by any previous frameworks.[47] For studying these types of problems, theorists still often turn to the techniques bundled together within

45. See esp. Hoddeson et al., *Rise of the Standard Model* (1997); Pickering, *Constructing Quarks* (1984); Crease and Mann, *Second Creation* (1986); Riordan, *Hunting of the Quark* (1987); and Fitch and Rosner, "Elementary particle physics" (1995).

46. Cf., e.g., Itzykson and Zuber, *Quantum Field Theory* (1980); Ryder, *Quantum Field Theory* (1985); Lowell Brown, *Quantum Field Theory* (1992); and Peskin and Schroeder, *Quantum Field Theory* (1995); with Jauch and Rohrlich, *Theory of Photons and Electrons* (1955); Mandl, *Quantum Field Theory* (1959); Schweber, *Relativistic Quantum Field Theory* (1961); and Bjorken and Drell, *Relativistic Quantum Fields* (1965).

47. David Gross emphasizes this point in Gross, "Asymptotic freedom" (1997), 223. See also Hartmann, "Models and stories" (1999).

Chew's program. Recent preprints exploiting dispersion relations, Regge poles, pomeron exchanges, and the like show that when making actual calculations, tools speak louder than theories. Chew's *S*-matrix program might have been declared dead, but many of its techniques live on.[48] The most telling sign of the *S*-matrix program's afterlife comes from the transmogrification of Gabriele Veneziano's 1968 "duality" model for nuclear-particle scattering—which he worked out as a concrete manifestation of Chew's bootstrap approach, combining diagram-based crossing symmetry with Regge trajectory behavior—into the first model of quantized strings.[49] Today's superstring theories owe their existence to the same kinds of paper-tool bricolage as did the earlier theoretical programs studied in this book. Feynman diagrams, suitably tweaked yet one more time, continue to prosper in the new work. (See fig. 10.11.) Feynman diagrams live on within the Standard Model and its more recent cousins, but not because one theory has defeated another.

Given the continuing relevance of Chew's paper tools, it is worth revisiting Kuhn's well-known observations about the functions of "exemplars" and "paradigms." Kuhn introduced the term "paradigm" in 1959 to mean what he later came to call "exemplars": typical problems, the completion of which helped the student to build up a cache of certain calculational tools, techniques, and skills.[50] Of course, this first sense of "paradigm" hardly exhausted Kuhn's use of the term. Margaret Masterman chided Kuhn as early as 1967, charging that no fewer than twenty-one distinct senses of "paradigm" lurk within Kuhn's famous *Structure of Scientific Revolutions.* Kuhn reminisced ten years later on the many uses of the term within his earlier work. In his recollections, the word "paradigm" assumed an eerie self-directing agency:

> The concept of paradigms proved to be the missing element I required in order to write the book [*Structure*], and a first full draft was prepared between the summer of 1959 and the end of 1960. Unfortunately, in that process, paradigms took on a life of their own, largely displacing the previous talk of consensus. Having begun

48. A sample of recent papers includes Aznauryan, "Multipole amplitudes" (2003); Baldini et al., "Form factors" (2000); Hurtado, Morales, and Quimbay, "Dispersion relations" (2000): 373–81; Brisudova et al., "Nonlinear Regge trajectories" (2003); Sibirtsev et al., "Systematic Regge theory analysis" (2003); Simonov, "Spectrum and Regge trajectories" (2002); Lengyel and Machado, "Two pomeron description" (2003); and Shuryak and Zahed, "Double pomeron production" (2003).

49. Veneziano, "Construction" (1968); similar work appeared in Dolen, Horn, and Schmid, "Finite-energy sum rules" (1968); my thanks to Denyse Chew for bringing this paper to my attention (e-mail to the author, 7 Sep 2003). See also Cushing, *Theory Construction* (1990), 193–96; Green, Schwarz, and Witten, *Superstring Theory* (1987), vol. 1, chap. 1; Polchinski, *String Theory* (1998), 1:178–84. For a popular account, see Greene, *Elegant Universe* (1999).

50. Kuhn, "Essential tension" (1977 [1959]), esp. 229.

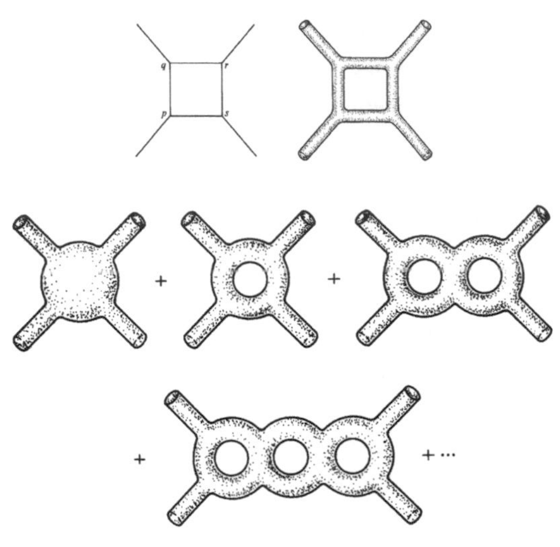

Figure 10.11. *Top row,* Feynman diagrams for point particles (*left*) and strings (*right*). *Bottom two rows,* expanding a string amplitude diagrammatically. (Source: Green, Schwarz, and Witten, *Superstring Theory* [1987], 1:30–31.)

> simply as exemplary problem solutions, they expanded their empire to include, first, the classic books in which these accepted examples initially appeared and, finally, the entire global set of commitments shared by the members of a particular scientific community. That more global use of the term is the only one most readers of the book have recognized, and the inevitable result has been confusion: many of the things there said about paradigms apply only to the original sense of the term. Though both senses seem to me important, they do need to be distinguished, and the word 'paradigm' is appropriate only to the first. Clearly, I have made unnecessary difficulties for many readers.[51]

Kuhn thought that the blurring of distinctions between "paradigm" as exemplar and "paradigm" as *Weltbild* was unfortunate because the two notions, though equally important, required differentiation. I find Kuhn's blurring of distinctions unfortunate for a different reason: it privileged a certain image of scientific practice—stemming from his paradigm-as-*Weltbild* line—that obscures more than it clarifies of the history of recent physics.

* * *

Feynman diagrams have become ubiquitous and routine today, and are applied in nearly every branch of physics. What gets lost from view in today's

51. Kuhn, "Preface," in Kuhn, *Essential Tension* (1977), ix–xxiii, on xix–xx; cf. Masterman, "Nature of a paradigm" (1970). Masterman's essay was first delivered in 1967. See also Rouse, *Knowledge and Power* (1987), chap. 2.

sea of Feynman diagrams is the diagrams' own historicity—and, even more important, the tremendous pedagogical apparatus which, springing up soon after World War II, allowed the new calculational techniques to spread and flourish. The diagrams' very success today hides both the variety of uses to which they had earlier been put and the work required for students to acquire, adopt, and subtly adapt these skills. Feynman diagrams do not occur in nature; and theoretical physicists are not born, they are made. During the middle decades of the twentieth century, both were fashioned as part of the same pedagogical process.

APPENDIX A

Feynman Diagrams in the *Physical Review,* 1949–54

Entries marked with an asterisk discussed elements of the diagrams, but did not include explicit figures. (L) indicates that the article was published as a brief letter to the editor.

1949

1. F. J. Dyson. "The radiation theories of Tomonaga, Schwinger, and Feynman." *PR* 75 (1 Feb 1949): 486–502.

2. F. J. Dyson. "The *S* matrix in quantum electrodynamics." *PR* 75 (1 June 1949): 1736–55.

*3. P. T. Matthews. "The application of Dyson's methods to meson interactions." *PR* 76 (1 Sep 1949): 684–85 (L).

4. R. P. Feynman. "Space-time approach to quantum electrodynamics." *PR* 76 (15 Sep 1949): 769–89.

5. K. M. Watson and J. V. Lepore. "Radiative corrections to nuclear forces in the pseudoscalar meson theory." *PR* 76 (15 Oct 1949): 1157–63.

6. J. Steinberger. "On the use of subtraction fields and the lifetimes of some types of meson decay." *PR* 76 (15 Oct 1949): 1180–86.

*7. D. Feldman. "On realistic field theories and the polarization of the vacuum." *PR* 76 (1 Nov 1949): 1369–75.

8. C. Morette. "On the production of π-mesons by nucleon-nucleon collisions." *PR* 76 (15 Nov 1949): 1432–39.

1950

*9. J. C. Ward. "The scattering of light by light." *PR* 77 (15 Jan 1950): 293 (L).

10. F. Rohrlich. "The self-stress of the electron." *PR* 77 (1 Feb 1950): 357–60.

11. R. Karplus and N. M. Kroll. "Fourth-order corrections in quantum electrodynamics and the magnetic moment of the electron." *PR* 77 (15 Feb 1950): 536–49.

*12. J. C. Ward. "An identity in quantum electrodynamics." *PR* 78 (15 Apr 1950): 182. (L)

13. J. Ashkin, T. Auerbach, and. R. Marshak. "Note on a possibile annihilation process for negative protons." *PR* 79 (15 July 1950): 266–71.

14. J. C. Ward. "A convergent non-linear field theory." *PR* 79 (15 July 1950): 406 (L).

15. A. Simon. "Bremsstrahlung in high energy nucleon-nucleon collisions." *PR* 79 (15 Aug 1950): 573–76.

16. M. N. Rosenbluth. "High energy elastic scattering of electrons on protons." *PR* 79 (15 Aug 1950): 615–19.

17. K. A. Brueckner. "The production of mesons by photons." *PR* 79 (15 Aug 1950): 641–50.

*18. A. Salam. "Differential identities in three-field renormalization problem." *PR* 79 (1 Sep 1950): 910–11 (L).

*19. C. N. Yang and D. Feldman. "The S-matrix in the Heisenberg representation." *PR* 79 (15 Sep 1950): 972–78.

20. R. Jost, J. M. Luttinger, and M. Slotnick. "Distribution of recoil nucleus in pair production by photons." *PR* 80 (15 Oct 1950): 189–96.

*21. G. C. Wick. "The evaluation of the collision matrix." *PR* 80 (15 Oct 1950): 268–72.

*22. P. T. Matthews. "Spinless mesons in the electromagnetic field." *PR* 80 (15 Oct 1950): 292 (L).

*23. P. T. Matthews. "Spinless mesons and nucleons in the electromagnetic field." *PR* 80 (15 Oct 1950): 292–93 (L).

24. Robert Karplus and Maurice Neuman. "Non-linear interactions between electromagnetic fields." *PR* 80 (1 Nov 1950): 380–85.

*25. C. B. van Wyk. "Selection rules for closed loop processes." *PR* 80 (1 Nov 1950): 487–88 (L).

26. F. Rohrlich. "Quantum electrodynamics of charged particles without spin." *PR* 80 (15 Nov 1950): 666–87.

*27. N. Hu. "The *S*-matrix in meson theory." *PR* 80 (15 Dec 1950): 1109–10 (L).

1951

*28. W. A. Newcomb and F. Rohrlich. "On the Lamb shift for spinless electrons." *PR* 81 (15 Jan 1951): 282–83 (L).

29. M. Peshkin. "Scattering and absorption of scalar and pseudoscalar mesons by nucleons." *PR* 81 (1 Feb 1951): 425–29.

*30. F. Coester. "On the evaluation of the *S*-matrix." *PR* 81 (1 Feb 1951): 455–56 (L).

*31. P. T. Matthews. "Renormalization of neutral mesons in three-field problems." *PR* 81 (15 Mar 1951): 936–39.

32. A. Salam. "Overlapping divergences and the *S*-matrix." *PR* 82 (15 Apr 1951): 217–27.

*33. F. J. Dyson. "Heisenberg operators in quantum electrodynamics." Pt. 1. *PR* 82 (1 May 1951): 428–39.

*34. J. S. R. Chisholm. "Calculation of matrix elements." *PR* 82 (1 May 1951): 448 (L).

*35. A. Petermann and E. C. G. Stueckelberg. "Restriction of possible interactions in quantum electrodynamics." *PR* 82 (15 May 1951): 548–49 (L).

36. K. Brueckner. "Production of π-mesons in nucleon-nucleon collisions." *PR* 82 (1 June 1951): 598–606.

*37. K. V. Roberts. "An equivalence theorem in meson theory." *PR* 83 (1 July 1951): 188–89 (L).

*38. F. J. Dyson. "Heisenberg operators in quantum electrodynanics." Pt. 2. *PR* 83 (1 Aug 1951): 608–27.

39. M. F. Kaplon. "The contribution of the Pauli moment to π-meson production by photons." *PR* 83 (15 Aug 1951): 712–15.

40. D. B. Beard and H. A. Bethe. "Field corections to neutron-proton scattering in a new mixed meson theory." *PR* 83 (15 Sep 1951): 1106–14.

41. R. M. Frank. "The fourth-order contribution to the self-energy of the electron." *PR* 83 (15 Sep 1951): 1189–93.

*42. F. J. Dyson. "The Schrödinger equation in quantum electrodynamics." *PR* 83 (15 Sep 1951): 1207–16.

43. M. Gell-Mann and F. Low. "Bound states in quantum field theory." *PR* 84 (15 Oct 1951): 350–54.

*44. A. Salam. "Divergent integrals in renormalizable field theories." *PR* 84 (1 Nov 1951): 426–31.

45. M. Baranger. "Relativistic corrections to the Lamb shift." *PR* 84 (15 Nov 1951): 866–67 (L).

46. J. C. Ward. "Renormalization theory of the interactions of nucleons, mesons, and photons." *PR* 84 (1 Dec 1951): 897–901.

47. E. E. Salpeter and H. A. Bethe. "A relativistic equation for bound-state problems." *PR* 84 (15 Dec 1951): 1232–42.

1952

48. M. Neuman. "Eigenvalue problem in quantum electrodynamics." *PR* 85 (1 Jan 1952): 129–33.

49. M. Ruderman. "Decay of the π-meson." *PR* 85 (1 Jan 1952): 157 (L).

50. L. M. Brown and R. P. Feynman. "Radiative corrections to Compton scattering." *PR* 85 (15 Jan 1952): 231–44.

51. F. J. Belinfante. "Problems connected with the prohibition of self-interactions in integro-causal quantum field-theory." *PR* 85 (1 Feb 1952): 468–73.

*52. C. A. Hurst. "The graphs for the kernel of the Bethe-Salpeter equation." *PR* 85 (1 Mar 1952): 920 (L).

*53. M. F. Kaplon. "Electromagnetic interference effects in charged meson-proton scattering." *PR* 85 (15 Mar 1952): 1059–60 (L).

54. F. Rohrlich and R. L. Gluckstern. "Forward scattering of light by a Coulomb field." *PR* 86 (1 Apr 1952): 1–9.

55. A. Pais. "Some remarks on the V-particles." *PR* 86 (1 June 1952): 663–72.

56. A. Salam. "Renormalized S-matrix for scalar electrodynamics." *PR* 86 (1 June 1952): 731–44.

*57. M. M. Lévy. "The symmetrical pseudoscalar meson theory of nuclear forces." *PR* 86 (1 June 1952): 806 (L).

58. N. M. Kroll and F. Pollock. "Second-order radiative corrections to hyperfine structure." *PR* 86 (15 June 1952): 876–88.

59. E. E. Salpeter. "Mass corrections to the fine structure of hydrogen-like atoms." *PR* 87 (15 July 1952): 328–43.

60. H. P. Noyes. "Decay of a neutral scalar heavy meson." *PR* 87 (15 July 1952): 344–51.

61. J. S. Levinger. "Small angle coherent scattering of gammas by bound electrons." *PR* 87 (15 Aug 1952): 656–62.

62. S. D. Drell. "Recoil correction to Bremsstrahlung cross section." *PR* 87 (1 Sep 1952): 753–55.

63. R. Karplus and A. Klein. "Electrodynamic displacement of atomic energy levels." Pt. 3, "The hyperfine structure of positronium." *PR* 87 (1 Sep 1952): 848–58.

64. A. Pais and R. Jost. "Selection rules imposed by charge conjugation and charge symmetry." *PR* 87 (1 Sep 1952): 871–75.

65. F. Low. "Natural line shape." *PR* 88 (1 Oct 1952): 53–57.

66. M. M. Lévy. "Non-adiabatic treatment of the relativistic two-body problem." *PR* 88 (1 Oct 1952): 72–82.
67. D. R. Yennie. "Quantum corrections to classical nonlinear meson theory." *PR* 88 (1 Nov 1952): 527–36.
68. M. Baranger, F. J. Dyson, and E. E. Salpeter. "Fourth-order vacuum polarization." *PR* 88 (1 Nov 1952): 680 (L).
69. M. M. Lévy. "Meson theory of nuclear forces and low energy properties of the neutron-proton system." *PR* 88 (15 Nov 1952): 725–39.
70. B. D. Fried. "The electron-neutron interaction as deduced from pseudoscalar meson theory." *PR* 88 (1 Dec 1952): 1142–49.

1953

*71. E. E. Salpeter. "The Lamb shift for hydrogen and deuterium." *PR* 89 (1 Jan 1953): 92–97.
72. P. Lindenfeld, A. Sachs, and J. Steinberger. "The internal pair production of γ-rays of mesonic origin: Alternate modes of π^0 decay." *PR* 89 (1 Feb 1953): 531–37.
73. G. Wentzel. "Three-nucleon interactions in Yukawa theory." *PR* 89 (15 Feb 1953): 684–88.
74. R. G. Moorhouse. "Pair creation in intermediate coupling theory." *PR* 89 (1 Mar 1953): 958–65.
*75. A. Petermann. "Divergence of perturbation expansion." *PR* 89 (1 Mar 1953): 1160–61 (L).
76. M. Ruderman. "Nuclear forces from pseudoscalar meson theory." *PR* 90 (15 Apr 1953): 183–85.
*77. S. F. Edwards. "A nonperturbation approach to quantum electrodynamics." *PR* 90 (15 Apr 1953): 284–91.
78. K. A. Brueckner, M. Gell-Mann, and M. Goldberger. "On the damping of virtual nucleon-pair formation in pseudoscalar meson theory." *PR* 90 (1 May 1953): 476–78.
*79. A. Salam and P. T. Matthews. "Fredholm theory of scattering in a given time-dependent field." *PR* 90 (15 May 1953): 690–95.
80. R. Chisholm and B. Touschek. "Spin orbit coupling and the mesonic Lamb shift." *PR* 90 (1 June 1953): 763–65.
81. A. Lenard. "Inner Bremsstrahlung in μ-meson decay." *PR* 90 (1 June 1953): 968–73.
*82. R. L. Gluckstern, M. H. Hull, and G. Breit. "Polarization of Bremsstrahlung radiation." *PR* 90 (15 June 1953): 1026–29.
83. J. S. Blair and G. F. Chew. "Fourth-order corrections to the scattering of pions by nonrelativistic nucleons." *PR* 90 (15 June 1953): 1065–67.
84. S. Deser and P. Martin. "A covariant meson-nucleon equation." *PR* 90 (15 June 1953): 1075–78.
85. A. Klein. "The Tamm-Dancoff formalism and the symmetric pseudoscalar theory of nuclear forces." *PR* 90 (15 June 1953): 1101–15.
86. L. Dresner. "Spin-orbit coupling in pseudoscalar meson theory." *PR* 91 (1 July 1953): 201–2 (L).
87. A. Klein. "Convergence of the adiabatic nuclear potential." *PR* 91 (1 Aug 1953): 740–48.
88. K. A. Brueckner. "Radiative corrections to nuclear forces in pseudoscalar theory." *PR* 91 (1 Aug 1953): 761–62 (L).
*89. R. J. Riddell, Jr.. "The number of Feynman diagrams." *PR* 91 (1 Sep 1953): 1243–48.
90. J. Weneser, R. Bersohn, and N. M. Kroll. "Fourth-order radiative corrections to atomic energy levels." *PR* 91 (1 Sep 1953): 1257–62.

*91. J. Hamilton. "The Fredholm theory of the S matrix." *PR* 91 (15 Sep 1953): 1524–26.

92. S. D. Drell and K. Huang. "Many-body forces and nuclear saturation." *PR* 91 (15 Sep 1953): 1527–42.

93. M. Baranger, H. A. Bethe, and R. P. Feynman. "Relativistic correction to the Lamb shift." *PR* 92 (15 Oct 1953): 482–501.

*94. J. G. Valatin. "Coupled-field Green's functions." *PR* 92 (15 Oct 1953): 522 (L).

*95. A. Klein. "Convergence of the adiabatic nuclear potential." Pt. 2. *PR* 92 (15 Nov 1953): 1017–20.

96. K. A. Brueckner and K. M. Watson. "Nuclear forces in pseudoscalar meson theory." *PR* 92 (15 Nov 1953): 1023–35.

97. E. M. Henley and M. A. Ruderman. "Nuclear forces from p-wave mesons." *PR* 92 (15 Nov 1953): 1036–44.

*98. B. Kursunoglu. "Derivation and renormalization of the Tamm-Dancoff equations." *PR* 92 (15 Nov 1953): 1069–70 (L).

*99. W. Macke. "The single-time Bethe-Salpeter equation." *PR* 92 (15 Nov 1953): 1072 (L).

100. R. D. Lawson. "Photoproduction of π-meson pairs." *PR* 92 (1 Dec 1953): 1272–79.

101. R. Finkelstein and P. Kaus. "Note on the beta interaction." *PR* 92 (1 Dec 1953): 1316–19.

1954

102. N. M. Kroll and M. A. Ruderman. "A theorem on photomeson production near threshold and the suppression of pairs in pseudoscalar meson theory." *PR* 93 (1 Jan 1954): 233–38.

103. J. S. Kovacs. "The angular correlation of mesons produced in inelastic meson-nucleon collisions." *PR* 93 (1 Jan 1954): 254 (L).

*104. S. Deser. "Radiative effects in meson-nucleon scattering." *PR* 93 (1 Feb 1954): 612–15.

105. T. Fulton and R. Karplus. "Bound state corrections in two-body systems." *PR* 93 (1 Mar 1954): 1104–16.

106. A. Aitken, H. Mahmoud, E. M. Henley, M. A. Ruderman, and K. M. Watson. "Some possible relationships between π-meson nucleon scattering and π-meson production in nucleon-nucleon collisions." *PR* 93 (15 Mar 1954): 1349–55.

*107. P. T. Matthews and A. Salam. "Renormalization." *PR* 94 (1 Apr 1954): 185–91.

108. M. Ross. "A connection between pion photoproduction and scattering phase shifts." *PR* 94 (15 Apr 1954): 454–60.

109. M. M. Lévy. "A covariant treatment of meson-nucleon scattering." *PR* 94 (15 Apr 1954): 460–68.

*110. D. Fox. "Production of antiprotons in p-p collisions." *PR* 94 (15 Apr 1954): 499 (L).

*111. T. Kinoshita and Y. Nambu. "The collective description of many-particle systems: A generalized theory of Hartree fields." *PR* 94 (1 May 1954): 598–617.

112. J. L. Anderson. "Green's functions in quantum electrodynamics." *PR* 94 (1 May 1954): 703–11.

113. S. Deser, W. E. Thirring, and M. L. Goldberger. "Low-energy limits and renormalization in meson theory." *PR* 94 (1 May 1954): 711–23.

114. A. Klein. "Single-time formalisms from covariant equations." *PR* 94 (15 May 1954): 1052–56.

*115. R. Arnowitt and S. Gasiorowicz. "Effect of negative energy components in the two-nucleon system." *PR* 94 (15 May 1954): 1057–62.

116. L. Spruch and G. Goertzel. "Magnetic internal Compton coefficients in the Born approximation." *PR* 94 (15 June 1954): 1671–78.

117. G. F. Chew. "Renormalization of meson theory with a fixed extended source." *PR* 94 (15 June 1954): 1748–54.
118. G. F. Chew. "Method of approximation for the meson-nucleon problem when the interaction is fixed and extended." *PR* 94 (15 June 1954): 1755–59.
119. H. Enatsu, H. Hasegawa, and P. Y. Pac. "Theory of unstable heavy particles." *PR* 95 (1 July 1954): 263–70.
120. R. Arnowitt and S. Gasiorowicz. "Covariant approximation scheme for Green's functions of coupled fields." *PR* 95 (15 July 1954): 538–45.
*121. H. S. Green. "Integral equations of quantized field theory." *PR* 95 (15 July 1954): 548–56.
122. A. Klein. "Suppression of pair coupling in nuclear forces." *PR* 95 (15 Aug 1954): 1061–64.
123. R. E. Cutkosky. "Internal Bremsstrahlung." *PR* 95 (1 Sep 1954): 1222–25.
124. M. Gell-Mann and F. E. Low. "Quantum electrodynamics at small distances." *PR* 95 (1 Sep 1954): 1300–1312.
125. J. C. Taylor. "Tamm-Dancoff method." *PR* 95 (1 Sep 1954): 1313–17.
*126. T. D. Lee. "Some special examples in renormalizable field theory." *PR* 95 (1 Sep 1954): 1329–34.
127. F. J. Dyson, M. Ross, E. E. Salpeter, S. S. Schweber, M. K. Sundaresan, W. M. Visscher, and H. A. Bethe. "Meson-nucleon scattering in the Tamm-Dancoff approximation." *PR* 95 (15 Sep 1954): 1644–58.
128. A. Klein. "New Tamm-Dancoff formalism." *PR* 95 (15 Sep 1954): 1676–82.
129. R. J. Finkelstein and S. A. Moszowski. "Mesonic corrections to the beta-decay coupling constants." *PR* 95 (15 Sep 1954): 1695–97 (L).
130. G. Feldman. "Antiproton production." *PR* 95 (15 Sep 1954): 1697 (L).
131. A. N. Mitra. "Modified nucleon propagators." *PR* 95 (15 Sep 1954): 1697–98 (L).
132. C. N. Yang and R. L. Mills. "Conservation of isotopic spin and isotopic gauge invariance." *PR* 96 (1 Oct 1954): 191–95.
133. W. M. Visscher. "Self-energy effects on meson-nucleon scattering according to the Tamm-Dancoff method." *PR* 96 (1 Nov 1954): 788–93.
134. A. M. Sessler. "Mesonic corrections to the quadrupole moment of the deuteron." *PR* 96 (1 Nov 1954): 793–96.
*135. M. Gell-Mann and M. L. Goldberger. "Scattering of low-energy photons by particles of spin 1/2." *PR* 96 (1 Dec 1954): 1433–38.
*136. J. C. Taylor. "Calculation of potentials from the new Tamm-Dancoff equation." *PR* 96 (1 Dec 1954): 1438–41.
137. H. W. Wyld, Jr. "Fourth order corrections to meson-nucleon scattering in pseudoscalar meson theory." *PR* 96 (15 Dec 1954): 1661–78.
*138. R. J. N. Phillips. "Renormalization of a neutral vector meson interaction." *PR* 96 (15 Dec 1954): 1678–79.
139. B. Kursunoglu. "Tamm-Dancoff methods and nuclear forces." *PR* 96 (15 Dec 1954): 1690–1701.

APPENDIX B

Feynman Diagrams in *Proceedings of the Royal Society*, 1950–54

Entries marked with an asterisk discussed elements of the diagrams, but did not include explicit figures. (L) indicates that the article was published as a brief letter to the editor.

1950

*1. K. V. Roberts. "On the quantum theory of the elementary particles." Pt. 1, "Introduction and classical field dynamics." *PRSA* 204 (22 Nov 1950): 123–44.

1951

*2. R. H. Dalitz. "On higher Born approximations in potential scattering." *PRSA* 206 (22 May 1951): 509–20.

3. R. H. Dalitz. "On radiative corrections to the angular correlation in internal pair creation." *PRSA* 206 (22 May 1951): 521–38.

*4. K. V. Roberts. "On the quantum theory of the elementary particles." Pt. 2, "Quantum field dynamics." *PRSA* 207 (22 Jun 1951): 228–51.

*5. F. J. Dyson. "The renormalization method in quantum electrodynamics." *PRSA* 207 (6 Jul 1951): 395–401.

*6. E. A. Power. "On a phenomenological approach to meson production in nucleon-nucleon collisions." *PRSA* 210 (7 Dec 1951): 85–98.

1952

*7. R. J. Eden. "Threshold behaviour in quantum field theory." *PRSA* 210 (7 Jan 1952): 388–404.

*8. A. Salam. "Renormalization of scalar electrodynamics using β-formalism." *PRSA* 211 (21 Feb 1952): 276–84.

9. C. A. Hurst. "The enumeration of graphs in the Feynman-Dyson technique." *PRSA* 214 (7 Aug 1952): 44–61.

*10. R. J. Eden. "Quantum field theory of bound states." Pt. 1, "Bound states in weak interaction." *PRSA* 215 (6 Nov 1952): 133–46.

11. G. E. Brown. "Bound-state perturbation theory in four-dimensional momentum representation." *PRSA* 215 (5 Dec 1952): 371–84.

*12. F. Mandl and T. H. R. Skyme. "The theory of the double Compton effect." *PRSA* 215 (22 Dec 1952): 497–507.

1953

13. R. J. Eden. "Quantum field theory of bound states." Pt. 2, "Relativistic theory of resonance reactions." *PRSA* 217 (7 May 1953): 390–408.

*14. E. A. Power. "A new proof of the perturbation expansions in quantum mechanics." *PRSA* 218 (7 Jul 1953): 384–91.

15. R. J. Eden and G. Rickayzen. "Quantum field theory of bound states." Pt. 3, "Bound states in strong interaction." *PRSA* 219 (11 Aug 1953): 109–19.

*16. R. J. Eden. "Quantum field theory of bound states." Pt. 4, "Relativistic theory of excited states of hydrogen." *PRSA* 219 (7 Oct 1953): 516–26.

17. M. L. G. Redhead. "Radiative corrections to the scattering of electrons and positrons by electrons." *PRSA* 220 (10 Nov 1953): 219–39.

1954

*18. P. T. Matthews and Abdus Salam. "Covariant Fock equations." *PRSA* 221 (7 Jan 1954): 128–34.

*19. J. G. Valatin. "On the Dirac-Heisenberg theory of vacuum polarization." *PRSA* 222 (9 Mar 1954): 228–39.

20. G. Feldman. "Modified propagators in field theory (with application to the anomalous magnetic moment of the nucleon)." *PRSA* 223 (7 Apr 1954): 112–29.

*21. J. G. Valatin. "On the propagation functions of quantum electrodynamics." *PRSA* 225 (22 Sep 1954): 535–48.

*22. J. G. Valatin. "On the definition of finite operator quantities in quantum electrodynamics." *PRSA* 226 (9 Nov 1954): 254–65.

23. J. C. Polkinghorne. "Renormalization of the transformation operators of quantum electrodynamics." *PRSA* 227 (21 Dec 1954): 94–102.

APPENDIX C

Feynman Diagrams in *Progress of Theoretical Physics*, 1949–54

Entries marked with an asterisk discussed elements of the diagrams, but did not include explicit figures. (L) indicates that the article was published as a brief letter to the editor.

1949

1. G. Takeda. "Note on the vacuum polarization." *PTP* 4 (Oct–Dec 1949): 573–75 (L).

1950

2. H. Fukuda and Y. Miyamoto. "The decay of a $\tau^{\pm}$ meson into a $\pi^{\pm}$ meson and a photon." *PTP* 5 (Jan–Feb 1950): 148–50 (L).
3. K. Baba, D. Itō, T. Miyazima, and M. Sasaki. "On the concept of the nuclear potential." *PTP* 5 (Jan–Feb 1950): 159–60 (L).
4. S. Ōneda, S. Sasaki, and S. Ozaki. "On the decay of heavy mesons." Pt. 3. *PTP* 5 (Mar–Apr 1950): 165–76.
5. K. Sawada. "Note on the self-energy and self-stress." Pt. 2. *PTP* 5 (Mar–Apr 1950): 236–51.
6. Y. Nambu. "Derivation of the interaction potential from field theory." *PTP* 5 (Mar–Apr 1950): 321–23 (L).
7. T. Kinoshita. "A note on the C-meson hypothesis." *PTP* 5 (Mar–Apr 1950): 335–36 (L).
8. Y. Katayama and S. Takagi. "Note on five-dimensional space and the self-energy of electron." *PTP* 5 (Mar–Apr 1950): 336–38 (L).
9. H. Fukuda, S. Hayakawa, and Y. Miyamoto. "On the nature of τ-mesons." Pt. 2. *PTP* 5 (May–Jun 1950): 352–72.
10. S. Ozaki. "On the decay of a heavy meson into lighter mesons." *PTP* 5 (May–Jun 1950): 373–94.
11. T. Kinoshita. "On the interaction of mesons with the electromagnetic field." Pt. 1. *PTP* 5 (May–Jun 1950): 473–88.
12. Y. Nambu. "Force potentials in quantum field theory." *PTP* 5 (Jul–Aug 1950): 614–33.

13. J. Ashkin, A. Simon, and R. Marshak. "On the scattering of π-mesons by nucleons." *PTP* 5 (Jul–Aug 1950): 634–68.
14. T. Kinoshita and Y. Nambu. "On the interaction of mesons with the electromagnetic field." Pt. 2, *PTP* 5 (Sep–Oct 1950): 749–68.
15. H. Fukuda and G. Takeda. "Production of π-mesons in nucleon-nucleon collisions near the threshold energy." *PTP* 5 (Sep–Oct 1950): 800–812.
16. N. Fukuda and T. Miyazima. "The covariant theory of radiation damping." Pt. 1, "General formalism." *PTP* 5 (Sep–Oct 1950): 849–60.
17. Y. Ōno. "On the anomalous magnetic moment of meson." *PTP* 5 (Sep–Oct 1950): 861–69.
18. T. Nakano, Y. Watanabe, S. Hanawa, and T. Miyazima. "Radiative corrections to decay processes." Pt. 2, "Beta disintegration of nucleon." *PTP* 5 (Nov–Dec 1950): 1014–23.
19. T. Kinoshita. "Note on the infrared catastrophe." *PTP* 5 (Nov–Dec 1950): 1045–47 (L).

1951

20. J. Yukawa and H. Umezawa. "On the problem of covariance in quantum electrodynamics." Pt. 1. *PTP* 6 (Jan–Feb 1951): 112–21.
21. J. Yukawa and H. Umezawa. "On the problem of covariance in quantum electrodynamics." Pt. 2. *PTP* 6 (Mar–Apr 1951): 197–201.
22. S. Ogawa, E. Yamada, and Y. Nagahara. "On the absorption of the negative π-meson by deuteron." *PTP* 6 (Mar–Apr 1951): 227–37.
23. K. Nakabayasi and I. Sato. "Radiative corrections to anomalous magnetic moment of nucleon in pseudoscalar meson theory." *PTP* 6 (Mar–Apr 1951): 252–55 (L).
24. K. Ida. "Production of vector π-mesons by high energy nucleon-nucleon collisions." *PTP* 6 (Mar–Apr 1951): 258–59 (L).
25. S. Tani. "Connection between particle models and field theories." Pt. 1, "The case spin 1/2." *PTP* 6 (May–Jun 1951): 267–85.
26. Z. Koba, N. Mugibayashi, and S. Nakai. "On gauge invariance and equivalence theorems." *PTP* 6 (May–Jun 1951): 322–41.
27. K. Husimi and R. Utiyama. "Note on Belinfante's new theory." *PTP* 6 (May–Jun 1951): 432–34 (L).
28. M. Taketani and S. Machida. "On the production of negative protons." *PTP* 6 (Jul–Aug 1951): 559–71.
29. R. Utiyama, T. Imamura, S. Sunakawa, and T. Dodo. "Note on the longitudinal and scalar photons." *PTP* 6 (Jul–Aug 1951): 587–603.
30. K. Nishijima. "Generalized Furry's theorem for closed loops." *PTP* 6 (Jul–Aug 1951): 614–15 (L).
31. Y. Nambu, K. Nishijima, and Y. Yamaguchi. "On the nature of V-particles." Pt. 1. *PTP* 6 (Jul–Aug 1951): 615–19 (L).
32. S. Ōneda. "Note on the theory of V-particles and τ-mesons." *PTP* 6 (Jul–Aug 1951): 633–35 (L).
33. H. Fukuda, Y. Fujimoto, and M. Koshiba. "Nuclear interaction of μ-meson." *PTP* 6 (Sep–Oct 1951): 788–800.
34. K. Nishijima. "On the adiabatic nuclear potential." Pt. 1. *PTP* 6 (Sep–Oct 1951): 815–28.
35. Z. Koba, T. Kotani, and S. Nakai. "Production of charged π-meson by γ-ray: Higher order corrections." *PTP* 6 (Sep–Oct 1951): 849–90.

36. T. Kotani, S. Machida, S. Nakamura, H. Takebe, M. Umezawa, and T. Yoshimura. "On the mesonic correction to the β-decay." *PTP* 6 (Nov–Dec 1951): 1007–12.
37. D. Itō. "'Infra-red catastrophe'-like divergency in meson-decay process." *PTP* 6 (Nov–Dec 1951): 1020–22 (L).
38. D. Itō. "On the divergence of the transition probability due to energy conservation in intermediate states." *PTP* 6 (Nov–Dec 1951): 1022–23 (L).
39. T. Nakano and K. Nishijima. "Divergences arising from nuclear forces." *PTP* 6 (Nov–Dec 1951): 1024–25 (L).

1952

40. M. Taketani, S. Machida, and S. O-Numa. "The meson theory of nuclear forces." Pt. 1, "The deuteron ground state and low energy neutron-proton scattering." *PTP* 7 (Jan 1952): 45–56.
41. S. Machida and K. Nishijima. "Remarks on the adiabatic nuclear potential." *PTP* 7 (Jan 1952): 57–68.
42. S. Minami. "Neutral-meson production by gamma-ray." *PTP* 7 (Jan 1952): 69–92.
43. D. Itō, H. Tanaka, Y. Watanabe, and M. Yamazaki. "Group theoretical aspects in S-matrix theory." *PTP* 7 (Jan 1952): 128–30 (L).
44. H. Miyazawa. "Non-additivity of nucleon magnetic moment in the deuteron." *PTP* 7 (Feb 1952): 207–16.
45. H. Kita. "Relativistic two-body problem." *PTP* 7 (Feb 1952): 217–24.
*46. R. Utiyama, S. Sunakawa, and T. Imamura. "On the Green-function of the quantum electrodynamics." *PTP* 7 (Mar 1952): 328 (L).
47. G. Takeda. "On the renormalization theory of the interaction of electrons and photons." *PTP* 7 (Mar 1952): 359–66.
48. C. Hayashi and Y. Munakata. "On a relativistic integral equation for bound states." *PTP* 7 (May 1952): 481–516.
49. S. Hori. "On the well-ordered S-matrix." *PTP* 7 (Jun 1952): 578–84.
50. S. Goto. "Note on the non-relativistic limit of the Compton scattering." *PTP* 7 (Jun 1952): 585–6 (L).
51. K. Sawada. "On the model of V-particle and meson-nucleon scattering." *PTP* 7 (Jun 1952): 592–95 (L).
52. N. Shōno and N. Oda. "Note on the non-local interaction." *PTP* 8 (Jul 1952): 28–38.
53. T. Nakano and K. Nishijima. "The *S* matrix method in pion reactions." *PTP* 8 (Jul 1952): 53–76.
54. R. Utiyama, S. Sunakawa, and T. Imamura. "On the theory of the Green-function in quantum-electrodynamics." *PTP* 8 (Jul 1952): 77–110.
*55. S. Kamefuchi. "Radiative correction to β-decay matrix element." *PTP* 8 (Jul 1952): 137–38 (L).
56. Y. Takahashi and H. Umezawa. "On the self-stress." *PTP* 8 (Aug 1952): 193–204.
57. R. Majumdar and A. N. Mitra. "The effect of damping on radiative corrections to electron scattering and the problem of infra-red catastrophe." *PTP* 8 (Oct 1952): 479–92.
58. S. Minami, T. Nakano, K. Nishijima, H. Okonogi, and E. Yamada. "Pion reactions in one nucleon system and nucleon isobars." *PTP* 8 (Nov 1952): 531–48.
59. S. Kamefuchi and H. Umezawa. "On the structure of the interactions of the elementary particles." Pt. 3, "On the renormalizable field theory." *PTP* 8 (Dec 1952): 579–98.

60. Z. Tokuoka and H. Tanaka. "On the equivalence of the particle formalism and the wave formalism of meson." *PTP* 8 (Dec 1952): 599–614.

1953

61. S. Minami. "The scattering of gamma-ray by nucleon and nucleon isobar." *PTP* 9 (Feb 1953): 108–16.
62. S. Tanaka and M. Itō. "On the natural decay of the free neutron." *PTP* 9 (Feb 1953): 169–80.
63. S. Ōneda. "On some properties of the interactions between elementary particles." *PTP* 9 (Apr 1953): 327–44.
64. R. Utiyama and T. Imamura. "Difficulty of divergence of the perturbation method in the quantum field theory." *PTP* 9 (Apr 1953): 431–54.
65. S. Kamefuchi and H. Umezawa. "On the structure of the interaction of the elementary particles." Pt. 4, "On the interaction of the second kind." *PTP* 9 (May 1953): 529–49.
66. T. Hamada and M. Sugawara. "Remarks on Lévy's fourth order potential." *PTP* 9 (May 1953): 555–57 (L).
67. R. Utiyama. "On the convergence of the perturbation method in quantum field theory." *PTP* 9 (Jun 1953): 593–606.
68. Z. Tokuoka. "On the equivalence of the particle formalism and the wave formalism of meson." Pt. 2, "Case of interacting meson and nucleon fields." *PTP* 10 (Aug 1953): 137–57.
69. T. Matsumoto, T. Hamada, and M. Sugawara. "On the effects of excited states of nucleons upon static nuclear potential in symmetrical pseudoscalar meson theory." Pt. 1, "Derivation of nuclear potential and qualitative conclusions." *PTP* 10 (Aug 1953): 199–226.
70. T. Hamada. "Effect of nucleon excited state on magnetic moment anomaly." *PTP* 10 (Sep 1953): 309–22.
71. I. Sato. "Nuclear forces in pseudoscalar meson theory." *PTP* 10 (Sep 1953): 323–58.
72. A. Kanazawa and M. Sugawara. "Pion-nucleon scattering and nucleon isobar." *PTP* 10 (Oct 1953): 399–414.
73. T. Tati. "On the many-body problem in the intermediate coupling theory." Pt. 1. *PTP* 10 (Oct 1953): 421–30.
74. K. Nishijima. "Many-body problem in quantum field theory." *PTP* 10 (Nov 1953): 549–74.
75. S. Tanaka and H. Umezawa. "On the transition matrix and the Green function in the quantum field theory." *PTP* 10 (Dec 1953): 617–29.
76. S. Okubo. "Non-perturbation approach method by edwards." *PTP* 10 (Dec 1953): 692–94 (L).
77. K. Nakabayasi, K. Hasegawa, and I. Yamamura. "Fourth order calculations of meson-proton scattering in the symmetrical Ps (Ps) theory." *PTP* 10 (Dec 1953): 694–95 (L).

1954

78. S. Okubo. "Note on the second kind interaction." *PTP* 11 (Jan 1954): 80–94.
79. S. Nakai. "On the Green-functions of many-electron problem." *PTP* 11 (Feb 1954): 155–78.
80. A. Kanazawa and M. Sugawara. "Anomalous magnetic moment of nucleon and nucleon isobar." *PTP* 11 (Mar 1954): 231–43.
81. S. Kamefuchi. "On the structure of the interaction of the elementary particles." Pt. 5, "Interaction of the second kind and non-local interaction." *PTP* 11 (Mar 1954): 273–87.

82. M. Hamaguchi. "The generalization of Stueckelberg formalism in the theory of quantized field." *PTP* 11 (Apr–May 1954): 461–75.
83. S. Chiba. "Renormalization in the covariant treatment of pion-nucleon scattering." *PTP* 11 (Jun 1954): 494–96 (L).
84. D. Itō and H. Tanaka. "Covariant subtraction of 'overlapping divergences' appearing in the pion-nucleon scattering." *PTP* 11 (Jun 1954): 501–3 (L).
85. D. Itō and H. Tanaka. "On the variational solution of Bethe-Salpeter equation in pion-nucleon scattering." *PTP* 12 (Jul 1954): 105–6 (L).
86. N. Fukuda, K. Sawada, and M. Taketani. "On the construction of potential in field theory." *PTP* 12 (Aug 1954): 156–66.
87. S. Hatano, T. Kaneno, and M. Shindo. "The effects of heavy particles on π-meson-proton scattering." *PTP* 12 (Aug 1954): 167–76.
88. Y. Miyachi. "Meson production in meson-nucleon collisions." *PTP* 12 (Aug 1954): 243–44 (L).
89. S. Chiba. "Renormalization in the covariant treatment of pion-nucleon scattering." *PTP* 12 (Oct 1954): 481–502.
90. K. Itabashi. "On the renormalization in the Tamm-Dancoff approximation for one-nucleon problem." Pt. 1, "A covariant generalization of the Tamm-Dancoff method." *PTP* 12 (Oct 1954): 494–502.
91. K. Hasegawa, S. Matsuyama, and T. Akiba. "On the analysis of the anomalous magnetic moment of the nucleon in the second order calculations." *PTP* 12 (Oct 1954): 548–49 (L).
92. K. Itabashi. "On the renormalization in Tamm-Dancoff approximation for one-nucleon problem." Pt. 2, "Subtraction of divergences in the generalized Tamm-Dancoff equations." *PTP* 12 (Nov 1954): 585–602.
93. S. Ōkubo. "Diagonalization of Hamiltonian and Tamm-Dancoff equation." *PTP* 12 (Nov 1954): 603–22.
94. S. Sunakawa, T. Imamura, and R. Utiyama. "Renormalization of two-electron Green-function." *PTP* 12 (Nov 1954): 642–52.
95. S. Ogawa, B. Sakita, and Y. Taguchi. "Notes on the unstable particles." *PTP* 12 (Nov 1954): 691–93 (L).
96. S. Goto. "The cut-off method in meson theory." *PTP* 12 (Dec 1954): 699–712.
97. S. Chiba, M. Yamazaki, and N. Fukuda. "Pion-nucleon scattering in the Tamm-Dancoff approximation." *PTP* 12 (Dec 1954): 767–88.

APPENDIX D

Feynman Diagrams in *Soryūshi-ron Kenkyū*, 1949–52

These articles and preprints appeared in Japanese; the following translations were completed by Kenji Ito. *Soken* was an informal mimeographed newsletter whose founders did not originally foresee long-term publication; hence, the volume-numbers for the first year or two were not always consistent.

1949

1. Mituo Taketani, Shigeru Machida, and Minoru Umezawa. "Creation of a nucleon pair by a pi-meson." *Soken* 1 (5 Aug 1949): 86–89.
2. Mituo Taketani, Shigeru Machida, and Taro Tamura. "Corrections of the nucleon's static electric field by means of the meson field." *Soken* 1 (5 Aug 1949): 89–99.
3. N. Fukuda and D. Ito. "Some generalizations of Dyson's results." *Soken* 1 (5 Aug 1949): 161–74.
4. Hiroshi Fukuda and Yoneji Miyamoto. "Tau meson." *Soken* 1 (15 Oct 1949): 179–213.
5. Mamoru Ono, Kazuo Baba, and Dasuke Ito. "Møller interaction of electrons." *Soken* 1 (25 Dec 1949): 44–64.
6. Suteo Kanazawa and Takao Tanabe. "On the self-energy of the photon." *Soken* 1 (25 Dec 1949): 155–58.
7. Toichiro Kinoshita and Yoichiro Nambu. "On the interaction of mesons with the electromagnetic field." Pt. 2. *Soken* 1 (25 Dec 1949): 176–81.
8. Toichiro Kinoshita. "A few remarks on the S-matrix." *Soken* 1 (25 Dec 1949): 182–89.

1950

9. Ziro Koba, Tsuneyuki Kotani, and Shinzo Nakai. "Production of the pi-meson by x-rays." *Soken* 1 (10 Jan 1950): 76–83.
10. Eiji Yamada. "Anomalous magnetic moment of the nucleon." *Soken* 1 (10 Jan 1950): 96–104.
11. Ziro Koba and Norimichi Mugibayashi. "Scattering of the photon by the neutron, preliminary report." *Soken* 1 (10 Jan 1950): 128–29.

12. Yoichiro Nambu. "Relativistic two-body problem." *Soken* 1 (10 Jan 1950): 133–41.
13. Nobuyuki Fukuda, Yoshikatsu Iida, Tatsuoki Miyajima, Muneo Sasaki, Ryoji Suzuki, and Takao Tati. "Relativistic formualtion of a damping theory, II: Reaction of the field in the meson theory." Pt. 1. *Soken* 2 (1 Mar 1950): 36–46.
14. Nobuyuki Fukuda and Yoshikatsu Iida. "Influence of $K^{(4)}$ on Compton scattering." *Soken* 2 (1 Mar 1950): 47–54.
15. Ziro Koba, Tsuneyuki Kotani, and Shinzo Nakai. "Production of the meson by x-rays." Pt. 2. *Soken* 2 (1 Mar 1950): 74–81.
16. Yasuhisa Katayama. "On the finite extension of the electron: Concerning Mr. Koba's spring theory." *Soken* 2 (1 Mar 1950): 157–69.
17. Tatsuoki Miyajima, Muneo Sasaki, Ryoji Suzuki, Takao Tati, Yoshikatsu Iida, Nobuyuki Fukuda, and Tsuneo Nagata. "On the theory of damping." *Soken* 2 (1 Apr 1950): 39–41.
18. Gyo Takeda and Hiroshi Fukuda. "Production of the pi-meson." Pt. 1. *Soken* 2 (1 Apr 1950): 110–31.
19. Yoichiro Nambu. "On the potential of the higher orders." *Soken* 2 (1 Apr 1950): 164–70.
20. Ziro Koba, Norimichi Mugibayashi, and Shinzo Nakai. "On gauge invariance." *Soken* 2 (June 1950): 14–21.
21. Ziro Koba, Norimichi Mugibayashi, and Shinzo Nakai. "On the equivalence theorem." *Soken* 2 (June 1950): 64–78.
22. Tatsuoki Miyajima, Nobuyuki Fukuda, and Tsuneo Nagata. "Relativistic formulation of the damping theory." Pt. 3, "Scattering of the meson by nucleons." *Soken* 2 (June 1950): 131–40.
23. Hiroshi Enatsu, Yasushi Ataka, and Yoshiro Takano. "Theory of mixed interactions." Pt. 4. *Soken* 2 (June 1950): 162–78.
24. Takao Nakabayashi and Iwao Sato. "Higher order corrections of the nucleon's anomalous magnetic moment." *Soken* 2 (29 Aug 1950): 118–24.
25. Hideji Kita and Yasuo Munakata. "Relativistic eigenvalue problem." *Soken* 2 (17 Sep 1950): 48–57.
26. Ziro Koba, Tsuneyuki Kotani, and Shinzo Nakai. "Production of charged pi-meson by x-rays." *Soken* 2 (17 Sep 1950): 63–84.
27. Hideo Goto. "On S-matrix and the third quantization." Pt. 2. *Soken* 2 (17 Sep 1950): 143–54.
28. Ichiei Watanabe. "On subtraction in the Heisenberg picture." *Soken* 2 (Oct 1950): 55–75.

1951

29. Jiro Yukawa and Hiroomi Umezawa. "Problem of covariance in quantum electrodynamics." *Soken* 3 (5 Feb 1951): 47–52.
30. Kazuhiko Nishijima. "On adiabatic potential." *Soken* 3 (5 Feb 1951): 75–84.
31. Ziro Koba, Tsuneyuki Kotani, Shigeo Minami, and Shinzo Nakai. "Production of neutral π-meson by γ-ray." Pt. 1. *Soken* 3 (5 Feb 1951): 143–52.
32. Suzo Ogawa, Eiji Yamada, and Yukio Nagahara. "Absorption of the pi-meson by the heavy proton." *Soken* 3 (5 Feb 1951): 153–67.
33. Takao Nakabayashi and Iwao Sato. "Anomalous magnetic moment of nucleons in the pseudo-scalar theory." *Soken* 3 (Apr 1951): 127–33.
34. Kojiro Ida. "On the production of vector pi-meson." *Soken* 3 (Apr 1951): 133–37.
35. Watanabe Ichiei. "On the expectation value of the Feynman-Dyson theory." *Soken* 3 (20 June 1951): 51–55.
36. Sadao Ōneda, "On the V-particle and meson." *Soken* 3 (Sep 1951): 271–79.

37. Shoichi Hori. "On the notion of the free field." *Soken* 3 (20 Nov 1951): 20–27.
38. Daisuki Ito. "On the phenomena similar to the infrared catastrophe in the process of the meson decay." *Soken* 3 (20 Nov 1951): 28–33.
39. Shigeru Machida and Kazuhiko Nishijima. "Remarks on the adiabatic potential." *Soken* 3 (20 Nov 1951): 41–55.
40. Masahiko Matsumoto, Shigeru Fujinaga, and Wataro Watari. "Corrections of the meson field in electromagnetic scattering between nucleons." *Soken* 3 (20 Nov 1951): 65–69.
41. Sadahiko Matsuyama, Hironari Miyazawa, and Yoshimoto Nakata. "V-particle." *Soken* 3 (20 Nov 1951): 89–91.
42. Tsuneyuki Kotani, Shigeru Machida, Hisao Takebe, and Minoru Umezawa. "Mesonic correction to the β-decay." *Soken* 3 (20 Nov 1951): 143–49.
43. Toichiro Kinoshita and Hironari Miyazawa. "On saturation of the nuclear force." *Soken* 3 (20 Nov 1951): 154–57.
44. Susumu Kamefuchi. "On 'charge renormalization' in quantum electrodynamics." *Soken* 3 (20 Nov 1951): 175–80.

1952

45. Gyo Takeda. "On the divergence of S-matrices." Pt. 1, "Cases of the electron and electromagnetic field." *Soken* 4 (10 Feb 1952): 1–8.
46. Gyo Takeda. "On the divergence of S-matrices." Pt. 2, "Cases of the meson and nucleon system." *Soken* 4 (10 Feb 1952): 8–13.
47. Taro Momo. "On the S-matrix and the damping theory." *Soken* 4 (10 Feb 1952): 43–49.
48. Kazuo Yamazaki. "Application of the Feynman-Fujiwara operator calculus on the second-quantized Dirac field." *Soken* 4 (Mar 1952): 1–21.
49. Shoichi Hori. "On the well-ordered S-matrix." *Soken* 4 (Mar 1952): 31–37.
50. Kazuhiko Nishijima. "Formulation of the heavy proton reaction in field theory." *Soken* 4 (10 Apr 1952): 55–66.
51. Gyo Takeda. "On the Dyson transformation (Equivalence theory in the pseudo-scalar meson)." *Soken* 4 (10 Apr 1952): 89–92.
52. Minoru Hamaguchi. "Interaction of spinor particles." *Soken* 4 (10 May 1952): 21–31.
53. Yasushi Takahashi and Hiroomi Umezawa. "On the self-stress of elementary particles." *Soken* 4 (10 May 1952): 62–76.
54. Yoichiro Nambu. "On scattering of the meson and nuclear force." *Soken* 4 (June 1952): 17–33.
55. Katuro Sawada. "V-particle model and pi-nucleon scattering." *Soken* 4 (June 1952): 33–38.
56. Yasushi Ataka. "Nucleon scattering in meson theory." *Soken* 4 (June 1952): 49–57.
57. Daisuke Ito, Yasutaro Takahashi, Yutaka Tanaka, and Miwae Yamazaki. "Double decay process of mesons." *Soken* 4 (Aug 1952): 113–28.
58. Shinzo Nakai. "On the Green function of many electron problems." *Soken* 4 (Sep 1952): 74–97.
59. Daisuke Ito. "On the meson-nucleon interaction by means of the ladder approximation." *Soken* 4 (Oct 1952): 6–19.
60. Susumu Kamefuchi and Hiroomi Umezawa. "On the interaction of the second kind." *Soken* 4 (Nov 1952): 1–27.
61. Chikashi Iso. "On the heavy meson." *Soken* 4 (Nov 1952): 40–51.
62. Masayuki Muto. "On the bound state of many bodies." *Soken* 4 (Nov 1952): 51–80.

63. Harufumi Sasaki. "On unstable particles." *Soken* 4 (Nov 1952): 80–91.
64. Shigeo Goto. "π-N scattering." *Soken* 4 (Dec 1952): 36–48.
65. Taro Shintomi. "On the interaction of nucleons exchanging mesons: A semi-classical treatment." *Soken* 4 (Dec 1952): 80–95.
66. Sadao Ōneda. "On the interaction of elementary particles." *Soken* 4 (Dec 1952): 121–32.

APPENDIX E

Feynman Diagrams in *Zhurnal eksperimental'noi i teoreticheskoi fiziki*, 1952–59

Beginning in 1955, all articles from *ZhETF* were translated and published in the journal *Sov Phys JETP*. For articles published in *ZhETF* between 1952–54, all transliterations and translations are by Karl Hall; following 1955, all translations are from the published versions in *Sov Phys JETP*. Entries marked with an asterisk discussed the diagrams in words, but did not include explicit figures. (L) indicates that the article was published as a brief letter to the editor.

1952

1. A. D. Galanin. "Radiatsionnye popravki v kvantovoi elektrodinamike" [Radiation corrections in quantum electrodynamics]. Pt. 1. *ZhETF* 22 (Apr 1952): 448–61.
2. A. D. Galanin. "Radiatsionnye popravki v kvantovoi elektrodinamike" [Radiation corrections in quantum electrodynamics]. Pt. 2. *ZhETF* 22 (Apr 1952): 462–70.
3. G. F. Zharkov. "Obrazovanie par pi-mezonov fotonami na nuklonakh" [Pair formation of pi-mesons by photons in nucleons]. *ZhETF* 22 (June 1952): 677–86.
4. A. D. Galanin. "Reliativistskoe uravnenie vzaimodeistvuiushchikh chastits" [Relativistic equation of interacting particles]. *ZhETF* 23 (Nov 1952): 488–92.

1953

5. D. D. Ivanenko and A. M. Brodskii. "O nelineinoi kvanotovoi teorii elektrona" [On nonlinear quantum theory of the electron]. *ZhETF* 24 (Apr 1953): 383–88.
6. G. M. Gandel'man. "Komptonovskoe rasseianie poliarizovannogo po krugu kvanta na elektrone s zadannym napravleniem spina" [Compton scattering of a circularly polarized quantum on an electron with a set spin direction]. *ZhETF* 25 (Oct 1953): 429–34.

1954

7. A. D. Galanin. "O parametre razlozheniia v psevdoskaliarnoi mezonnoi teorii s psevdoskaliarnoi sviaz'iu" [On the expansion parameter in pseudoscalar meson theory with pseudoscalar coupling]. *ZhETF* 26 (Apr 1954): 417–22.

8. A. D. Galanin. "Nekotorye zamechaniia o raskhodimostiakh v toerii psevdoskaliarnogo mezona s psevdovektornoi sviaz'iu" [Several remarks on the divergences in the theory of the pseudoscalar meson with pseudoscalar coupling]. *ZhETF* 26 (Apr 1954): 423–29.

9. I. E. Tamm, Iu. A. Gol'fand, and V. Ia. Fainberg. "Polufenomenologicheskaia teoriia vzaimodeistviia pi-mezonov s nuklonami" [Semi-phenomenological theory of the interaction of pi-mesons wtih nucleons]. Pt. 1. *ZhETF* 26 (June 1954): 649–67.

10. A. D. Galanin and V. G. Solov'ev. "Radiatsionnaia popravka k vremeni zhizni π^0-mezona" [Radiative correction to the lifetime of a π^0-meson]. *ZhETF* 27 (July 1954): 112–114 (L).

11. V. T. Khoziainov. "Obrazovanie mezonov pri perifericheskikh stolknoveniiakh nuklonov" [Meson formation during peripheral collisions of nucleons]. *ZhETF* 27 (Oct 1954): 445–57.

12. V. I. Ritus. "Obrazovanie π-mezonov fotonami i nuklonnye izobary" [π-meson formation by photons[,] and nucleonic isobars]. *ZhETF* 27 (Dec 1954): 660–76.

1955

*13. Iu. A. Gol'fand. "Construction of a distribution function by the method of quasi-fields." *ZhETF* 28 (Feb 1955): 140–50. [*Sov Phys JETP* 1 (July 1955): 118–25.]

*14. E. S. Fradkin. "The asymptote of Green's function in quantum electrodynamics." *ZhETF* 28 (June 1955): 750–52 (L). [*Sov Phys JETP* 1 (Nov 1955): 604–6 (L).]

15. V. P. Silin, I. E. Tamm, and V. Ia. Fainberg. "Method of terminated field equations and its applications to scattering of mesons by nucleons." *ZhETF* 29 (July 1955): 6–19. [*Sov Phys JETP* 2 (Jan 1956): 3–13.]

16. A. D. Galanin, B. L. Ioffe, and I. Ia. Pomeranchuk. "The asymptotic Green's function of nucleon and meson in pseudo-scalar theory with weak interaction." *ZhETF* 29 (July 1955): 51–63. [*Sov Phys JETP* 2 (Jan 1956): 37–45.]

17. E. S. Fradkin. "The quantum theory of fields." Pt. 1. *ZhETF* 29 (July 1955): 121–34. [*Sov Phys JETP* 2 (Jan 1956): 148–58.]

*18. E. S. Fradkin. "Concerning some general relations of quantum electrodynamics." *ZhETF* 29 (Aug 1955): 258–61 (L). [*Sov Phys JETP* 2 (Mar 1956): 361–63 (L).]

19. V. B. Berestetskii. "Asymptotic behavior of electromagnetic vacuum-polarization in the presence of meson interactions." *ZhETF* 29 (Nov 1955): 585–98. [*Sov Phys JETP* 2 (May 1956): 540–49.]

20. S. S. Gershtein and Ia. B. Zel'dovich. "Meson corrections in the theory of beta decay." *ZhETF* 29 (Nov 1955): 698–99 (L). [*Sov Phys JETP* 2 (May 1956): 576–78 (L).]

*21 A. A. Rukhadze. "Interaction of nucleons through a pseudoscalar meson field." *ZhETF* 29 (Nov 1955): 709–11 (L). [*Sov Phys JETP* 2 (May 1956): 570–71 (L).]

22. V. B. Berestetskii and I. Ia. Pomeranchuk. "Formation of a μ-meson pair in positron annihilation." *ZhETF* 29 (Dec 1955): 864 (L). [*Sov Phys JETP* 2 (May 1956): 580–81 (L).]

23. I. Ia. Pomeranchuk. "Solution of the equations of pseudo-scalar meson theory with pseudo-scalar coupling." *ZhETF* 29 (Dec 1955): 869–71 (L). [*Sov Phys JETP* 2 (July 1956): 739–41. (L).]

1956

*24. N. N. Bogoliubov and D. V. Shirkov. "The multiplicative renormalization group in the quantum theory of fields." *ZhETF* 30 (Jan 1956): 77–86. [*Sov Phys JETP* 3 (Aug 1956): 57–64.]

*25. V. V. Sudakov. "Vertex parts at very high energies in quantum electrodynamics." *ZhETF* 30 (Jan 1956): 87–95. [*Sov Phys JETP* 3 (Aug 1956): 65–71.]

26. A. A. Abrikosov. "The infrared catastrophe in quantum electrodynamics." *ZhETF* 30 (Jan 1956): 96–108. [*Sov Phys JETP* 3 (Aug 1956): 71–80.]

27. A. A. Abrikosov. "The Compton effect at high energies." *ZhETF* 30 (Feb 1956): 386–98. [*Sov Phys JETP* 3 (Nov 1956): 474–83.]

28. I. T. Diatlov and K. A. Ter-Martirosian. "Asymptotic meson-meson scattering theory." *ZhETF* 30 (Feb 1956): 416–19 (L). [*Sov Phys JETP* 3 (Oct 1956): 454–56 (L).]

29. A. A. Abrikosov. "Scattering of high-energy electrons and positrons by electrons." *ZhETF* 30 (Mar 1956): 544–50. [*Sov Phys JETP* 3 (Oct 1956): 379–84.]

30. Iu. M. Lomsadze. "On the singularity of the electromagnetic potential in the higher approximations of perturbation theory." *ZhETF* 30 (Apr 1956): 707–12. [*Sov Phys JETP* 3 (Nov 1956): 554–58.]

31. L. M. Afrikian. "Contribution to the theory of production and annihilation of antiprotons." *ZhETF* 30 (Apr 1956): 734–45. [*Sov Phys JETP* 3 (Nov 1956): 503–11.]

*32. E. N. Avrorin and E. S. Fradkin. "Renormalizability of pseudoscalar meson theory with pseudovector coupling." *ZhETF* 30 (Apr 1956): 756–60. [*Sov Phys JETP* 3 (Jan 1957): 862–65.

*33. A. A. Logunov. "Concerning a certain generalization of a renormalization group." *ZhETF* 30 (Apr 1956): 793–95 (L). [*Sov Phys JETP* 3 (Dec 1956): 766–68 (L).]

*34. D. A. Kirzhnits. "On the mass of the photon in quantum electrodynamics." *ZhETF* 30 (Apr 1956): 796–97 (L). [*Sov Phys JETP* 3 (Dec 1956): 768–70 (L).]

*35. V. I. Ritus. "Renormalization in the equations of the new Tamm-Dancoff method." *ZhETF* 30 (May 1956): 965–67 (L). [*Sov Phys JETP* 3 (Dec 1956): 805–7 (L).]

*36. D. A. Kirzhnits. "On mass renormalization in the Tamm-Dancoff method." *ZhETF* 30 (May 1956): 971–73 (L). [*Sov Phys JETP* 3 (Dec 1956): 809–12 (L).]

*37. V. I. Ritus. "The scattering of photons by nucleons and nuclear isobars." *ZhETF* 30 (June 1956): 1070–78. [*Sov Phys JETP* 3 (Jan 1957): 926–34.]

*38. V. I. Karpman. "On the S-matrix for particles with arbitrary spin." *ZhETF* 30 (June 1956): 1104–11. [*Sov Phys JETP* 3 (Jan 1957): 934–40.]

39. I. E. Dzialoshinskii. "Account of retardation in the interaction of neutral atoms." *ZhETF* 30 (June 1956): 1152–54 (L). [*Sov Phys JETP* 3 (Jan 1957): 977–79 (L).]

40. A. M. Korolev. "The dynamical magnetic moment of the deuteron." *ZhETF* 31 (Aug 1956): 211–17. [*Sov Phys JETP* 4 (Feb 1957): 73–79.]

41. A. V. Svidzinskii. "Determination of the Green's function in the Bloch-Nordsieck model by functional integration." *ZhETF* 31 (Aug 1956): 324–29. [*Sov Phys JETP* 4 (Mar 1957): 179–83.]

42. R. V. Polovin. "Radiative corrections to the scattering of electrons by electrons and positrons." *ZhETF* 31 (Sep 1956): 449–58. [*Sov Phys JETP* 4 (Apr 1957): 385–92.]

*43. B. L. Ioffe. "Dispersion relations for scattering and photoproduction." *ZhETF* 31 (Oct 1956): 583–95. [*Sov Phys JETP* 4 (May 1957): 534–44.]

44. G. N. Vialov. "Anomalous magnetic moments of nucleons." *ZhETF* 31 (Oct 1956): 620–24. [*Sov Phys JETP* 4 (May 1957): 562–65.]

*45. Iu. M. Lomsadze. "Concerning a certain possibility in quantum field theory." *ZhETF* 31 (Nov 1956): 887–89 (L). [*Sov Phys JETP* 4 (June 1957): 754–55 (L).]

46. V. V. Sudakov and K. A. Ter-Martirosian. "Consequences of renormalizability of pseudoscalar meson theory with two interaction constants." *ZhETF* 31 (Nov 1956): 899–901 (L). [*Sov Phys JETP* 4 (June 1957): 763–64 (L).]

*47. A. I. Alekseev. "The photoproduction cross section for positronium in an external field taking into account radiative corrections." *ZhETF* 31 (Nov 1956): 909–10 (L). [*Sov Phys JETP* 4 (June 1957): 771–73 (L).]

48. L. P. Gor'kov and I. M. Khalatnikov. "Electrodynamics of charged scalar particles." *ZhETF* 31 (Dec 1956): 1062–78. [*Sov Phys JETP* 4 (July 1957): 777–89.]

49. E. G. Melikian. "The internal Compton effect." *ZhETF* 31 (Dec 1956): 1088–90 (L). [*Sov Phys JETP* 4 (July 1956): 930–31 (L).]

1957

50. G. D'erdi. "A physical model of the hyperon." *ZhETF* 32 (Jan 1957): 152–54 (L). [*Sov Phys JETP* 5 (Aug 1957): 152–54 (L).]

*51. Iu. M. Popov. "Scattering of π-mesons on nucleons in higher approximations of the Tamm-Dancoff method." *ZhETF* 32 (Jan 1957): 169–71 (L). [*Sov Phys JETP* 5 (Aug 1957): 131–33 (L).]

*52. L. P. Gor'kov. "Two limiting momenta in scalar electrodynamics." *ZhETF* 32 (Feb 1957): 359–62. [*Sov Phys JETP* 5 (Sep 1957): 167–69.]

*53. V. S. Barashenkov. "The construction of a phenomenological scattering matrix with non-local interaction." *ZhETF* 32 (Feb 1957): 368–69 (L). [*Sov Phys JETP* 5 (Sep 1957): 313–15 (L).]

*54. E. G. Melikian. "Internal Compton effect in pair conversion." *ZhETF* 32 (Feb 1957): 384–85 (L). [*Sov Phys JETP* 5 (Sep 1957): 331–33 (L).]

*55. D. A. Kirzhnits. "Contribution to field theory involving a cut-off factor." *ZhETF* 32 (Mar 1957): 534–41. [*Sov Phys JETP* 5 (Oct 1957): 445–51.]

56. A. D. Galanin. "On the possibility of formulating a meson theory with several fields." *ZhETF* 32 (Mar 1957): 552–58. [*Sov Phys JETP* 5 (Oct 1957): 460–64.]

57. A. V. Romankevich. "Concerning the lifetime of the two forms of the π^0 meson." *ZhETF* 32 (Mar 1957): 615 (L). [*Sov Phys JETP* 5 (Oct 1957): 509 (L).]

*58. A. M. Brodskii. "On the derivation of the Low equation in the theory of meson scattering." *ZhETF* 32 (Mar 1957): 616–17 (L). [*Sov Phys JETP* 5 (Oct 1957): 509–11 (L).]

59. I. T. Diatlov, V. V. Sudakov, and K. A. Ter-Martirosian. "Asymptotic meson-meson scattering theory." *ZhETF* 32 (Apr 1957): 767–80. [*Sov Phys JETP* 5 (Nov 1957): 631–42.]

60. A. I. Alekseev. "Covariant equation for two annihilating particles." *ZhETF* 32 (Apr 1957): 852–62. [*Sov Phys JETP* 5 (Nov 1957): 696–704.]

*61. V. Z. Blank. "Application of a renormalized group to different scattering problems in quantum electrodynamics." *ZhETF* 32 (Apr 1957): 932–33 (L). [*Sov Phys JETP* 5 (Nov 1957): 759–61 (L).]

*62. V. G. Solov'ev. "Investigation of a model in quantum field theory." *ZhETF* 32 (May 1957): 1050–57. [*Sov Phys JETP* 5 (Dec 1957): 859–66.]

63. B. T. Geilikman. "Magnetic interaction of electrons and anomalous diamagnetism." *ZhETF* 32 (May 1957): 1206–11. [*Sov Phys JETP* 5 (Dec 1957): 981–85.]

*64. M. I. Riazanov. "Phenomenological study of the effect of nonconducting medium in quantum electrodynamics." *ZhETF* 32 (May 1957): 1244–46 (L). [*Sov Phys JETP* 5 (Dec 1957): 1013–15 (L).]

*65. S. N. Sokolov. "Photon Green function accurate to e^4." *ZhETF* 32 (May 1957): 1261–62. [*Sov Phys JETP* 5 (Dec 1957): 1029–30.]

66. A. A. Loganov and A. N. Tavkhelidze. “Dispersion relations for photoproduction of pions on nucleons.” *ZhETF* 32 (June 1957): 1393–1403. [*Sov Phys JETP* 5 (Dec 1957): 1134–44.]

*67. A. G. Galanin and Iu. N. Lokhov. “Convergence of the perturbation-theory series for a non-relativistic nucleon.” *ZhETF* 33 (July 1957): 285–86 (L). [*Sov Phys JETP* 6 (Jan 1958): 221–22 (L).]

68. Iu. I. Kulakov. “Application of matrix polynomials to determine scattering phases.” *ZhETF* 33 (Aug 1957): 501–13. [*Sov Phys JETP* 6 (Feb 1958): 391–400.]

69. I. Iu. Kobzarev. “On the possibility of $\pi \rightarrow e + \nu + \gamma$ decay.” *ZhETF* 33 (Aug 1957): 551–53 (L). [*Sov Phys JETP* 6 (Feb 1958): 431–32 (L).]

*70. Iu. A. Tarasov. “Bound states in positronium.” *ZhETF* 33 (Sep 1957): 706–9. [*Sov Phys JETP* 6 (Mar 1958): 542–44.]

71. A. I. Akhiezer, L. N. Rozentsveig, and I. M. Shmushkevich. “Scattering of electrons by protons.” *ZhETF* 33 (Sep 1957): 765–72. [*Sov Phys JETP* 6 (Mar 1958): 588–94.]

*72. Chou Guan-Chao. “Concerning a symmetry property of the new Gell-Mann theory.” *ZhETF* 33 (Oct 1957): 1058–59 (L). [*Sov Phys JETP* 6 (Apr 1958): 815–16 (L).]

*73. V. N. Gribov. “Determination of phases of matrix elements of the S-matrix.” *ZhETF* 33 (Dec 1957): 1431–36. [*Sov Phys JETP* 6 (June 1958): 1102–7.]

1958

74. N. N. Bogoliubov. “A new method in the theory of superconductivity.” Pt. 1. *ZhETF* 34 (Jan 1958): 58–65. [*Sov Phys JETP* 7 (July 1958): 41–46.]

75. V. V. Tolmachev and S. V. Tiablikov. “A new method in the theory of superconductivity.” Pt. 2. *ZhETF* 34 (Jan 1958): 66–72. [*Sov Phys JETP* 7 (July 1958): 46–50.]

76. N. N. Bogoliubov. “A new method in the theory of superconductivity.” Pt. 3. *ZhETF* 34 (Jan 1958): 73–79. [*Sov Phys JETP* 7 (July 1958): 51–55.]

77. V. M. Galitskii and A. B. Migdal. “Application of quantum field theory methods to the many body problem.” *ZhETF* 34 (Jan 1958): 139–50. [*Sov Phys JETP* 7 (July 1958): 96–104.]

78. V. M. Galitskii. “The energy spectrum of a non-ideal Fermi gas.” *ZhETF* 34 (Jan 1958): 151–62. [*Sov Phys JETP* 7 (July 1958): 104–12.]

79. V. I. Mamasakhlisov, S. G. Matinian, and M. E. Perel’man. “Photoproduction of strange particles on protons.” *ZhETF* 34 (Jan 1958): 195–97. [*Sov Phys JETP* 7 (July 1958): 133–34.]

80. S. T. Beliaev. “Application of the methods of quantum field theory to a system of bosons.” *ZhETF* 34 (Feb 1958): 417–32. [*Sov Phys JETP* 7 (Aug 1958): 289–99.]

81. S. T. Beliaev. “Energy-spectrum of a non-ideal Bose gas.” *ZhETF* 34 (Feb 1958): 433–46. [*Sov Phys JETP* 7 (Aug 1958): 299–307.]

82. L. B. Okun’. “Some remarks on a compound model of elementary particles.” *ZhETF* 34 (Feb 1958): 469–76. [*Sov Phys JETP* 7 (Aug 1958): 322–27.]

83. S. M. Bilen’kii. “Contributions to the theory of dispersion relations.” *ZhETF* 34 (Feb 1958): 518–19 (L). [*Sov Phys JETP* 7 (Aug 1958): 357–18 (L).]

84. I. Iu. Kobzarev and I. E. Tamm. “Strange-particle decays in the theory of Feynman and Gell-Mann.” *ZhETF* 34 (Apr 1958): 899–901. [*Sov Phys JETP* 7 (Oct 1958): 622–24.]

85. V. M. Galitskii. “Sound excitations in Fermi systems.” *ZhETF* 34 (Apr 1958): 1011–13 (L). [*Sov Phys JETP* 7 (Oct 1958): 698–99 (L).]

86. A. I. Alekseev. “Two-photon annihilation of positronium in the p-state.” *ZhETF* 34 (May 1958): 1195–1201. [*Sov Phys JETP* 7 (Nov 1958): 826–30.]

*87. S. V. Tiablikov and V. V. Tolmachev. "The interaction of electrons with lattice vibrations." *ZhETF* 34 (May 1958): 1254–57. [*Sov Phys JETP* 7 (Nov 1958): 867–69.]

*88. M. I. Riazanov. "Radiative corrections to Compton scattering taking into account polarization of the surrounding medium." *ZhETF* 34 (May 1958): 1258–66. [*Sov Phys JETP* 7 (Nov 1958): 869–75.]

89. A. B. Migdal. "Interaction between electrons and lattice vibrations in a normal metal." *ZhETF* 34 (June 1958): 1438–46. [*Sov Phys JETP* 7 (Dec 1958): 996–1001.]

90. A. M. Brodskii. "Dispersion relations and the derivation of the equations for K-meson scattering." *ZhETF* 34 (June 1958): 1531–38. [*Sov Phys JETP* 7 (Dec 1958): 1056–61.]

91. Chen Chun-Sian and Chow Shih-Hsun. "Energy spectrum of a high density electron gas." *ZhETF* 34 (June 1958): 1566–73. [*Sov Phys JETP* 7 (Dec 1958): 1080–85.]

92. L. D. Landau. "On the theory of the Fermi liquid." *ZhETF* 35 (July 1958): 97–103. [*Sov Phys JETP* 8 (Jan 1959): 70–74.]

93. I. T. Diatlov. "Photoproduction of electron and μ-meson pairs on nucleons." *ZhETF* 35 (July 1958): 154–58. [*Sov Phys JETP* 8 (Jan 1959): 108–11.]

94. Iu. A. Gol'fand. "On the theory of the weak interactions." Pt. 1. *ZhETF* 35 (July 1958): 170–77. [*Sov Phys JETP* 8 (Jan 1958): 118–23.]

95. V. G. Vaks and B. L. Ioffe. "On the $\pi \rightarrow e + \nu + \gamma$ decay." *ZhETF* 35 (July 1958): 221–27. [*Sov Phys JETP* 8 (Jan 1959): 151–56.]

96. I. Zlatev and P. S. Isaev. "Bremsstrahlung and pair production from protons with allowance for form factor." *ZhETF* 35 (July 1958): 309–10 (L). [*Sov Phys JETP* 8 (Jan 1959): 213–15 (L).]

*97. V. N. Gribov. "The spectral representation of the two-meson Green's function." *ZhETF* 35 (Aug 1958): 416–27. [*Sov Phys JETP* 8 (Febr 1959): 287–95.]

*98. I. M. Dremin. "Investigation of Ke_3-decay with the emission of a gamma photon." *ZhETF* 35 (Aug 1958): 515–17 (L). [*Sov Phys JETP* 8 (Feb 1959): 355–57 (L).]

99. N. F. Nelipa. "Elastic scattering of photons on excited nucleons." *ZhETF* 35 (Sep 1958): 662–67. [*Sov Phys JETP* 8 (Mar 1959): 460–63.]

100. I. Ia. Pomeranchuk. "Isotopic effect in the residual electrical resistance of metals." *ZhETF* 35 (Oct 1958): 992–95. [*Sov Phys JETP* 8 (Apr 1959): 693–95.]

101. D. A. Kirzhnits. "Correlation energy of an inhomogenous electron gas." *ZhETF* 35 (Nov 1958): 1198–1208. [*Sov Phys JETP* 8 (May 1959): 838–45.]

*102. Chen Chun-Sian. "A new method in statistical perturbation theory." *ZhETF* 35 (Dec 1958): 1518–21. [*Sov Phys JETP* 8 (June 1959): 1062–64.]

103. A. A. Ansel'm. "Asymptotic theory of a one-dimensional four-fermion interaction." *ZhETF* 35 (Dec 1958): 1522–31. [*Sov Phys JETP* 8 (June 1959): 1065–71.]

104. A. A. Abrikosov and L. P. Gor'kov. "On the theory of superconducting alloys." Pt. 1, "The electrodynamics of alloys at absolute zero." *ZhETF* 35 (Dec 1958): 1558–71. [*Sov Phys JETP* 8 (July 1959): 1090–98.]

1959

105. L. B. Okun' and I. Ya. Pomeranchuk. "Peripheral interactions between elementary particles." *ZhETF* 36 (Jan 1959): 300–12. [*Sov Phys JETP* 9 (July 1959): 207–15.]

*106. D. V. Shirkov. "On the compensation equation in superconductivity theory." *ZhETF* 36 (Feb 1959): 607–12. [*Sov Phys JETP* 9 (Aug 1959): 421–24.]

107. V. I. Ogievetskii. "Interaction between K and π mesons." *ZhETF* 36 (Feb 1959): 642–43 (L). [*Sov Phys JETP* 9 (Aug 1959): 447–48 (L).]

*108. A. A. Ansel'm. "A model of a field theory with nonvanishing renormalized charge." *ZhETF* 36 (Mar 1959): 863–68. [*Sov Phys JETP* 9 (Sep 1959): 608–11.]

109. A. A. Abrikosov, L. P. Gor'kov, and I. E. Dzyaloshinskii. "On the application of quantum-field-theory methods to problems of quantum statistics at finite temperatures." *ZhETF* 36 (Mar 1959): 900–908. [*Sov Phys JETP* 9 (Sep 1959): 636–41.]

110. W. Zoellner. "Dispersion relations and Chew-Low type equations for inelastic meson processes in the fixed source case." *ZhETF* 36 (Apr 1959): 1103–9. [*Sov Phys JETP* 9 (Oct 1959): 784–88.]

111. A. A. Vedenov and A. I. Larkin. "Equation of state of a plasma." *ZhETF* 36 (Apr 1959): 1133–42. [*Sov Phys JETP* 9 (Oct 1959): 806–11.]

112. L. P. Pitaevskii. "Properties of the spectrum of elementary excitations near the disintegration threshold of the excitations." *ZhETF* 36 (Apr 1959): 1168–78. [*Sov Phys JETP* 9 (Oct 1959): 830–37.]

113. Chen Chun-Sian and Chow Shih-Hsun. "The basic compensation equation in superconductivity theory when the Coulomb interaction is taken into account." *ZhETF* 36 (Apr 1959): 1246–53. [*Sov Phys JETP* 9 (Oct 1959): 885–89.]

*114. E. S. Fradkin. "The Green's function method in quantum statistics." *ZhETF* 36 (Apr 1959): 1286–98. [*Sov Phys JEPT* 9 (Oct 1959): 912–19.]

115. Ya. B. Zel'dovich. "The transformation $K_2^0 \rightarrow K_1^0$ by electrons." *ZhETF* 36 (May 1959): 1381–86. [*Sov Phys JETP* 9 (Nov 1959): 984–87.]

116. Yu. V. Tsekhmistrenko. "Some applications in the theory of metals of the method of summation of the main Feynman diagrams." *ZhETF* 36 (May 1959): 1546–49. [*Sov Phys JETP* 9 (Nov 1959): 1097–99.]

*117. A. F. Grashin. "Nucleon-nucleon scattering in high angular momentum states." *ZhETF* 36 (June 1959): 1717–24. [*Sov Phys JETP* 9 (Dec 1959): 1223–28.]

118. I. E. Dzyaloshinskii and L. P. Pitaevskii. "Van Der Waals forces in an inhomogenous dielectric." *ZhETF* 36 (June 1959): 1797–1805. [*Sov Phys JETP* 9 (Dec 1959): 1282–87.]

119. A. I. Alekseev. "Three-photon annihilation of positronium in the p state." *ZhETF* 36 (June 1959): 1839–44. [*Sov Phys JETP* 9 (Dec 1959): 1312–15.]

*120. V. G. Vaks. "Radiative deviations from the Coulomb law at small differences." *ZhETF* 36 (June 1959): 1882–89. [*Sov Phys JETP* 9 (Dec 1959): 1340–44.]

121. L. D. Landau. "On the analytic properties of vertex parts in quantum field theory." *ZhETF* 37 (July 1959): 62–70. [*Sov Phys JETP* 10 (Jan 1960): 45–50.]

*122. D. V. Shirkov. "On taking the Coulomb effects into account in the theory of superconductivity." *ZhETF* 37 (July 1959): 179–86. [*Sov Phys JETP* 10 (Jan 1960): 127–31.]

123. A. I. Larkin. "Passage of particles through a plasma." *ZhETF* 37 (July 1959): 264–72. [*Sov Phys JETP* 10 (Jan 1960): 186–91.]

*124. M. I. Shirokov. "On a symmetry in τ^0 decay." *ZhETF* 37 (July 1959): 328–29 (L). [*Sov Phys JETP* 10 (Jan 1960): 232–33 (L).]

125. Ya. I. Granovskii. "The electromagnetic interaction in the Heisenberg theory." *ZhETF* 37 (Aug 1959): 442–51. [*Sov Phys JETP* 10 (Feb 1960): 314–20.]

126. Yu. A. Gol'fand. "On the introduction of an 'elementary length' in the relativistic theory of elementary particles." *ZhETF* 37 (Aug 1959): 504–9. [*Sov Phys JETP* 10 (Feb 1960): 356–60.]

127. I. S. Zlatev and P. S. Isaev. "Dispersion relations for the virtual Compton effect." *ZhETF* 37 (Sep 1959): 728–34. [*Sov Phys JETP* 10 (Mar 1960): 519–23.]

*128. N. N. Bogolyubov, A. A. Logunov, and D. V. Shirkov. "The method of dispersion relations and perturbation theory." *ZhETF* 37 (Sep 1959): 805–15. [*Sov Phys JETP* 10 (Mar 1960): 574–81.]

*129. G. V. Avakov. "Electron-electron scattering and quantum electrodynamics at small distances." *ZhETF* 37 (Sep 1959): 848–49 (L). [*Sov Phys JETP* 10 (Mar 1960): 604 (L).]

130. L. I. Podlubnyi. "The non-additivity of London-Van Der Waals forces." *ZhETF* 37 (Sep 1959): 888–89 (L). [*Sov Phys JETP* 10 (Mar 1960): 633–34 (L).]

131. E. D. Zhizhin. "Scattering of a photon by a nucleon in the one-meson approximation." *ZhETF* 37 (Oct 1959): 994–99. [*Sov Phys JETP* 10 (Apr 1960): 707–10.]

132. K. A. Ter-Martirosyan. "Incompatibility of the conditions of analyticity and unitarity in the Lee model." *ZhETF* 37 (Oct 1959): 1005–9. [*Sov Phys JETP* 10 (Apr 1960): 714–17.]

133. G. M. Gandel'man and V. S. Pinaev. "Emission of neutrino pairs by electrons and the role played by it in stars." *ZhETF* 37 (Oct 1959): 1072–78. [*Sov Phys JETP* 10 (Apr 1960): 764–68.]

134. V. P. Kuznetsov. "On electromagnetic corrections in μ-e decay." *ZhETF* 37 (Oct 1959): 1102–5. [*Sov Phys JETP* 10 (Apr 1960): 784–86.]

135. I. S. Zlatev and P. S. Isaev. "Use of dispersion relations for a test of quantum electrodynamics at small distances." *ZhETF* 37 (Oct 1959): 1161–62 (L). [*Sov Phys JETP* 10 (Apr 1960): 826–27 (L).]

*136. Yu. V. Tsekhmistrenko. "Two-particle excitations of superfluid Fermi-systems." *ZhETF* 37 (Oct 1959): 1164–66 (L). [*Sov Phys JETP* 10 (Apr 1960): 829–30 (L).]

137. I. T. Dyatlov. "Expansion of the amplitude for a reaction with production of three low-energy particles in powers of the threshold momenta." *ZhETF* 37 (Nov 1959): 1330–36. [*Sov Phys JETP* 10 (May 1960): 947–51.]

138. Yu. M. Lomsadze, V. I. Lend'el, and B. M. Érnst. "Anomalous magnetic moments of nucleons in the Chew method." *ZhETF* 37 (Nov 1959): 1342–45. [*Sov Phys JETP* 10 (May 1960): 955–57.]

139. V. Ya. Fainberg. "Application of the dispersion relations method in quantum electrodynamics." *ZhETF* 37 (Nov 1959): 1361–71. [*Sov Phys JETP* 10 (May 1960): 968–74.]

140. L. P. Gor'kov. "Theory of superconducting alloys in a strong magnetic field near the critical temperature." *ZhETF* 37 (Nov 1959): 1407–16. [*Sov Phys JETP* 10 (May 1960): 998–1004.]

141. V. A. Lyul'ka and V. A. Filimonov. "The contribution of three-particle forces to the binding energies of hypernuclei." *ZhETF* 37 (Nov 1959): 1431–33. [*Sov Phys JETP* 10 (May 1960): 1015–17.]

142. V. S. Barashenkov. "Angular asymmetry in (π N) collisions and ($\pi\pi$) collisions." *ZhETF* 37 (Nov 1959): 1464–66 (L). [*Sov Phys JETP* 10 (May 1960): 1038–39 (L).]

*143. A. P. Bukhvostov and I. M. Shmushkevich. "Capture of polarized μ^- mesons by deuterons." *ZhETF* 37 (Nov 1959): 1471–73 (L). [*Sov Phys JETP* 10 (May 1960): 1042–44 (L).]

144. B. B. Dotsenko. "On the pionic and electromagnetic structure of nucleons." *ZhETF* 37 (Nov 1959): 1478–79 (L). [*Sov Phys JETP* 10 (May 1960): 1048–49 (L).]

145. Yu. A. Gol'fand. "On the masses of leptons." *ZhETF* 37 (Nov 1959): 1493–94 (L). [*Sov Phys JETP* 10 (May 1960): 1059–60 (L).]

146. A. D. Galanin, A. F. Grashin, B. L. Ioffe, and I. Ya. Pomeranchuk. "On the collisions of nucleons with large overall angular momentum." *ZhETF* 37 (Dec 1959): 1663–79. [*Sov Phys JETP* 10 (June 1960): 1179–89.]

147. V. L. Lyuboshitz. "Polarization phenomena in radiative collisions of two electrons." *ZhETF* 37 (Dec 1959): 1727–40. [*Sov Phys JETP* 10 (June 1960): 1221–28.]

*148. V. A. Shakhbazyan. "On the two-charge renormalization group in scalar quantum electrodynamics." *ZhETF* 37 (Dec 1959): 1789–93. [*Sov Phys JETP* 10 (June 1960): 1263–66.]

149. L. P. Pitaevskii. "On the superfluidity of liquid He3." *ZhETF* 37 (Dec 1959): 1794–1807. [*Sov Phys JETP* 10 (June 1960): 1267–75.]

APPENDIX F

Feynman Diagrams in Other Journals, 1950–54

Entries marked with an asterisk discussed elements of the diagrams, but did not include explicit figures. (L) indicates that the article was published as a brief letter to the editor.

1950

1. P. Benoist Gueutal, J. Prentki, and J. Ratier. "Sur la production de mésons nucléaires de spin 0 par les photons." *Le Journal de Physique et le Radium* 11 (Oct 1950): 553–63.

1951

2. W. Thirring. "Zum Kopplungsschema der Mesonen." *Zeitschrift für Naturforschung* 6 (1951): 53–54 (L).
3. H. Salecker. "Zur Frage eines Massenunterschiedes zwischen Elektron und Positron." *Zeitschrift für Naturforschung* 6 (1951): 484–86.
4. M. Cini and L. Radicati. "Sullo scattering dei mesoni da parte dei nucleoni." *Il Nuovo Cimento* 8 (May 1951): 317–25.
5. M. Cini and L. Radicati. "On the double scattering of mesons by nucleons." *Il Nuovo Cimento* 8 (Aug 1951): 542–51.
6. G. Petiau. "Sur la création de paires de corpuscules dans les processus de collisions entre corpuscules de spin $h/2$." *Le Journal de Physique et le Radium* 12 (Dec 1951): 911–19.

1952

7. Reinhard Oehme. "Zerfall neutraler Mesonen." *Zeitschrift für Naturforschung* 7 (1952): 55–60.
8. G. Lüders, R. Oehme, and W. Thirring. "π-Mesonen und quantisierte Feldtheorien." *Zeitschrift für Naturforschung* 7 (1952): 213–34.
9. P. Urban and H. Mitter. "Über die Streuung von schnellen Elektronen Vorläufige Mitteilung." *Zeitschrift für Naturforschung* 7 (1952): 818 (L).

10. R. Ascoli. "Interazioni non localizzabili: Esempio dell'effetto Compton." *Il Nuovo Cimento* 9 (Sep 1952): 757–63.
11. P. Havas. "Sur la création de paires de corpuscules dans les processus de collisions entre corpuscules de Spin $h/2$." *Le Journal de Physique et le Radium* 13 (Oct 1952): 438–40.

1953

12. W. Macke. "Zum relativistischen Zweikörperproblem der Quantenmechanik." Pt. 1. *Zeitschrift für Naturforschung* 8 (1953): 599–615.
13. H. Jordan and W. Frahn. "Nichtlokale Feldtheorie auf der Grundlage der Salpeter-Bethe-Gleichung." Pt. 1, "'Freie' Teilchen." *Zeitschrift für Naturforschung* 8 (1953): 620–28.
14. E. Freese. "Gebundene Teilchen und Streuprobleme in der Quantenfeldtheorie." *Zeitschrift für Naturforschung* 8 (1953): 776–90.
15. M. Cini. "A covariant formulation of the non-adiabatic method for the relativistic two-body problem." Pt. 1. *Il Nuovo Cimento* 10 (May 1953): 526–39.
16. S. Fubini. "A covariant non-adiabatic equation for nucleon-pion scattering." *Il Nuovo Cimento* 10 (June 1953): 851–54 (L).
17. P. Budini and C. Villi. "Teoria non locale de l'interazzione tra particelle di Fermi." *Il Nuovo Cimento* 10 (Aug 1953): 1172–86.
18. W. Macke. "Renormalised two particle potential from field theory." *Il Nuovo Cimento* 10 (Aug 1953): 1198–1200.
19. P. Budini. "Processi del second'ordine dell'interazione alla Fermi." *Il Nuovo Cimento* 10 (Sep 1953): 1299–1310.
20. H. Corben. "Long range nuclear forces." *Il Nuovo Cimento* 10 (Oct 1953): 1485.
21. P. Budini. "Processi del quart'ordine dell'interazione alla Fermi." *Il Nuovo Cimento* 10 (Oct 1953): 1486–88.

1954

22. K. Hain. "Über gebundene Zustände von π-Mesonen." *Zeitschrift für Naturforschung* 9 (1954): 495–508.
23. G. Eder. "Das Potential zwischen Nukleonen." *Zeitschrift für Naturforschung* 9 (1954): 565–72.
24. H. Jordan and W. Frahn. "Nichtlokale Feldtheorie auf der Grundlage der Salpeter-Bethe-Gleichung." *Zeitschrift für Naturforschung* 9 (1954): 572–78.
25. K. Symanzik. "Zur renormierten einzeitigen Bethe-Salpeter-Gleichung." *Il Nuovo Cimento* 11 (Jan 1954): 88–91 (L).
26. M. Lévy and R. Marshak. "The *S*-wave in pion-nucleon scattering." *Il Nuovo Cimento* 11 (Apr 1954): 366–71.
27. E. Caianiello. "On quantum field theory." Pt. 2, "Non-perturbative equations and methods." *Il Nuovo Cimento* 11 (May 1954): 492–529.
28. W. Zimmermann. "Yang-Feldmanformlismus und einzeitige Wellenfunktionen." *Il Nuovo Cimento* 11 (Jun 1954): 577–89.
29. L. Sartori and V. Wataghin. "Pion-nucleon scattering by variational method." *Il Nuovo Cimento* 12 (Aug 1954): 260–69.

30. E. Power. "No-pair terms in the adiabatic nuclear potential." *Il Nuovo Cimento* 12 (Sep 1954): 323–34.
31. W. Królikowski. "The wave function of photons and their configurational equation." *Il Nuovo Cimento* 12 (Dec 1954): 852–58.
32. J. Rayski. "On a systematization of heavy mesons and hyperons." *Il Nuovo Cimento* 12 (Dec 1954): 945–47.

Interviews

Unless otherwise indicated, all interviews were conducted by the author.

Adler, Stephen. 16 February 1999, via telephone.
Baranger, Michel. 18 January 2001, MIT.
Bethe, Hans. 8 May 1972. Interview with Charles Weiner. Transcript in *NBL,* call number OH39.
Case, Kenneth. 22 April 2002, La Jolla, California.
Chew, Geoffrey. 10 February 1998, Berkeley, California.
Dyson, Freeman. 8 January 2001, Princeton, New Jersey.
Feynman, Richard. 1966. Interview with Charles Weiner. Transcript in *NBL,* call number OH30140.
Finkelstein, Jerry. 24 July 1998, Berkeley, California.
Frazer, William. 7 July 1998, Berkeley, California.
Goldberger, Marvin. 23 April 2002, La Jolla, California.
Helmholz, A. Carl. 14 July 1998, Berkeley, California.
Kroll, Norman. 20 April 2002, La Jolla, California.
Low, Francis. 11 April 2001, MIT.
Mandelstam, Stanley. 6 August 1998, Berkeley, California.
Marshak, Robert. 1970. Interview with Charles Weiner. Transcript in *NBL,* call number OH308.
Salpeter, Edwin. 30 March 1978. Interview with Spencer Weart. Transcript in *NBL,* call number OH415.
Schrieffer, John. 26 September 1974. Interview with Joan Warnow and Robert Williams. Transcript in *NBL,* call number OH426.
Shugart, Howard. 29 July 1998, Berkeley, California.
Stapp, Henry. 21 August 1998, Berkeley, California.
Wichmann, Eyvind. 13 August 1998, Berkeley, California.

Bibliography

Aaserud, Finn. 1995. "Sputnik and the 'Princeton three': The national security laboratory that was not to be." *HSPS* 25:185–239.

Abrikosov, A. A. 1989. "Recollections of L. D. Landau." In *Landau: The Physicist and the Man,* ed. I. M. Khalatnikov, 29–35. New York: Pergamon.

Adler, Stephen, and Roger Dashen. 1968. *Current Algebras and Applications to Particle Physics.* New York: W. A. Benjamin.

Ahmadzadeh, Akbar. 1963. *A Numerical Study of the Regge Parameters in Potential Scattering.* Ph.D. dissertation, University of California, Berkeley.

Akhiezer, Alexander I. 1989. "Teacher and friend." In *Landau: The Physicist and the Man,* ed. I. M. Khalatnikov, 36–56. New York: Pergamon.

———. 1994. "Recollections of Lev Davidovich Landau." *PT* 47 (Jun): 35–42.

Akhiezer, A. I., and V. B. Berestetskii. 1953. *Kvantovaia Elekrodinamika* [Quantum Electrodynamics]. Moscow: Gosudartsvennoe Izdatelstvo Tekhniko-Teoreticheskoi Literatury.

———. 1957. "Quantum electrodynamics." Mimeographed Technical Report TR-2876. Oak Ridge, TN: U.S. Atomic Energy Commission.

———. 1965 [1959]. *Quantum Electrodynamics.* Trans. G. M. Volkoff. New York: Interscience.

Akhiezer, A., and R. Polovin. 1953. "Radiatsionnye popravki k rasseianiiu elektrona electronom" [Radiative corrections to electron-electron scattering]. *Doklady Akademii Nauk SSSR* 90:55–57.

Alborn, Timothy. 1989. "Negotiating notation: Chemical symbols and British society, 1831–1835." *Annals of Science* 46:437–60.

Alvarez, Luis. 1957. "Excerpts from a Russian diary." *PT* 10 (May): 24–32.

———. 1957. "Further excerpts from a Russian diary." *PT* 10 (Jun): 22–32.

———. 1987. *Alvarez: Adventures of a Physicist.* New York: Basic Books.

Anderson, H. L., W. C. Davidon, and U. E. Kruse. 1955. "Causality in the pion-proton scattering." *PR* 100:339–43.

Anderson, Philip. 2000. "Brainwashed by Feynman?" *PT* 53 (Feb): 11–12.

Anderton, Keith. 1993. *The Limits of Science: A Social, Political, and Moral Agenda for Epistemology in Nineteenth Century Germany.* Ph.D. dissertation, Harvard University.

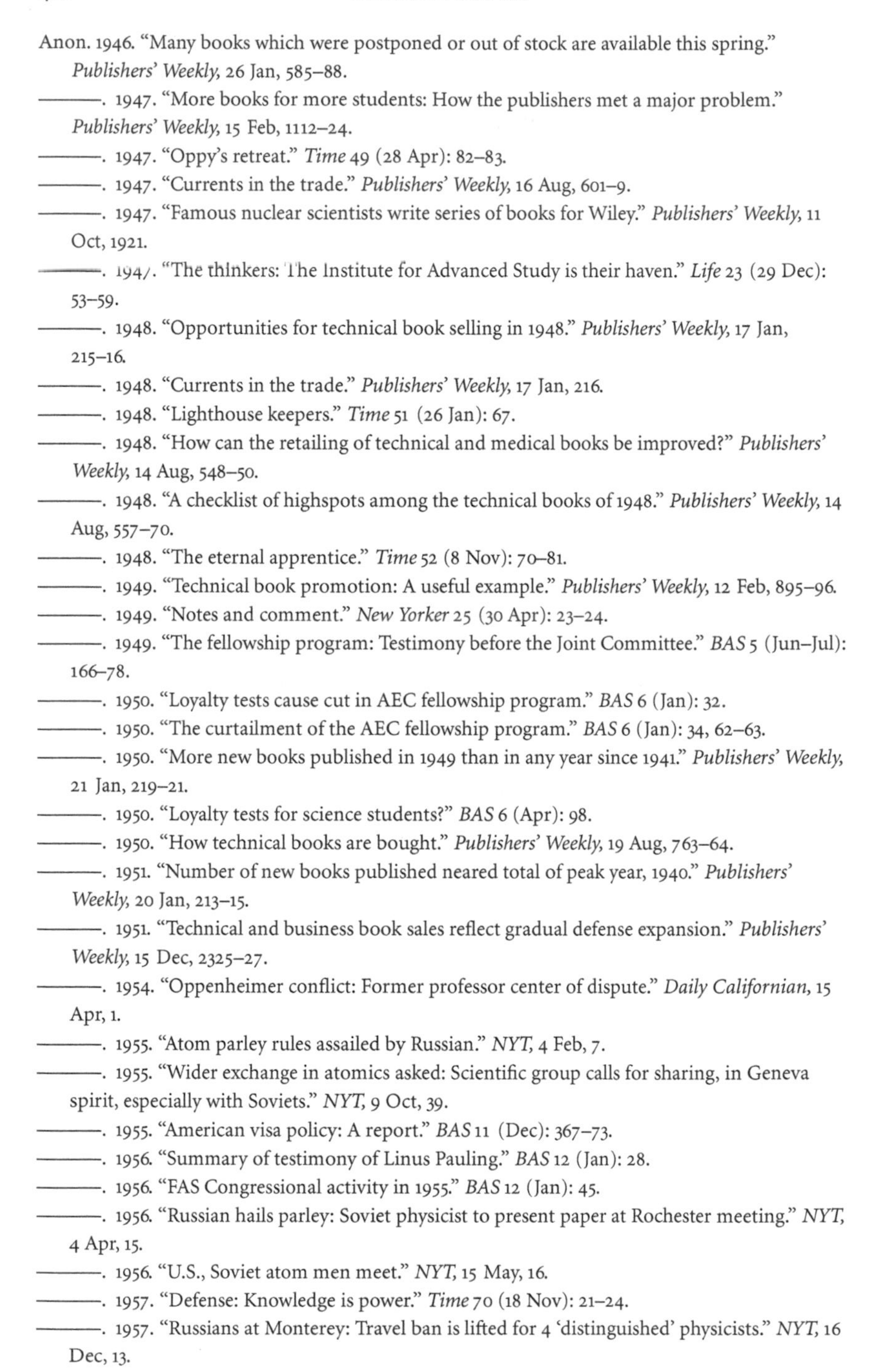

Anon. 1946. "Many books which were postponed or out of stock are available this spring." *Publishers' Weekly,* 26 Jan, 585–88.

———. 1947. "More books for more students: How the publishers met a major problem." *Publishers' Weekly,* 15 Feb, 1112–24.

———. 1947. "Oppy's retreat." *Time* 49 (28 Apr): 82–83.

———. 1947. "Currents in the trade." *Publishers' Weekly,* 16 Aug, 601–9.

———. 1947. "Famous nuclear scientists write series of books for Wiley." *Publishers' Weekly,* 11 Oct, 1921.

———. 1947. "The thinkers: The Institute for Advanced Study is their haven." *Life* 23 (29 Dec): 53–59.

———. 1948. "Opportunities for technical book selling in 1948." *Publishers' Weekly,* 17 Jan, 215–16.

———. 1948. "Currents in the trade." *Publishers' Weekly,* 17 Jan, 216.

———. 1948. "Lighthouse keepers." *Time* 51 (26 Jan): 67.

———. 1948. "How can the retailing of technical and medical books be improved?" *Publishers' Weekly,* 14 Aug, 548–50.

———. 1948. "A checklist of highspots among the technical books of 1948." *Publishers' Weekly,* 14 Aug, 557–70.

———. 1948. "The eternal apprentice." *Time* 52 (8 Nov): 70–81.

———. 1949. "Technical book promotion: A useful example." *Publishers' Weekly,* 12 Feb, 895–96.

———. 1949. "Notes and comment." *New Yorker* 25 (30 Apr): 23–24.

———. 1949. "The fellowship program: Testimony before the Joint Committee." *BAS* 5 (Jun–Jul): 166–78.

———. 1950. "Loyalty tests cause cut in AEC fellowship program." *BAS* 6 (Jan): 32.

———. 1950. "The curtailment of the AEC fellowship program." *BAS* 6 (Jan): 34, 62–63.

———. 1950. "More new books published in 1949 than in any year since 1941." *Publishers' Weekly,* 21 Jan, 219–21.

———. 1950. "Loyalty tests for science students?" *BAS* 6 (Apr): 98.

———. 1950. "How technical books are bought." *Publishers' Weekly,* 19 Aug, 763–64.

———. 1951. "Number of new books published neared total of peak year, 1940." *Publishers' Weekly,* 20 Jan, 213–15.

———. 1951. "Technical and business book sales reflect gradual defense expansion." *Publishers' Weekly,* 15 Dec, 2325–27.

———. 1954. "Oppenheimer conflict: Former professor center of dispute." *Daily Californian,* 15 Apr, 1.

———. 1955. "Atom parley rules assailed by Russian." *NYT,* 4 Feb, 7.

———. 1955. "Wider exchange in atomics asked: Scientific group calls for sharing, in Geneva spirit, especially with Soviets." *NYT,* 9 Oct, 39.

———. 1955. "American visa policy: A report." *BAS* 11 (Dec): 367–73.

———. 1956. "Summary of testimony of Linus Pauling." *BAS* 12 (Jan): 28.

———. 1956. "FAS Congressional activity in 1955." *BAS* 12 (Jan): 45.

———. 1956. "Russian hails parley: Soviet physicist to present paper at Rochester meeting." *NYT,* 4 Apr, 15.

———. 1956. "U.S., Soviet atom men meet." *NYT,* 15 May, 16.

———. 1957. "Defense: Knowledge is power." *Time* 70 (18 Nov): 21–24.

———. 1957. "Russians at Monterey: Travel ban is lifted for 4 'distinguished' physicists." *NYT,* 16 Dec, 13.

———. 1963. "Physics professor wins prize." *Daily Californian,* 3 Jan, 5.
———. 1966. "Scholars: Paradise in Princeton." *Time* 88 (8 July): 70–71.
———. 1970. "The Institute advances." *Newsweek* 46 (6 Apr): 54–55.
———. 2003. "Wired superstrings." *Scientific American* 288 (May): 38–39.
Aramaki, S. 1987. "Formation of the renormalization theory in quantum electrodynamics." *Historia Scientiarum* 32:1–42.
———. 1989. "Development of the renormalization theory in quantum electrodynamics." Pt. 2. *Historia Scientiarum* 37:91–113.
Arnowitt, Richard. 1952. *The Hyperfine Structure of Hydrogen.* Ph.D. dissertation, Harvard University.
Ashkin, J. 1960. "Experimental information on the strong interactions of pions and nucleons (photoproduction, pion scattering, pion production)." In *Proceedings of the 1960 Annual International Conference on High Energy Physics at Rochester,* ed. E. C. G. Sudarshan, J. H. Tinlot, and A. C. Melissinos, 634–36. Rochester: University of Rochester.
Asimov, Isaac. 1976. *Asimov on Physics.* Garden City, NY: Doubleday.
Assmus, Alexi. 1992. "The Americanization of molecular physics." *HSPS* 23:1–34.
———. 1993. "The creation of postdoctoral fellowships and the siting of American scientific research." *Minerva* 31:151–83.
Atomic Energy Commission. 1954. *In the Matter of J. Robert Oppenheimer: Transcript of Hearing before Personnel Security Board.* Washington, DC: Atomic Energy Commission.
Aznauryan, I. G. 2003. "Multipole amplitudes of pion photoproduction on nucleons up to 2 GeV within dispersion relations and unitary isobar model." *Physical Review C* 67:015209.
Badash, Lawrence. 2000. "Science and McCarthyism." *Minerva* 38:53–80.
Baigrie, Brian S., ed. 1996. *Picturing Knowledge: Historical and Philosophical Problems Concerning the Use of Art in Science.* Toronto: University of Toronto Press.
Balázs, Louis. 1966. "Lectures on bootstraps." In *Lectures on Mathematics and Physics.* Bombay: Tata Institute of Fundamental Research.
Baldin, A. M., and V. V. Mikhailov. 1950. "Obrazovanie edinichnykh mezonov γ-kvantami" [Formation of individual mesons by γ-quanta]. *ZhETF* 20:1057–63.
Baldini, R., et al. 2000. "Determination of nucleon and pion form factors via dispersion relations." *Nuclear Physics A* 666:38–43.
Ball, James S. 1960. *The Application of the Mandelstam Representation to Photoproduction of Pions from Nucleons.* Ph.D. dissertation, University of California, Berkeley.
Barnes, Barry, David Bloor, and John Henry. 1996. *Scientific Knowledge: A Sociological Analysis.* Chicago: University of Chicago Press.
Barnett, Lincoln. 1951. "Physicist Oppenheimer." In *Writing on Life: Sixteen Close-Ups,* by Lincoln Barnett, 345–83. New York: William Sloane.
Barthes, Roland. 1977. "The death of the author." In *Image, Music, Text,* ed. and trans. Stephen Heath, 142–48. New York: Hill and Wang.
Barton, Gabriel. 1965. *Introduction to Dispersion Techniques in Field Theory.* New York: W. A. Benjamin.
Bassett, Bruce, Fabrizio Tamburini, David Kaiser, and Roy Maartens. 1999. "Metric preheating and limitations of linearized gravity." *Nuclear Physics B* 561:188–240.
Beck, Carl. 1959. *Contempt of Congress.* New Orleans: Hauser.
Belinfante, F. J. 1948. "Supermultatempa Teorio por Mezonaj Kampoj." *PTP* 3:460–62.
———. 1949. "Pri la Kalkulado de Elektromagnetaj Fenomenoj per Kampo de Neutraj Vektor-Mezonoj kun Neglektebla Maso." *PTP* 4:165–70.

Beller, Mara. 1999. *Quantum Dialogue: The Making of a Revolution.* Chicago: University of Chicago Press.

Berelson, Bernard. 1960. *Graduate Education in the United States.* New York: McGraw-Hill.

———. 1962. "Postdoctoral work in American universities: A recent survey." *Journal of Higher Education* 33 (Mar): 119–30.

Berestetskii, V. B. 1952. "Teoriia vozmushchenii v kvantovoi elektrodinamike" [Perturbation theory in quantum electrodynamics]. *Uspekhi fizicheskikh nauk* 46:231–78.

Berestetskii, V. B., and L. D. Landau. 1949. "O vzaimodeistvii mezhdu elektronom i pozitronom" [On the electron positron interaction]. *ZhETF* 19:673–79.

Bernardini, G., E. T. Booth, L. Lederman, and J. Tinlot. 1950. "Nuclear interaction of π^--mesons." *PR* 80:924–25.

Bernstein, Barton J. 1982. "In the matter of J. Robert Oppenheimer." *HSPS* 12:195–252.

———. 2001. "Interpreting the elusive Robert Serber: What Serber says and what Serber does not explicitly say." *SHPMP* 32:443–86.

Bernstein, Jeremy. 1983. "The need to know." In *Asymptotic Realms of Physics: Essays in Honor of Francis E. Low,* ed. Alan Guth, Kerson Huang, and Robert Jaffe, xvii-xxiv. Cambridge: MIT Press.

———. 1987. *The Life It Brings: One Physicist's Beginnings.* New York: Ticknor and Fields.

Bess, Demaree. 1955. "Where the biggest ideas are born." *Saturday Evening Post* 228 (22 Oct): 25, 119–20.

Bethe, Hans. 1937. "Nuclear physics B: Nuclear dynamics, theoretical." *Reviews of Modern Physics* 9:69–244.

———. 1947. "The electromagnetic shift of energy levels." *PR* 72:339–41.

———. 1947. *Elementary Nuclear Theory: A Short Course on Selected Topics.* New York: John Wiley and Sons.

———. 1948. "Electromagnetic shift in energy levels." Unpublished report prepared for the eighth Solvay conference, held autumn 1948 in Brussels. Copy available in *HAB,* Folder 16:1.

———. 1953. "What holds the nucleus together?" *Scientific American* 189 (Sep): 58–63.

———. 1954. "Mesons and nuclear forces." *PT* 7 (Feb): 5–11.

———. 1956. "Nuclear many-body problem." *PR* 103:1353–90.

———. 1968. "J. Robert Oppenheimer, 1904–1967." *Biographical Memoirs of Fellows of the Royal Society* 14:391–416.

Bethe, Hans, and Robert Bacher. 1936. "Nuclear physics A: Stationary states of nuclei." *Reviews of Modern Physics* 8:82–229.

Bethe, Hans, and Enrico Fermi. 1932. "Über die Wechselwirkung von Zwei Electronen." *Zeitschrift für Physik* 77:296–306.

Bethe, Hans, and Jeffrey Goldstone. 1957. "Effect of a repulsive core in the theory of complex nuclei." *PRSA* 238:551–67.

Bethe, Hans, and J. Robert Oppenheimer. 1946. "Reaction of radiation on electron scattering and Heitler's theory of radiation damping." *PR* 70:451–58.

Bethe, H. A., and E. E. Salpeter. 1951. "A relativistic equation for bound state problems." *PR* 82:309–10.

Biagioli, Mario. 1993. *Galileo, Courtier: The Practice of Science in the Culture of Absolutism.* Chicago: University of Chicago Press.

Birge, Raymond T. Ca. 1970. *History of the Physics Department, University of California, Berkeley.* 5 vols. Copies in the Bancroft and Physics Department Libraries, University of California, Berkeley.

Bitboul, Roger, ed. 1976. *Retrospective Index to Theses of Great Britain and Ireland, 1716–1950.* Vol. 4: *Physical Sciences.* Santa Barbara: American Bibliographic Center, Clio Press.

Bjorken, James, and Sidney Drell. 1964. *Relativistic Quantum Mechanics.* New York: McGraw-Hill.

———. 1965. *Relativistic Quantum Fields.* New York: McGraw-Hill.

Blankenbecler, R., and M. L. Goldberger. 1962. "Behavior of scattering amplitudes at high energies, bound states, and resonances." *PR* 126:766–86.

Bleuler, K. M. 1950. "Eine neue Methode zur Behandlung der longitudinalen und skalaren Photonen." *Helvetica Physica Acta* 23:567–86.

Bogoliubov, N. N., and D. V. and Shirkov. 1959 [1957]. *Introduction to the Theory of Quantized Fields.* Trans. G. M. Volkoff. New York: Interscience.

Bogoljubow, N. N., and D. W. Schirkow. 1955. "Probleme der Quantentheorie der Felder." *Fortschrtitte der Physik* 3:439–95. Originally published in Russian in *Uspekhi fizicheskikh nauk* 55:149–214.

Bohm, David. 1952. "A suggested interpretation of the quantum theory in terms of 'hidden' variables." Pt. 1. *PR* 85:166–79.

———. 1952. "A suggested interpretation of the quantum theory in terms of 'hidden' variables." Pt. 2. *PR* 85:180–93.

Bohr, Niels. 1961. "The Solvay meetings and the development of quantum physics." In *The Quantum Theory of Fields,* ed. R. Stoop, 13–36. New York: Interscience.

Bohr, Niels, and Léon Rosenfeld. 1979 [1933]. "On the question of the measurability of electromagnetic field quantities." In *Selected Papers of Léon Rosenfeld,* ed. R. S. Cohen and J. J. Stachel, 357–400. Boston: Reidel.

Born, M., W. Heisenberg, and P. Jordan. 1926. "Zur Quantenmechanik." Pt. 2. *Zeitschrift für Physik* 35:557–615.

Boyd, Richard. 1991. "Confirmation, semantics, and the interpretation of scientific theories." In *The Philosophy of Science,* ed. Richard Boyd, Philip Gasper, and J. D. Trout, 3–23. Cambridge: MIT Press.

Brain, Robert, and M. Norton Wise. 1994. "Muscles and engines: Indicator diagrams and Helmholtz's graphical methods." In *Universalgenie Helmholtz: Rueckblick nach 100 Jahren,* ed. Lorenz Krueger, 124–49. Berlin: Akademie Verlag.

Breit, Gregory. 1947. "Relativistic corrections to magnetic moments of nuclear particles." *PR* 71:400–402.

———. 1947. "Does the electron have an intrinsic magnetic moment?" *PR* 72:984.

Breit, Gregory, and M. H. Hull. 1953. "Advances in knowledge of nuclear forces." *AJP* 21:183–220.

Brisudova, M. M., et al. 2003. "Nonlinear Regge trajectories and glueballs." Preprint nucl-th/0303012 (Mar), available from http://www.arXiv.org.

Brode, Bernice. 1980. "Tales of Los Alamos." In *Reminiscences of Los Alamos, 1943–1945,* ed. L. Badash, J. Hirschfelder, and H. Broida, 133–59. Boston: D. Reidel.

Bromberg, Joan. 1976. "The concept of particle creation before and after quantum mechanics." *HSPS* 7:161–91.

Brooke, John Hedley. 2000. "Introduction: The study of chemical textbooks." In *Communicating Chemistry: Textbooks and Their Audiences, 1789–1939,* ed. Anders Lundgren and Bernadette Bensaude-Vincent, 1–18. Canton, MA: Science History Publications.

Brown, Laurie. 1981. "Yukawa's prediction of the meson." *Centaurus* 25:71–132.

———, ed. 1993. *Renormalization: From Lorentz to Landau (and Beyond).* New York: Springer.

Brown, Laurie, Max Dresden, and Lillian Hoddeson, eds. 1989. *Pions to Quarks: Particle Physics in the 1950s.* New York: Cambridge University Press.

Brown, Laurie, Rokuo Kawabe, Michiji Konuma, and Zirō Maki, eds. 1988. *Elementary Particle Theory in Japan, 1935–1960*. Kyoto: Yukawa Hall Archival Library.

Brown, Laurie, and Yoichiro Nambu. 1998. "Physicists in wartime Japan." *Scientific American* 279 (Dec): 96–103.

Brown, Laurie, and Helmut Rechenberg. 1990. "Landau's work on quantum field theory and high-energy physics (1930–61)." In *Frontiers of Physics: Proceedings of the Landau Memorial Conference*, ed. Errol Gotsman, Yuval Ne'eman, and Alexander Voronel, 53–81. New York: Pergamon.

———. 1996. *The Origin of the Concept of Nuclear Forces*. Philadelphia: Institute of Physics Publishing.

Brown, Laurie, and John Rigden, eds. 1993. *Most of the Good Stuff: Memories of Richard Feynman*. New York: American Institute of Physics.

Brown, Laurie, Michael Riordan, Max Dresden, and Lillian Hoddeson. 1997. "The Rise of the Standard Model: 1964–1979." In *The Rise of the Standard Model: Particle Physics in the 1960s and 1970s*, ed. Lillian Hoddeson, Laurie Brown, Michael Riordan, and Max Dresden, 3–35. New York: Cambridge University Press.

Brown, Lowell S. 1992. *Quantum Field Theory*. New York: Cambridge University Press.

Brown, Sanborn. 1966. "Post-doctoral training." In *The Education of a Physicist*, ed. Sanborn Brown and Norman Clarke, 47–56. Cambridge: MIT Press.

Brueckner, K. A. 1955. "Many-body problem for strongly interacting particles." Pt. 2, "Linked cluster expansion." *PR* 100:36–45.

Brueckner, K. A., and C. A. Levinson. 1955. "Approximate reduction of the many-body problem for strongly interacting particles to a problem of self-consistent fields." *PR* 97:1344–52.

Bryan, Ronald. 1994. "Marshak's nucleon-nucleon theory program at Rochester in the 1950s." In *A Gift of Prophecy: Essays in Celebration of the Life of Robert Eugene Marshak*, ed. E. C. G. Sudarshan, 50–55. Singapore: World Scientific.

Buchwald, Jed, ed. 1995. *Scientific Practice: Theories and Stories of Doing Physics*. Chicago: University of Chicago Press.

Calvert, Jack, James Pitts, and George Dorion. 1972. *Graduate School in the Sciences: Entrance, Survival, and Careers*. New York: Wiley-Interscience.

Cambrosio, Alberto, Daniel Jacobi, and Peter Keating. 1993. "Ehrlich's 'beautiful pictures' and the controversial beginnings of immunological imagery." *Isis* 84:662–99.

Cao, Tian Yu. 1991. "The Reggeization program, 1962–82: Attempts at reconciling quantum field theory with *S*-matrix theory." *Archive for the History of the Exact Sciences* 41:239–83.

———. 1993. "New philosophy of renormalization: From the renormalization group to effective field theories." In *Renormalization: From Lorentz to Landau (and Beyond)*, ed. Laurie Brown, 87–133. New York: Springer.

———. 1997. *Conceptual Developments of 20th Century Field Theories*. New York: Cambridge University Press.

Cao, Tian Yu, and Silvan Schweber. 1993. "The conceptual foundations and the philosophical aspects of renormalization theory." *Synthèse* 97:33–108.

Capella, A., et al. 1985. "The pomeron story." In *A Passion for Physics: Essays in Honor of Geoffrey Chew*, ed. Carleton DeTar, J. Finkelstein, and Chung-I Tan, 79–87. Singapore: World Scientific.

Carson, Cathryn. 1995. *Particle Physics and Cultural Politics: Werner Heisenberg and the Shaping of a Role for the Physicist in Postwar West Germany*. Ph.D. dissertation, Harvard University.

———. 1996. "The peculiar notion of exchange forces." Pt. 1, "Origins in quantum mechanics, 1926–1928." *SHPMP* 27:23–45.

———. 1996. "The peculiar notion of exchange forces." Pt. 2, "From nuclear force to QED, 1929–1950." *SHPMP* 27:99–131.

Cartwright, Nancy. 1983. *How the Laws of Physics Lie*. New York: Oxford University Press.

———. 1997. "Models: The blueprints for laws." *Philosophy of Science* 64, suppl.:S292–S303.

———. 1999. *The Dappled World: A Study of the Boundaries of Science*. New York: Cambridge University Press.

Cartwright, Nancy, Towfic Shomar, and Mauricio Suárez. 1995. "The tool box of science: Tools for the building of models with a superconductivity example." *Poznan Studies in the Philosophy of Science and the Humanities* 44:137–49.

Case, Kenneth. 1949. "Equivalence theorems for meson-nucleon couplings." *PR* 76:14–17.

Cassidy, David. 1981. "Cosmic ray showers, high energy physics, and quantum field theories: Programmatic interactions in the 1930s." *HSPS* 12:1–39.

Castillejo, L., R. H. Dalitz, and F. J. Dyson. 1956. "Low's scattering equation for the charged and neutral scalar theories." *PR* 101:453–58.

Chamberlain, Owen. 1985. "Interactions with Geoff Chew." In *A Passion for Physics: Essays in Honor of Geoffrey Chew*, ed. Carleton DeTar, J. Finkelstein, and Chung-I Tan, 11–13. Singapore: World Scientific.

Chartier, Roger. 1992. *The Order of Books: Readers, Authors, and Libraries in Europe between the Fourteenth and Eighteenth Centuries*, trans. Lydia Cochrane. Stanford: Stanford University Press.

Cheston, W. B. 1952. "The interaction of a charged pi-meson with the deuteron." *PR* 85:952–61.

Chevalley, Catherine. 1996. "Physics as an art: The German tradition and the symbolic turn in philosophy, history of art and natural science in the 1920s." In *The Elusive Synthesis: Aesthetics and Science*, ed. A. I. Tauber, 227–49. Boston: Kluwer.

Chew, Geoffrey. 1950. "Academic freedom on trial at the University of California." *BAS* 6 (Nov): 333–36.

———. 1956. "Passport problems." *BAS* 12 (Jan): 26–28.

———. 1958. "Proposal for determining the pion-nucleon coupling constant from the angular distribution for nucleon-nucleon scattering." *PR* 112:1380–83.

———. 1958. "The nucleon and its interactions with pions, photons, nucleons, and antinucleons: Theoretical." Pt. 1. In *Proceedings of the 1958 Annual Conference on High Energy Physics at CERN*, ed. B. Ferretti, 93–113. Geneva: CERN.

———. 1959. "The pion-nucleon interaction and dispersion relations." *Annual Review of Nuclear Science* 9:29–60.

———. 1960. "Double dispersion relations and unitarity as the basis for a dynamical theory of strong interactions." In *Relations de dispersion et particules élémentaires*, ed. C. DeWitt and R. Omnès, 455–514. Paris: Hermann.

———. 1961. "Double dispersion relations and unitarity as the basis for a dynamical theory of strong interactions." In *Dispersion Relations: Scottish Universities' Summer School, 1960*, ed. G. R. Screaton, 167–258. New York: Interscience.

———. 1961. *S-Matrix Theory of Strong Interactions*. New York: W. A. Benjamin.

———. 1962. "S-matrix theory of strong interactions without elementary particles." *Reviews of Modern Physics* 34:394–401.

———. 1963. "The dubious role of the space-time continuum in microscopic physics." *Science Progress* 51:529–539.

———. 1963. "Strong-interaction S-matrix theory without elementary particles." In *1962 Cargèse Lectures in Theoretical Physics*, ed. Maurice Lévy, chap. 11, pp. 1–37. New York: W. A. Benjamin.

———. 1964. "Nuclear democracy and bootstrap dynamics." In *Strong-Interaction Physics: A Lecture Note Volume,* by Maurice Jacob and Geoffrey Chew, 103–52. New York: W. A. Benjamin.

———. 1966. *The Analytic S Matrix: A Basis for Nuclear Democracy.* New York: W. A. Benjamin.

———. 1968. "Bootstrap: A scientific idea?" *Science* 161 (23 Aug): 762–65.

———. 1970. "Hadron bootstrap: Triumph or frustration?" *PT* 23 (Oct): 23–28.

———, ed. 1971. *Multiperipheral Dynamics.* New York: Gordon and Breach.

———. 1974. "Impasse for the elementary-particle concept." In *The Great Ideas Today, 1974,* ed. Robert Hutchins and Mortimer Adler, 92–125. Chicago: Encyclopedia Britannica.

———. 1989. "Particles as S-matrix poles: Hadron democracy." In *Pions to Quarks: Particle Physics in the 1950s,* ed. Laurie Brown, Max Dresden, and Lillian Hoddeson, 600–607. New York: Cambridge University Press.

Chew, Geoffrey, and Fritjof Capra. 1985. "Bootstrap physics: A conversation with Geoffrey Chew." In *A Passion for Physics: Essays in Honor of Geoffrey Chew,* ed. Carleton DeTar, J. Finkelstein, and Chung-I Tan, 247–86. Singapore: World Scientific.

Chew, G. F., and S. C. Frautschi. 1960. "Unified approach to high- and low-energy strong interactions on the basis of the Mandelstam representation." *PRL* 5:580–83.

———. 1961. "Principle of equivalence for all strongly-interacting particles within the S-matrix framework." *PRL* 7:394–97.

Chew, G. F., S. C. Frautschi, and S. Mandelstam. 1962. "Regge poles in $\pi\pi$ scattering." *PR* 126:1202–8.

Chew, Geoffrey, Murray Gell-Mann, and Arthur Rosenfeld. 1964. "Strongly interacting particles." *Scientific American* 210:74–93.

Chew, G. F., M. L. Goldberger, F. E. Low, and Y. Nambu. 1957. "Application of dispersion relations to low-energy meson-nucleon scattering." *PR* 106:1337–44.

———. 1957. "Relativistic dispersion relation approach to photomeson production." *PR* 106:1345–55.

Chew, G. F., and F. E. Low. 1959. "Unstable particles as targets in scattering experiments." *PR* 113:1640–48.

Chew, G. F., and S. Mandelstam. 1960. "Theory of low-energy pion-pion interaction." *PR* 119:467–77.

———. 1961. "Theory of low-energy pion-pion interaction." Pt. 2. *Il Nuovo Cimento* 19:752–76.

Chomsky, Noam, et al. 1997. *The Cold War and the University: Toward an Intellectual History of the Postwar Years.* New York: New Press.

Chu, Shu-yuan. 1966. *A Study of Multi-channel Dynamics in the New Strip Approximation.* Ph.D. dissertation, University of California, Berkeley.

Cini, Marcello. 1980. "The history and ideology of dispersion relations: The pattern of internal and external factors in a paradigm shift." *Fundamenta Scientiae* 1:157–72.

Clarke, Adele, and Joan Fujimura, eds. 1992. *The Right Tools for the Job: At Work in Twentieth-Century Life Sciences.* Princeton: Princeton University Press.

Coben, Stanley. 1971. "The scientific establishment and the transmission of quantum mechanics to the United States, 1919–32." *American Historical Review* 76:442–66.

Collins, Harry. 1974. "The TEA set: Tacit knowledge and scientific networks." *Social Studies of Science* 4:165–86.

———. 1992 [1985]. *Changing Order: Replication and Induction in Scientific Practice.* 2nd ed. Chicago: University of Chicago Press.

———. 1990. *Artificial Experts: Social Knowledge and Intelligent Machines.* Cambridge: MIT Press.

———. 2001. "Tacit knowledge, trust and the Q of sapphire." *Social Studies of Science* 31:71–85.
Collins, H. M., G. H. de Vries, and W. E. Bijker. 1997. "Ways of going on: An analysis of skill applied to medical practice." *Science, Technology, and Human Values* 22:267–85.
Collins, Randall. 1998. *The Sociology of Philosophies: A Global Theory of Intellectual Change.* Cambridge: Harvard University Press.
Corry, John. 1966. "Visit to 'an intellectual hotel.'" *NYT Magazine,* 15 May, 50–70.
Creager, Angela. 2002. *The Life of a Virus: Tobacco Mosaic Virus as an Experimental Model, 1930–1965*. Chicago: University of Chicago Press.
Crease, Robert, and Charles Mann. 1986. *The Second Creation: Makers of the Revolution in Twentieth-Century Physics.* New York: Macmillan.
Cressy, David. 1986. "Books as totems in seventeenth-century England and New England." *Journal of Library History* 21:92–106.
Cushing, James. 1985. "Is there just one possible world? Contingency vs. the bootstrap." *SHPS* 16:31–48.
———. 1986. "The importance of Heisenberg's *S*-matrix program for the theoretical high-energy physics of the 1950s." *Centaurus* 29:110–49.
———. 1990. *Theory Construction and Selection in Modern Physics: The S Matrix.* New York: Cambridge University Press.
———. 1994. *Quantum Mechanics: Historical Contingency and the Copenhagen Hegemony.* Chicago: University of Chicago Press.
Cutkosky, Richard E. 1960. "Singularities and discontinuities of Feynman amplitudes." *Journal of Mathematical Physics* 1:429–33.
Cziffra, Peter. 1960. *The Two-Pion Exchange Contribution to the Higher Partial Waves of Nucleon-Nucleon Scattering.* Ph.D. dissertation, University of California, Berkeley.
Cziffra, P., M. H. MacGregor, M. J. Moravcsik, and H. P. Stapp. 1959. "Modified analysis of nucleon-nucleon scattering." Pt. 1, "Theory and *p-p* scattering at 310 Mev." *PR* 114:880–86.
Cziffra, P., and M. J. Moravcsik. 1959. "Determination of the pion-nucleon coupling constant from *n-p* scattering angular distribution." *PR* 116:226–30.
Dalitz, R. H. 1962. *Strange Particles and Strong Interactions.* New York: Oxford University Press.
Dalitz, Richard H., and Frank J. Duarte. 2000. "Obituary: John Clive Ward." *PT* 53 (Sep): 99.
Darnton, Robert. 1984. *The Great Cat Massacre, and Other Episodes in French Cultural History.* New York: Vintage.
Darrigol, Olivier. 1986. "The origin of quantized matter waves." *HSPS* 16:197–253.
———. 1988. "The quantum electrodynamical analogy in early nuclear theory; or, The roots of Yukawa's theory." *Revue d'Histoire des Sciences* 41:225–97.
———. 1988. "Elements of a scientific biography of Tomonaga Sin-itiro." *Historia Scientiarum* 35:1–29.
———. 1992. *From c-Numbers to q-Numbers: The Classical Analogy in the History of Quantum Theory.* Berkeley: University of California Press.
Davidon, W. C., and M. L. Goldberger. 1956. "Comparison of spin-flip dispersion relations with pion-nucleon scattering data." *PR* 104:1119–21.
Davies, Lawrence. 1957. "Soviet physicists at coast meeting: Attend the opening session of nuclear conference—3 to speak on Friday." *NYT,* 18 Dec, 6.
———. 1957. "Russian gives U.S. pure science lead: But top physicist declares that Soviet is closing gap—Stanford work hailed." *NYT,* 22 Dec, 2.
Davis, Nuel Pharr. 1968. *Lawrence and Oppenheimer.* New York: Simon and Schuster.
Davis, Watson. 1962. "Thinkers unlimited." *Américas* 14 (Apr): 14–17.

de Grazia, Margreta, and Peter Stallybrass. 1993. "The materiality of the Shakespearean text." *Shakespeare Quarterly* 44:255–83.

Delamont, Sara, and Paul Atkinson. 2001. "Doctoring uncertainty: Mastering craft knowledge." *Social Studies of Science* 31:87–107.

De Maria, M., and A. Russo. 1985. "The discovery of the positron." *Rivista di Storia della Scienza* 2:237–86.

Desai, Bipin. 1961. *Low-Energy Pion-Photon Interaction: The* $(2\pi, 2\gamma)$ *Vertex.* Ph.D. dissertation, University of California, Berkeley.

Deser, Stanley. 1953. *Relativistic Two-Body Interactions.* Ph.D. dissertation, Harvard University.

Deser, S., et al., eds. 1979. *Themes in Contemporary Physics: Essays in Honour of Julian Schwinger's 60th birthday.* Amsterdam: North-Holland.

DeTar, Carleton. 1985. "What are the quark and gluon poles?" In *A Passion for Physics: Essays in Honor of Geoffrey Chew,* ed. Carleton DeTar, J. Finkelstein, and Chung-I Tan, 71–78. Singapore: World Scientific.

DeTar, Carleton, J. Finkelstein, and Chung-I Tan, eds. 1985. *A Passion for Physics: Essays in Honor of Geoffrey Chew.* Singapore: World Scientific.

DeWitt, C., and R. Omnès, eds. 1960. *Relations de dispersion et particules élémentaires.* Paris: Hermann.

Diamond, Sigmund. 1992. *Compromised Campus: The Collaboration of Universities with the Intelligence Community, 1945–1955.* New York: Oxford University Press.

Dirac, P. A. M. 1959 [1927]. "The quantum theory of the emission and absorption of radiation." In *Selected Papers on Quantum Electrodyanmics,* ed. Julian Schwinger, 1–23. New York: Dover.

———. 1928. "The quantum theory of the electron." Pts. 1, 2. *PRSA* 117:610–24; 118:351–61.

———. 1930. "A theory of electrons and protons." *PRSA* 126:360–65.

———.1994 [1934]. "Theory of the positron." In *Early Quantum Electrodyanmics: A Source Book,* ed. Arthur I. Miller, 136–44. New York: Cambridge University Press.

Dolen, R., D. Horn, and C. Schmid. 1968. "Finite-energy sum rules and their application to πN charge exchange." *PR* 166:1768–81.

Douglass, D. H., and M. W. P. Strandberg. 1963. "Stages in the education of a physicist: An attempted solution of a pedagogical problem." *AJP* 31 (Sep): 707–12.

Dower, John. 1999. *Embracing Defeat: Japan in the Wake of World War II.* New York: Norton.

Drell, Sidney. 1961. "Peripheral contributions to high-energy interaction processes." *Reviews of Modern Physics* 33:458–66.

Dresden, Max. 1987. *H. A. Kramers: Between Tradition and Revolution.* New York: Springer.

———. 1993. "Renormalization in historical perspective: The first stage." In *Renormalization: From Lorentz to Landau (and Beyond),* ed. Laurie Brown, 31–55. New York: Springer.

Dumit, Joseph. 2004. *Picturing Personhood: Brain Scans and Biomedical Identity.* Princeton: Princeton University Press.

Dyson, Freeman. 1947. "The approximation to algebraic numbers by rationals." *Acta Mathematica* 89:225–40. Reprinted in *Selected Papers of Freeman Dyson,* ed. Freeman Dyson (Providence: American Mathematical Society, 1996), 65–80.

———. 1948. "The electromagnetic shift of energy levels." *PR* 73:617–26.

———. 1951. "Advanced quantum mechanics." Mimeographed notes (by Stan Cohen, Don Edwards, and Carl Greifinger) from lectures delivered at Cornell during autumn 1951. Copy available in the Physics Department Library, University of California, Berkeley.

———. 1952. "Quantum electrodynamics." *PT* 5 (Sep): 6–9.

———. 1952. "Divergence of perturbation theory in quantum electrodynamics." *PR* 85:631–32.

———. 1953. "Field theory." *Scientific American* 188 (Apr): 57–64.

———. 1953. "The dynamics of a disordered linear chain." *PR* 92:1331–38.

———. 1954. "Advanced quantum mechanics." Rev. ed. Mimeographed notes from lectures delivered at Cornell during autumn 1951 and later updated. Copy available in the Physics Department Library, University of California, Berkeley.

———. 1955. Review of *Mesons and Fields*, vol. 2, by Silvan S. Schweber, Hans A. Bethe, and Frederic de Hoffmann. *PT* 8 (Jun): 27.

———. 1955. Review of *Einführung in die Quantenelektrodynamik*, by Walter E. Thirring. *PT* 8 (Dec): 22.

———. 1956. Review of *Mesons and Fields*, vol. 1, by Silvan S. Schweber, Hans A. Bethe, and Frederic de Hoffmann. *PT* 9 (May): 32–34.

———. 1956. "General theory of spin-wave interactions." *PR* 102:1217–30.

———. 1965. "Old and new fashions in field theory." *PT* 18 (June): 21–24.

———. 1970. "Reflections: The sellout." *New Yorker* 46 (21 Feb): 44–59.

———. 1979. *Disturbing the Universe*. New York: Basic Books.

———, ed. 1996. *Selected Papers of Freeman Dyson with Commentary*. Providence: American Mathematical Society.

———. 1996. "Comments on selected papers." In *Selected Papers of Freeman Dyson with Commentary*, ed. Freeman Dyson, 2–49. Providence: American Mathematical Society.

Dzyaloshinskii, I. E. 1989. "Landau through a pupil's eyes." In *Landau: The Physicist and the Man*, ed. I. M. Khalatnikov, 89–96. New York: Pergamon.

Eberhard, Mark. 1999. "Why things break." *Scientific American* 281 (Oct): 66–73.

Eden, Richard. 1961. "Lectures on the use of perturbation methods in dispersion theory." Mimeographed notes from lectures delivered at the University of Maryland during March and April 1961. University of Maryland Physics Department Technical Report 211. Copy in the possession of Sam Schweber.

———. 1967. *High Energy Collisions of Elementary Particles*. New York: Cambridge University Press.

Eden, R., P. Landshoff, D. Olive, and J. Polkinghorne. 1966. *The Analytic S-Matrix*. Cambridge: Cambridge University Press.

Edgerton, Samuel. 1984. "Galileo, Florentine 'disegno,' and the 'strange spottednesse' of the moon." *Art Journal* 44:225–32.

Edgerton, Samuel, and Michael Lynch. 1988. "Aesthetics and digital image processing: Representational craft in contemporary astronomy." In *Picturing Power: Visual Depiction and Social Relations*, ed. Gordon Fyfe and John Law, 184–219. London: Routledge.

Edwards, S. F. 1955. "Nucleon propagators in functional form." In *The Proceedings of the 1954 Glasgow Conference on Nuclear and Meson Physics*, ed. E. H. Bellamy and R. G. Moorhouse, 299–302. New York: Pergamon.

———. 1979. "Disordered systems as quantum field theories." In *Themes in Contemporary Physics: Essays in honour of Julian Schwinger's 60th birthday*, ed. S. Deser et al., 212–22. Amsterdam: North-Holland.

Eisenstadt, Jean. 1989. "The low water mark of general relativity, 1925–1955." In *Einstein and the History of General Relativity*, ed. Don Howard and John Stachel, 277–92. Boston: Birkhäuser.

Eisenstein, Elizabeth. 1979. *The Printing Press as an Agent of Change: Communications and Cultural Transformaitons in Early-Modern Europe*. 2 vols. New York: Cambridge University Press.

———. 2002. "An unacknowledged revolution revisited." *American Historical Review* 107:87–105.

Elkins, James. 1995. "Art history and images that are not art." *Art Bulletin* 77:553–72.

———. 1999. *The Domain of Images.* Ithaca: Cornell University Press.

Endō, Shinji, Tōichirō Kinoshita, and Zirō Koba. 1948. "Reactive corrections for the elastic scattering of an electron." *PTP* 3:320–21.

———. 1949. "Errata." *PTP* 4:100–101.

———. 1949. "Reactive corrections for the elastic scattering of an electron." *PTP* 4:218–29.

Erwin, A. R., R. March, W. D. Walker, and E. West. 1961. "Evidence for a π-π resonance in the $I = 1, J = 1$ state." *PRL* 6:628–30.

Euler, Hans. 1936. "Über die Streuung von Licht an Licht nach der Diracschen Theorie." *Annalan der Physik* 26:398–448.

Euler, H., and B. Kockel. 1935. "Über die Streuung von Licht an Licht nach der Diracschen Theorie." *Die Naturwissenschaften* 15:246–47.

Fairlie, D. B., P. V. Landshoff, J. Nuttall, and J. C. Polkinghorne. 1962. "Singularities of the second type." *Journal of Mathematical Physics* 3:594–602.

Fermi, Enrico, and C. N. Yang. 1949. "Are mesons elementary particles?" *PR* 76:1739–43.

Fermi, Laura. 1957. *Atoms for the World: United States Participation in the Conference on the Peaceful Uses of Atomic Energy.* Chicago: University of Chicago Press.

———. 1968. *Illustrious Immigrants: The Intellectual Migration from Europe, 1930–41.* Chicago: University of Chicago Press.

Feyerabend, Paul. 1975. *Against Method: Outline of an Anarchistic Theory of Knowledge.* Atlantic Highlands: Humanities Press.

Feynman, Michelle. 1995. *The Art of Richard P. Feynman.* New York: Gordon and Breach.

Feynman, Richard. 1948. "Pocono conference." *PT* 1 (Jun): 8–10.

———. 1948. "A relativistic cut-off for classical electrodynamics." *PR* 74:939–46.

———. 1948. "Relativistic cut-off for quantum electrodynamics." *PR* 74:1430–38.

———. 1948. "Space-time approach to non-relativistic quantum mechanics." *Reviews of Modern Physics* 20:367–87.

———. 1949. "The theory of positrons." *PR* 76:749–59.

———. 1949. "Quantum electrodynamics." Mimeographed notes (by H. L. Brode) from lectures delivered at Cornell during autumn 1949. Copy in the possession of Sam Schweber.

———. 1950. "Mathematical formulation of the quantum theory of electromagnetic interaction." *PR* 80:440–57.

———. 1950. "Quantum electrodynamics and meson theories." Mimeographed notes (by Carl Helstrom and Malvin Ruderman) from lectures delivered at California Institute of Technology between 6 February and 2 March 1950. Copy in the possession of Sam Schweber.

———. 1951. "An operator calculus having applications in quantum electrodynamics." *PR* 84:108–28.

———. 1951. "High energy phenomena and meson theories." Mimeographed notes (by William Karzas) from lectures delivered at Caltech between January and March 1951. Copy in the possession of Sam Schweber.

———. 1953. "Atomic theory of the λ transition in helium." *PR* 91:1291–1301.

———. 1953. "Atomic theory of liquid helium near absolute zero." *PR* 91:1301–8.

———. 1953. "Quantum Mechanics III." Mimeographed notes (by A. R. Hibbs) from lectures delivered at California Institute of Technology during autumn 1953. Copy in the possession of Elisha Huggins.

———. 1957. "Superfluidity and superconductivity." *Reviews of Modern Physics* 29:205–12.

———. 1961. "The present status of quantum electrodynamics." In *The Quantum Theory of Fields,* ed. R. Stoop, 61–91. New York: Interscience.

———. 1961. *Quantum Electrodynamics: A Lecture Note and Reprint Volume.* New York: W. A. Benjamin.

———. 1966. "The development of the space-time view of quantum field theory." *Science* 153:699–708.

———. 1985. *QED: The Strange Theory of Light and Matter.* Princeton: Princeton University Press.

———. 1987. "The reason for antiparticles." In *Elementary Particles and the Laws of Physics: The 1986 Dirac Memorial Lectures,* by Richard Feynman and Steven Weinberg, 1–59. New York: Cambridge University Press.

Feynman, Richard, with Ralph Leighton. 1985. *"Surely You're Joking, Mr. Feynman!" Adventures of a Curious Character.* New York: W. W. Norton.

———. 1988. *"What Do You Care What Other People Think?" Further Adventures of a Curious Character.* New York: W. W. Norton.

Feynman, Richard, Robert Leighton, and Matthew Sands. 1965. *The Feynman Lectures on Physics.* 3 vols. Reading, MA: Addison-Wesley.

Fish, Stanley. 1980. *Is There a Text in This Class? The Authority of Interpretive Communities.* Cambridge: Harvard University Press.

———. 1989. *Doing What Comes Naturally: Change, Rhetoric, and the Practice of Theory in Literary and Legal Studies.* Durham: Duke University Press.

Fitch, Val, and Jonathan Rosner. 1995. "Elementary particle physics in the second half of the twentieth century." In *Twentieth Century Physics,* ed. Laurie Brown, Abraham Pais, and Brian Pippard, 2:635–794. New York: AIP Press.

Fleck, Ludwik. 1979 [1935]. *Genesis and Development of a Scientific Fact.* Trans. F. Bradley and T. J. Trenn. Chicago: University of Chicago Press.

Fleming, Donald, and Bernard Bailyn, eds. 1969. *The Intellectual Migration: Europe and America, 1930–1960.* Cambridge: Harvard University Press.

Flexner, Abraham. 1930. *Universities: American, English, German.* New York: Oxford University Press.

Fogelin, Robert. 1987 [1976]. *Wittgenstein.* 2nd ed. New York: Routledge and Kegan Paul.

Foldy, L. L., and R. E. Marshak. 1949. "Production of π^- mesons in nucleon-nucleon collisions." *PR* 75:1493–99.

Ford, Kenneth W. 1963. *The World of Elementary Particles.* New York: Blaisdell.

Forman, Paul. 1971. "Weimar culture, causality and quantum theory, 1918–1927: Adaptation by German physicists and mathematicians to a hostile intellectual environment." *HSPS* 3:1–115.

———. 1987. "Behind quantum electronics: National security as basis for physical research in the United States, 1940–1960." *HSPS* 18:149–229.

Fortun, Michael, and Silvan Schweber. 1993. "Scientists and the legacy of World War II: The case of operations research." *Social Studies of Science* 23:595–642.

Francis, N. C. 1953. "Photoproduction of π^0 mesons in hydrogen and deuterium." *PR* 89:766–74.

Francis, N. C., and K. M. Watson. 1953. "The elastic scattering of particles by atomic nuclei." *PR* 92:291–303.

Frautschi, Steven. 1963. *Regge Poles and S-Matrix Theory.* New York: W. A. Benjamin.

———. 1985. "My experiences with the S-matrix program." In *A Passion for Physics: Essays in Honor of Geoffrey Chew,* ed. Carleton DeTar, J. Finkelstein, and Chung-I Tan, 44–48. Singapore: World Scientific.

Frazer, William R. 1959. *A Proposal for Determining the Electromagnetic Form Factor of the Pion.* Ph.D. dissertation, University of California, Berkeley.

———. 1966. *Elementary Particles.* Englewood Cliffs, NJ: Prentice Hall.

———. 1985. "The analytic and unitary S-matrix." In *A Passion for Physics: Essays in Honor of Geoffrey Chew,* ed. Carleton DeTar, J. Finkelstein, and Chung-I Tan, 1–8. Singapore: World Scientific.

Frazer, William, and José Fulco. 1959. "Effect of a pion-pion scattering resonance on nucleon structure." *PRL* 2:365–68.

———. 1960. "Partial-wave dispersion relations for the process $\pi + \pi \rightarrow N + \overline{N}$." *PR* 117:1603–8.

Freedberg, David. 1989. *The Power of Images: Studies in the History and Theory of Response.* Chicago: University of Chicago Press.

Freundlich, Yehudah. 1980. "Theory evaluation and the bootstrap hypothesis." *SHPS* 11:267–77.

Fujimura, Joan. 1996. *Crafting Science: A Sociohistory of the Quest for the Genetics of Cancer.* Cambridge: Harvard University Press.

Fulton, Thomas. 1954. *Energy Levels of Positronium.* Ph.D. dissertation, Harvard University.

———. 1963. "Resonances in strong interaction physics." In *Elementary Particle Physics and Field Theory: Brandeis Summer Institute 1962,* ed. K. W. Ford, 1:1–122. New York: W. A. Benjamin.

Furry, W. 1937. "A symmetry theorem in the positron theory." *PR* 51:125–29.

———. 1951. "On bound states and scattering in positron theory." *PR* 81:115–24.

Fyfe, Gordon, and John Law, eds. 1988. *Picturing Power: Visual Depictions and Social Relations.* New York: Routledge.

Gakushikai Shimeiroku [Directory of the Society of Bachelors]. 1940–. Tokyo: Gakushikai.

Galanin, A. D. 1951. "Radiatsionnye popravki v kvantovoi elektrodinamike" [Radiative corrections in quantum electrodynamics]. *Doklady Akademii Nauk SSSR* 79:229–32.

Galanin, A. D., B. L. Ioffe, and I. Ia. Pomeranchuk. 1954. "Perenormirovka massy i zariada v kovariantnykh uravneniiakh kvantovoi teorii polia" [Mass and charge renormalization in the covariant equations of quantum field theory]. *Doklady Akademii Nauk SSSR* 98:361–64.

Galen. 1978. *On the Doctrines of Hippocrates and Plato.* Ed. and trans. Phillip de Lacy. Berlin: Akademie Verlag.

Galison, Peter. 1979. "Minkowski's space-time: From visual thinking to the absolute world." *HSPS* 10:85–121.

———. 1987. *How Experiments End.* Chicago: University of Chicago Press.

———. 1997. *Image and Logic: A Material Culture of Microphysics.* Chicago: University of Chicago Press.

———. 1998. "Feynman's war: Modelling weapons, modelling nature." *SHPMP* 29:391–434.

———. 1998. "Judgment against objectivity." In *Picturing Science, Producing Art,* ed. Caroline Jones and Peter Galison, 327–59. New York: Routledge.

Galison, Peter, Michael Gordin, and David Kaiser, eds. 2001. *Science and Society: The History of Modern Physical Science in the Twentieth Century.* 4 vols. New York: Routledge.

Galison, Peter, and Andrew Warwick, eds. 1998. *The Cultures of Theory.* Special issue of *SHPMP* 29:287–434.

García-Belmar, Antonio, José Ramón Bertomeu-Sánchez, and Bernadette Bensaude-Vincent. 2005. "The power of didactic writings: French chemistry textbooks in the nineteenth century." In *Pedagogy and the Practice of Science: Historical and Contemporary Perspectives,* ed. David Kaiser, 219–51. Cambridge: MIT Press.

Gardner, David. 1967. *The California Oath Controversy.* Berkeley: University of California Press.

Gasiorowicz, Stephen. 1966. *Elementary Particle Physics.* New York: John Wiley and Sons.

Geddes, Robert. 1972. "Theory in practice." Pt. 2. *Architectural Forum* 137 (Oct): 52–59.

Geiger, Roger. 1993. *Research and Relevant Knowledge: American Research Universities since World War II.* New York: Oxford University Press.

Geisen, Gerald, and Frederic L. Holmes, eds. 1993. *Research Schools: Historical Reappraisals.* Special issue of *Osiris,* vol. 8.

Gell-Mann, Murray. 1961. "Symmetry properties of fields." In *The Quantum Theory of Fields,* ed. R. Stoop, 131–42. New York: Interscience.

———. 1964. "A schematic model of baryons and mesons." *Physics Letters* 8:214–15.

———. 1964. "The symmetry group of vector and axial vector currents." *Physics* 1:63–75.

———. 1987. "Particle theory from *S*-matrix to quarks." In *Symmetries in Physics, 1600–1980,* ed. M. G. Doncel, A. Hermann, L. Michel, and A. Pais, 479–97. Barcelona: Bellaterra.

Gell-Mann, Murray, and Keith Brueckner. 1957. "Correlation energy of an electron at high density." *PR* 106:364–68.

Gell-Mann, Murray, Marvin Goldberger, and Walter Thirring. 1954. "Use of causality conditions in quantum theory." *PR* 95:1612–27.

Giere, Ronald. 1988. *Explaining Science: A Cognitive Approach.* Chicago: University of Chicago Press.

———. 1999. *Science without Laws.* Chicago: University of Chicago Press.

Gingerich, Owen. 2004. *The Book Nobody Read: Chasing the Revolutions of Nicolaus Copernicus.* New York: Walker.

Ginsparg, Paul. 1992. "Computopia, here we come." *PT* 45 (June): 13–15, 100.

Ginzburg, Vitaly. 1989. "Landau's attitude toward physics and physicists." *PT* 42 (May): 54–61.

Gitlin, Todd. 1993 [1987]. *The Sixties: Years of Hope, Days of Rage.* Rev. ed. New York: Bantam.

Gleick, James. 1992. *Genius: The Life and Science of Richard Feynman.* New York: Pantheon.

Goldberg, Stanley. 1984. *Understanding Relativity: Origin and Impact of a Scientific Revolution.* Boston: Birkhäuser.

Goldberger, M. L. 1955. "Causality conditions and dispersion relations." Pt. 1, "Boson fields." *PR* 99:979–85.

———. 1960. "Introduction to the theory and applications of dispersion relations." In *Relations de dispersion et particules élémentaires,* ed. C. de Witt and R. Omnès, 15–157. Paris: Hermann.

———. 1961. "Theory and applications of single variable dispersion relations." In *The Quantum Theory of Fields,* ed. R. Stoop, 179–96. New York: Interscience.

———. 1970. "Fifteen years in the life of dispersion theory." In *Subnuclear Phenomena,* ed. A. Zichichi, 685–93. New York: Academic Press.

———. 1983. "Francis E. Low: A sixtieth birthday tribute." In *Asymptotic Realms of Physics,* ed. Alan Guth, Kerson Huang, and Robert Jaffe, xi–xv. Cambridge: MIT Press.

———. 1985. "A passion for physics." In *A Passion for Physics: Essays in Honor of Geoffrey Chew,* ed. Carleton DeTar, J. Finkelstein, and Chung-I Tan, 241–45. Singapore: World Scientific.

Goldberger, M. L., H. Miyazawa, and R. Oehme. 1955. "Application of dispersion relations to pion-nucleon scattering." *PR* 99:986–88.

Goldberger, M. L., and S. B. Treiman. 1958. "Decay of the pi meson." *PR* 110:1178–84.

Goldfarb, L. J. B., and D. Feldman. 1952 "High-energy proton-proton scattering and associated polarization effects." *PR* 88:1099–1109.

Goldstein, Jack S. 1953. *Properties of Bound State Solutions to the Relativistic Two-Body Equation.* Ph.D. dissertation, Cornell University.

Goldstone, Jeffrey. 1957. "Derivation of the Brueckner many-body theory." *PRSA* 239:267–79.

———. 1961. "Field theories with 'superconductor' solutions."*Il Nuovo Cimento* 19:154–64.

Gombrich, Ernst. 1969 [1960]. *Art and Illusion: A Study in the Psychology of Pictorial Representation.* Princeton: Princeton University Press.

———. 1981. "Image and code: Scope and limits of conventionalism in pictorial representation." In *Image and Code*, ed. Wendy Steiner, 11–42. Ann Arbor: University of Michigan Press.

Goncharev, German. 1996. "Thermonuclear milestones." *Physics Today* 49 (Nov): 44–61.

Gooday, Graeme. 1990. "Precision measurement and the genesis of physics teaching laboratories in Victorian Britain." *British Journal for the History of Science* 23:25–51.

———. 1991. "Teaching telegraphy and electrotechnics in the physics laboratory: William Ayrton and the creation of an academic space for electrical engineering, 1873–84." *History of Technology* 13:73–111.

Gooding, David. 1986. "How do scientists reach agreement about novel observations?" *SHPS* 17:205–30.

———. 1990. "Mapping experiment as a learning process: How the first electromagnetic motor was invented." *Science, Technology, and Human Values* 15:165–201.

———. 1990. "Theory and observation: The experimental nexus." *International Studies in the Philosophy of Science* 4:131–48.

———. 1994. "Imaginary science." *British Journal for the Philosophy of Science* 45:1029–46.

Gooding, David, Trevor Pinch, and Simon Schaffer, eds. 1989. *The Uses of Experiment: Studies in the Natural Sciences.* New York: Cambridge University Press.

Goodman, Nelson. 1976 [1968]. *Languages of Art: An Approach to a Theory of Symbols.* Indianapolis: Hackett.

———. 1978. *Ways of Worldmaking.* Indianapolis: Hackett.

Goodstein, David. 1998. "Feynmaniacs should read this review, skip lecture collection, save 22 simoleans." *American Scientist* 86 (Jul–Aug): 374–77.

Goodstein, David, and Judith Goodstein. 2000. "Richard Feynman and the history of superconductivity." *Physics in Perspective* 2:30–47.

Gordon, Stephen. 1998. *Strong Interactions: Particles, Passion and the Rise and Fall of Nuclear Democracy.* A.B. honors thesis, Harvard University.

Gorelik, Gennady. 1999. "The metamorphosis of Andrei Sakharov." *Scientific American* 280 (Mar): 98–101.

Gottfried, Kurt, and Victor Weisskopf. 1984. *Concepts of Particle Physics.* New York: Oxford University Press.

Gouiran, Robert. 1967. *Particles and Accelerators.* Trans. W. F. G. Crozier. New York: McGraw-Hill.

Graham, Loren. 1998. *What Have We Learned about Science and Technology from the Russian Experience?* Stanford: Stanford University Press.

Green, M. B., J. H. Schwarz, and E. Witten. 1987. *Superstring Theory.* 2 vols. New York: Cambridge University Press.

Greene, Brian. 1999. *The Elegant Universe: Superstrings, Hidden Dimensions, and the Quest for the Ultimate Theory.* New York: Norton.

Gribbin, John, and Mary Gribbin. 1997. *Richard Feynman: A Life in Science.* New York: Dutton.

Griesemer, James. 1991. "Must scientific diagrams be eliminable? The case of path analysis." *Biology and Philosophy* 6:155–80.

Griesemer, James, and William Wimsatt. 1989. "Picturing Weismannism: A case study of conceptual evolution." In *What the Philosophy of Biology Is*, ed. Michael Ruse, 75–137. Dordrecht: Kluwer.

Gross, David. 1985. "On the uniqueness of physical theories." In *A Passion for Physics: Essays in Honor of Geoffrey Chew*, ed. Carleton DeTar, J. Finkelstein, and Chung-I Tan, 128–36. Singapore: World Scientific.

———. 1990. "Chasing the Landau ghost." In *Frontiers of Physics: Proceedings of the Landau Memorial Conference,* ed. Errol Gotsman, Yuval Ne'eman, and Alexander Voronel, 97–111. New York: Pergamon.

———. 1997. "Asymptotic freedom and the emergence of QCD." In *The Rise of the Standard Model: Particle Physics in the 1960s and 1970s,* ed. Lillian Hoddeson, Laurie Brown, Michael Riordan, and Max Dresden, 199–232. New York: Cambridge University Press.

Gross, David, and Frank Wilczek. 1973. "Ultraviolet behavior of non-Abelian gauge theories." *PRL* 30:1343–46.

Grythe, I. 1982. "Some remarks on the early *S*-matrix." *Centaurus* 26:198–203.

Gupta, S. N. 1950. "Theory of longitudinal photons in quantum electrodynamics." *PRSA* 63:681–91.

Hacking, Ian. 1983. *Representing and Intervening: Introductory Topics in the Philosophy of Natural Science.* New York: Cambridge University Press.

Hagedorn, R. 1961. "Introduction to field theory and dispersion relations." CERN preprint 61-6 (14 February). Copy in the possession of Sam Schweber.

Hahn, T. 1999. "Generating and calculating one-loop Feynman diagrams with *FeynArts, FormCalc,* and *LoopTools.*" Preprint hep-ph/9905354 (May), available from http://www.arXiv.org.

Hall, Karl. 1999. *Purely Practical Revolutionaries: A History of Stalinist Theoretical Physics.* Ph.D. dissertation, Harvard University.

———. 2005. "'Think less about foundations': A short course on Landau and Lifshitz's *Course of Theoretical Physics.*" In *Pedagogy and the Practice of Science: Historical and Contemporary Perspectives,* ed. David Kaiser, 253–86. Cambridge: MIT Press.

Halpern, O. 1933. "Scattering processes produced by electrons in negative energy states." *PR* 44:855–56.

Hamilton, John. 1959. *The Theory of Elementary Particles.* Oxford: Clarendon Press.

———. 1960. "Dispersion relations for elementary particles." *Progress of Nuclear Physics* 8:143–95.

———. 1964–65. *Elementary Introduction to Elementary Particle Physics.* Mimeographed lectures from NORDITA.

Hanson, Norwood Russell. 1958. *Patterns of Discovery: An Inquiry into the Conceptual Foundations of Science.* New York: Cambridge University Press.

———. 1963. *The Concept of the Positron: A Philosophical Analysis.* New York: Cambridge University Press.

Hartmann, Stephan. 1999. "Models and stories in hadron physics." In *Models as Mediators: Perspectives on Natural and Social Science,* ed. Mary Morgan and Margaret Morrison, 326–46. New York: Cambridge University Press.

Hayakawa, Satio. 1983. "The development of meson physics in Japan." In *The Birth of Particle Physics,* ed. Laurie Brown and Lillian Hoddeson, 82–107. New York: Cambridge University Press.

———. 1988. "Sin-itiro Tomonaga and his contributions to quantum electrodynamics and high energy physics." In *Elementary Particle Theory in Japan, 1935–1960,* ed. Laurie Brown, Rokuo Kawabe, Michiji Konuma, and Zirō Maki, 43–60. Kyoto: Yukawa Hall Archival Library.

Heisenberg, W. 1927. "Über den anschaulichen Inhalt der Quantentheoretischen Kinematik u. Mechanik." *Zeitschrift für Physik* 43:172–98.

Heisenberg, W., and W. Pauli. 1929–30. "Zur Quantenelektrodynamik der Wellenfelder." Parts 1, 2. *Zeitschrift für Physik* 56:1–61; 59:168–90.

Heilbron, John. 1985. "The earliest missionaries of the Copenhagen spirit." *Revue d'histoire des sciences* 38:195–230.

Heitler, Walter. 1936. *The Quantum Theory of Radiation*. Oxford: Clarendon Press.

———. 1944. *The Quantum Theory of Radiation*. 2nd ed. Oxford: Clarendon Press.

Helmholz, A. Carl, with Graham Hale and Ann Lage. 1993. *Faculty Governance and Physics at the University of California, Berkeley, 1937–1990*. Berkeley: Regional Oral History Office, Bancroft Library.

Henley, Ernest, and Walter Thirring. 1962. *Elementary Quantum Field Theory*. New York: McGraw-Hill.

Herken, Gregg. 2002. *Brotherhood of the Bomb: The Tangled Lives and Loyalties of Robert Oppenheimer, Ernest Lawrence, and Edward Teller*. New York: Henry Holt.

Hershberg, James. 1993. *James B. Conant: Harvard to Hiroshima and the Making of the Nuclear Age*. Stanford: Stanford University Press.

Hess, Karl. 1956. "Dr. Oppenheimer's Institute." *American Mercury* 83 (July): 5–12.

Hewlett, Richard, and Francis Duncan. 1969. *A History of the United States Atomic Energy Commission*. Vol. 2, *Atomic Shield, 1947–1952*. University Park: Pennsylvania State University Press.

Hirokawa, Shunkichi, and Shūzō Ogawa. 1989. "Shōichi Sakata: His physics and methodology." *Historia Scientiarum* 36:67–81.

Hoddeson, Lillian, Ernest Braun, Jürgen Teichmann, and Spencer Weart, eds. 1992. *Out of the Crystal Maze: Chapters from the History of Solid-State Physics*. New York: Oxford University Press.

Hoddeson, Lillian, Laurie Brown, Michael Riordan, and Max Dresden, eds. 1997. *The Rise of the Standard Model: Particle Physics in the 1960s and 1970s*. New York: Cambridge University Press.

Hoddeson, Lillian, and Vicki Daitch. 2002. *True Genius: The Life and Science of John Bardeen*. Washington, DC: Joseph Henry Press.

Hofstadter, Douglas. 1979. *Gödel, Escher, Bach: An Eternal Golden Braid*. New York: Basic Books.

Holloway, David. 1994. *Stalin and the Bomb*. New Haven: Yale University Press.

———. 1996. "New light on early Soviet bomb secrets." *Physics Today* 49 (Nov): 26–27.

Holmes, Frederic L., and Trevor Levere, eds. 2000. *Instruments and Experimentation in the History of Chemistry*. Cambridge: MIT Press.

Hornby, Catharine M. 1997. *Harvard Astronomy in the Age of McCarthyism*. A.B. honors thesis, Harvard University.

House Un-American Activities Committee. 1949. *Hearings Regarding Communist Infiltration of Radiation Laboratory and Atomic Bomb Project at the University of California, Berkeley, Calif.* 3 vols. Washington, DC: Government Printing Office.

Hu, Ning. 1949. "On the treatment of quantum electrodynamics without eliminating the longitudinal field." *PR* 76:391–96.

Hugenholtz, N. M. 1957. "Perturbation theory of large quantum systems." *Physica* 23:481–532.

———. 1957. "Perturbation approach to the Fermi gas model of heavy nuclei." *Physica* 23:533–45.

———. 2000. "Perturbation theory of large quantum systems." In *The Legacy of Léon van Hove*, ed. Alberto Giovannini, 65–67. Singapore: World Scientific.

Hugenholtz, N. M., and David Pines. 1959. "Ground-state energy and excitation spectrum of a system of interacting bosons." *PR* 116:489–506.

Hugenholtz, N. M., and L. van Hove. 1958. "A theorem on the single particle energy in a Fermi gas with interaction." *Physica* 24:363–76.

Hughes, Thomas. 1983. *Networks of Power: Electrification in Western Society, 1880–1930*. Baltimore: Johns Hopkins University Press.

Hughes, V. W., and T. Kinoshita. 1999. "Anomalous *g* values of the electron and muon." *Review of Modern Physics* 71:S133–S139.

Hurst, C. Angas. 2001. "Herbert Sydney Green, 1920–1999." *Historical Records of Australian Science* 13, no. 3:301–22.

Hurtado, Rafael, John Morales, and Carlos Quimbay. 2000. "Dispersion relations at finite temperature and density for nucleons and pions." *Heavy Ion Physics* 11:373–81.

Index to Theses Accepted for Higher Degrees in the Universities of Great Britain and Ireland. 1951–. London: Aslib.

Infeld, Leopold. 1941. *Quest: The Evolution of a Scientist*. New York: Doubleday, Doran.

Institute for Advanced Study. 1955. *Publications of Members, 1930–1954*. Princeton: Princeton University Press.

———. 1980. *A Community of Scholars: The Institute for Advanced Study Faculty and Members, 1930–1980*. Princeton: Institute for Advanced Study.

Ioffe, Boris L. 1954. "O raskhodimosti riada teorii vozmyshchenii v kvantovoi elektrodinamike" [On the divergence of the perturbation theory series in quantum electrodynamics]. *Doklady Akademii Nauk SSSR* 94:437–38.

———. 1954. "Sistemy kovariantnykh uravnenii v teorii kvantovykh polei" [Systems of covariant equations in quantum field theory]. *Doklady Akademii Nauk SSSR* 95:761–64.

———. 1989. "If Landau were alive now." In *Landau: The Physicist and the Man*, ed. I. M. Khalatnikov, 153–56. New York: Pergamon.

———. 2001. "A top secret assignment." In *At the Frontiers of Particle Physics: Handbook of QCD, Boris Ioffe Festschrift*, ed. M. Shifman, vol. 1, 33–35. New Jersey: World Scientific.

———. 2002. "Landau's theoretical minimum, Landau's seminar, ITEP in the beginning of the 1950s." Preprint hep-ph/0204295 (April), available from http://www.arXiv.org.

Ioffe, B. L., L. B. Okun', and A. P. Rudik. 1957. "The problem of parity non-conservation in weak interactions." *Soviet Physics JETP* 5 (Sep): 328–30. Originally published in Russian in *ZhETF* 32 (Feb): 396–97].

Iser, Wolfgang. 1978 [1976]. *The Act of Reading: A Theory of Aesthetic Response*. Baltimore: Johns Hopkins University Press.

Israeli, Nathan. 1944. "American postdoctoral education." *Journal of Higher Education* 15 (Nov): 428–30.

Itō, Daisuke. 1988. "My positive and negative contribution to the development of Tomonaga's theory of renormalization." In *Elementary Particle Theory in Japan, 1935–1960*, ed. Laurie Brown, Rokuo Kawabe, Michiji Konuma, and Zirō Maki, 61–62. Kyoto: Yukawa Hall Archival Library.

Ito, Kenji. 2002. *Making Sense of* Ryōshiron *(Quantum Theory): Introduction of Quantum Mechanics into Japan, 1920–1940*. Ph.D. dissertation, Harvard University.

Itzykson, Claude, and Jean-Bernard Zuber. 1980. *Quantum Field Theory*. New York: McGraw-Hill.

IUPAP SUN Commission. 1956. "Symbols and units." *PT* 9 (Nov): 23–27.

Ivanter, I. G. and L. B. Okun'. 1957. "On the theory of scattering of particles by nuclei." *Soviet Physics JETP* 5 (Sep): 340–41. Originally published in Russian in *ZhETF* 32 (Feb): 402–3.

Jackson, J. D. 1961. "Introduction to dispersion relation techniques." In *Dispersion Relations: Scottish Universities' Summer School, 1960*, ed. G. R. Screaton, 1–63. New York: Interscience.

Jackson, Myles. 2000. *Spectrum of Belief: Joseph von Fraunhofer and the Craft of Precision Optics*. Cambridge: MIT Press.

Jacob, Maurice. 1964. "An introduction to the analysis of strong-interaction processes." In *Strong-Interaction Physics: A Lecture Note Volume,* by Maurice Jacob and Geoffrey Chew, 1–102. New York: W. A. Benjamin.

Jacob, Maurice, and Geoffrey Chew. 1964. *Strong-Interaction Physics: A Lecture Note Volume.* New York: W. A. Benjamin.

Jammer, Max. 1966. *The Conceptual Development of Quantum Mechanics.* New York: McGraw-Hill.

Jauch, Josef, and Fritz Rohrlich. 1955. *The Theory of Photons and Electrons.* Cambridge, MA: Addison-Wesley.

Jauss, Hans Robert. 1982 [1974]. *Toward an Aesthetic of Reception.* Trans. Timothy Bahti. Minneapolis: University of Minnesota Press.

Jerome, Fred. 2002. *The Einstein File: J. Edgar Hoover's Secret War against the World's Most Famous Scientist.* New York: St. Martin's Press.

Johns, Adrian. 1998. *The Nature of the Book: Print and Knowledge in the Making.* Chicago: University of Chicago Press.

———. 2000. "The past, present, and future of the scientific book." In *Books and the Sciences in History,* ed. Marina Frasca-Spada and Nick Jardine, 408–26. New York: Cambridge University Press.

———. 2002. "How to acknowledge a revolution." *American Historical Review* 107:106–25.

Johnson, George. 1999. *Strange Beauty: Murray Gell-Mann and the Revolution in Twentieth-Century Physics.* New York: Knopf.

Jones, Caroline, and Peter Galison, eds. 1998. *Picturing Science, Producing Art.* New York: Routledge.

Jones, C. Edward. 1985. "Deducing T, C, and P invariance for strong interactions in topological particle theory." In *A Passion for Physics: Essays in Honor of Geoffrey Chew,* ed. Carleton DeTar, J. Finkelstein, and Chung-I Tan, 189–94. Singapore: World Scientific.

Jordan, Kathleen, and Michael Lynch. 1992. "The sociology of a genetic engineering technique: Ritual and rationality in the performance of the 'plasmid prep.'" In *The Right Tools for the Job: At Work in Twentieth-Century Life Sciences,* ed. Adele Clarke and Joan Fujimura, 77–114. Princeton: Princeton University Press.

Jordan, P. 1927. "Zur Quantenmechanik des Gasentartung." *Zeitschrift für Physik* 44:473–80.

Josephson, Paul. 1991. *Physics and Politics in Revolutionary Russia.* Berkeley: University of California Press.

———. 2000. *Red Atom: Russia's Nuclear Power Program from Stalin to Today.* New York: W. H. Freeman.

Jungnickel, Christa, and Russell McCormmach. 1986. *Intellectual Mastery of Nature: Theoretical Physics from Ohm to Einstein.* 2 vols. Chicago: University of Chicago Press.

Kaempffer, F. A. 1965. *Concepts in Quantum Mechanics.* New York: Academic Press.

Kaiser, David. 1998. "A ψ is just a ψ? Pedagogy, practice, and the reconstitution of general relativity, 1942–1975." *SHPMP* 29:321–38.

———. 2000. *Making Theory: Producing Physics and Physicists in Postwar America.* Ph.D. dissertation, Harvard University.

———. 2001. "Francis E. Low: Coming of age as a physicist in postwar America." *Physics@MIT* 14:24–31, 70–77.

———. 2002. "Cold war requisitions, scientific manpower, and the production of American physicists after World War II." *HSPS* 33:131–59.

———. 2004. "The postwar suburbanization of American physics." *American Quarterly* 56: 851–88.

———, ed. 2005. *Pedagogy and the Practice of Science: Historical and Contemporary Perspectives.* Cambridge: MIT Press.

Källén, Gunnar. 1952. "On the definition of the renormalization constants in quantum electrodynamics." *Helvetica Physica Acta* 25:417–34.

———. 1964. *Elementary Particle Physics.* Reading, MA: Addison-Wesley.

Kamefuchi, Susumu. 1988. "Study of the divergence problem at Nagoya around 1950: Personal recollections." In *Elementary Particle Theory in Japan, 1935–1960*, ed. Laurie Brown, Rokuo Kawabe, Michiji Konuma and Zirō Maki, 126–32. Kyoto: Yukawa Hall Archival Library.

———. 2001. "Quantum field theory in postwar Japan: Early works of H. Umezawa." In *Selected Papers of Hiroomi Umezawa*, ed. A. Arimitsu et al., 3–19. Tokyo: Syokabo.

Kaneseki, Yoshinori. 1974 [1950]. "The elementary particle theory group." In *Science and Society in Modern Japan: Selected Historical Sources*, ed. Shigeru Nakayama, David Swain, and Eri Yagi, 221–52. Tokyo: University of Tokyo Press. This article originally appeared in Japanese in 1950.

Kaplon, Morton. 1951. *Meson Production.* Ph.D. dissertation, University of Rochester.

Karplus, Robert, and Abraham Klein. 1952. "Electrodynamic displacement of atomic energy levels." Pt. 1, "Hyperfine structure." *PR* 85:972–84.

Karplus, Robert, Abraham Klein, and Julian Schwinger. 1952. "Electrodynamic displacement of atomic energy levels." Pt. 2, "Lamb shift." *PR* 86:288–301.

Karplus, Robert, and Norman Kroll. 1949. "Fourth-order corrections in quantum electrodynamics and the magnetic moment of the electron." *PR* 76:846–47.

Karplus, Robert, and Malvin Ruderman. 1955. "Applications of causality to scattering." *PR* 98:771–74.

Katzenstein, J. 1950. "The radiative collisions of positrons and electrons." *PR* 79:481–86.

Keller, Evelyn Fox. 2000. "Models of and models for: Theory and practice in contemporary biology." *Philosophy of Science* 67, suppl.:S72–S86.

Keller, Joseph. 1949. "Mesons old and new." *AJP* 17:356–67.

Kemmer, N. 1937. "Field theory of nuclear interaction." *PR* 52:906–10.

———. 1938. "The charge-dependence of nuclear forces." *Proceedings of the Cambridge Philosophical Society* 34:354–64.

Kennefick, Daniel. 2000. "Star crushing: Theoretical practice and the theoreticians' regress." *Social Studies of Science* 30:5–40.

Kevles, Daniel. 1977. "The National Science Foundation and the debate over postwar research policy, 1942–1945." *Isis* 68:5–26.

———. 1995 [1978]. *The Physicists: The History of a Scientific Community in Modern America.* 3rd ed. Cambridge: Harvard University Press.

Khalatnikov, I. M., ed. 1989. *Landau: The Physicist and the Man.* New York: Pergamon.

———. 1989. "Reminiscences of Landau." *PT* 42 (May): 34–41.

Khariton, Yuli, Viktor Adamskii, and Yuri Smirnov. 1996. "The way it was." *BAS* 52 (Nov/Dec): 53–59.

Khariton, Yuli, and Yuri Smirnov. 1993. "The Khariton version." *BAS* 49 (May): 20–31.

Kibble, T. B. W. 1988. "Paul Taunton Matthews, 19 November 1919–26 February 1987." *Biographical Memoirs of Fellows of the Royal Society* 34:555–80.

———. 1998. "Muhammed Abdus Salam, K.B.E." *Biographical Memoirs of Fellows of the Royal Society* 44:387–401.

Kim, Yongduk. 1961. *Production of Pion Pairs by a Photon in the Coulomb Field of a Nucleus.* Ph.D. dissertation, University of California, Berkeley.

Kinoshita, Tōichirō. 1962. "Mass singularities of Feynman amplitudes." *Journal of Mathematical Physics* 3:650–77.

———. 1988. "Personal recollections, 1944–1952." In *Elementary Particle Theory in Japan, 1935–1960,* ed. Laurie Brown, Rokuo Kawabe, Michiji Konuma, and Zirō Maki, 7–10. Kyoto: Yukawa Hall Archival Library.

Kinoshita, T., and Y. Nambu. 1950. "On the electromagnetic properties of mesons." *PTP* 5:307–11.

Kinoshita, T., and M. Nio. 2003. "Revised α^4 term of lepton $g-2$ from the Feynman diagrams containing an internal light by light scattering subdiagram." *PRL* 90:021803.

Klein, Abraham. 1950. *On the Relativistic Theory of Meson Fields.* Ph.D. dissertation, Harvard University.

———. 1993. "Autobiographical notes." In *Symposium on Contemporary Physics: Celebrating the 65th Birthday of Professor Abraham Klein,* ed. Michel Vallières and Da Hsuan Feng, 3–60. Singapore: World Scientific.

———. 1996. "Recollections of Julian Schwinger." In *Julian Schwinger: The Physicist, the Teacher, and the Man,* ed. Y. Jack Ng, 1–7. Singapore: World Scientific.

Klein, Howard. 1951. "Promoting technical books." *Publishers' Weekly,* 15 Dec, 2238–40.

Klein, Ursula. 1999. "Techniques of modelling and paper-tools in classical chemistry." In *Models as Mediators: Perspectives on Natural and Social Science,* ed. Mary Morgan and Margaret Morrison, 146–67. New York: Cambridge University Press.

———. 2001. "Paper tools in experimental cultures." *SHPS* 32:265–302.

———, ed. 2001. *Tools and Modes of Representation in the Laboratory Sciences.* Boston: Kluwer.

———. 2003. *Experiments, Models, Paper Tools: Cultures of Organic Chemistry in the Nineteenth Century.* Stanford: Stanford University Press.

Klepikov, N. P. 1954. "K teorii vakuumnogo funktsionala" [On the theory of vacumm functional]. *Doklady Akademii Nauk SSSR* 98:937–40.

Knuth, Donald. 1979. "Mathematical typography." *American Mathematical Society Bulletin,* nos. 1, 2 (Mar): 337–72.

Koba, Zirō, and Gyō Takeda. 1948. "Radiative corrections in e^2 for an arbitrary process involving electrons, positrons, and light quanta." *PTP* 3:203–5.

———. 1948. "Radiation reaction in collision processes, II." *PTP* 3:407–21.

———. 1949. "Radiation reaction in collision processes, III." Pt. 1. *PTP* 4:60–70.

———. 1949. "Radiation reaction in collision processes, III." Pt. 2. *PTP* 4:130–41.

Koba, Zirō, and Sin-itiro Tomonaga. 1948. "On radiation reactions in collision processes." Pt. 1. *PTP* 3:290–303.

Kobzarev, I. Iu., and L. B. Okun'. 1957. "The spin of the Λ-particle." *Soviet Physics JETP* 3 (Jan): 954–55. Originally published in Russian in *ZhETF* 30 (Apr 1956): 798–99.

———. 1957. "Simultaneous creation of Λ and θ-particles." *Soviet Physics JETP* 5 (Nov): 761–62. Originally published in Russian in *ZhETF* 32 (Apr): 933–34.

———. 1958. "Decay probabilities of the Σ-hyperon with parity nonconservation." *Soviet Physics JETP* 6 (Jan): 230–31. Originally published in Russian in *ZhETF* 33 (Jul 1957): 296–97.

Koester, D., D. Sullivan, and D. H. White. 1982. "Theory selection in particle physics: A quantitative case study of the evolution of weak-electromagnetic unification theory." *Social Studies of Science* 12:73–100.

Kohler, Robert. 1990. "The Ph.D. machine: Building on the collegiate base." *Isis* 81:638–62.

———. 1994. *Lords of the Fly:* Drosophila *Genetics and the Experimental Life.* Chicago: University of Chicago Press.

Kojevnikov, Alexei. 1996. "President of Stalin's Academy: The mask and responsibility of Sergei Vavilov." *Isis* 87:18–50.

———. 1999. "Freedom, collectivism, and quasiparticles: Social metaphors in quantum physics." *HSPS* 29:295–331.

———. 2002. "David Bohm and collective movement." *HSPS* 33:161–92.

Konuma, Michiji. 1988. "Start of the interuniversity institutes in Japan in the 1950s." In *Elementary Particle Theory in Japan, 1935–1960*, ed. Laurie Brown, Rokuo Kawabe, Michiji Konuma, and Zirō Maki, 24–27. Kyoto: Yukawa Hall Archival Library.

———. 1989. "Social aspects of Japanese particle physics in the 1950s." In *Pions to Quarks: Particle Physics in the 1950s*, ed. Laurie Brown, Max Dresden, and Lillian Hoddeson, 536–48. New York: Cambridge University Press.

Kragh, Helge. 1981. "The genesis of Dirac's relativistic theory of electrons." *Archive for the History of the Exact Sciences* 24:31–67.

———. 1990. *Dirac: A Scientific Biography.* New York: Cambridge University Press.

———. 1996. *Cosmology and Controversy: The Historical Development of Two Theories of the Universe.* Princeton: Princeton University Press.

———. 1999. *Quantum Generations: A History of Physics in the Twentieth Century.* Princeton: Princeton University Press.

Kramers, Hendrik. 1994 [1938]. "The interaction between charged particles and the radiation field." In *Early Quantum Electrodynamics: A Source Book*, ed. Arthur I. Miller, 254–58. New York: Cambridge University Press.

Krementsov, Nikolai. 2002. *The Cure: A Story of Cancer and Politics from the Annals of the Cold War.* Chicago: University of Chicago Press.

Krieger, Martin. 1992. *Doing Physics: How Physicists Take Hold of the World.* Bloomington: Indiana University Press.

Kroll, Norman, and Willis Lamb, Jr. 1949. "On the self-energy of a bound electron." *PR* 75:388–98.

Kroll, Norman, and Franklin Pollock. 1951. "Radiative corrections to the hyperfine structure and the fine structure constant." *PR* 84:594–95.

Kuhn, Thomas. 1977 [1959]. "The essential tension: Tradition and innovation in scientific research." In *The Essential Tension: Selected Studies in Scientific Tradition and Change*, 225–39. Chicago: University of Chicago Press.

———. 1977 [1961]. "The function of measurement in modern physical science." In *The Essential Tension: Selected Studies in Scientific Tradition and Change*, 178–224. Chicago: University of Chicago Press.

———. 1977. *The Essential Tension: Selected Studies in Scientific Tradition and Change.* Chicago: University of Chicago Press.

———. 1996 [1962]. *The Structure of Scientific Revolutions.* 3rd ed. Chicago: University of Chicago Press.

Kulakov, Iu. I. 1957. "Inelastic proton-proton scattering." *Soviet Physics JETP* 5 (Oct): 477–82. Originally published in Russian in *ZhETF* 32 (Mar): 576–83.

Kuznick, Peter J. 1987. *Beyond the Laboratory: Scientists as Political Activists in 1930s America.* Chicago: University of Chicago Press.

Landau, Lev. 1959. "On analytic properties of vertex parts in quantum field theory." *Nuclear Physics* 13:181–92.

———. 1960. "On analytical properties of vertex parts in quantum field theory." In *Ninth International Annual Conference on High Energy Physics*, 2:95–101. Moscow: International Union of Pure and Applied Physics.

———. 1965. *Collected Papers of L. D. Landau.* Ed. D. Ter Haar. New York: Gordon and Breach.

Landau, L. D., A. A. Abrikosov, and I. M. Khalatnikov. 1954. "The removal of infinities in quantum electrodynamics." *Doklady Akademii Nauk SSSR* 95:497–500. Reprinted in *Collected Papers of L. D. Landau,* ed. and trans. D. Ter Haar (New York: Gordon and Breach, 1965), 607–10.

———. 1954. "An asymptotic expression for the electron Green function in quantum electrodynamics." *Doklady Akademii Nauk SSSR* 95:773–77. Reprinted in *Collected Papers of L. D. Landau,* ed. and trans. D. Ter Haar (New York: Gordon and Breach, 1965), 611–15.

———. 1954. "An asymptotic expression for the photon Green function in quantum electrodynamics." *Doklady Akademii Nauk SSSR* 95:1177–81. Reprinted in *Collected Papers of L. D. Landau,* ed. and trans. D. Ter Haar (New York: Gordon and Breach, 1965), 616–20.

———. 1954. "The electron mass in quantum electrodynamics." *Doklady Akademii Nauk SSSR* 96:261–65. Reprinted in *Collected Papers of L. D. Landau,* ed. and trans. D. Ter Haar (New York: Gordon and Breach, 1965), 621–25.

Landau, Lev, and Rudolf Peierls. 1931. "Erweiterung des Unbestimmtheitsprinzips für die relativistische Quantentheorie." *Zeitschift für Physik* 69:56–67.

Latour, Bruno. 1983. "Give me a laboratory and I will raise the world." In *Science Observed: Perspectives on the Social Study of Science,* ed. Karin Knorr-Cetina and Michael Mulkay, 141–70. London: Sage.

———. 1988 [1984]. *The Pasteurization of France.* Trans. Alan Sheridan and John Law. Cambridge: Harvard University Press.

———. 1990. "Drawing things together." In *Representation in Scientific Practice,* ed. Michael Lynch and Steve Woolgar, 19–68. Cambridge: MIT Press. Originally published as "Visualization and cognition: Thinking with eyes and hands," *Knowledge and Society: Studies in the Sociology of Culture Past and Present* 6 (1986): 1–40.

———. 1987. *Science in Action: How to Follow Scientists and Engineers through Society.* Cambridge: Harvard University Press.

———. 1996. *Aramis; or, The Love of Technology.* Trans. Catherine Porter. Cambridge: Harvard University Press.

Latour, Bruno, and Steve Woolgar. 1986 [1979]. *Laboratory Life: The Construction of Scientific Facts.* 2nd ed. Princeton: Princeton University Press.

Laurence, William. 1948. "New guide offered on atom research: Top scientists crowd meeting as Prof. Schwinger outlines way to pass obstacles." *NYT,* 1 Feb, 50.

Lautrup, B., and H. Zinkernagel. 1999. "$g - 2$ and the trust in experimental results." *SHPMP* 30:85–110.

Lave, Jean. 1988. *Cognition in Practice: Mind, Mathematics, and Culture in Everyday Life.* New York: Cambridge University Press.

Layzer, Arthur. 1960. *The Free-Propagator Expansion in the Evaluation of the Lamb Shift.* Ph.D. dissertation, Columbia University.

———. 1960. "New theoretical value for the Lamb shift." *PRL* 4:580–82.

Lee, Jennifer. 2001. "Postdoc trail: Long and filled with pitfalls." *NYT,* 21 Aug.

Lee, T. D., and M. Nauenberg. 1964. "Degenerate systems and mass singularities." *PR* 133:B1549–B1562.

Leighton, Ralph. 1991. *Tuva or Bust! Richard Feynman's Last Journey.* New York: W. W. Norton.

Leighton, Robert B. 1959. *Principles of Modern Physics.* New York: McGraw-Hill.

Lengyel, A., and M. V. T. Machado. 2003. "On a two pomeron description of the $F(2)$ structure function." Preprint hep-ph/0304079 (Apr), available from http://www.arXiv.org.

Lenoir, Timothy. 1988. "Social interests and the organic physics of 1847." In *Science in Reflection*, ed. Edna Ullmann-Margalit, 169–91. Dordrecht: Kluwer.

Leskov, Sergei. 1993. "Dividing the glory of the fathers." *BAS* 49 (May): 37–9.

Levere, Trevor. 2000. "Measuring gases and measuring goodness." In *Instruments and Experimentation in the History of Chemistry*, ed. Frederic L. Holmes and Trevor Levere, 105–35. Cambridge: MIT Press.

Levi Setti, Riccardo. 1963. *Elementary Particles*. Chicago: University of Chicago Press.

Lévi-Strauss, Claude. 1966 [1962]. *The Savage Mind*. London: Weidenfeld and Nicolson.

Linde, Andrei. 1992. "How physics fostered freedom in the USSR." *PT* 45 (Jun): 13.

Littlewood, Sandra, and Skip Garretson. 1953. "Espionage in the Rad Lab—Naw!" *Daily Californian*, 11 Dec, 8.

Livingston, M. Stanley. 1968. *Particle Physics: The High-Energy Frontier*. New York: McGraw-Hill.

Livingston, M. S., and Hans Bethe. 1937. "Nuclear physics C: Nuclear dynamics, experimental." *Reviews of Modern Physics* 9:245–390.

Low, Francis E. 1963. "Dispersion relations." Mimeographed notes from lectures delivered at the Theoretical Physics Division of the Canadian Association of Physicists during August 1963. Copy in the possession of Sam Schweber.

———. 1985. "Complete sets of wave-packets." In *A Passion for Physics: Essays in Honor of Geoffrey Chew*, ed. Carleton DeTar, J. Finkelstein, and Chung-I Tan, 17–22. Singapore: World Scientific.

Low, F. E., and E. E. Salpeter. 1951. "On the hyperfine structure of hydrogen and deuterium." *PR* 83:478.

Lubkin, Gloria. 1989. "Special Issue: Richard Feynman." *PT* 42 (Feb): 22–23.

Lundgren, Anders, and Bernadette Bensaude-Vincent. 2000. "Preface." In *Communicating Chemistry: Textbooks and Their Audiences, 1789–1939*, ed. Anders Lundgren and Bensaude-Vincent, vii. Canton, MA: Science History Publications.

Lüthy, Christoph. 2002. "The invention of atomist iconography." In *The Emergence of Scientific Imagery*, ed. J. Renn, W. Lefevre, and U. Schöpflin, 117–38. Basil: Birkhäuser.

Lynch, Michael. 1985. *Art and Artifact in Laboratory Science: A Study of Shop Work and Shop Talk in a Research Laboratory*. Boston: Routledge and Kegan Paul.

———. 1985. "Discipline and the material form of images: An analysis of scientific visibility." *Social Studies of Science* 15:37–66.

———. 1991. "Science in the age of mechanical reproduction: Moral and epistemic relations between diagrams and photographs." *Biology and Philosophy* 6:205–26.

Lynch, Michael, and Steve Woolgar, eds. 1990. *Representation in Scientific Practice*. Cambridge: MIT Press.

MacGregor, M. H., Michael J. Moravcsik, and Henry P. Stapp. 1960. "Nucleon-nucleon scattering experiments and their phenomenological analysis." *Annual Review of Nuclear Science* 10:291–352.

MacKenzie, Donald. 2003. "An equation and its worlds: Bricolage, exemplars, disunity, and performativity in financial economics." *Social Studies of Science* 33:831–68.

MacKenzie, Donald, and Graham Spinardi. 1995. "Tacit knowledge, weapons design, and the uninvention of nuclear weapons." *American Journal of Sociology* 101:44–99.

Major, John. 1971. *The Oppenheimer Hearing*. New York: Stein and Day.

Maki, Zirō. 1988. "Tomonaga and the meson theory." In *Elementary Particle Theory in Japan, 1935–1960*, ed. Laurie Brown, Rokuo Kawabe, Michiji Konuma, and Zirō Maki, 69–77. Kyoto: Yukawa Hall Archival Library.

———. 1989. "The development of elementary particle theory in Japan: Methodological aspects of the formation of the Sakata and Nagoya models." *Historia Scientiarum* 36:83–95.

Mandelstam, S. 1958. "Determination of the pion-nucleon scattering amplitude from dispersion relations and unitarity: General theory." *PR* 112:1344–60.

———. 1959. "Analytic properties of transition amplitudes in perturbation theory." *PR* 115:1741–51.

———. 1961. "Two-dimensional representations of scattering amplitudes and their applications." In *The Quantum Theory of Fields*, ed. R. Stoop, 209–25. New York: Interscience.

Mandl, Franz. 1959. *Introduction to Quantum Field Theory*. New York: Interscience.

Mandl, Franz, and Graham Shaw. 1984. *Quantum Field Theory*. New York: Wiley.

———. 1993. *Quantum Field Theory*. Rev. ed. New York: Wiley.

Marling, Karal Ann. 1994. *As Seen on TV: Visual Culture in the 1950s*. Cambridge: Harvard University Press.

Marshak, Robert E. 1939. *Contributions to the Theory of the Internal Constitution of Stars*. Ph.D. dissertation, Cornell University.

———. 1952. *Meson Physics*. New York: McGraw-Hill.

———. 1970. "The Rochester conferences: The rise of international cooperation in high energy physics." *BAS* 26:92–98.

———. 1983. "Particle physics in rapid transition: 1947–1952." In *The Birth of Particle Physics*, ed. Laurie Brown and Lillian Hoddeson, 376–401. New York: Cambridge University Press.

———. 1985. "Origin of the two-meson theory." In *Shelter Island II*, ed. Roman Jackiw, Nicola Khuri, Steven Weinberg, and Edward Witten, 355–62. Cambridge: MIT Press.

———. 1989. "Scientific impact of the first decade of the Rochester conferences (1950–1960)." In *Pions to Quarks: Particle Physics in the 1950s*, ed. Laurie Brown, Max Dresden, and Lillian Hoddeson, 645–67. New York: Cambridge University Press.

Marshak, R. E., and H. A. Bethe. 1947. "On the two-meson hypothesis." *PR* 72:506–9.

Martin, Paul. 1954. *Bound State Problems in Electrodynamics*. Ph.D. dissertation, Harvard University.

———. 1979. "Schwinger and statistical physics: A spin-off success story and some challenging sequels." In *Themes in Contemporary Physics: Essays in Honour of Julian Schwinger's 60th Birthday*, ed. S. Deser et al., 70–88. Amsterdam: North-Holland.

———. 1996. "Julian Schwinger: Personal recollections." In *Julian Schwinger: The Physicist, the Teacher, and the Man*, ed. Y. Jack Ng, 83–89. Singapore: World Scientific.

Masterman, Margaret. 1970. "The nature of a paradigm." In *Criticism and the Growth of Knowledge*, ed. Imre Lakatos and Alan Musgrave, 59–89. New York: Cambridge University Press.

Matthews, P. T. 1949. "The *S*-matrix for meson-nucleon interactions." *PR* 76:1254–55.

———. 1949. "Application of the Tomonaga-Schwinger theory to the interaction of nucleons with neutral scalar and vector mesons." *PR* 76:1657–74.

———. 1950. "The *S*-matrix for meson-nucleon interactions." *Philosophical Magazine* 41:185–95.

———. 1962. "Some particles are more elementary than others." Inaugural lecture. London: Imperial College.

Matthews, P. T., and Abdus Salam. 1951. "The renormalization of meson theories." *Reviews of Modern Physics* 23:311–14.

Mattuck, Richard. 1967. *A Guide to Feynman Diagrams in the Many-Body Problem*. New York: McGraw-Hill.

———. 1976. *A Guide to Feynman Diagrams in the Many-Body Problem*. 2nd ed. New York: Dover.

Mayer, Martin. 1963. "The leisure of the theory class." *Esquire* 59 (Mar): 16–26.

Mayer, Meinhard. 1990. "Reminiscences of the Landau seminar, 1957–58." In *Frontiers of Physics: Proceedings of the Landau Memorial Conference,* ed. Errol Gotsman, Yuval Ne'eman, and Alexander Voronel, 31–41. New York: Pergamon.

McConnell, J. 1951. "Diverging integrals in the self-charge problem." *PR* 81:275.

McCumber, John. 2001. *Time in the Ditch: American Philosophy and the McCarthy Era.* Evanston: Northwestern University Press.

McDonald, W. S., V. Z. Peterson, and D. R. Corson. 1957. "Photoproduction of neutral pions from hydrogen at forward angles from 240 to 480 MeV." *PR* 107:577–85.

McMahon, Allan. 1948. "How to expand the sale of technical books." *Publishers' Weekly,* 28 Feb, 1129–31.

McMillen, J. H. 1958. "Our universities' research-associate positions in physics." *PT* 11 (Aug): 14–15.

Mehra, Jagdish. 1975. *The Solvay Conferences on Physics.* Dordrecht: Reidel.

———. 1994. *The Beat of a Different Drum: The Life and Science of Richard Feynman.* New York: Oxford University Press.

Mehra, Jagdish, and Kimball Milton. 2000. *Climbing the Mountain: The Scientific Biography of Julian Schwinger.* New York: Oxford University Press.

Melcher, Frederic. 1946. "Editorial: Keeping books in print." *Publishers' Weekly,* 8 June, 3007.

[———]. 1950. "Again the demand increases for technical books." *Publishers' Weekly,* 19 Aug, 775.

Mermin, N. David. 1968. *Space and Time in Special Relativity.* New York: McGraw-Hill.

———. 1991. "Publishing in computopia." *PT* 44 (May): 9–11.

Merz, Martina. 1998. "'Nobody can force you when you are across the ocean': Face to face and e-mail exchanges between theoretical physicists." In *Making Space for Science: Territorial Themes in the Shaping of Knowledge,* ed. Crosbie Smith and Jon Agar, 313–29. New York: St. Martin's Press.

Miller, Arthur I. 1984. *Imagery in Scientific Thought: Creating 20th-Century Physics.* Boston: Birkhäuser.

———, ed. 1994. *Early Quantum Electrodynamics: A Source Book.* New York: Cambridge University Press.

———. 1994. "Frame-setting essay." In *Early Quantum Electrodynamics: A Source Book,* ed. Arthur I. Miller, 1–104. New York: Cambridge University Press.

———. 1996. *Insights of Genius: Imagery and Creativity in Science and Art.* New York: Springer.

Mills, Robert. 1955. *Fourth-Order Radiative Corrections to Atomic Energy Levels, II.* Ph.D. dissertation, Columbia University.

———. 1993. "Tutorial on infinities in QED." In *Renormalization: From Lorentz to Landau (and Beyond),* ed. Laurie Brown, 59–85. New York: Springer.

Mitchell, W. J. T. 1986. *Iconology: Image, Text, Ideology.* Chicago: University of Chicago Press.

———. 1994. *Picture Theory: Essays on Verbal and Visual Representation.* Chicago: University of Chicago Press.

———. 1998. *The Last Dinosaur Book: The Life and Times of a Cultural Icon.* Chicago: University of Chicago Press.

Miyamoto, Yoneji. 1988. "Personal recollections." In *Elementary Particle Theory in Japan, 1935–1960,* ed. Laurie Brown, Rokuo Kawabe, Michiji Konuma, and Zirō Maki, 63–67. Kyoto: Yukawa Hall Archival Library.

Mlodinow, Leonard. 2003. *Feynman's Rainbow: A Search for Beauty in Physics and in Life.* New York: Warner Books.

Møller, Christian. 1952. *The Theory of Relativity.* Oxford: Clarendon Press.

Moravcsik, Michael. 1961. "Practical utilisation of the nearest singularity in dispersion relations." In *Dispersion Relations: Scottish Universities' Summer School, 1960*, ed. G. R. Screaton, 117–65. New York: Interscience.

———. 1963. *The Two-Nucleon Interaction*. Clarendon: Oxford University Press.

———. 1985. "Thirty years of one-particle exchange." In *A Passion for Physics: Essays in Honor of Geoffrey Chew*, ed. Carleton DeTar, J. Finkelstein, and Chung-I Tan, 26–34. Singapore: World Scientific.

Moreton, David. 1951. "Jackets sell technical books." *Publishers' Weekly*, 15 Dec, 2240–42.

Morgan, Mary, and Margaret Morrison, eds. 1999. *Models as Mediators: Perspectives on Natural and Social Science*. New York: Cambridge University Press.

Morrison, Harry, ed. 1962. *The Quantum Theory of Many-Particle Systems*. New York: Gordon and Breach.

Mueller, Al. 1985. "Renormalons and phenomenology in QCD." In *A Passion for Physics: Essays in Honor of Geoffrey Chew*, ed. Carleton DeTar, J. Finkelstein, and Chung-I Tan, 137–42. Singapore: World Scientific.

Mukherji, V. 1974. "A history of the meson theory of nuclear forces from 1935–52." *Archives for the History of the Exact Sciences* 13:27–102.

Mullet, Shawn. 1999. *Political Science: The Red Scare as the Hidden Variable in the Bohmian Interpretation of Quantum Theory*. B.A. thesis, University of Texas, Austin.

Nakayama, Shigeru. 2001. "Introduction: occupation period, 1945–52." In *A Social History of Science and Technology in Contemporary Japan*, ed. Shigeru Nakayama, vol. 1, 23–56. Melbourne: Trans Pacific Press.

———. 2001. "Destruction of cyclotrons." In *A Social History of Science and Technology in Contemporary Japan*, ed. Shigeru Nakayama, 1:108–18. Melbourne: Trans Pacific Press.

———. 2001. "The international exchange of scientific information." In *A Social History of Science and Technology in Contemporary Japan*, ed. Shigeru Nakayama, 1:237–48. Melbourne: Trans Pacific Press.

———. 2001. "Sending scientists overseas." In *A Social History of Science and Technology in Contemporary Japan*, ed. Shigeru Nakayama, 1:249–60. Melbourne: Trans Pacific Press.

Nambu, Yoichiro. 1949. "The level shift and the anomalous magnet moment of the electron." *PTP* 4:82–94.

———. 1955. "Structure of Green's functions in quantum field theory." *PR* 100:394–411.

———. 1988. "Summary of personal recollections of the Tokyo group." In *Elementary Particle Theory in Japan, 1935–1960*, ed. Laurie Brown, Rokuo Kawabe, Michiji Konuma, and Zirō Maki, 3–5. Kyoto: Yukawa Hall Archival Library.

National Research Council. 1966. *Physics: Survey and Outlook*. Washington, DC: National Academy of Sciences.

———. 1969. *The Invisible University: Postdoctoral Education in the United States*. Washington, DC: National Academy of Sciences.

———. 1981. *Postdoctoral Appointments and Disappointments*. Washington, DC: National Academy Press.

National Science Foundation. 1988. *Career Progression of NATO Postdoctoral Fellows*. Washington, DC: National Science Foundation.

Nelipa, N. F. 1958. "On the problem of the excited states of nucleons." *Soviet Physics JETP* 6 (May): 981–84. Originally published in Russian in *ZhETF* 33 (Nov 1957): 1277–81.

Nersessian, Nancy. 1988. "Reasoning from imagery and analogy in scientific concept formation." *Philosophy of Science Association* 1:41–47.

———. 1993. "In the theoretician's laboratory: Thought experimenting as mental modeling." *Philosophy of Science Association* 2:291–301.

———. 1999. "Model-based reasoning in conceptual change." In *Model-Based Reasoning in Scientific Discovery,* ed. L. Magnani, N. J. Nersessian, and P. Thagard, 5–22. New York: Kluwer.

Ng, Y. Jack, ed. 1996. *Julian Schwinger: The Physicist, the Teacher, and the Man.* Singapore: World Scientific.

Nihon Butsuri Gakkai Meibo [Directory of the Japanese Physical Society]. 1956, 1963. Tokyo: Nihon Butsuri Gakkai.

Nihon Hakushi Roku [Directory of Japanese Doctorate Holders]. 1985. 9 vols. Tokyo: Nihon Tosho Senta.

Norris, Robert, and William Arkin. 1993. "Russian/Soviet weapons secrets revealed." *BAS* 49 (Apr): 48.

Novozhilov, Y. V. 1961. *Elementary Particles.* Delhi: Hindustan.

Noyes, H. P., E. M. Hafner, J. Klarmann, and A. E. Woodruff, eds. 1954. *Proceedings of the Fourth Annual Rochester Conference on High Energy Nuclear Physics.* Rochester: University of Rochester.

Nye, Mary Jo. 1993. *From Chemical Philosophy to Theoretical Chemistry.* Berkeley: University of California Press.

Ochs, Elinor, Sally Jacoby, and Patrick Gonzales. 1994. "Interpretive journeys: How physicists talk and travel through graphic space." *Configurations* 2:151–71.

Okun', L. B. 1956. "K-meson charge exchange in hydrogen and deuterium." *Soviet Physics JETP* 3 (Aug): 142–43. Originally published in Russian in *ZhETF* 30 (Jan): 218–19.

———. 1957. "Isotopic invariance and 'strange' particles." *Soviet Physics JETP* 3 (Jan): 994–96. Originally published in Russian in *ZhETF* 30 (June 1956): 1172–73.

———. 1957. "On the probabilities of Σ-particle decay." *Soviet Physics JETP* 4 (Mar): 284–86. Originally published in Russian in *ZhETF* 31 (Aug 1956): 333–35.

———. 1957. "The μ-decay of K-particles and hyperons." *Soviet Physics JETP* 5 (Sep): 334–35. Originally published in Russian in *ZhETF* 32 (Feb): 400–402.

———. 1958. "On K_{e3} decay." *Soviet Physics JETP* 6 (Feb): 409–10. Originally published in Russian in *ZhETF* 33 (Aug 1957): 525–26.

Okun', L. B., and I. Ia. Pomeranchuk. 1956. "The conservation of isotopic spin and the cross section of the interaction of high-energy π-mesons and nucleons with nucleons." *Soviet Physics JETP* 3 (Sep): 307–8. Originally published in Russian in *ZhETF* 30 (Feb): 424.

Okun', L. B., and B. Pontecorvo. 1957. "Some remarks on slow processes of transformation of elementary particles." *Soviet Physics JETP* 5 (Dec): 1297–99. Originally published in Russian in *ZhETF* 32 (June): 1587–88.

Okun', L. B., and A. P. Rudik. 1957. "Possible correlations in π-μ-e decay." *Soviet Physics JETP* 5 (Oct): 520–21. Originally published in Russian in *ZhETF* 32 (Mar): 627–28.

———. 1960. "On a method of finding singularities of Feynman graphs." *Nuclear Physics* 15:261–88.

Okun', L. B., and M. I. Shmushkevich. 1956. "Capture of K^- mesons by deuterium and hyperon-nucleon interaction." *Soviet Physics JETP* 3 (Dec): 792–93. Originally published in Russian in *ZhETF* 30 (May): 979–81.

Olesko, Kathryn. 1991. *Physics as a Calling: Discipline and Practice in the Königsberg Seminar for Physics.* Ithaca: Cornell University Press.

———. 1993. "Tacit knowledge and school formation." *Osiris* 8:16–29.

Olson, Keith. 1974. *The G.I. Bill, the Veterans, and the Colleges.* Lexington: University of Kentucky Press.

Olwell, Russell. 1999. "Physical isolation and marginalization in physics: David Bohm's cold war exile." *Isis* 90:738–56.

Omnès, R., and M. Froissart. 1963. *Mandelstam Theory and Regge Poles: An Introduction for Experimentalists.* New York: W. A. Benjamin.

Ōneda, Sadao. 1988. "Personal recollections on weak interactions during the 1950s and early 1960s." In *Elementary Particle Theory in Japan, 1935–1960,* ed. Laurie Brown, Rokuo Kawabe, Michiji Konuma, and Zirō Maki, 15–19. Kyoto: Yukawa Hall Archival Library.

Oppenheimer, J. Robert. 1950. "Electron theory." In *Rapports du 8e Conseil de Physique, Solvay,* ed. R. Stoop, 1–11. Brussels: Solvay.

———. 1966. "Thirty years of mesons." *PT* 18 (Nov): 51–58.

Oreskes, Naomi. 1999. *The Rejection of Continental Drift: Theory and Method in American Earth Science.* New York: Oxford University Press.

———. 2003. "The role of quantitative models in science." In *Models in Ecosystem Science,* ed. Charles D. Canham, Jonathan J. Cole, and William K. Lauenroth, 13–31. Princeton: Princeton University Press.

Oreskes, Naomi, and Ronald Rainger. 2000. "Science and security before the atomic bomb: The loyalty case of Harald U. Sverdrup." *SHPMP* 31:309–69.

Oreskes, Naomi, Kristin Shrader-Frechette, and Kenneth Belitz. 1994. "Verification, validation, and confirmation of numerical models in the earth sciences." *Science* 263 (4 Feb): 641–46.

Overbye, Dennis. 2001. "On stage, a day in the life of an idiosyncratic physicist." *NYT,* 13 Nov, D5.

Pais, Abraham. 1945. "On the theory of the electron and the nucleon." *PR* 68:227–28.

———. 1946. "On the theory of elementary particles." *Verhandelingen der Koninklijke Nederlandsche Akademie van Wetenschappen, Afdeeling Natuurkunde* 19:1–91.

———. 1986. *Inward Bound: Of Matter and Forces in the Physical World.* New York: Clarendon.

———. 1997. *A Tale of Two Continents: A Physicist's Life in a Turbulent World.* Princeton: Princeton University Press.

Pang, Alex Soojung-Kim. 1994. "Victorian observing practices, printing technology, and representations of the solar corona." Pt. 1, "The 1860s and 1870s." *Journal for the History of Astronomy* 25:249–74.

———. 1995. "Victorian observing practices, printing technology, and representations of the solar corona." Pt. 2, "The age of photomechanical reproduction." *Journal for the History of Astronomy* 26:63–75.

———. 1997. "Visual representation and post-constructivist history of science." *HSPS* 28:139–71.

———. 1998. "Technology, aesthetics, and the development of astrophotography at the Lick Observatory." In *Inscribing Science: Scientific Texts and the Materiality of Communication,* ed. Timothy Lenoir, 223–48. Stanford: Stanford University Press.

Panofsky, Erwin. 1991 [1927]. *Perspective as Symbolic Form.* Trans. Christopher S. Wood. New York: Zone Books.

Park, Buhm Soon. 2005. "In the 'context of pedagogy': Teaching strategy and theory change in quantum chemistry." In *Pedagogy and the Practice of Science: Historical and Contemporary Perspectives,* ed. David Kaiser, 287–319. Cambridge: MIT Press.

Pauli, Wolfgang. 1946. *Meson Theory of Nuclear Forces.* New York: Interscience.

Pauli, W., and F. Villars. 1949. "On the invariant regularization in relativistic quantum theory." *Reviews of Modern Physics* 21:434–44.

Pauli, W., and V. Weisskopf. 1994 [1934]. "The quantization of the scalar relativistic wave equation." In *Early Quantum Electrodynamics: A Source Book,* ed. Arthur I. Miller, 188–205. New York: Cambridge University Press.

Peaslee, D. C. 1951. "Boson current corrections to second order." *PR* 81:94–106.

———. 1951. "Infinite integrals in quantum electrodynamics." *PR* 81:107–9.

Peat, F. David. 1997. *Infinite Potential: The Life and Times of David Bohm.* Reading: Addison-Wesley.

Peierls, Rudolf. 1985. *Bird of Passage: Recollections of a Physicist.* Princeton: Princeton University Press.

Perry, Georgella. 1985. "My years with Professor Chew." In *A Passion for Physics: Essays in Honor of Geoffrey Chew,* ed. Carleton DeTar, J. Finkelstein, and Chung-I Tan, 14–16. Singapore: World Scientific.

Peskin, Michael, and Daniel Schroeder. 1995. *Introduction to Quantum Field Theory.* Reading, MA: Addison-Wesley.

Petermann, A. 1957. "Magnetic moment of the electron." *Nuclear Physics* 3:689–90.

———. 1957. "Fourth order magnetic moment of the electron." *Helvetica Physica Acta* 30:407–8.

———. "Fourth order magnetic moment of the electron." *Nuclear Physics* 5:677–83.

Pickering, Andrew. 1984. *Constructing Quarks: A Sociological History of Particle Physics.* Chicago: University of Chicago Press.

———. 1984. "Against putting the phenomena first: The discovery of the weak neutral current." *SHPS* 15:85–117.

———. 1989. "From field theory to phenomenology: The history of dispersion relations." In *Pions to Quarks: Particle Physics in the 1950s,* ed. Laurie Brown, Max Dresden, and Lillian Hoddeson, 579–99. New York: Cambridge University Press.

———. 1995. *The Mangle of Practice: Time, Agency, and Science.* Chicago: University of Chicago Press.

Pickering, Andrew, and Adam Stephanides. 1992. "Constructing quaternions: On the analysis of conceptual practice." In *Science as Practice and Culture,* ed. Andrew Pickering, 139–67. Chicago: University of Chicago Press.

Pickup, E., and L. Voyvodic. 1951. "Cosmic-ray nucleon-nucleon collisions in photographic emulsions." *PR* 82:265.

Pinch, Trevor. 1997. "Old habits die hard: Retrieving practices from social theory." *SHPS* 28:203–8.

Pinch, Trevor, H. M. Collins, and Larry Carbone. 1996. "Inside knowledge: Second order measures of skill." *Sociological Review* 44:163–86.

Pines, David. 1961. *The Many-Body Problem.* New York: W. A. Benjamin.

———. 1963. *Elementary Excitations in Solids.* New York: W. A. Benjamin.

———. 1989. "Richard Feynman and condensed matter physics." *PT* 42 (Feb): 61–66.

Pines, David, and Philippe Nozières. 1966. *The Theory of Quantum Liquids.* New York: W. A. Benjamin.

Polanyi, Michael. 1958. *Personal Knowledge.* London: Routledge and Kegan Paul.

———. 1967. *The Tacit Dimension.* New York: Anchor.

Polchinski, Joseph. 1998. *String Theory.* 2 vols. New York: Cambridge University Press.

Politzer, H. David. 1973. "Reliable perturbative results for strong interactions?" *PRL* 30:1346–49;

Polkinghorne, John. 1956. Review of *The Theory of Photons and Electrons,* by J. M. Jauch and F. Rohrlich. *PT* 9 (Aug): 34.

———. 1956. Review of *Quantum Field Theory,* by H. Umezawa. *PT* 9 (Nov): 38–40.

———. 1961. "The analytic properties of perturbation theory." In *Dispersion Relations: Scottish Universities' Summer School, 1960,* ed. G. R. Screaton, 65–93. New York: Interscience.

———. 1962. "Analyticity and unitarity." Pt. 2. *Il Nuovo Cimento* 25:901–11.

———. 1985. "Salesman of ideas." In *A Passion for Physics: Essays in Honor of Geoffrey Chew*, ed. Carleton DeTar, J. Finkelstein, and Chung-I Tan, 23–25. Singapore: World Scientific.

———. 1989. *Rochester Roundabout: The Story of High Energy Physics*. San Francisco: W. H. Freeman.

Pollock, Franklin. 1952. *The Fourth Order Corrections to the Propagation Function and Mass of the Electron*. Ph.D. dissertation, Columbia University.

Pontecorvo, B. M. 1956. "The processes of production of heavy mesons and V_1^0-particles." *Soviet Physics JETP* 2 (Jan): 135–39. Originally published in Russian in *ZhETF* 29 (Aug 1955): 140–46.

Porter, Laura. 1988. *From Intellectual Sanctuary to Social Responsibility: The Founding of the Institute for Advanced Study, 1930–1933*. Ph.D. dissertation, Princeton University.

Prange, R. E. 1958. "Dispersion relations for Compton scattering." *PR* 110:240–52.

Priestley, Joseph. 1775. *Experiments and Observations on Different Kinds of Air*. London: J. Johnson.

Pyenson, Lewis. 1977. "'Who the guys were': Prosopography in the history of science." *History of Science* 15:155–88.

Quigg, Chris. 1983. *Gauge Theories of the Strong, Weak, and Electromagnetic Interactions*. Reading, MA: Benjamin/Cummings.

[Rabinowitch, Eugene]. 1949. "The 'cleansing' of AEC fellowships." *BAS* 5 (Jun–Jul): 161–62.

[———]. 1952. "How to lose friends." *BAS* 8 (Jan): 2–5.

Rader, Karen. 1998. "The mouse people: Murine genetics work at the Bussey Institution, 1909–1936." *Journal for the History of Biology* 31:327–54.

———. 1999. "Of mice, medicine, and genetics: C. C. Little's creation of the inbred laboratory mouse, 1909–1918." *Studies in History and Philosophy of Biological and Biomedical Sciences* 30:319–43.

Radetsky, Peter. 1994. "The modern postdoc: Prepping for the job market." *Science* 265 (23 Sep): 1909–10.

Ramond, Pierre. 1981. *Field Theory: A Modern Primer*. Reading, MA: Benjamin-Cummings.

Rand, Myrton. 1951. "The National Research fellowships." *Scientific Monthly* 73 (Aug): 71–80.

Rasmussen, Nicolas. 1997. *Picture Control: The Electron Microscope and the Transformation of Biology in America, 1940–1960*. Stanford: Stanford University Press.

Ravetz, Jerome. 1971. *Scientific Knowledge and Its Social Problems*. Oxford: Clarendon.

Raymond, Jack. 1956. "Americans in Soviet report science for peace stressed." *NYT*, 22 May, 1, 8.

Rechenberg, Helmut. 1989. "The early *S*-matrix theory and its propagation (1942–1952)." In *Pions to Quarks: Particle Physics in the 1950s*, ed. L. Brown, M. Dresden, and L. Hoddeson, 551–78. New York: Cambridge University Press.

Reed, Thomas, and Arnold Kramish. 1996. "Trinity at Dubna." *Physics Today* 49 (Nov): 30–35.

Regge, Tullio. 1958. "Analytic properties of the scattering matrix." *Il Nuovo Cimento* 8:671–79.

———. 1958. "On the analytic behavior of the eigenvalue of the *S*-matrix in the complex plane of the energy." *Il Nuovo Cimento* 9:295–302.

———. 1959. "Introduction to complex orbital momenta." *Il Nuovo Cimento* 14:951–76.

———. 1960. "Bound states, shadow states and Mandelstam representation." *Il Nuovo Cimento* 18:947–56.

Regis, Ed. 1987. *Who Got Einstein's Office? Eccentricity and Genius at the Institute for Advanced Study*. Reading, MA: Perseus Books.

Reingold, Nathan. 1987. "Vannevar Bush's New Deal for research: Or the triumph of the old order." *HSPS* 17:299–344.

Rider, Robin. 1984. "Alarm and opportunity: Emigration of mathematicians and physicists to Britain and the United States, 1933–1945." *HSPS* 15:107–76.

———. 1998. "Shaping Information: Mathematics, Computing, and Typography." In *Inscribing Science: Scientific Texts and the Materiality of Communication*, ed. Timothy Lenoir, 39–54. Stanford: Stanford University Press.

Rigden, John. 1987. *Rabi: Scientist and Citizen*. New York: Basic Books.

Riordan, Michael. 1987. *The Hunting of the Quark: A True Story of Modern Physics*. New York: Simon and Schuster.

Rocke, Alan. 1993. *The Quiet Revolution: Hermann Kolbe and the Science of Organic Chemistry*. Berkeley: University of California Press.

Rohrlich, Fritz. 1953. "Applied quantum electrodynamics." Mimeographed notes (by Carroll Alley et al.) from lectures delivered at Princeton University during spring 1953. Copy available at http://hrst.mit.edu/hrs/renormalization/public.

Roman, Paul. 1960. *Theory of Elementary Particles*. Amsterdam: North-Holland.

Rorabaugh, W. J. 1989. *Berkeley at War: The 1960s*. New York: Oxford University Press.

Rosenbaum, E. P., with Robert Marshak and Robert Wilson. 1956. "Physics in the U.S.S.R." *Scientific American* 195 (Aug): 29–35.

Rosenfeld, Léon. 1948. *Nuclear Forces*. New York: Interscience.

———. 1950. "Meson fields and nuclear forces." *PTP* 5:519–22.

Rouse, Joseph. 1987. *Knowledge and Power: Toward a Political Philosophy of Science*. Ithaca: Cornell University Press.

Rudolph, John. 2002. *Scientists in the Classroom: The Cold War Reconstruction of American Science Education*. New York: Palgrave.

Rudwick, Martin. 1976. "The emergence of a visual language for geological science, 1760–1840." *History of Science* 14:149–95.

Rueger, Alexander. 1992. "Attitudes towards infinities: Responses to anomalies in quantum electrodynamics, 1927–1947." *HSPS* 22:309–37.

Ryder, Lewis. 1985. *Quantum Field Theory*. New York: Cambridge University Press.

Sagdeev, Roald. 1993. "Russian scientists save American secrets." *BAS* 49 (May): 32–36.

Salam, Abdus. 1989. "Physics and the excellences of life it brings." In *Pions to Quarks: Particle Physics in the 1950s*, ed. Laurie Brown, Max Dresden, and Lillian Hoddeson, 525–35. New York: Cambridge University Press.

Salisbury, Harrison. 1956. "Curtains are parted on science in Soviet." *NYT*, 3 June, 1, 12.

Salpeter, E. E., and J. S. Goldstein. 1953. "Momentum space wave functions." Pt. 2, "The deuteron bound state." *PR* 90:983–86.

Samuels, Gertrude. 1950. "Where Einstein surveys the cosmos." *NYT Magazine*, 19 Nov, 14–37.

Schaffer, Simon. 1989. "Glass works: Newton's prisms and the uses of experiment." In *The Uses of Experiment: Studies in the Natural Sciences*, ed. David Gooding, Trevor Pinch, and Simon Schaffer, 67–104. New York: Cambridge University Press.

———. 1992. "Late Victorian metrology and its instrumentation: A manufactory of ohms." In *Invisible Connections: Instruments, Institutions, and Science*, ed. Robert Bud and Susan Cozzens, 23–56. Bellingham, WA: SPIE Optical Engineering Press.

Schmidt, Stanley. 1963. "The 'basic' particle—it's out of date." *Daily Californian*, 23 Oct, 1, 12.

Schrecker, Ellen. 1986. *No Ivory Tower: McCarthyism and the Universities*. New York: Oxford University Press.

———. 1994. *The Age of McCarthyism: A Brief History with Documents*. Boston: Bedford Books.

———. 1998. *Many Are the Crimes: McCarthyism in America*. New York: Little, Brown.

Schwartz, Harry. 1955. "U.S. barred visits to Soviet parley: 2 physicists tell why they accepted, then rejected, invitation by Academy." *NYT*, 7 Apr, 1, 16.

Schweber, Silvan. 1961. *Introduction to Relativistic Quantum Field Theory.* Evanston, IL: Row, Peterson.

———. 1986. "The empiricist temper regnant: Theoretical physics in the United States, 1920–1950." *HSPS* 17:55–98.

———. 1986. "Feynman and the visualization of space-time processes." *Reviews of Modern Physics* 58:449–508.

———. 1989. "Some reflections on the history of particle physics in the 1950s." In *Pions to Quarks: Particle Physics in the 1950s*, ed. Laurie Brown, Max Dresden, and Lillian Hoddeson, 668–93. New York: Cambridge University Press.

———. 1993. "Changing conceptualization of renormalization theory." In *Renormalization: From Lorentz to Landau (and Beyond)*, ed. Laurie Brown, 135–66. New York: Springer.

———. 1994. *QED and the Men who Made It: Dyson, Feynman, Schwinger, and Tomonaga.* Princeton: Princeton University Press.

———. 2000. *In the Shadow of the Bomb: Oppenheimer, Bethe, and the Moral Responsibility of the Scientist.* Princeton: Princeton University Press.

Schweber, Silvan, Hans Bethe, and Frederic de Hoffmann. 1955. *Mesons and Fields.* 2 vols. White Plains, NY: Row and Peterson.

Schwinger, Julian. 1948. "On quantum-electrodynamics and the magnetic moment of the electron." *PR* 73:416–17.

———. 1948. "Quantum electrodynamics." Pt. 1, "A covariant formulation." *PR* 74:1439–61.

———. 1951. "On the Green's functions of quantized fields." Pts. 1, 2. *Proceedings of the National Academy of Sciences* 37:452–59.

———, ed. 1958. *Selected Papers on Quantum Electrodynamics.* New York: Dover.

———. 1983. "Renormalization theory of quantum electrodynamics: An individual view." In *The Birth of Particle Physics*, ed. Laurie Brown and Lillian Hoddesson, 329–53. New York: Cambridge University Press.

———. 1983. "Two shakers of physics: Memorial lecture for Sin-itiro Tomonaga." In *The Birth of Particle Physics*, ed. Laurie Brown and Lillian Hoddeson, 354–75. New York: Cambridge University Press.

———. 1989. "A path to quantum electrodynamics." *PT* 42 (Feb): 42–48.

———. 1996. "The Greening of quantum field theory: George and I." In *Julian Schwinger: The Physicist, the Teacher, and the Man*, ed. Y. Jack Ng, 13–27. Singapore: World Scientific.

Science Citation Index. 1961–. Philadelphia: Institute for Scientific Information.

Screaton, G. R., ed. 1961. *Dispersion Relations: Scottish Universities' Summer School, 1960.* New York: Interscience.

Secord, James. 2000. *Victorian Sensation: The Extraordinary Publication, Reception, and Secret Authorship of "Vestiges of the Natural History of Creation."* Chicago: University of Chicago Press.

Segrè, Emilio. 1993. *A Mind Always in Motion: The Autobiography of Emilio Segrè.* Berkeley: University of California Press.

Seidel, Robert. 1986. "A home for big science: The Atomic Energy Commission's laboratory system." *HSPS* 16:135–75.

Serber, Robert. 1967. "The early years." *PT* 20 (Oct): 35–39.

Serber, Robert, with Robert Crease. 1998. *Peace and War: Reminiscences of a Life on the Frontiers of Science.* New York: Columbia University Press.

Shankar, R. 1999. "Effective field theory in condensed matter physics." In *Conceptual Foundations of Quantum Field Theory*, ed. Tian Yu Cao, 47–55. New York: Cambridge University Press.

Shapin, Steven, and Simon Schaffer. 1985. *Leviathan and the Air-Pump: Hobbes, Boyle, and the Experimental Life*. Princeton: Princeton University Press.

Shapin, Steven, and Arnold Thackray. 1974. "Prosopography as a research tool in history of science: The British scientific community, 1700–1900." *History of Science* 12:1–28.

Shirkov, Dmitri. 1993. "Historical remarks on the renormalization group." In *Renormalization: From Lorentz to Landau (and Beyond)*, ed. Laurie Brown, 169–86. New York: Springer.

Shuryak, E., and I. Zahed. 2003. "Semiclassical double pomeron production of glueballs and eta-prime." Preprint hep-ph/0302231 (Feb), available from http://www.arXiv.org.

Sibirtsev, A., et al. 2003. "Systematic Regge theory analysis of Omega photoproduction." Preprint nucl-th/0301015 (Jan), available from http://www.arXiv.org.

Simon, Albert. 1950. *Bremsstrahlung in High Energy Nucleon-Nucleon Collisions and Radiative Corrections to Meson-Nucleon Scattering*. Ph.D. dissertation, University of Rochester.

Simonov, Yu. A. 2002. "Spectrum and Regge trajectories in QCD." Preprint hep-ph/0210309 (Oct), available from http://www.arXiv.org.

Simpson, J. A., and E. S. C. Weiner, eds. 1989. *Oxford English Dictionary*. 2nd ed. New York: Oxford University Press.

Sismondo, Sergio, ed. 1999. *Modeling and Simulation*. Special issue of *Science in Context*, vol. 12 (Summer).

Sklar, Lawrence, ed. 2000. *The Nature of Scientific Theory*. New York: Garland.

Skorniakov, G. V. 1953. "Izluchenie γ-kvantov pri raspade μ-mezona" [Radiation of γ-quanta during μ-meson decay]. *Doklady Akademii Nauk SSSR* 89:431–34.

Smith, Alice Kimball. 1965. *A Peril and a Hope: The Scientists' Movement in America, 1945–7*. Chicago: University of Chicago Press.

Smith, Alice Kimball, and Charles Weiner, eds. 1980. *Robert Oppenheimer: Letters and Recollections*. Cambridge: Harvard University Press.

Smith, Crosbie, and M. Norton Wise. 1989. *Energy and Empire: A Biographical Study of Lord Kelvin*. New York: Cambridge University Press.

Smorodinskii, I. 1949. "Smeshchenie termov vodorodopodobnykh atomov i anomal'nyi magnitnyi moment elektrona" [Shift of terms of hydrogen-like atoms and the anomolous magnetic moment of the electron]. *Uspekhi fizicheskikh nauk* 39:325–58.

Snow, C. P. 1959. *The Two Cultures and the Scientific Revolution: The Rede Lecture, 1959*. New York: Cambridge University Press.

Snow, George, and Hartland Snyder. 1950. "On the self-energies of quantum field theory." *PR* 80:987–89.

Snyder, Hartland. 1950. "Quantum electrodynamics: The self-energy problem." *PR* 78:98–103.

———. 1950. "Quantum field theory." *PR* 79:520–25.

Snyder, Joel. 1980. "Picturing vision." In *The Languages of Images*, ed. W. J. T. Mitchell, 219–46. Chicago: University of Chicago Press.

Sommerfield, Charles. 1957. "Magnetic dipole moment of the electron." *PR* 107:328–29.

———. 1958. "The magnetic moment of the electron." *Annals of Physics* 5:26–57.

Sopka, Katherine. 1988 [1980]. *Quantum Physics in America: The Years through 1935*. New York: Tomash.

Soto, Maximillian, Jr. 1968. *Fourth-Order Radiative Corrections to Atomic Energy Levels*. Ph.D. dissertation, Columbia University.

Squires, E. J. 1963. *Complex Angular Momenta and Particle Physics*. New York: W. A. Benjamin.

Stachel, John. 1995. "History of relativity." In *Twentieth Century Physics*, ed. Laurie Brown, Abraham Pais, and Brian Pippard, vol. 1, 249–456. Philadelphia: Institute of Physics Publishers.

Stapp, Henry P. 1962. "Derivation of the CPT theorem and the connection between spin and statistics from postulates of the *S*-matrix theory." *PR* 125:2139–62.

———. 1962. "Axiomatic *S*-matrix theory." *Review of Modern Physics* 34:390–94.

———. 1965. "Analytic S-matrix theory." In *High-Energy Physics and Elementary Particles*, ed. Abdus Salam, 3–54. Vienna: International Atomic Energy Agency.

———. 1965. "Space and time in *S*-matrix theory." *PR* 139:B257–B270.

Steinberger, Jack. 1989. "A particular view of particle physics in the fifties." In *Pions to Quarks: Particle Physics in the 1950s*, ed. Laurie Brown, Max Dresden, and Lillian Hoddeson, 307–30. New York: Cambridge University Press.

———. 1997. "Early particles." *Annual Review of Nuclear and Particle Science* 47:xiii-xlii.

Stern, A. W. 1964. "The third revolution in 20th century physics." *PT* 17:42–45.

Stern, Beatrice. 1964. "A History of the Institute for Advanced Study, 1930–1950." Unpublished typescript. Copy available in the Hoover Library, McDaniel College, Westminster, MD.

Stern, Philip. 1969. *The Oppenheimer Case: Security on Trial.* New York: Harper and Row.

Stewart, George. 1950. *The Year of the Oath.* New York: Doubleday.

Stoop, R., ed. 1961. *The Quantum Theory of Fields.* New York: Interscience.

Stuckey, William. 1975. "The garden of lonely wise: A profile of the Institute for Advanced Study." *Science Digest* 77 (Feb): 28–37.

Stueckelberg, E. C. G. 1942. "La mécanique du point matériel en théorie de relativité et en théorie des quanta." *Helvetica Physica Acta* 15:23–37.

Stueckelberg, E. C. G., and A. Petermann. 1953. "La normalisation des constantes dans la theories des quanta." *Helvetica Physica Acta* 26:499–520.

Suits, C. G. 1960. "The postgraduate training of physicists." In *International Education in Physics*, ed. Sanborn Brown and Norman Clarke, 88–95. Cambridge: MIT Press.

Sullivan, D., D. Koester, D. H. White, and K. Kern. 1980. "Understanding rapid theoretical change in particle physics: A month-by-month co-citation analysis." *Scientometrics* 2:309–19.

Suppe, Frederick. 1977. "The search for philosophic understanding of scientific theories." In *The Structure of Scientific Theories*, ed. Frederick Suppe, 2nd ed., 3–241. Urbana: University of Illinois Press.

Sykes, Christopher, ed. 1994. *No Ordinary Genius: The Illustrated Richard Feynman.* New York: W. W. Norton.

Tanikawa, Yasutaka. 1947. "On the quantum theory of radiation damping and the lifetime of the meson." *PTP* 3:38–53.

Taylor, J. C. 1960. "Analytic properties of perturbation expansions." *PR* 117:261–65.

———. 1960. "Special topics in dispersion relations." Mimeographed notes from lectures delivered at the University of Rochester during November 1960. AEC preprint NYO-9364. Copy in the possession of Sam Schweber.

Taylor, Peter, and Ann Blum, eds. 1991. *Pictorial Representation in Biology*. Special issue of *Biology and Philosophy*, vol. 6.

———. 1991. "Pictorial representation in biology." *Biology and Philosophy* 6:125–34.

Tentyukov, M., and J. Fleischer. 1999. "DIANA, a program for Feynman diagram evaluation." Preprint hep-ph/9905560 (May), available from http://www.arXiv.org.

———. 2000. "A Feynman diagram analyzer: DIANA." *Computational Physics Communications* 132:124–41.

Terent'ev, M. V. 1963. "Application of the dispersion relations technique to the calculation of the magnetic moment of the electron." *Soviet Physics JETP* 16:444–54.

't Hooft, G. 1971. "Renormalization of massless Yang-Mills fields." *Nuclear Physics B* 33:173–99.

———. 1971. "Renormalizable Lagrangians for massive Yang-Mills fields." *Nuclear Physics B* 35:167–88.

———. 1997. "Renormalization of gauge theories." In *The Rise of the Standard Model: Particle Physics in the 1960s and 1970s,* ed. Lillian Hoddeson, Laurie Brown, Michael Riordan, and Max Dresden, 179–98. New York: Cambridge University Press.

't Hooft, G., and M. Veltman. 1972. "Renormalization and regularization of gauge fields." *Nuclear Physics B* 44:189–213.

———. 1973. "Diagrammar." CERN Yellow Report 73-9. Reprinted in *Under the Spell of the Gauge Principle,* by G. 't Hooft (Singapore: World Scientific, 1994), 29–173.

Thorpe, Charles, and Steven Shapin. 2000. "Who was J. Robert Oppenheimer? Charisma and complex organization." *Social Studies of Science* 30:545–90.

Toll, John. 1958. "Scientists urge lifting travel restrictions." *BAS* 14 (Oct): 326–28.

Tomonaga, S. 1946. "On a relativistically invariant formulation of the quantum theory of wave fields." *PTP* 1:27–42.

———. 1948. "On infinite reactions in quantum field theory." *PR* 74:224–25.

Topham, Jonathan. 2000. "A textbook revolution." In *Books and the Sciences in History,* ed. Marina Frasca-Spada and Nick Jardine, 317–37. New York: Cambridge University Press.

Travis, John. 1992. "Postdocs: Tales of woe from the 'invisible university.'" *Science* 257 (18 Sep): 1738–40.

Traweek, Sharon. 1988. *Beamtimes and Lifetimes: The World of High-Energy Physics.* Cambridge: Harvard University Press.

Treiman, Sam. 1963. "Analyticity in particle physics." In *Proceedings of the Eastern Theoretical Physics Conference, October 26–27, 1962,* ed. M. E. Rose, 127–74. New York: Gordon and Breach.

———. 1989. "A connection between the strong and weak interactions." In *Pions to Quarks: Particle Physics in the 1950s,* ed. Laurie Brown, Max Dresden, and Lillian Hoddeson, 384–89. New York: Cambridge University Press.

———. 1996. "A life in particle physics." *Annual Review of Nuclear and Particle Science* 46:1–30.

Treiman, Sam, Roman Jackiw, and David Gross. 1972. *Lectures on Current Algebra and Its Applications.* Princeton: Princeton University Press.

Turner, Stephen. 1994. *The Social Theory of Practices: Tradition, Tacit Knowledge, and Presuppositions.* Chicago: The University of Chicago Press.

Uehling, Edwin. 1935. "Polarization effects in the positron theory." *PR* 48:55–63.

Umezawa, H. 1956. *Quantum Field Theory.* New York: Interscience.

Umezawa, Hiroomi, and Rokuo Kawabe. 1949. "Some general formulae relating to vacuum polarization." *PTP* 4:423–42.

Uretsky, Jack, Robert Kenney, Edward Knapp, and Victor Perez-Mendez. 1958. "Photoproduction of positive pions from protons." *PRL* 1:12–14.

Valatin, J. G. 1947. "On molecular coupling effects." *PR* 71:458.

———. 1951. "On a formulation of quantum electrodynamics." *PR* 83:850–51.

———. 1954. "Singularities of electron kernel functions in an external electromagnetic field." *PRSA* 222:93–108.

Van Helden, Albert, and Thomas Hankins, eds. 1994. *Instruments.* Special issue of *Osiris,* vol. 9.

van Hove, Léon. 1955. "Energy corrections and persistent perturbation effects in continuous spectra." *Physica* 21:901–23.

———. 1956. "Energy corrections and persistent perturbation effects in continuous spectra." Pt. 2, "The perturbed stationary states." *Physica* 22:343–54.

———. 2000. "Introduction: An autobiographical sketch." In *The Legacy of Léon van Hove*, ed. Alberto Giovannini, xv–xvii. Singapore: World Scientific.

van Hove, L., N. M. Hugenholtz, and L. P. Howland, 1961. *Problems in Quantum Theory of Many-Particle Systems.* New York: W. A. Benjamin.

Veltman, M. 1994. *Diagrammatica: The Path to Feynman Diagrams.* New York: Cambridge University Press.

———. 1997. "The path to renormalizability." In *The Rise of the Standard Model: Particle Physics in the 1960s and 1970s*, ed. Lillian Hoddeson, Laurie Brown, Michael Riordan, and Max Dresden, 145–78. New York: Cambridge University Press.

Veneziano, G. 1968. "Construction of a crossing-symmetric, Regge-behaved amplitude for linearly rising trajectories." *Il Nuovo Cimento* 57A:190–97.

Villars, F. 1950. "On the energy-momentum tensor of the electron." *PR* 79:122–28.

———. 1951. "Quantum electrodynamics." Mimeographed notes from lectures delivered at MIT during July 1951. Copy in the possession of Sam Schweber.

Visher, Stephen. 1939. "The education of the younger starred scientists." *Journal of Higher Education* 10 (Mar): 124–32.

Wang, Jessica. 1995. "Liberals, the progressive Left, and the political economy of postwar American science: The National Science Foundation debate revisited." *HSPS* 26:139–66.

———. 1999. *American Science in an Age of Anxiety: Scientists, Anticommunism, and the Cold War.* Chapel Hill: University of North Carolina Press.

Warwick, Andrew. 1992. "Cambridge mathematics and Cavendish physics: Cunningham, Campbell, and Einstein's relativity, 1905–1911." Pt. 1, "The uses of theory." *SHPS* 23:625–56.

———. 1993. "Cambridge mathematics and Cavendish physics: Cunningham, Campbell, and Einstein's relativity, 1905–1911." Pt. 2, "Comparing traditions in Cambridge physics." *SHPS* 24:1–25.

———. 2003. *Masters of Theory: Cambridge and the Rise of Mathematical Physics.* Chicago: University of Chicago Press.

Watson, Kenneth. 1953. "Multiple scattering and the many-body problem: Applications to photomeson production in complex nuclei." *PR* 89:575–87.

Weinberg, Steven. 1964. "Photons and gravitons in *S*-matrix theory: Derivation of charge conservation and equality of gravitational and inertial mass." *PR* 135:B1049–B1056.

———. 1964. "Derivation of gauge invariance and the equivalence principle from Lorentz invariance of the *S*-matrix." *Physics Letters* 9:357–59.

———. 1964. "Feynman rules for any spin." *PR* 133:B1318–B1332.

———. 1977. "The search for unity: Notes for a history of quantum field theory." *Daedalus* 106:17–35.

———. 1995. *The Quantum Theory of Fields.* 3 vols. New York: Cambridge University Press.

———. 1997. "Changing attitudes and the Standard Model." In *The Rise of the Standard Model: Particle Physics in the 1960s and 1970s*, ed. Lillian Hoddeson, Laurie Brown, Michael Riordan, and Max Dresden, 36–44. New York: Cambridge University Press.

Weiner, Charles. 1969. "A new site for the seminar: The refugees and American physics in the thirties." In *The Intellectual Migration: Europe and America, 1930–1960*, ed. Donald Fleming and Bernard Bailyn, 190–234. Cambridge: Harvard University Press.

———. 1975. "Cyclotrons and internationalism: Japan, Denmark, and the United States, 1935–1945." In *Proceedings of 14th International Congress of the History of Science, 1974*, 353–65. Tokyo: Science Council of Japan.

———. 1978. "Retroactive saber rattling?" *BAS* 34 (Apr): 10–12.

Weisskopf, Victor. 1994 [1936]. "The electrodynamics of the vacuum based on the quantum theory of the electron." In *Early Quantum Electrodynamics: A Source Book*, ed. Arthur I. Miller, 206–26. New York: Cambridge University Press.

———. 1958 [1939]. "On the self-energy and the electromagnetic field of the electron." In *Selected Papers in Quantum Electrodynamics*, ed. Julian Schwinger, 68–81. New York: Dover.

———. 1952. "Report on the visa situation." *BAS* 8 (Oct): 221–22.

———. 1954. "Visas for foreign scientists." *BAS* 10 (Mar): 68–69, 112.

Welton, Theodore. 1948. "Some observable effects of the quantum-mechanical fluctuations of the electromagnetic field." *PR* 74:1157–67.

Weneser, Joseph. 1952. *Fourth-Order Radiative Corrections to Atomic Energy Levels.* Ph.D. dissertation, Columbia University.

Wentzel, Gregor. 1947. "Recent research in meson theory." *Reviews of Modern Physics* 19:1–18.

———. 1949 [1943]. *Quantum Theory of Fields.* Trans. C. Houtermans and J. M. Jauch. New York: Interscience.

———. 1950. "μ-pair theories of the π-meson." *PR* 79:710–17.

Westwick, Peter. 2003. *The National Labs: Science in an American System, 1947–74.* Cambridge: Harvard University Press.

Wheeler, John. 1989. "The young Feynman." *PT* 42 (Feb): 24–32.

Wheeler, J. A., and R. P. Feynman. 1945. "Interaction with the absorber as a mechanism of radiation." *Reviews of Modern Physics* 17:157–81.

———. 1949. "Classical electrodynamics in terms of direct interparticle action." *Reviews of Modern Physics* 21:235–433.

Wheeler, John, with Kenneth Ford. 1998. *Geons, Black Holes, and Quantum Foam: A Life in Physics.* New York: W. W. Norton.

White, Bebo. 1997. "The World Wide Web and high-energy physics." *Annual Reviews of Nuclear and Particle Science* 47:1–26.

Wichmann, Eyvind. 1956. *Vacuum Polarization in a Strong Coulomb Field.* Ph.D. dissertation, Columbia University.

Wichmann, Eyvind, and Norman Kroll. 1954. "Vacuum polarization in a strong Coulomb field." *PR* 96:232–34.

Wick, Gian Carlo. 1955. "Introduction to some recent work in meson theory." *Reviews of Modern Physics* 27:339–62.

Wightman, Arthur. 1989. "The general theory of quantized fields in the 1950s." In *Pions to Quarks: Particle Physics in the 1950s*, ed. Laurie Brown, Max Dresden, and Lillian Hoddeson, 608–29. New York: Cambridge University Press.

Williams, Dudley. 1954. "Coherent Systems of Physical Units." *PT* 7 (Apr): 8–11.

Williams, P. K., and V. Hagopian, eds. 1973. *π-π Scattering—1973 (Tallahassee Conference).* New York: American Institute of Physics.

Wilson, Richard. 1994. "Commemoration of the life of R. E. (Bob) Marshak." In *A Gift of Prophecy: Essays in Celebration of the Life of Robert Eugene Marshak*, ed. E. C. G. Sudarshan, 536–41. Singapore: World Scientific.

Wittgenstein, Ludwig. 1958. *Philosophical Investigations: The English Text of the Third Edition.* Ed. G. E. M. Anscombe. New York: Macmillan, 1958.

Yang, C. N. 1962. *Elementary Particles: The Vanuxem Lectures.* Princeton: Princeton University Press.

Yarmolinsky, Adam, ed. 1955. *Case Studies in Personnel Security.* Washington, DC: Bureau of National Affairs.

Young, James A. 1961. *Two Problems of Structure in the Theory of Weak Interactions.* Ph.D. dissertation, University of California, Berkeley.

Yukawa, Hideki. 1935. "On the interaction of elementary particles." Pt. 1. *Proceedings of the Physical-Mathematical Society of Japan* 17:48–57.

———. 1982 [1957]. *The Traveller (Tabibito).* Trans. L. Brown and R. Yoshida. Singapore: World Scientific.

Zachariasen, Frederik. 1961. "Self-consistent calculation of the mass and width of the $J = 1$, $T = 1, \pi\pi$ Resonance." *PRL* 7:112–13.

———. 1961. Erratum. *PRL* 7:268.

———. 1963. "The theory and application of Regge poles." In *1962 Cargèse Lectures in Theoretical Physics,* ed. Maurice Lévy, chap. 10, pp. 1–42. New York: W. A. Benjamin.

———. 1966. "Lectures on bootstraps." In *Recent Developments in Particle Physics,* ed. Michael Moravcsik, 86–151. New York: Gordon and Breach.

Zachariasen, Frederik, and Charles Zemach. 1962. "Pion resonances." *PR* 128:849–58.

Zel'dovich, Ia. B. 1954. "K teorii π-mezonov" [On the theory of π-mesons]. *Doklady Akademii Nauk SSSR* 97:225–28.

———. 1954. "O raspade zariazhennykh π-mezonov" [On the decay of charged π-mesons]. *Doklady Akademii Nauk SSSR* 97:421–24.

Zwanziger, Daniel. 1961. *α^3 Corrections to Hyperfine Structure in Hydrogenic Atoms.* Ph.D. dissertation, Columbia University.

———. 1961. "α^3 corrections to hyperfine structure in hydrogenic atoms." *PR* 121:1128–42.

Zweig, George. 1964. "An SU(3) model for strong interaction symmetry and its breaking." CERN preprint 8419/TH.142.

Index

CPSIA information can be obtained
at www.ICGtesting.com
Printed in the USA
LVHW03s0509130718
583439LV00003BA/17/P

9 780226 422671